The Photosynthetic Bacterial Reaction Center

Structure and Dynamics

NATO ASI Series

Advanced Science Institutes Series

A series presenting the results of activities sponsored by the NATO Science Committee, which aims at the dissemination of advanced scientific and technological knowledge, with a view to strengthening links between scientific communities.

The series is published by an international board of publishers in conjunction with the NATO Scientific Affairs Division

A	**Life Sciences**	Plenum Publishing Corporation
B	**Physics**	New York and London
C	**Mathematical and Physical Sciences**	Kluwer Academic Publishers Dordrecht, Boston, and London
D	**Behavioral and Social Sciences**	
E	**Applied Sciences**	
F	**Computer and Systems Sciences**	Springer-Verlag
G	**Ecological Sciences**	Berlin, Heidelberg, New York, London,
H	**Cell Biology**	Paris, and Tokyo

Recent Volumes in this Series

Volume 143—Antiviral Drug Development: A Multidisciplinary Approach
edited by Erik De Clercq and Richard T. Walker

Volume 144—H-2 Antigens: Genes, Molecules, Function
edited by Chella S. David

Volume 145—Pharmacokinetics: Mathematical and Statistical Approaches to Metabolism and Distribution of Chemicals and Drugs
edited by A. Pecile and A. Rescigno

Volume 146—Reactive Oxygen Species in Chemistry, Biology, and Medicine
edited by Alexandre Quintanilha

Volume 147—The Biophysics of Organ Cryopreservation
edited by David E. Pegg and Armand M. Karow, Jr.

Volume 148—Cerebellum and Neuronal Plasticity
edited by Mitchell Glickstein, Christopher Yeo, and John Stein

Volume 149—The Photosynthetic Bacterial Reaction Center: Structure and Dynamics
edited by Jacques Breton and André Verméglio

Series A: Life Sciences

The Photosynthetic Bacterial Reaction Center

Structure and Dynamics

Edited by

Jacques Breton

CEN Saclay
Gif-sur-Yvette, France

and

André Verméglio

ARBS
CEN Cadarache
Saint Paul lez Durance, France

Springer Science+Business Media, LLC

Proceedings of a NATO Advanced Research Workshop on the Structure of the Photosynthetic Bacterial Reaction Center: X-Ray Crystallography and Optical Spectroscopy with Polarized Light, held September 20-25, 1987, at the Centre d' Études Nucléaires de Cadarache, France

Library of Congress Cataloging in Publication Data

NATO Advanced Research Workshop on the Structure of the Photosynthetic Bacterial Reaction Center (1987: Cadarache, France)
The photosynthetic bacterial reaction center: structure and dynamics / edited by Jacques Breton and André Verméglio.
p cm.—(NATO ASI series. Series A, Life sciences; v. 149)
"Proceedings of NATO Advanced Research Workshop on the Structure of the Photosynthetic Bacterial Reaction Center . . . held September 20-25, 1987, Cadarache, France."
"Published in cooperation with NATO Scientific Affairs Division."
Bibliography: p.
Includes index.

ISBN 978-1-4899-0817-9 ISBN 978-1-4899-0815-5 (eBook)
DOI 10.1007/978-1-4899-0815-5

1. Photosynthesis—Congresses. 2. Bacteria, Photosynthetic—Congresses 3. Bacterial pigments—Congresses. I. Breton, Jacques. II. Verméglio, André. III. North Atlantic Treaty Organization. Scientific Affairs Division. IV. Title. V. Series.
QR88.5 1987 88-12581
581.1′3342—dc19 CIP

Originally published by Plenum Press, New York in 1988
Softcover reprint of the hardcover 1st edition 1988

PREFACE

This volume contains the contributions from the speakers at the NATO Advanced Research Workshop on "Structure of the Photosynthetic Bacterial Reaction Center : X-ray Crystallography and Optical Spectroscopy with Polarized Light" which was held at the "Maison d'Hôtes" of the Centre d'Etudes Nucléaires de Cadarache in the South of France, 20-25 September, 1987. This meeting continued in the spirit of a previous workshop which took place in Feldafing (FRG), March 1985.

Photosynthetic reaction centers are intrinsic membrane proteins which, by performing a photoinduced transmembrane charge separation, are responsible for the conversion and storage of solar energy. Since the pioneering work of Reed and Clayton (1968) on the isolation of the reaction center from photosynthetic bacteria, optical spectroscopy with polarized light has been one of the main tools used to investigate the geometrical arrangement of the various chromophores in these systems. The recent elucidation by X-ray crystallography of the structure of several bacterial reaction centers, a breakthrough initiated by Michel and Deisenhofer, has provided us with the atomic coordinates of the pigments and some details about their interactions with neighboring aminoacid residues. This essential step has given a large impetus both to experimentalists and to theoreticians who are now attempting to relate the X-ray structural model to the optical properties of the reaction center and ultimately to its primary biological function. The initial photosynthetic reactions result in an ultrafast transfer of an electron from the primary donor, a dimer of bacteriochlorophylls, to a bacteriopheophytin located some 17 Å away. In this electron transfer process the role of a monomeric bacteriochlorophyll molecule, which is intriguingly positioned almost in between the primary donor and the bacteriopheophytin, is the object of considerable scrutinity and of much controversy. The importance of the protein for fine-tuning the energy levels of the chromophores as wells as for controlling the relaxation steps which accompany the electron transfer and permit the remarkable efficiency of the transmembrane charge stabilization process is increasingly recognized. Most of these issues have been addressed during the workshop and are discussed in these proceedings.

The NATO Scientific Affairs Division is gratefully acknowledged for awarding a grant that made the organization of the workshop possible. Additional funding support was also provided by the Institut National de la Recherche Agronomique, the Département de Biologie du Commissariat à l'Energie Atomique, the Association pour la Recherche en Bioénergie Solaire and the SMC-TBT.

We are greatly indebted to Ms. P. Belle for her dedicated assistance with the secretarial correspondence and the preparation of these proceedings and also to Mrs. R. Vallerie for her efficient contribution to the smooth running of the meeting. We would also like to express our sincere thanks to F. Bel, M. Bombal, P. Druzian, F. Guichod, J. Marcesse, R. Rabette and R. Siadoux for their assistance with the local organization of the conference which was held in the pleasant and stimulating atmosphere of the medieval castle of Cadarache.

Finally we wish to express our gratitude to the participants to the workshop for their enthusiastic cooperation which has resulted in the success of the meeting and the high scientific standard of this volume. There is clearly a need for meetings devoted to the photosynthetic bacterial reaction center at regular intervals in the future.

January, 1988

Jacques BRETON

André VERMEGLIO

CONTENTS

THE CRYSTAL STRUCTURE OF THE PHOTOSYNTHETIC REACTION CENTER FROM *RHODOPSEUDOMONAS VIRIDIS*

Johann Deisenhofer and Hartmut Michel

Max-Planck-Institut fuer Biochemie
D-8033 Martinsried, FRG

The photosynthetic reaction center (RC) from *Rhodopseudomonas viridis* (*R. viridis*) was one of the first integral membrane proteins which could be crystallized (Michel, 1982). The crystals turned out to be suitable for X-ray structure analysis at atomic resolution. In a study at 3Å resolution, using phases determined from multiple isomorphous replacement experiments with heavy atom compounds, the arrangement of the major prosthetic groups (Deisenhofer et al., 1984), and the folding of the protein subunits (Deisenhofer et al., 1985) were determined. Sequencing of the RC's protein subunits (Michel et al., 1985; Michel et al., 1986a; Weyer et al., 1987) facilitated interpretation of the electron density map and model building; information from X-ray diffraction and from sequencing together allowed a detailed description of pigment-protein interactions in the RC (Michel et al., 1986b).

These studies resulted in an atomic model of the RC, a simplified representation of which is shown in figure 1. The membrane spanning core of the RC is formed by the subunits L and M, hydrophobic polypeptides of 273 and 323 amino acids, repectively. Large portions of these subunits are folded in a similar way; outstanding elements of secondary structure are 5 membrane spanning helices in each subunit. The arrangement of L and M shows a remarkable local 2-fold symmetry. On both sides of the membrane a peripheral subunit is noncovalently attached to the L-M complex: The 4-heme cytochrome with 336 amino acids at the periplasmic surface, and the globular part of the H-subunit at the cytoplasmic surface. The N-terminal segment of the H-subunit forms a membrane spanning helix; in total, this subunit consists of 258 amino acids.
Imbedded into the L-M complex, and exhibiting largely the same local 2-fold symmetry, is the group of pigments more or less involved in the primary charge separation: 4 BChl-b, 2 BPh-b, a menaquinone, and the non-heme iron. These pigments form 2 symmetrically arranged branches, both originating from the tightly interacting pair of BChl-bs, the special pair, near the periplasmic surface of the membrane. Each branch leads via an accessory BChl-b and a BPh-b to a quinone binding site near the

membrane's cytoplasmic surface. Only the binding site of Qa, the primary quinone acceptor, is fully occupied by a menaquinone molecule; most of the Qb molecule, a ubiquinone, appears to have been lost during isolation or crystallization of the RC. From symmetry considerations, and from binding studies with o-phenanthroline and terbutryn, the Qb binding site could be identified. The non-heme iron is bound by 4 histidines and one glutamic acid; it sits on the local twofold axis between both quinone binding sites.
This model (without the cytochrome subunit) was used as a starting point for the crystal structure analysis of RC from *Rb. sphaeroides* (see Feher et al, this volume; Tiede et al., this volume, and references therein). These studies confirmed the close structural similarity between RCs from *R. viridis* and *Rb. sphaeroides*.

The model of the *R. viridis* RC was improved by crystallographic refinement, first at 2.9Å resolution, later at 2.3Å resolution (Deisenhofer, Epp, Michel, in preparation). Apart from the correction of local errors in the model, the refinement process led to the discovery of a number of ordered constituents of the RC structure which, due to large phase errors and lack of resolution, could not be recognized during the early stages of the X-ray analysis. These new features include a long, bent molecule near one of the accessory BChl-bs, which most probably represents the carotenoid found by chemical analysis in the RC (Sinning, Michel, Eugster, unpublished). Another new feature in the refined model is the Qb head group, found with low occupancy in the Qb binding site. About 200 bound water molecules were located, some of them in positions that indicate probable functional importance. A few electron density features, too big to be water, were interpreted as bound ions, most probably sulfate. The list of ordered solvent is completed by one molecule of the detergent LDAO; the vast majority of the detergent in the crystal is disordered.
With its improved accuracy, and with the new features mentioned, the refined model of the *R. viridis* RC can contribute to a deeper understanding of the primary reactions in bacterial photosynthesis, and to the solution of some of the problems discussed in other articles within this issue.

References

Deisenhofer, J., Epp, O., Miki, K., Huber, R. and Michel, H. (1985). Structure of the protein subunits in the photosynthetic reaction centre of Rhodopseudomonas viridis at 3 Å resolution, *Nature* 318, 618-624.

Deisenhofer, J., Epp, O., Miki, K., Huber, R. and Michel, H. (1984). X-Ray Structure Analysis of a Membrane Protein Complex: Electron Density Map at 3 Å Resolution and a Model of the Chromophores of the Photosynthetic Reaction Center from Rhodopseudomonas viridis, *J. Mol. Biol.* 180, 385-398.

Michel, H. (1982). Three-Dimensional Crystals of a Membrane Protein Complex the Photosynthetic Reaction Centre from Rhodopseudomonas viridis, *J. Mol. Biol.* Vol 158, 567-572.

Michel, H., Weyer, K.A., Gruenberg, H. and Lottspeich, F. (1985). The 'heavy' subunit of the photosynthetic reaction centre from Rhodopseudomonas viridis: isolation of the gene, nucleotide and amino acid sequence, *EMBO J.* 4, 1667-1672.

Michel, H., Weyer, K.A., Gruenberg, H., Dunger, I., Oesterhelt, D. and Lottspeich, F. (1986a). The 'light' and 'medium'

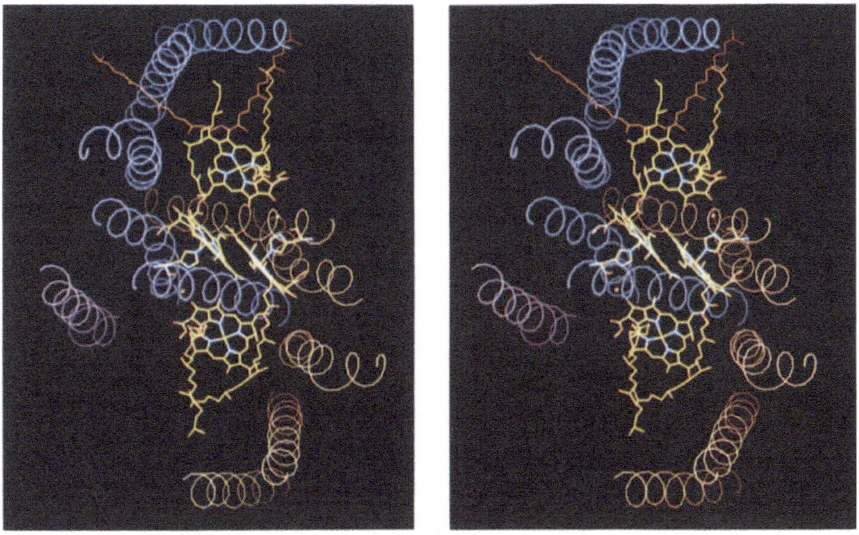

Figure 1 (stereo pair): View from the cytochrome binding site (periplasmic membrane surface) showing smoothed backbone models of the transmembrane helices of subunits L (light brown), M (blue), and H (purple), together with cofactors of the RC from *R. viridis*. The cofactors shown are, from top to bottom, the putative carotenoid (brown), the accessory BChl-b BCMA, the special pair of BChl-bs with histidine ligands and bound water molecules, and the accessory BChl-b BCLA). BChl-bs, histidines, and waters are colored according to atom types (carbon and magnesium yellow, nitrogen blue, oxygen red).

subunits of the photosynthetic reaction centre from Rhodopseudomonas viridis: isolation of the genes, nucleotide and amino acid sequence, *EMBO J.* 5, 1149-1158.
Michel, H., Epp, O. and Deisenhofer, J. (1986b). Pigment-protein interactions in the photosynthetic reaction centre from Rhodopseudomonas viridis, *EMBO J.* 5, 2445-2451.
Weyer, K.A., Lottspeich, F., Gruenberg, H., Lang, F., Oesterhelt, D. and Michel, H. (1987). Amino acid sequence of the cytochrome subunit of the photosynthetic reaction centre from the purple bacterium Rhodopseudomonas viridis, *EMBO J.* 6, 2197-2202.

STRUCTURE OF THE REACTION CENTER FROM *RHODOBACTER SPHAEROIDES* R-26 AND 2.4.1

J. P. Allen and G. Feher
University of California, San Diego, La Jolla, California 92093

T. O. Yeates, H. Komiya, and D. C. Rees
University of California, Los Angeles, Los Angeles, California 90024

I. INTRODUCTION

Detailed theories of electron transfer in reaction centers (RCs) require knowledge of their three dimensional structure. We have determined the structure of RCs from *Rb. sphaeroides* by x-ray diffraction of single crystals. Diffraction data of RCs from both the carotenoidless mutant, R-26, and the wild type strain, 2.4.1 were analyzed at resolutions of 2.8 Å and 3.5 Å respectively. These structures have been refined to current R-factors of 25% (for the R26 data) and 22% (for the 2.4.1. data). Details concerning data collection and analysis, with descriptions of the structure of the RC from the R-26 strain have been presented elsewhere (1-4). In this report we shall focus on the general features of the structure and compare them with those reported for the RC from *R. viridis* (5). The structures from these two species have been shown by the molecular replacement method to be homologous (6,7). We shall emphasize the importance of the structural features to the function of electron transfer.

II. SYMMETRY FEATURES OF THE RC STRUCTURE

One of the striking features of the RC is the symmetry exhibited by the cofactors and the protein subunits. The 8 cofactors, the bacteriochlorophyll dimer, (D), the bacteriochlorphyll monomers, (B), the bacteriopheophytin (ϕ), and the quinones (Q),* form two branches, A and B, that are approximately related to each other by a two fold symmetry axis (see Fig. 1). The single Fe^{2+} lies approximately on the symmetry axis but the carotenoid, (C), lies far from the axis, thus breaking the two-fold symmetry. The two subunits L and M are also approximately related by this two fold axis while the H subunit has no such obvious symmetry properties.

To quantitate the symmetry of the cofactors, we determined the rotation matrix that optimized the superposition of the cofactors of the A branch onto the cofactors of the B branch (8). The rotation axis obtained from this transformation is specified by the polar angles $\phi = 81°$,

*The nomenclature for the cofactors was the subject of an extended discussion at this conference. It was agreed that a simple two letter code would be desirable, i.e. one letter denoting the cofactor with a subscript, A or B, denoting the appropriate branch. It was also agreed that the bacteriochlorophyll monomers should be designated by B and the quinones by Q. There was no unanimity concerning the nomenclature for the bacteriochlorophyll dimer and the bacteriopheophytin. We suggest D for the dimer (it also stands for donor, in case that P_{680} or P_{700} in green plants should turn out to be monomer) and ϕ for the bacteriopheophytin. ϕ is pronounced either "fee" or "fie", both of which are phonetically close to *pheophytin*.

$\psi = 52°$, $\kappa = 183°$ (where ϕ is the azimuth from the x axis, ψ is the inclination from the y axis, and κ is the angle of rotation) (for details see 1). We also determined the rotation matrix that optimized the superposition of equivalent C^{α} atoms of the membrane spanning helices of the L and M subunits; this transformation has a rotation axis given by $\phi = 81°$, $\psi = 53°$, $\kappa = 180°$ (2). Since the angles specifying the rotations of the cofactors and the C^{α} atoms of the helices are close, we shall consider the symmetry axes of the cofactors and the L and M subunits to be equivalent.

The transformation that best relates the cofactors and the L and M subunits deviates slightly from an exact two fold rotation. Considering the cofactors alone, the transformation results in a rms deviation between equivalent atoms of 0.7Å for D, 1.3Å for B, 1.4Å for ϕ, and 2.2Å for Q. In terms of function the cofactors that are involved in the more primary electron transfers obey the two fold symmetry better than those involved in the later stages. The rms deviation of equivalent C^{α} atoms of the transmembrane helices from the symmetry axis is 1.2 Å. Much larger differences exist between the other regions of the L and M subunits, most notably the regions of the amino termini.

III. ELECTRON TRANSFER

The factors contributing to electron transfer are the electronic matrix elements and the nuclear (vibronic) Frank-Condon factors (9,10). The three dimensional structure determined by x-ray diffraction is essential but not sufficient to calculate these factors as the electronic structure is also needed (11,12). Futhermore, only the time-averaged, or static, structure is determined; whereas dynamical features (eg. light induced conformational changes) are believed to play an important role in electron transfer. Thus x-ray diffraction studies must be complemented by other spectroscopic techniques (eg. optical, FTIR, EPR, etc.) as were discussed at this conference.

The three dimensional structure has the striking feature of the two fold symmetry that divides the cofactors into two branches, A and B (as discussed in Sec. II). In suprising contrast to this symmetry is the preferential direction of electron transfer along the cofactors of the A branch. Clearly, the environments of the cofactors must be asymmetric, as we shall discuss in this section. In a later section (V), we will discuss a possible role of B_B, a cofactor that apparently is not involved in the electron transfer process.

A. Cofactors

The electronic matrix elements depend critically on the overlap of the electronic wave functions of the the cofactors. These are related to the spatial overlap, which is obtained from the three dimensional structure by the following procedure: The surface areas of the individual, isolated cofactor ring systems were calculated by the method of Lee and Richard (13). The surface areas were also calculated for the cofactors in their appropriate position in the RC. The differences between the two surface areas were used to determine the overlaps. The overlap between D_B and B_A is larger by a factor of approximately 1.5 than the corresponding overlap between D_A and B_B, as illustrated in Fig. 2. To visualize the overlap, the surfaces were generated by spheres centered on the atoms with radii 10% larger than the standard van der Waals distance (see Fig. 2).

The refined structures of D, B and ϕ show deviations from planarity with differences observed between corresponding cofactors of the two branches. Small deviations from planarity of atoms in the ring conjugation may result in large changes in the electronic properties of the cofactors. Details of these deviations awaits further refinement of the structure.

Interactions between the cofactors and the protein are also likely to have important consequences on the electron transfer rates. Different arrangements of hydrogen bonds are found for equivalent cofactors of the two branches. For example, ring V of ϕ_A appears to be hydrogen bonded to Glu L104 (see Fig. 3) while ring V of ϕ_B does not appear to form a hydrogen bond (14). This may be partly responsible for the difference in the Q_X transition moments of ϕ_A and ϕ_B leading to the splitting of the absorption bands at 532nm and 545nm. The non-planar

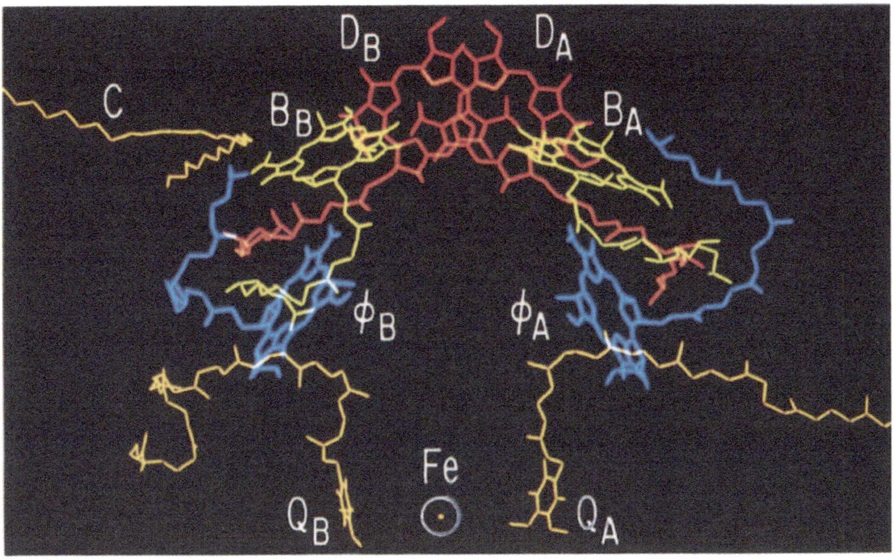

Fig. 1. *The structure of the 10 cofactors of the RC from Rb. sphaeroides (see text for nomenclature). The two-fold symmetry axis is approximately aligned in the plane of the paper with the A branch on the right and the B branch on the left. The carotenoid C is represented as a simple polyene chain (See Sec. V) and is present only in the wild type species; all other cofactors are present in both the 2.4.1 and R-26 strains.*

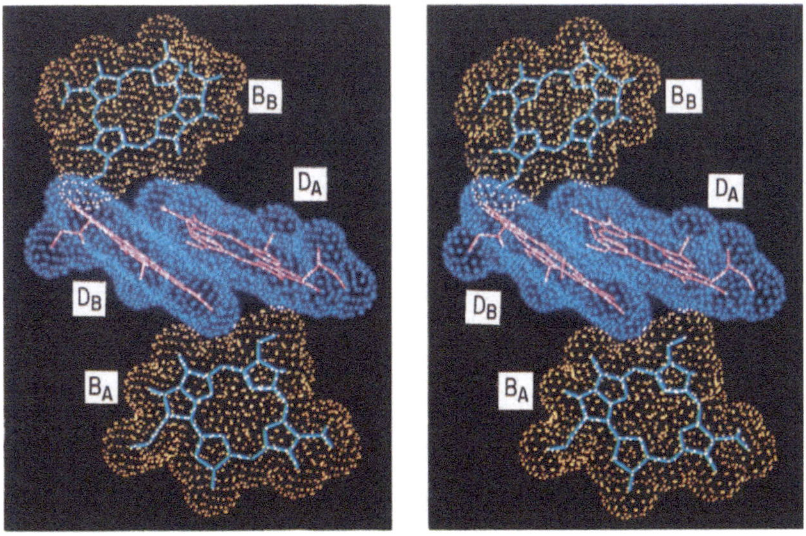

Fig. 2. *Stereo view of the bacteriochlorophyll dimer, D, and the bacteriochlorophyll monomers,* B_A *and* B_B *with their van der Waals surfaces enlarged by 10% to visualize the overlap. The surface contact between* D_B *and* B_A *is more extensive than between* D_A *and* B_B. *For ease of viewing the ester carbonyl and phytyl chains of ring IV have been removed.*

features of the ester carbonyl group of ring V of ϕ_A (see Fig. 3), is consistent with a keto form of the ring. Recent ENDOR experiments on ϕ_A^- corroborate this assignment (15).

B. Aromatic Residues

Many aromatic residues of the RC are positioned near the cofactors, in particular, the dimer. Residue Tyr M210 is in van der Waals contact with both D_B and ϕ_A (see Fig. 4) and may serve as a conduit for the transfer of electrons from D to ϕ_A. The symmetry related residue, Phe L181, is not as favorably positioned for electron transfer from D to ϕ_B.

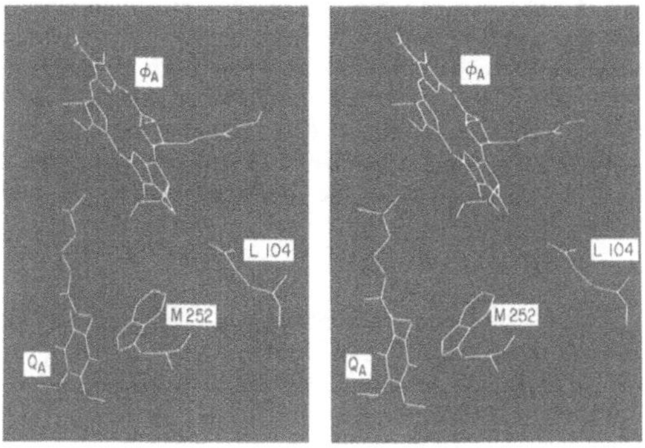

Fig. 3. *Stereo view showing the bacteriopheophytin ϕ_A, primary quinone Q_A, and amino acid residues Glu L104 and Trp M252.*

Electrons are transferred from ϕ_A to Q_A in approximately 200ps (at 295K (16,17)) through a distance of 13Å[‡]. Bridging these cofactors is the aromatic residue Trp M252 (see Fig. 3). This arrangement is basically conserved in the RC of *R. viridis*, although in that case the aromatic group of the corresponding Trp is more nearly parallel to the ring system of Q_A (a menaquinone). In *Rb. sphaeroides*, the symmetry related residue, Phe L216 is near Q_B but the planes of the two rings are approximately perpendicular (14).

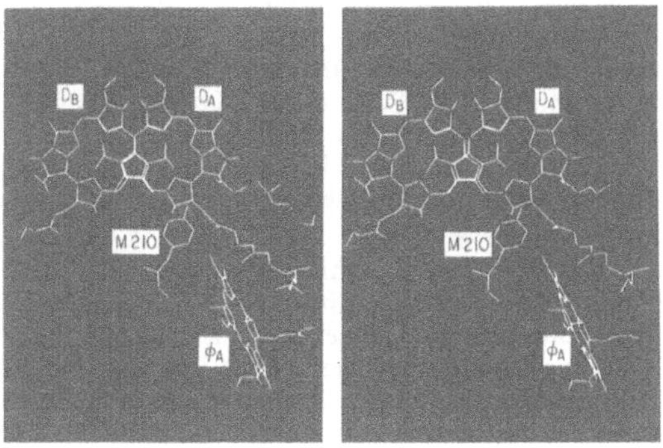

Fig. 4. *Stereo view showing the bacteriochlorophyll dimer D, bacteriopheophytin ϕ_A, and amino acid residue Tyr M210.*

Recent experiments have shown that aromatic residues participate in the electron transfer of the reaction center core of Photosystem II, which is believed to be homologous to the bacterial systems. Both D^+, which is characterized by "Signal II" and Z , which transfers electrons from the site of water oxidation to the primary donor have been identified as tyrosine residues (18,19).

C. Charged Residues

Since the RC is an integral membrane protein the distribution of its residues differs greatly from that found in water soluble proteins (3). Charged residues are located only in the two outer regions of the RC with none in the central region that is formed by the 11 transmembrane helices. The cofactors of the RC lie in this central, hydrophobic region. No charged residues are found near D, ϕ, or B (or C in RCs from the 2.4.1 species) with the exception of Glu L104 (see Sec. III.A). Since the iron-quinone complex is located at the edge of the hydrophobic region, charged residues are present near these cofactors. A noticeable asymmetry arises from the larger number of charged residues near Q_B compared to Q_A. The ionization state of the charged residues cannot be determined directly from the crystallographic data, but only deduced from an analysis of the structure. From electrostatic considerations (3), Glu L212 and Asp L213 are not expected to be negatively charged but are believed to be neutral (protonated). These residues may be involved in the protonation of Q_B (see Sec. IV).

---

‡ distances are quoted between ring centers.

Collectively, the charged residues create an electrostatic field throughout the RC. The electrostatic potential was numerically evaluated by solving the finite difference form of Poisson's equation (20). The results show that from electrostatic energy considerations, electron transfer from D to Q_A is preferred along the A branch over that of the B branch (3). Although it is difficult to draw quantitative conclusions concerning the electron transfer kinetics from these calculations, the results identify electrostatic effects as a possible source of the asymmetry in the electron transfer rates for the two branches.

D. Iron Quinone Complex

The iron atom is coordinated to 4 histidines (L230, L190, M219, and M266) and to the bidentate ligands of Glu M234 (see Fig. 5). The ligands form a distorted octahedron as had been predicted from spectroscopic measurements (21 and references therein). This arrangement provides a major polar interaction that contributes to the stabilization of the tertiary structure (3). Such interactions are necessary for the stabilization of the tertiary structure of RC since hydrophobic interactions, which stabilize the structures of water soluble proteins play a small role in stabilizing integral membrane proteins.

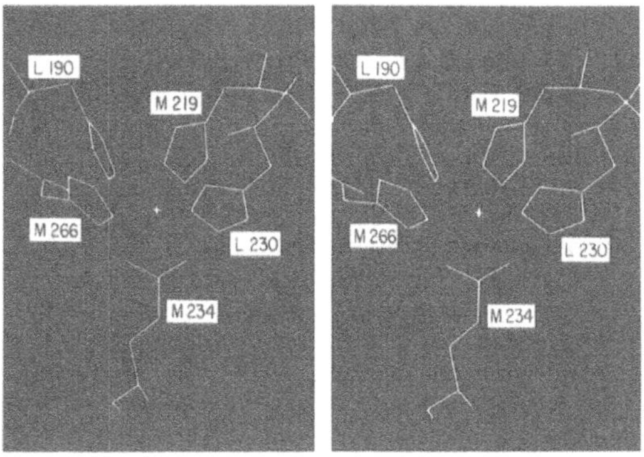

Fig. 5. *Stereo view of the* Fe^{2+} *(center) surrounded by four histidines (L190, L230, M219, and M266) and one glutamic acid (M266) that coordinate to the iron atom.*

The Fe^{2+} lies approximately midway between the quinones, being approximately 2Å closer to Q_B than Q_A. The rings of both quinones form hydrogen bonds to nearby amino acid residues and are well protected from the solvent (analysis of the structure shows that the rings have neglible exposure to the surface of the protein). Since the charge of the Fe^{2+} is only partially compensated by Glu M234, a net positive charge remains. This electrostatic energy may be important in stabilizing the electron on the closer quinone, Q_B.

Electron transfer from Q_A to Q_B occurs in approximately 100 μs (at 295K (16,17)) over a distance of 18.5Å . The Fe^{2+} forms an anti-ferromagnetically coupled system with the anion radicals Q_A^- or Q_B^-. However, removal of Fe^{2+} does not significantly affect the rate of electron transfer; possible roles of Fe^{2+} (in addition to the structural role discussed above) are presented elsewhere (22). Located between the quinone rings are the imidizole rings of His M219 and His L190. The separation between neighboring rings is 3-5Å. This arrangement suggests that these two histidines play a role in the electron transfer from Q_A to Q_B.

IV. PROTONATION

A transfer of protons accompanies the transfer of electrons in the RC. The one electron, semiquinone species, Q_A^- and Q_B^- are not protonated directly but interact with amino acid residues whose pK's are changed by electrostatic interactions giving rise to the observed proton uptake (23,24). Work is in progress by McPherson et al. to calculate from the structure the electrostatic effects of the semiquinones (25). Preliminary results indicate that not only nearby residues (eg. Tyr H40 near Q_A and Glu L212 and Asp L213 near Q_B) but a large number of distant residues must be considered in modeling the protonation (25).

When Q_B is doubly reduced, $Q_B^=$ accepts two protons through a process that is not well understood. If $Q_B^=$ is released to the surface of the RC before protonation, electrostatic forces presumably opened the structure to allow the release of the quinone (see Sec III.D). If $Q_B^=$ is protonated inside the RC, two different mechanisms may accomplish this. In one mechanism, a channel connects Q_B to the surface of the RC exposing it to the outside solvent. No solvent channels were found after examination of the RC structure. Alternatively, protonation can occur through a "bucket brigade". In this mechanism protons are transferred from the surface to Q_B through a chain of protonated residues and bound solvent. We are currently examining the structure to identify residues that could form such a bridge; they should include acidic groups near Q_B.

V. CAROTENOID

Crystals of RCs from *Rb. sphaeroides* 2.4.1 were grown under the same conditions as those from *Rb. sphaeroides* R-26. The 2.4.1 crystals had the same space group as the R26 crystals. Although the cell constant for the two forms were similar, the a axis of the 2.4.1 crystals was 1% smaller than the a axis of the R-26 crystals. Data from crystals of RC- 2.4.1 were collected to 3.5Å and the RC-R26 structure was refined against the 2.4.1 data in two stages. Since the cell constants were slightly different, the orientation (rotation, translation, and cell constants) were refined by treating the RC as a rigid body. The repositioned RC was then refined using the restrained least squares program PROLSQ of Hendrickson and Konnert (26). After refinement, the R factor was 22% between 8 and 3.5Å resolution.

An electron density map was generated by taking the Fourier transform of the difference between observed and calculated structure factor amplitudes. This difference map has a prominent extended region of electron density near B_B. Additional regions of electron density are located near the two ends of this prominent region. Together, these regions of electron density are large enough to accommodate most of the carotenoid. For a preliminary fit of the carotenoid we have modeled the carotenoid as a simple all trans polyene chain omitting such structural features as the cis bond and the out of plane twists. We shall add these features after we model the bound solvent (and possibly detergent) that is observed in this region, reposition the amino acid residues in the carotenoid region, and finish data collection at higher resolution.

The carotenoid is located near the cofactor B_B (see Fig. 6). It passes between the B and C helices of the M subunit in the central, hydrophobic region of the RC. The protein environment, which consists of a large number of aromatic residues, places strong steric constraints on the conformation of the carotenoid. This is consistent with the instability, and rapid isomerization of the carotenoid to the all trans form, upon extraction from the RC (27). The carotenoid's absorption spectra is red shifted relative to its spectra in organic solvents. This has been modeled by assuming the presence of a few charged amino acid residues near the carotenoid (28). No charged residues are found near C (see Sec. III.C) though several are located at distances of 10 Å and beyond. Alternatively, the conformation of the carotenoid may cause this spectral shift (29).

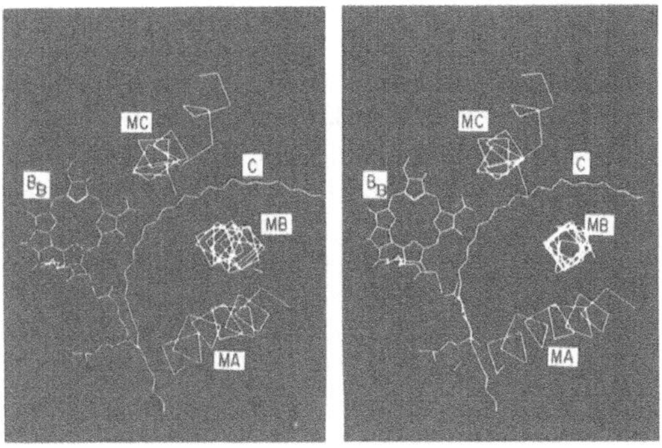

Fig. 6. *Stereo view showing the carotenoid, C, (using a simple polyene chain as described in Sec. V), bacteriochlorophyll monomer* B_B, *and backbone models of three of the transmembrane helices of the M subunit, A, B, and C.*

The positioning of the carotenoid suggests a role for the cofactor B_B in protecting the RC against photo-oxidation damage. B_B does not appear to play an important role in the electron transfer processes of the RC. However, the structural arrangement suggests that B_B serves as a conduit for the triplet energy transfer from the primary donor to the carotenoid. The involvement of a Bchl monomer during this transfer had been suggested from spectroscopic measurements (30).

ACKNOWLEDGMENTS

We thank E. Abresch for preparation of the RCs. The work was supported by grants from the National Institutes of Health (DK36053, GM31299, and GM13191).

REFERENCES

1. J. P. Allen, G. Feher, T. O. Yeates, H. Komiya, and D. C. Rees, Structure of the reaction center from *Rhodobacter sphaeroides* R-26: The cofactors, *Proc. Natl. Acad. Sci. USA* **84**:5730-5734 (1987).
2. J. P. Allen, G. Feher, T. O. Yeates, H. Komiya, and D. C. Rees, Structure of the reaction center from *Rhodobacter sphaeroides* R-26: The protein subunits, *Proc. Natl. Acad. Sci. USA* **84**:6162-6166 (1987).
3. T. O. Yeates, H. Komiya, D. C. Rees, J. P. Allen, and G. Feher, Structure of the reaction center from *Rhodobacter sphaeroides* R-26: Membrane protein interactions, *Proc. Natl. Acad. Sci. USA* **84**:6438-6442 (1987).
4. J. P. Allen, G. Feher, T. O. Yeates, H. Komiya, and D. C. Rees, Structure of the reaction center from *Rhodobacter sphaeroides* R-26: Pigment protein interactions, *Proc. Natl. Acad. Sci. USA*, in preparation.
5. H. Michel, O. Epp, and J. Deisenhofer, Pigment-protein interactions in the photosynthetic reaction center from *Rhodopseudomonas viridis*, *EMBO J.* **5**:2445-2451 (1986).
6. J. P. Allen, G. Feher, T. O. Yeates, D. C. Rees, J. Deisenhofer, H. Michel, and R. Huber, Structural homology of reaction centers from *Rhodopseudomonas sphaeroides* and *Rhodopseudomonas viridis* as determined by x-ray diffraction, *Proc. Natl. Acad. Sci. USA*, **83**:8589-8593 (1986).
7. C. H. Chang, D. Tiede, J. Tang, U. Smith, J. Norris, M. Schiffer, Structure of *Rhodopseudomonas sphaeroides* R-26 reaction center, *FEBS Letters* **205**:82-86 (1986).
8. W. Kabsch, A solution for the best rotation function to relate two sets of vectors, *Acta Crystallogr.* **A32**:922-923 (1976).
9. D. D. DeVault, "Quantum Mechanical Tunnelling in Biological Systems," Cambridge University Press (1984).
10. R. A. Marcus and N. Sutin, Electron transfers in chemistry and biology, *Biochim. Biophys. Acta* **811**:265-302 (1985).
11. W. W. Parson, S. Creighton, and A. Warshel. Calculations of Spectroscopic Properties and Electron transfer kinetics of Photosynthetic reaction centers, *J. Phys. Chem.*, in press.
12. M. E. Michel-Beyerle, M. Plato, J. Deisenhofer, H. Michel, M. Bixon, J. Jortner, Unidirectionality of charge separation in reaction centers of photosynthetic bacteria, *Biochem. Biophys. Acta* , in press.
13. B. Lee and F. M. Richards, The interpretation of protein structures: estimation of static accessibility, *J. Mol. Biol.* **55**:379-400 (1971).
14. J. P. Allen, G. Feher, T. O. Yeates, and D. C. Rees, Structure analysis of the reaction center from *Rhodopseudomonas sphaeroides*: electron density map at 3.5 Å resolution, *in*: "Progress in Photosynthesis Research," J. Biggins, ed., M. Nijohff/W. Junk, The Netherlands, Vol. 1, pp. 4-375-4-378 (1987).
15. G. Feher, R. A. Isaacson, M. Y. Okamura, and W. Lubitz, ENDOR of exchangeable protons of the reduced intermediate acceptor in reaction centers from *Rhodobacter sphaeroides* R-26. These proceedings.
16. W. W. Parson, and B. Ke, Primary photochemical reactions, *in* "Photosynthesis," Govindjee, ed., Academic Press, New York, pp. 331-385 (1982).
17. C. Kirmaier and D. Holten, Primary photochemistry of reaction centers from the photosynthetic purple bacteria, *Photosynth. Res.*, **13**:225-260 (1987).
18. B. A. Barry and G. T. Babcock, Tyrosine radicals are involved in the photosynthetic oxygen evolving system, *Proc. Natl. Acad. Sci. USA* **84**:7099-7103 (1987).
19. R. J. Debus, B. A. Barry, G. T. Babcock, L. McIntosh, Site-directed mutagenesis identifies a tyrosine radical involved in the photosynthetic oxygen evolving system, *Proc. Natl. Acad. Sci. USA*, in press.
20. W. Warwicker and H. C. Watson, Calculation of the electric potential in the active site cleft due to α-helix dipoles, *J. Mol. Biol.* **157**:671-679 (1982).

21. M. Y. Okamura, G. Feher, and N. Nelson, Reaction centers *in*: "Photosynthesis," Govindjee, ed. Academic Press, New York, pp. 195-272 (1982).

22. R. J. Debus, G. Feher, and M. Y. Okamura, Iron-depleted reaction centers from *Rhodopseudomonas sphaeroides* R-26.1: Characterization and reconstitution with Fe^{2+}, Mn2+, Co2+, Ni2+, Cu2+, Zn2+, *Biochemistry* **25**:2276-2287 (1986).

23. P. Marot, and C. A. Wraight, Light induced proton binding-unbinding dynamics in reaction centers from *Rhodobacter sphaeroides*, *in*: "Progress in Photosynthesis Research," J. Biggins, ed., M. Nijhoff/W. Junk, The Netherlands, Vol. 2, pp. 6.401-6.404 (1987).

24. P. H. McPherson, M. Y. Okamura, G. Feher, and M. Schonfeld, Light induced proton uptake by RCs from *Rb. sphaeroides* R-26.1 *Biophys. J.* (Abst.) **51**:225a (1987).

25. P. H. McPherson, M. Y. Okamura, and G. Feher, Light-induced proton uptake by photosynthetic reaction centers from *Rb. sphaeroides* R-26, manuscript in preparation.

26. W. A. Hendrickson, Stereochemically restrained refinement of macromolecules structures, *Methods Enzymol.* **115**:252-270 (1985).

27. M. Lutz, I. Agalidis, G. Hervo, R.J.Cogdell, and F. Reiss-Husson, On the state of carotenoids bound to reaction centers of photosynthetic bacteria: A Resonance Raman study, *Biochim. Biophys. Acta* **503**:287-303 (1987).

28. T. Kokitani, B. Honig, and A. R. Crofts, Theoretical studies of the electrochromic response of carotenoids in photosynthetic membranes, *Biophys. J.* **39**:57-63 (1982).

29. H. A. Frank, B. W. Chadwick, S. Taremi, S. Kolaczkowski, M. K. Bowman, Singlet and triplet absorption spectra of carotorenoids bound in the reaction centers of *Rhodopseudomonas sphaeroides* R26, *FEBS Letters* **203**:157-163 (1986).

30. C. C. Schenck, P. Mathias, M. Lutz, Triplet formation and triplet decay in reaction centers from the photosynthetic bacterium *Rhodopseudomonas sphaeroides*, *Photochem. Photobiol.* **39** 407-417 (1984).

SYMMETRY BREAKING STRUCTURES INVOLVED IN THE DOCKING OF CYTOCHROME C AND PRIMARY ELECTRON TRANSFER IN REACTION CENTERS OF RHODOBACTER SPHAEROIDES

D.M. Tiede, D.E. Budil, J. Tang, O. El-Kabbani and J.R. Norris
Chemistry Division, Argonne National Laboratory, Argonne, Illinois 60439

C.-H. Chang and M. Schiffer
Division of Biological and Medical Research, Argonne National Laboratory
Argonne, Illinois 60439

INTRODUCTION

A striking feature of Rb sphaeroides (1-4) and Rps viridis (5,6) reaction center structures is the approximate two-fold symmetry which relates the L and M protein subunits and the positions of the chromophores. Two sets of bacteriochlorophyll, B, bacteriopheophytin, H, and quinone, Q, molecules are positioned at nearly equivalent positions about the bacteriochlorophyll dimer, P. The chromophores are labelled L or M, according to whether they belong to the set predominately associated with the L or M subunit. Each set of chromophores potentially functions as an electron transfer pathway leading away from P. However the two-fold symmetry is only approximate. Significant differences have been noted in the protein environment surrounding the chromophores (2,6,7). These differences arise from changes in the type of amino acid residues at specific, symmetry equivalent positions on the L and M subunits. Small differences in the distances between chromophores on the L and M pathways were found in the Rps viridis structure after refinement using high resolution data (8).

The identification of structural asymmetries is important since spectroscopic evidence has indicated that both sets of chromophores are not photochemically equivalent. For example, low temperature formation of transient P^+H^- (9) and trapped PH^- (10,11) states involves only one bacteriopheophytin, presumably H_L. H_M can be photochemically reduced at room temperature following reduction of H_L with approximately 10-fold lower quantum yield (11). The comparison of photochemical reactivities and structural differences of the L and M pathways suggests that photosynthetic efficiency can be modified by relatively small changes in structure.

In this article we address three aspects of the correlation between symmetry breaking structures and photochemistry in reaction centers of Rb sphaeroides. First, we have asked whether a chemical asymmetry exists within P. This has been determined by measuring the orientation of the magnetic axes of the spin polarized triplet state in single crystals of Rps viridis and Rb sphaeroides. These directions were compared to the molecular coordinates of P. Second, we have compared the protein environments surrounding the pigments on the L and M pathways. Among the differences is a set of amino acid differences at symmetry related positions on the L and M subunits which are conserved in Rps viridis, Rb sphaeroides and capsulata. This suggests the possibility that some of these residues could contribute to the selectivity of electron transfer. Finally, we propose a site for the binding of c-cytochromes, and describe a model for the Rb sphaeroides reaction center-cytochrome c_2 complex.

Rb Sphaeroides REACTION CENTER STRUCTURE

The *Rb sphaeroides* R-26 reaction center structure was determined as described previously using the molecular replacement method (1,2). Previously, we reported a refinement of the *Rb sphaeroides* structure to 3.7Å resolution with a R-factor of 0.39 (1). At present we have extended the resolution to 3.2Å with a R-factor of 0.28. We have been able to fit 96% of the amino acid side chains into the electron density maps. To aid comparison of the *Rps viridis* and *Rb sphaeroides* structures, we have labeled the *Rb sphaeroides* amino acid residues according to the *Rps viridis* sequence and the alignment described by Michel et al (15).

I. CORRELATION OF PARAMAGNETIC STATES AND THE MOLECULAR STRUCTURE OF P

Whether both monomers of P contribute equally to the "working" electronic states of P can be determined by comparing the orientation of the spin polarized triplet axes to the molecular structure of P. The rationale for this experiment is straightforward. If both monomers are completely equivalent and the triplet state is delocalized over both molecules, then the triplet magnetic axes will be oriented in directions corresponding to the vector sum of triplet orientations for each monomer. Conversely, if the triplet state is localized on one molecule of the dimer, then the relationship between triplet and this monomer molecular axes will approximate that of a monomeric bacteriochlorophyll.

We have analyzed in detail (12) the spin polarized EPR signals measured in *Rps viridis* (13) and *Rb sphaeroides* (14) single crystals. A different, but striking correlation is found between magnetic and molecular coordinates of P in each case. In *Rps viridis* the x,y and z magnetic axes align within 5° of the molecular x,y and z directions for the monomer of P on the L pathway. In contrast, the triplet axes in the *Rb sphaeroides* crystals are found to coincide within 5° of an axis system derived by averaging the x,y and z molecular directions for each monomer of P. This near alignment of triplet magnetic and dimer molecular axes in *Rb sphaeroides* is shown in figure 1.

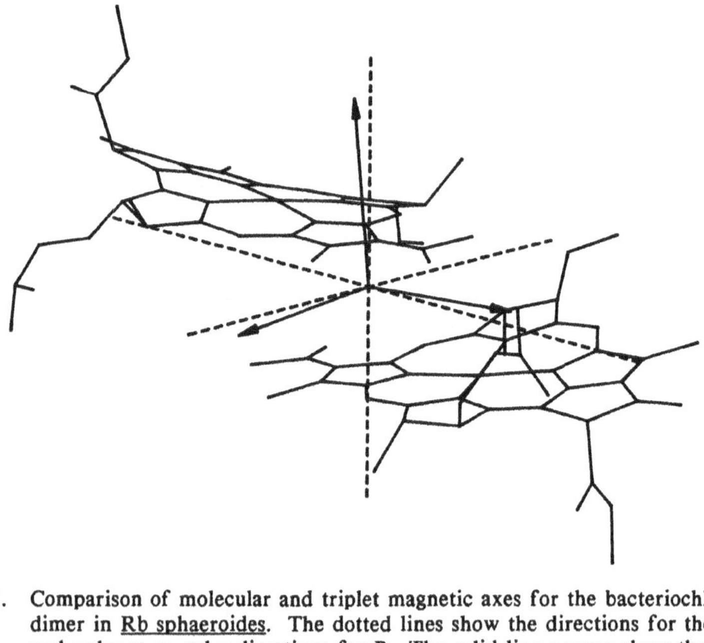

Figure 1. Comparison of molecular and triplet magnetic axes for the bacteriochlorophyll dimer in *Rb sphaeroides*. The dotted lines show the directions for the average molecular x, y and z directions for P. The solid line arrows show the magnetic x, y and z directions.

Figure 2. Aromatic and hydrogen bonding amino acid residues in the vicinity of the bacteriochlorophyll dimer in Rb sphaeroides.

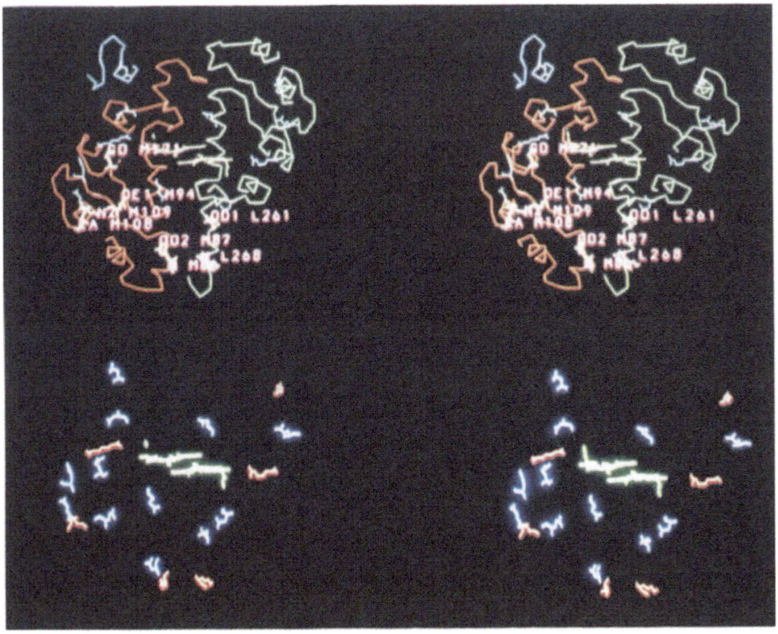

Figure 3. The cytochrome c binding site on the Rb sphaeroides reaction center. The top stereodiagram shows the periplasmic surface of the reaction center. The orange line shows the M subunit, green the L subunit and single helix of the H subunit is shown in blue. The bacteriochlorophyll dimer is seen edge-on in the center of the reaction center. The labels mark the locations of the symmetry breaking residues. The lower portion of the figure shows the same view, except the protein backbone has been removed. The patch of charges below the dimer is the proposed cytochrome c_2 binding site.

These results provide strong evidence that the triplet state is delocalized over both molecules of the dimer in Rb sphaeroides compared to localization on the L side bacteriochlorophyll in Rps viridis. The probability that these alignments could occur fortuitously is less than 0.1%. The accuracy of the angular determinations in the EPR measurement is at least equal to the certainty in molecular direction determination by x-ray diffraction. These results provide an experimental test for symmetry of an electronic state of P, unbiased by assumptions on the molecular wavefunctions. The greater delocalization of the triplet state in the Rb sphaeroides dimer predicts that the structural features which affect excited electronic state energies will be more symmetric than in Rps viridis.

Several asymmetries have been noted in the amino acid environment surrounding P in Rps viridis (5,6). In particular, differences in hydrogen bonding to the ring V keto group has been found for the L and M side monomers. A hydrogen bond is formed between this carbonyl and threonine L248 for the L side monomer. On the M-side, the threonine is replaced by a non-hydrogen bonding isoleucine. Conceivably, hydrogen bonding differences at this site could be significant, since these interactions could alter chemical reduction potentials and π-orbital distributions. Both L and M monomers show hydrogen bonding at the ring I acetyl carbonyl.

The amino acid environment surrounding P in Rb sphaeroides has fewer aromatic and hydrogen bonding amino acids than that in Rps viridis (2,6,7). In Rb sphaeroides the ring V keto carbonyls for both molecules of P are not hydrogen bonded. The residue equivalent to threonine L248 in the Rps viridis structure has be changed to a non-hydrogen bonding methionine. This means than unlike Rps viridis, both monomers of P are equivalently non-hydrogen bonded at the ring V keto position. Furthermore the tyrosine M195 which is hydrogen bonding to the acetyl carbonyl on the M-side monomer in Rps viridis has been changed to a phenylalanine in Rb sphaeroides, removing the possibility for hydrogen bonding at this position. The histidine equivalent to L168 in the Rps viridis structure is retained, and is positioned appropriately to hydrogen bond to the ring I acetyl carbonyl on the L-side monomer of P. If these hydrogen bonding interactions are significant in determining the triplet state energies, then the more delocalized state found in Rps sphaeroides would argue for a symmetric hydrogen bonding pattern. To preserve symmetry in the hydrogen bonding of the L and M-side monomers of P, we note that the ring I acetyl on the M-side monomer could be rotated by 180° from its position in the Rps viridis structure, which would put the carbonyl oxygen in position to hydrogen bond to tyrosine M208. Although this configuration would conserve symmetry in hydrogen bonding for the L and M monomers of P, it would destroy the possibility for maintaining strict C2 symmetry for the dimer. Figure 2 shows the hydrogen bonding and aromatic residues surrounding P in the Rb sphaeroides structure.

II. SYMMETRY BREAKING AMINO ACIDS IN THE CHROMOPHORE BINDING SITES

We have compared the amino acid residues on the L and M subunits which surround the chromophores P, B, and H in the Rb sphaeroides structure. These are tabulated in Tables 1,2,and 3 respectively. Simple aliphatic residues are listed by their one letter code; aromatic residues are listed in bold; and charged or polar residues are listed with their three letter code. Several notable differences are seen between residues on the L and M sides. An indication that some of these differences may contribute to the preferential involvement of the L pathway in photochemistry is suggested in those cases where the difference between the L and M side is conserved in three species of bacteria, Rps viridis, Rb sphaeroides, and Rb capsulata. These residues are marked by asterisks.

There is only one charged residue which is in the vicinity of the chromophores. This is glutamic L104, which is presumably protonated and in position to hydrogen bond to the H_L ring V carbonyl (6). Valine residues are at the corresponding position (M131) on the M-side in Rps viridis and Rb capsulata, ruling out hydrogen bonding at this position for H_M. This difference in hydrogen bonding for the two bacteriopheophytins has been proposed to contribute to the preferred use of H_L in electron transfer (6). In Rb sphaeroides M131 is a threonine, as is M144. Both residues are positioned near the H_M ring V carbonyl and could possibly form a hydrogen bond. At the present stage of refinement the oxygen-oxygen distances are a little over 3 Å for both residues. If hydrogen bonding at the ring V carbonyl of H is important in determining the direction of electron transfer it is expected that electron transfer in Rb sphaeroides may differ from that in Rb capsulata and Rps viridis. No such difference has been observed.

Table 1. Amino acid Residues in the vicinity of the special pair.

	L-side				M-side		
	sphaeroides	viridis	capsulata		sphaeroides	viridis	capsulata
L127	A	M	A	M154	L	F	L
L131	L	L	L	M158	L	I	L
L156	W	W	W	M183	W	W	W
L157	V	V	V	M184	Thr	L	Thr
L158	Ser	Asn	Ser	M185	Asn	Thr	Asn
L160	Thr	F	Thr	M187	F	F	F
L161	G	G	G	M188	Ser	Ser	Ser
*L162	Tyr	Tyr	Tyr	*M189	L	I	L
L167	F	W	F	M194	L	F	L
L168	His	His	His	M195	F	Tyr	F
L173	His	His	His	M200	His	His	His
L176	A	Ser	G	M203	Ser	Ser	Ser
L177	I	V	I	M204	I	I	I
L180	F	L	F	M207	L	A	L
*L181	F	F	F	*M208	Tyr	Tyr	Tyr
L240	A	I	A	M274	V	V	V
L244	Ser	G	Ser	M278	G	A	G
L247	Cys	G	Cys	M281	G	G	G
L248	M	Thr	M	M282	I	I	I
*M208	Tyr	Tyr	Tyr	*L181	F	F	F

Table 2. Amino acid residues in the vicinity of the bridging bacteriochlorophylls.

	L-side				M-side		
	sphaeroides	viridis	capsulata		sphaeroides	viridis	capsulata
*L97	F	F	F	*M124	V	L	V
L127	A	M	A	M154	L	F	L
L128	Tyr	F	Tyr	M155	W	V	W
L131	L	L	L	M158	L	I	L
*L146	F	F	F	*M173	V	V	P
L148	Tyr	Tyr	Tyr	M175	Tyr	F	Tyr
L150	I	I	I	M177	I	I	I
L151	W	L	W	M178	F	W	F
L153	His	His	His	M180	His	His	His
L154	L	L	L	M181	L	I	L
L156	W	W	W	M183	W	W	W
L157	V	V	V	M184	Thr	L	Thr
M195	F	Tyr	F	L168	His	His	His
M201	G	G	G	L174	M	M	M
M204	I	I	I	L177	I	V	I
M205	A	G	A	L178	Ser	Ser	Ser
*M208	Tyr	Tyr	Tyr	*L181	F	F	F

Table 3. Amino acid residues in the vicinity of the bacteriopheophytins.

	L-side				M-side		
	sphaeroides	viridis	capsulata		sphaeroides	viridis	capsulata
*L97	F	F	F	*M124	V	L	V
L100	W	W	W	M127	W	W	W
*L101	A	M	A	*M128	W	W	W
*L104	Glu	Glu	Glu	*M131	Thr	V	V
L117	I	V	I	M144	Thr	I	M
L118	P	P	P	M145	A	A	A
L121	F	F	F	M148	F	F	F
L124	A	P	A	M151	A	A	A
L237	Ser	A	A	M271	A	Ser	A
L238	L	Ser	L	M272	V	L	V
L240	A	I	A	M274	V	V	V
L241	V	F	V	M275	Thr	M	Thr
*M208	Tyr	Tyr	Tyr	*L181	F	F	F
M209	G	G	G	L182	Thr	V	Thr
M212	L	L	L	L185	L	M	W
M215	A	A	A	L188	A	G	A
M216	M	A	M	L189	L	L	M
*M250	W	W	W	*L216	F	F	F
*M253	Thr	Thr	Thr	*L219	L	V	L
M254	M	I	M	L220	V	V	M

The location of one of the conserved symmetry breaking residue pair, phenylalanine L181 and tyrosine M208, is striking. While both residues are aromatic only tyrosine is polar and has the potential to form a hydrogen bond. Phenylalanine L181 is within 4.5 Å (van der Waals distance) of four chromophores: P_L, P_M, B_M, H_M. Tyrosine M208 is within 4.5 Å of P_M, P_L, B_L, H_L. The conservation of these residues in three photosynthetic organisms and their central locations suggests that these residues may play a role in the primary electron transfer. Three conserved symmetry breaking pairs are in the vicinity of two chromophores. Residues, L97 and M124, are adjacent to both B and H. Histidine L168 forms a hydrogen bond with P_L and is close to B_M. The corresponding phenylalanine (tyrosine in Rps viridis) does not hydrogen bond, but is near to P_M and B_L. L216 and M250 are positioned directly between H and Q on the M and L sides respectively. The larger tryptophan ring M250 makes van der Waals contact with both H_L and Q_A, which could facilitate electron transfer between these components (6).

III. MODEL FOR THE CYTOCHROME C2-REACTION CENTER COMPLEX

We have also examined the binding of c-cytochromes to the Rb sphaeroides reaction center. Previous linear dichroism measurements showed that horse cytochrome c and Rps sphaeroides cytochrome c_2 orient differently in electron transfer complexes formed with the Rb sphaeroides reaction center (16). These experiments suggest that specific orientation of cytochrome in the complex will be dependent upon the distribution of charged residues on each protein. We have built a model of the Rb sphaeroides cyt c_2-reaction center complex, based upon the dichroism observed for this complex, and a complementary pairing of charges on each protein.

Examination of the reaction center periplasmic surface (figure 3) shows that the charged residues are not distributed symmetrically. Instead, a "patch" of charged residues is found. This patch is generated by appearance of a cluster of 7 charged residues which have uncharged residues at the symmetry related positions on the opposite protein subunit. These symmetry breaking charged residues occur on the surface ab segment (M86, M87, M94, M108), the L-terminus (L261, L268) and the surface cd segment (M171). A inversion of charge is also found by the replacement of glutamic M109 with lysine at the symmetry related position L82. In addition to these symmetry breaking charge residues, this patch also contains the charged residues (M99, M162, L257) which are conserved on both L and M subunits. We propose that this patch may function as the c-cytochrome binding site.

An x-ray structure of the Rb sphaeroides cyt c_2 is not currently available. Of the structures available in the Brookhaven National Laboratory Protein Data Bank, the cyt c_2 from Paracoccus denitrificans (17) has the closest homology to the Rb sphaeroides cytochrome (18). We have made a model of the Rb sphaeroides cytochrome from the Paracoccus structure by replacing the surface amino acid side chains with those that matched the sequence in Rb sphaeroides cytochrome. This model Rb sphaeroides cyt c_2 was used to build a model of the reaction center-cyt c_2 complex. Dichroism measurements previously determined the orientation of cyt c_2 with respect to the membrane normal (16). The model cyt c_2 was held at the same angle with respect to the reaction center two-fold symmetry axis. The cytochrome was then translated or rotated about a direction parallel to the reaction center symmetry axis in order to maximize charge pairing.

The largest number of charged pairs occurred with the cyt c_2 positioned over the reaction center patch described above. A model for the Rb sphaeroides cyt c_2-reaction center complex is shown in figure 4. This model complex allows formation of at least 11 salt bridges between residues on the reaction center and cytochrome. These are listed in Table 4. The symmetry of the reaction center residues indicates whether charge is conserved (+) or non-conserved (-) at the symmetry related position on the other reaction center subunit. This model uses 6 of the 8 symmetry breaking reaction center charged residues. The residues in parenthesis next to the RC column show the corresponding residues in the Rps viridis sequence when they differ from those in Rb sphaeroides. Several changes may be expected since Rb sphaeroides cyt c_2 and the Rps viridis cytochrome subunit are completely unrelated. This charged patch is almost completely conserved in Rb capsulata, except for lysine L268 which has been changed to an asparagine Rb capsulata and sphaeroides are similar in that both use a cyt c_2 as an electron donor to the RC. The first column gives the amino acid residues on the horse cyt c at positions homologous to those listed for the Rb sphaeroides cyt c_2. The residues in parenthesis show those positions where analogous charge residues are absent.

Table 4. Complementary Charge Pairs in the Rb sphaeroides Cytochrome c_2-Reaction Center Complex

cyt c	cyt c_2	RC	symmetry
lys 10	lys 10	asp M290	+
lys 25	arg 32	asp L155	+
(thr 47)	lys 55	asp L257 (arg)	+
glu 50	glu 59	lys L268 (asp)	-
glu 69	asp 82	lys M108 (his)	-
(------)	asp 93	arg M86	-
(------)	lys 95	asp M87 (gln)	-
(------)	lys 97	asp L261	-
lys 79	lys 99	asp M182	+
lys 85	lys 105	glu M99 (gln)	+
lys 86	lys 106	glu M94 (tyr)	-

Significantly, this model includes salt bridges formed between the reaction center and asp 93, lys 95, lys 97 on cyt c_2. These are part of an 8 amino acid insert which is present in cyt c_2 but is absent in cyt c. This model can provide an explanation for the difference in dichroism seen for the reaction center-cyt c and c_2 complexes. The absence of the insert segment may allow the bound cyt c to rotate away from this region, in accordance with the measured dichroism (16).

This model for the reaction center-cyt c_2 complex differs from that proposed by Allen et al (4). In the present case, the heme Fe^{+} to P Mg^{2+} is 20.5 Å, the binding site is determined by the patch of charges which is shifted slightly towards the M-side of the reaction center, and the angle between the heme normal and the average of the normals for the two bacteriochlorophylls of P is approximately 53°. The model described by Allen et al

(4), used the crystal structure of the R rubrum cyt c_2. Their model emphasized the formation of salt bridges with charged residues on the R rubrum cyt c_2 which are conserved in several species, attempted to maximize coplanarity between heme and bacteriochlorophyll rings, and positioned the binding site directly over the dimer. These may not be reliable criteria. Our previous experiments indicated that the orientation of the cytochrome will be sensitive to the specific distribution of charged residues on different cytochromes (16). This suggests that docking of the cytochrome will not be exclusively determined by the positions of charged residues which are conserved in several species. Furthermore, the structures of the Rps viridis and Rb sphaeroides reaction centers do not support the notion that coplanar alignment of donors and acceptors are required for efficient electron transfer. Clearly the model building approach is not definitive. However these models will be extremely important in designing experiments that identify how structural parameters regulate bimolecular electron transfer.

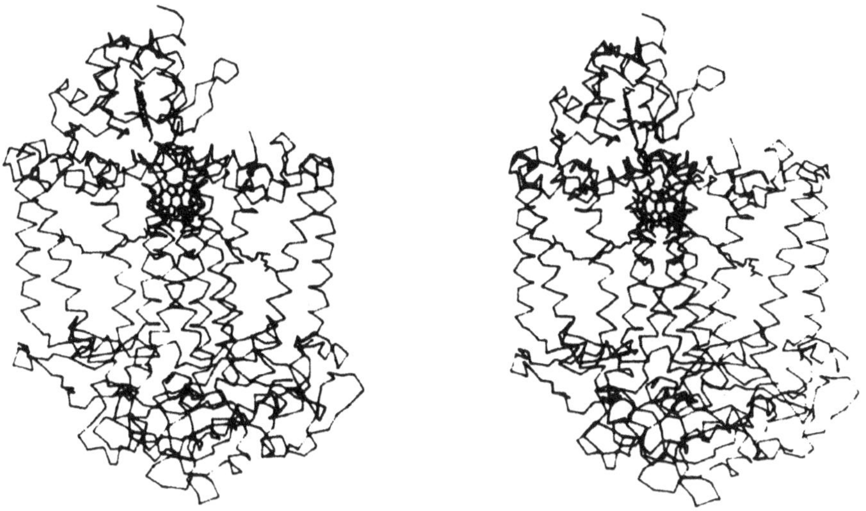

Figure 4. A model for the Rb sphaeroides cytochrome c_2-reaction center electron transfer complex.

ACKNOWLEDGEMENTS

D.M.T., D.E.B., O.E., J.T. and J.R.N. were supported by the U.S. Department of Energy, Office of Basic Energy Sciences, Division of Chemical Sciences under contract W-31-109-Eng-38; C.-H.C. and M.S. were supported by Public Health Service grant GM36598 and by the Office of Health and Environmental Research, U.S. Department of Energy under contract W-31-109-Eng-38.

REFERENCES

1. C.-H. Chang, D.M. Tiede, J. Tang, U. Smith, J.R. Norris, and M. Schiffer, Structure of Rhodobacter sphaeroides R-26 reaction center, FEBS Lett 205:82 (1986).
2. C.-H. Chang, D.M. Tiede, J. Tang, J. Norris and M. Schiffer, Crystallographic studies of the photosynthetic reaction center from R. sphaeroides, in:"Progress in Photosynthesis Research," J. Biggins, ed., Martinus Nijhoff Publishers, Dordrecht (1986).

3. J.P. Allen, G. Feher, T.O. Yeates, H. Komiya, and D. C. Rees, Structure of the reaction center from Rhodobacter sphaeroides R-26: The cofactors, PNAS 84:5730 (1987).
4. J.P. Allen, G. Feher, T.O. Yeates, H. Komiya, and D. C. Rees, Structure of the reaction center from Rhodobacter sphaeroides R-26: The protein subunits, PNAS 84:6162 (1987).
5. J. Deisenhofer, O. Epp, K. Miki, R. Huber and H. Michel, Structure of the protein subunits in the photosynthetic reaction centre of Rhodopseudomonas viridis at 3Å resolution, Nature 318:618 (1985).
6. H. Michel, O. Epp and J. Deisenhofer, Pigment-protein interactions in the photosynthetic reaction centre from Rhodopseudomonas viridis, EMBO J. 5:2245 (1986).
7. D. Budil, P. Gast, C.-H. Chang, M. Schiffer and J. R. Norris, Three-dimensional x-ray crystallography of membrane proteins: Insights into electron transfer, Ann. Rev. Phys. Chem. 38:561 (1987).
8. J. Deisenhofer, O. Epp and H. Michel, The crystal structure of the photosynthetic reaction centre from Rhodopseudomonas viridis, refined at 2.3Å resolution, These proceedings.
9. C. Kirmaier, D. Holten and W.W. Parson, Picosecond-photodichroism studies of the transient states in Rhodobacter sphaeroides reaction centers at 5 K, BBA 810:49 (1985).
10. D.M. Tiede, E. Kellogg and J. Breton, Conformational changes following reduction of the bacteriopheophytin electron acceptor in reaction centers of Rhodopseudomonas viridis, BBA 829:294 (1987).
11. S. Florin and D.M. Tiede, Photochemical reduction of either of the two bacteriopheophytins in bacterial photosynthetic reaction centers, in: "Progress in Photosynthesis Research," J. Biggins, ed., Martinus Nijhoff Publishers, Dordrecht (1987).
12. J.R. Norris, D.E. Budil, P. Gast, C.-H. Chang and M. Schiffer, unpublished information (1987).
13. P. Gast, M.R. Wasielewski, M. Schiffer and J.R. Norris, Orientation of the primary donor in single crystals of Rhodopseudomonas viridis reaction centers, Nature 305:451 (1983).
14. P. Gast and J.R. Norris, EPR detected triplet formation in a single crystal of reaction center protein from the photosynthetic bacterium Rhodobacter sphaeroides R-26, FEBS Lett 177:277 (1984).
15. H. Michel, K.A. Weyer, H. Gruenberg, I. Dunger, D. Oesterhelt, and F. Lottspeich, The "light" and "medium" subunits of the photosynthetic reaction centre from Rhodopseudomonas viridis: Isolation of the genes, nucleotide and amino acid sequence," EMBO J. 5:1149 (1986).
16. D.M. Tiede, Cytochrome c orientation in electron transfer complexes with photosynthetic reaction centers of Rhodobacter sphaeroides and when bound to the surface of negatively charged membranes: Characterization by optical linear dichroism, Biochemistry 26:397 (1987).
17. R. Timkovich and R.E. Dickerson, The structure of Paracoccus denitrificans cytochrome c_{550}, J. Biol. Chem. 251:4033 (1976).
18. T.E. Meyer and M.D. Kamen, New perspectives on c-type cytochromes, Adv. Protein Chem. 35:105 (1982).

CRYSTALLOGRAPHIC STUDIES OF THE PHOTOSYNTHETIC REACTION CENTER FROM WILD TYPE *RHODOBACTER SPHAEROIDES* (Y STRAIN)

Arnaud Ducruix, Bernadette Arnoux and Françoise Reiss-Husson*

Institut de Chimie des Substances Naturelles and *ER 307
Laboratoire de Photosynthèse, CNRS
91198 Gif sur Yvette, France

Reaction centers (RC) from various purple bacteria are expected to show many structural similarities. Indeed they fulfill analogous functions, and they are built in the same way by three polypeptide chains bearing a number of cofactors: four bacteriochlorophylls, two bacteriopheophytins, one carotenoid, two quinones and a metal ion (most often Fe). Thus the structure of the *Rps. viridis* RC, the first to be resolved at atomic resolution (1-3) could be considered as a model for this family of pigment-protein complexes. Very recently, the main structural features of *Rps. viridis* RC have been shown to be conserved in the R26 *Rb. sphaeroides* RC (4-7).

Here we report on the state of our current work on the structure of a RC from another *Rb. sphaeroides* strain (Y, wild type). The reason for carrying out such a structural study is two-fold. The RC from the *Rb. sphaeroides* strain Y differs from those of the R26 mutant in a few but important properties. Unlike the carotenoid - less R26 RC, this RC contains a firmly bound spheroidene molecule, present in a specific -cis conformation (8) ; its assignment has been reported at this meeting by Lutz and co-workers as a 15 cis configuration with an out of plane twisting of the chain. This spheroidene molecule plays a role in excitation and deexcitation processes (9) and should be in a precise position within the pigment array to fulfill its functions. Hopefully this localization will be determined in our study ; it could be compared to that reported by Deisenhofer and co-workers at this meeting for hydroxyneurosporene in *Rps. viridis*.

Another interesting difference between the two *Rb. sphaeroides* strains concerns the nature of the metal interacting with the quinones. Fe is found in R26 RC (10) (as in *Rps. viridis*). Mn is found predominantly in Y RC when cells are grown in a normal medium, and may totally replace Fe under iron deficiency ; this Mn interacts magnetically with the quinones (11). In the Y strain, *in vivo* exchange of this Mn for other metals such as Fe (1), Co or Zn (I. Agalidis, F. Reiss-Husson, A. W. Rutherford, unpublished results), and Cu (12) is readily performed ; in the R26 strain, Fe has been exchanged only *in vitro* after a rather drastic treatment of the RC (13). This difference in metal affinity is unexplained ; a description of the Mn binding site in Y RC crystals would be obviously interesting in this respect.

Characterization of RC crystals from Y *Rb. sphaeroides*

We have already reported the preliminary characterization by X ray diffraction of RC crystals from Y *Rb. sphaeroides* (14). Crystals were grown in the presence of β octylglucoside, with PEG 4000 and NaCl as precipitants, either by vapour diffusion or by microdialysis. Brown crystals were obtained within a few weeks ; the largest had dimensions of 1 to 3 mm x 0.15 mm x 0.15 mm. Only one form was observed, with $P2_12_12_1$ as the space group. The parameters of the orthorhombic cell are a = 143.2 Å, b = 139.8 Å, c = 77.6 Å ; they are indicative of the presence of one RC per asymmetric unit, and not very different from the parameters of R26 RCs in the same form (4, 5).

At that time (14) we measured the Mn and Fe contents of several washed crystals by atomic absorption spectrometry ; $\sim$ 1.1 $\pm$ 0.1 Mn and 0.3 $\pm$ 0.07 Fe per RC were found. The ratio Mn/Fe was higher in the crystals than in the starting solution, as expected if Fe was a contaminant. We verified also by solvent extraction of several washed crystals and absorption spectrophotometry of the extracts that the carotenoid content of the crystals was the same as that of freshly purified RCs.

Since then, in collaboration with Dr. W. Mäntele (University of Freiburg) we have been able to obtain linear dichroic spectra of thin single crystals in the visible and near infrared range. These crystals are photochemically active ; decay kinetics of P^+ have been measured (W. Mäntele and F. Reiss-Husson, unpublished results).

Crystallization condition and X ray studies

Since our first report (14) we have tried to optimize the conditions of crystallization, for improving the yield of crystals of a size (and quality) suitable for X ray studies. Some of our observations might be valuable in crystallization trials of other RCs or membrane proteins.

The first point is the requirement of purity ; by purity we mean not only the absence of proteic contaminants but also homogeneity and low lipid content. Indeed at the beginning of this work we isolated RCs by a published procedure (15) involving as the final step a DEAE cellulose (Whatman 52) chromatography with a linear NaCl gradient in presence of 0.1 % LDAO. The RC protein was eluted in a broad asymmetrical band, which might be considered as pure by the usual SDS-PAGE electrophoresis criteria, but which still contained a noticeable amount of phospholipids (14-17 per RC). With such a preparation, only microcrystals were occasionally observed using vapour diffusion (see below). An additional chromatofocalisation of the protein sample on a PB74 gel (Pharmacia) in 0.1 % LDAO with a linear pH gradient (from pH 7 to 5.5) revealed however that it was not homogeneous : a number of overlapping bands (up to 4) was observed, with apparent pIs in the range 6.15 to 6.40.

At that stage we tried other ion exchange resins instead of DEAE-cellulose : DEAE-Trisacryl (IBF), DEAE-Sephacel and DEAE-Sepharose (Pharmacia). The last one was the best ; the RC was eluted in 0.1 % LDAO with a linear NaCl gradient in two well separated bands, both with a satisfactory purity index (A 280/A 802 < 1.24). The major one contained only $\simeq$ 2 phospholipids per RC ; it was homogeneous both in a HPLC ion exchange chromatogram (Fig. 1) and by chromatofocusing (single apparent pI = 6.3). With this fraction the crystallization was clearly improved. The minor fraction was not homogeneous by these two criteria and was not used further for crystallization.

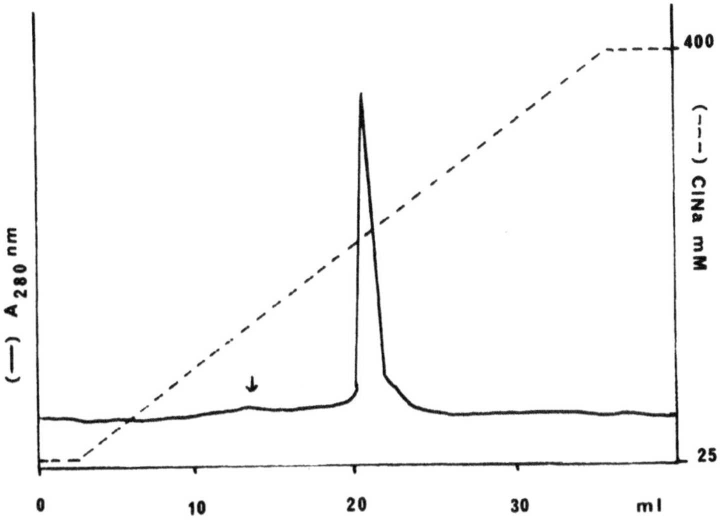

Fig. 1. Elution of RC major fraction from a TSK-5PW-DEAE ion exchange column, with a linear NaCl gradient from 25 to 400 mM in 10 mM Tris, 1 mM EDTA, 0.1 % LDAO buffer pH 8.0. (18° C, 1 ml/min). Arrow points towards baseline artefact.

The second point is the relative advantages of microdialysis versus vapour diffusion, at least for Y *Rb. sphaeroides* RC in the β octylglucoside-(PEG + NaCl) system. When vapour diffusion experiments were performed as already described (14), the behaviour of identical drops placed in the same chamber and equilibrated against a given reservoir was often not identical. In some of them, crystals grew without phase separation ; but most often, phase separation proceeded in the same time, and the growth of the

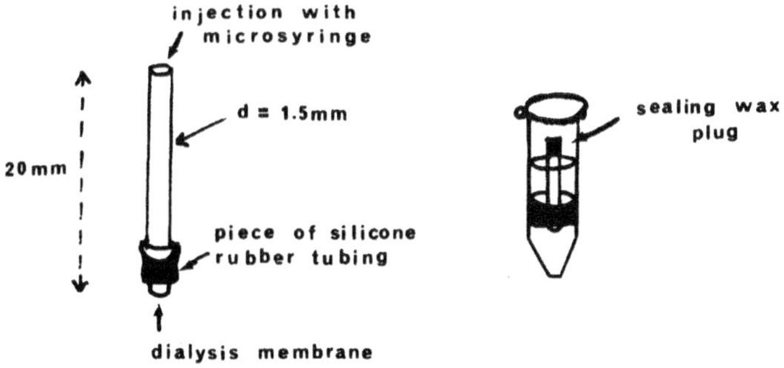

Fig. 2. Experimental set up (after (16)) for microdialysis in disposable glass capillaries (used for micropipettes), with 40 μl samples and a 500 μl reservoir in an Eppendorf conical tube.

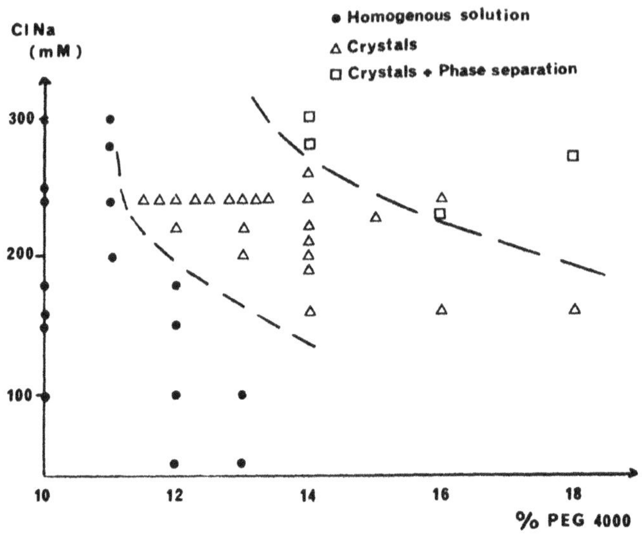

Fig. 3. Phase diagram observed after dialysis at 18° C of 40 µl samples containing : 2 mg RC protein/ml, 0.8 % β octylglucoside, 15 mM Tris-HCl pH 8.0 and initially NaCl and PEG 4000 at third of their final concentration. Ordinates are PEG and NaCl concentrations in the reservoir, which also contains 0.8 % β octylglucoside.

crystals was quite impaired. In such experiments, concentrations of all components of the drops are increasing in parallel; we turned therefore to microdialysis in order to better control each of them separately.

Keeping the detergent and the protein concentrations constant in a very simple dialysis set-up (Fig. 2), we could define a phase diagram giving a range of final PEG 4000 and NaCl concentrations where crystallization took place reproducibly without phase separation (Fig. 3). We should notice however that this diagram is dependent on the rate of the dialysis, that is on a number of factors, such as the initial PEG and NaCl concentrations, the dialysis membrane characteristics (thickness, cut-off value) and even the shape of the capillary (i.e. surface/volume ratio of the sample). Nevertheless, we now obtain consistently crystallization in ≃ 90 % of the samples ; although the majority of the crystals are still thin (< 150 x 150 µm in section), crystals with a size of 1 to 2 mm x 0.4 mm x 0.3 mm have been obtained.

We have now collected a complete set of data, obtained at 18° C from 5 crystals, on a rotation camera at the LURE Synchrotron facility (Orsay). The resolution extends to 2.6 Å ; processing of the films is in progress.

References

1. J. Deisenhofer, O. Epp, K. Miki, R. Huber, and H. Michel, J. Mol. Biol. 180:385 (1984).
2. J. Deisenhofer, O. Epp, K. Miki, R. Huber, and H. Michel, Nature 318:681 (1985).
3. H. Michel, O. Epp, and J. Deisenhofer, Embo J. 5:2445 (1986).

4. J. P. Allen, G. Feher, T. O. Yeates, D. C. Rees, J. Deisenhofer, H. Michel, and R. Huber, Proc. Natl. Acad. Sci. USA 83:8589 (1986).
5. C. H. Chang, D. Tiede, J. Tang, U. Smith, J. Norris, and M. Schiffer, FEBS Lett. 205:82 (1986).
6. J. P. Allen, G. Feher, T. O. Yeates, H. Komiya, and D. C. Rees, Proc. Natl. Acad. Sci. USA 84:5730 (1987).
7. J. P. Allen, G. Feher, T. O. Yeates, H. Komiya, and D. C. Rees, Proc. Natl. Acad. Sci. USA 84:6162 (1987).
8. M. Lutz, I. Agalidis, G. Hervo, R. J. Cogdell, and F. Reiss-Husson, Biochim. Biophys. Acta 503:287 (1978).
9. W. W. Parson, and T. G. Monger, Brookhaven Symp. Biol. 28:196 (1976).
10. G. Feher, R. A., Isaacson, J. D. Mc Elroy, L. Ackerson, and M. Y. Okamura, Biochim. Biophys. Acta 368:135 (1974).
11. A. W. Rutherford, I. Agalidis, and F. Reiss-Husson, FEBS Lett. 182:151 (1985).
12. S. K. Buchanan, and G. C. Dismukes, Biochemistry 26:5049 (1987).
13. R. J. Debus, G. Feher, and M. Y. Okamura, Biochemistry 25:2276 (1986).
14. A. Ducruix, and F. Reiss-Husson, J. Mol. Biol. 193:419 (1987).
15. E. Rivas, F. Reiss-Husson, and M. Le Maire, Biochemistry 19:2943 (1980).
16. S. E. Pronk, H. Hofstra, H. Groendiijk, J. Kingma, M. Swark, F. Dorner, J. Drenth, W. G. Hol, and B. Witholt, J. Biol. Chem. 260:13580 (1985).

SINGLE CRYSTALS OF THE PHOTOCHEMICAL REACTION CENTER FROM *Rhodobacter sphaeroides* WILD TYPE STRAIN 2.4.1 ANALYZED BY POLARIZED LIGHT

Harry A. Frank[a], Shahriar S. Taremi[a], James R. Knox[b] and Werner Mäntele[c]

Departments of [a]Chemistry and [b]Molecular and Cell Biology, University of Connecticut, Storrs, CT 06268 and [c]Institüt für Biophysik und Strahlenbiologie, D-7800 Freiburg FRG

Introduction

The crystal structures of the reaction centers from the carotenoidless mutant *Rb. sphaeroides* R26 and from *Rps. viridis,* have provided the means to explore the relationships between the structures of the chromophores and their spectroscopic properties (Deisenhofer *et al.*, 1985; Allen *et al.*, 1986; Chang *et al.*, 1986). Prompted by the fact that one specific relationship, that between the carotenoid molecule and its various spectroscopic properties has not been fully elucidated by the existing structures, we set out to crystallize and study the reaction center protein from *Rb. sphaeroides* 2.4.1. The abundance of biochemical, spectroscopic and dynamics information previously obtained (Cogdell *et al.*, 1976; Parson & Monger, 1976; Lutz *et al.*, 1978; Vermeglio *et al.*, 1978; McGann & Frank, 1985; Chadwick & Frank, 1986; Gagliano *et al.*, 1986) for this particular bacterial strain and its associated carotenoid, spheroidene, makes the elucidation of the structure of this reaction center complex of particular importance.

Recently, we have reported the crystallization and preliminary X-ray diffraction and optical spectroscopic characterization of the photochemical reaction centers from *Rhodobacter sphaeroides* 2.4.1 (Frank *et al., 1987*). The crystals were obtained in a solution of β-octylglucoside by the vapor diffusion technique using polyethylene glycol 4000 as the precipitant at 22 °C. The orthorhombic crystals (space group $P2_12_12_1$) have cell constants a = 142.5 Å, b = 136.1 Å, c = 78.5 Å and diffract to 3.7 Å. The X-ray data collection has now been completed at the CHESS synchrotron facility at Cornell University and the structural analysis of the reaction center protein is underway.

In the present work we present a detailed optical spectroscopic characterization of these reaction center crystals using polarized light. It was found that the crystals display pronounced linear dichroism throughout the absorption spectral region. The data taken in conjunction with results obtained from borohydride-treated *Rb. sphaeroides* 2.4.1 reaction centers (in solution) reveal the wavelengths of specific chromophore absorption bands. The data will allow a calculation of the angles the transition moments have with respect to the crystal axes. The crystals are shown via transient optical absorption changes to be photoactive.

Methods

Rb. sphaeroides 2.4.1 reaction centers were prepared according to the procedure described by Frank *et al.*, 1987. Crystallization was accomplished by vapor diffusion at 22°*C* in the dark. 50 μℓ droplets containing 30 μM reaction centers, 0.8% β-octylglucoside, 10% (w/v) PEG 4000, 1% heptanetriol, 0.15M NaCl, 10mM Tris buffer at pH 8.0, 1mM EDTA and 0.1% NaN_3 were equilibrated against 22% (w/v) PEG 4000, 0.25M NaCl, 10mM Tris buffer at pH 8.0, 1mM EDTA, 0.1% NaN_3 . After one week at 22°*C*, long prismatic crystals were observed. After about 2-4 weeks crystal growth was complete, and crystals with typical dimensions of 1.5mm x 0.05mm x 0.05mm were obtained.

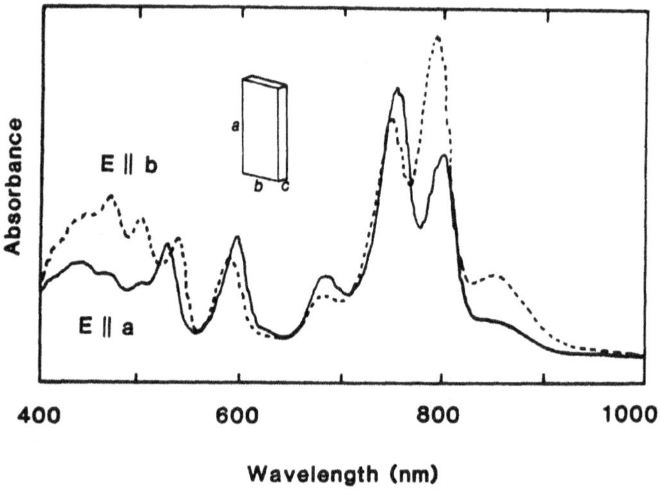

Figure 1. Polarized absorbance spectra of a single crystal of the photochemical reaction center from *Rhodopseudomonas sphaeroides* wild type strain 2.4.1.

Results

The detailed optical analysis was carried out using an absorption microspectrometer. (For a description of the apparatus, see Mäntele , W. *et al.*, this volume.) Figure 1 shows the polarized absorption spectra of the crystals. Figure 2 displays the linear dichroism calculated from the spectra taken in Figure 1.

The photoactivity of the crystallized reaction centers was tested by observing the light-induced absorption changes associated with charge separation in the reaction center. Upon photooxidation of the primary donor, the 860nm band bleaches and there is a shift of the 800nm reaction center band to shorter wavelengths (Vermeglio & Clayton, 1976). Precisely this behavior was observed in the crystals (Figure 3) confirming the integrity of the protein in its crystalline state.

Discussion

Linear dichroism of single crystals of reaction centers provide a means of determining the precise wavelengths and orientations of the chromophore transition moments. We proceed to discuss each spectral region in turn.

Primary donor Q_y : Figures 1 and 2 show small polarized absorption bands and negative linear dichroism for the primary donor around 860nm. The addition of sodium ascorbate to the mother liquor in which the crystals were suspended did not enhance these absorption bands or shift the bacteriochlorophyll monomeric Q_y bands. These changes would be expected if the primary donor was oxidized prior to treatment with ascorbate. Thus, these data indicate that the primary donor Q_y transition lies predominantly in the *bc* crystal plane, with some projection onto the *b* crystal axis and less projection onto the *a* crystal axis.

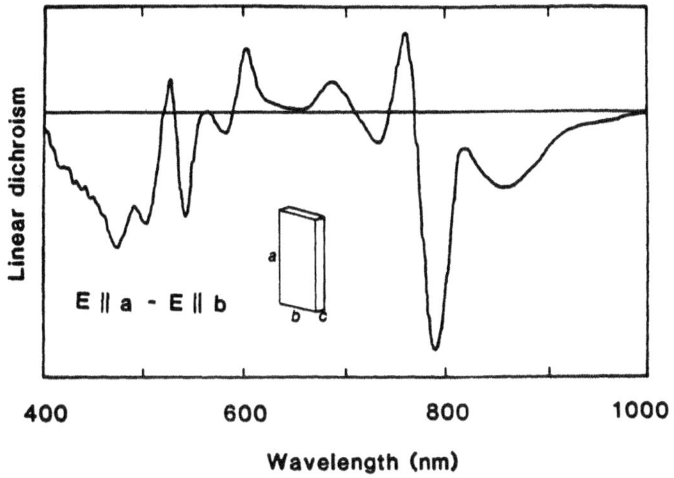

Figure 2. Linear dichroism of a single crystal of the photochemical reaction center from *Rhodopseudomonas sphaeroides* wild type strain 2.4.1.

Bacteriochlorophyll monomer Q_y : Upon rotation of the direction of the polarized measuring beam from parallel to *a* to parallel to *b* one observes a pronounced buildup of absorption intensity, a 7nm shift to shorter wavelength of the peak maximum and a marked broadening of the absorption feature around 800nm. This indicates that there are at least two transitions which contribute to this band. The peak positions indicate one transition at 803nm and another at 797nm, the former having its transition moment oriented close to the *b* crystal axis. This assignment is based on the extent of band broadening observed upon rotation of the measuring beam polarization from parallel to *a* to parallel to *b*. By a comparison with the absorption spectrum of borohydride-treated *Rb. sphaeroides* 2.4.1 reaction centers (Figure 4) this long wavelength (803nm) transition

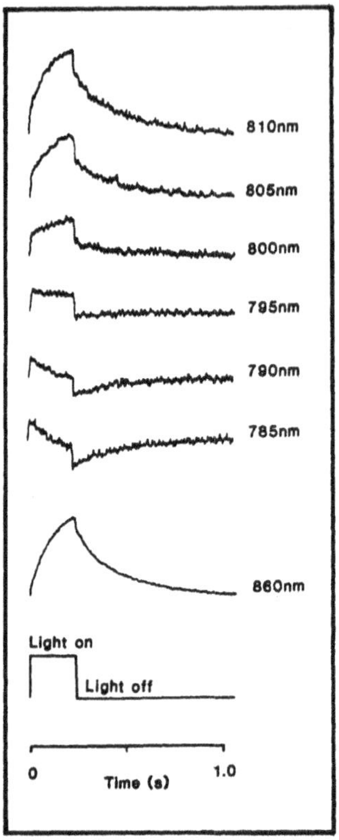

Figure 3. Transient absorption changes in a single crystal of the photochemical reaction center from *Rhodopseudomonas sphaeroides* wild type strain 2.4.1.

can be assigned to the monomeric BChl bound to the M protein subunit. The shorter wavelength (797nm) transition is thus associated with the L subunit monomeric BChl and is calculated from these data to have a large projection on the *b* crystal axis.

Bacteriopheophytin Q_y : Upon rotation of the direction of the polarized measuring beam from parallel to *a* to parallel to *b* one observes a small decrease of absorption intensity, an approximate 4nm shift to shorter wavelength of the peak maximum and a slight narrowing of the absorption feature around 760nm. As in the case for the monomeric BChl Q_y transitions this indicates that there are at least two transitions which contribute to this band. There is one at a slightly longer wavelength (probably associated with the L subunit (Michel *et al.*, 1986)) than the other. The longer wavelength transition moment is oriented with a large projection on the *a* crystal axis while the shorter wavelength transition probably has rather equal projections onto both the *a* and *b* crystal axes.

Bacteriochlorophyll monomer Q_x : In the region around 600nm there is a 5nm blue shift upon rotation of the direction of light polarization from parallel to *a* to parallel to *b*. The blue shift suggests that the long wavelength absorbing BChl (assigned above to BChl M) is oriented with its Q_x transition predominantly along the *a* crystal axis. The above conclusion that the Q_y of this molecule has a large projection along *b* is consistent with this assignment and places the plane of the BChl M chromophore in the *ab* crystal plane. The shorter wavelength BChl Q_x (assigned to the BChl L) displays a relatively small change in the linear dichroism experiment. Because the Q_y of this molecule is clearly oriented along the *b* crystal axis, its Q_x is probably positioned out of the *ab* plane.

Bacteriopheophytin Q_x : In the region around 535nm there is a large (approximately 9nm) shift to longer wavelength of the absorption maximum upon rotation of the direction of light polarization from parallel to *a* to parallel to *b*. Part of the shift may be attributable to enhanced carotenoid absorption in the spectrum taken with the polarizer parallel to *b*. Nevertheless, the shift to longer wavelength suggests that the long wavelength absorbing BPheo is oriented with its Q_x transition predominantly along the *b* crystal axis. The above conclusion that the Q_y of this molecule has a large projection along *a* places the plane of this BPheo L chromophore in the *ab* crystal plane. The shorter wavelength BPheo Q_x (assigned to the BPheo M) displays only small changes in the linear dichroism experiment suggesting that it has equal projections onto the *a* and *b* crystal axes. Taken together with the above conclusion that the Q_y of the M subunit BPheo also has similar projections onto these axes, the ring system of this short wavelength absorbing BPheo (like the L subunit BPheo) can be placed close to the *ab* crystal plane.

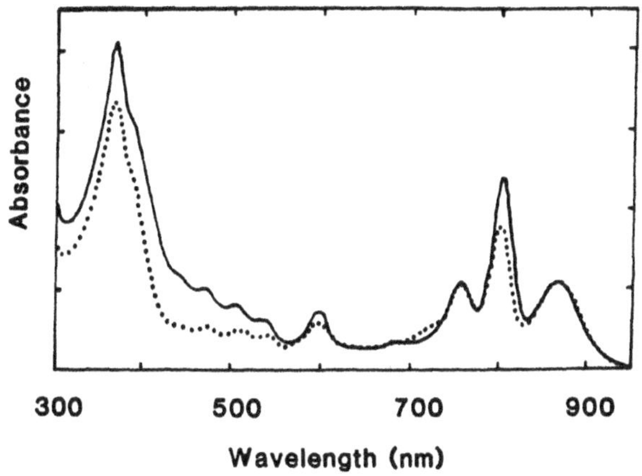

Figure 4. Reaction centers of *Rhodopseudomonas sphaeroides* wild type strain 2.4.1 treated with potassium borohydride essentially according to the procedure outlined by Ditson *et al.*, 1984, and Maroti *et al.*, 1985, with some minor modifications. The procedure is known to remove one of the monomeric bacteriochlorophylls (presumably the bacteriochlorophyll on M subunit side). The solid line represents the absorption spectrum of the reaction centers before treatment. The dotted line represents the absorption spectrum after treatment with borohydride. A difference spectrum generated by subtracting these two spectra (data not shown) has a peak at 803nm. The carotenoid absorption bands are greatly diminished in the treated material indicating the loss of a large portion of the bound carotenoid in those reaction centers.

Carotenoid: The carotenoid displays a strong negative linear dichroism consistent with the previous qualitative assignment (Frank *et al.*, 1987) that this molecule's transition moment is predominantly projected along the *b* crystal axis. Previous photoselection (Vermeglio *et al.*, 1978) and magnetophotoselection (McGann & Frank, 1985) studies have shown that the carotenoid transition moment makes an angle of about 75° with respect to the primary donor Q_y. This is consistent with the above assignment that the primary donor Q_y is positioned out of the *ab* plane.

Acknowledgements

This work was supported by grants from the Competitive Research Grant Office of the U. S. Department of Agriculture (86-CRCR-1-2016) and the National Institutes of Health (GM-30353) to H. A. F. and (GM-37742) to J. R. K.

References

Allen, J. P., Feher, G., Yeates, T. O., Rees, D. C., Deisenhofer, J., Michel, H. & Huber, R. (1986) *Proc. Nat. Acad. Sci. USA.* **83,** 8589-8593.

Chadwick, B. & Frank, H. A. (1986) *Biochim. Biophys. Acta* **851,** 257-266.

Chang, C.-H., Tiede, G., Tang, J., Smith, U., Norris, J. & Schiffer, M. (1986) *FEBS Letters* **205,** 82-86.

Cogdell, R. J., Parson, W. W. & Kerr, M. A. (1976) *Biochim. Biophys. Acta* **430,** 83-93.

Deisenhofer, J., Epp, O., Miki, K., Huber, R. & Michel, H. (1985) *Nature* **318,** 618-624.

Ditson, S. L., Davis, R. C. & Pearlstein, R. M. (1984) *Biochim. Biophys. Acta* **766,** 623-629.

Frank, H. A., Taremi, S. S. & Knox, J. R. (1987) *J. Mol. Biol.* **197,** in press.

Gagliano, A. G., Breton, J. & Geacintov, N. E. (1986) *Photobiochem. Photobiophys.* **10,** 213-221.

Krinsky, N.I. (1971) In *Carotenoids* (Isler, O., Gutmann, H. & Solms, U., eds), pp. 48-51, Birkhauser, Basel.

Lutz, M., Agalidis, I., Hervo, G., Cogdell, R. J. & Reiss-Husson, F. (1978) *Biochim. Biophys. Acta* **503,** 287-303.

Maroti, P., Kirmaier, C., Wraight, C., Holten, D. & Pearlstein, R. M. (1985) *Biochim. Biophys. Acta* **810,** 132-139.

McGann, W. J. & Frank, H. A. (1985) *Biochim. Biophys. Acta* **807,** 101-109.

Michel, H., Epp, O. & Deisenhofer, J. (1986) *EMBO J.* **5,** 2445-2451.

Parson, W.W. & Monger, T.G. (1976) *Brookhaven Symp. Biol.* **28,** 196-212.

Vermeglio, A., Breton, J., Paillotin, G. & Cogdell, R. (1978) *Biochim. Biophys. Acta* **501,** 514-530.

Vermeglio, A. & Clayton, R. K. (1976) *Biochim. Biophys. Acta* **449,** 500-515.

SPECTROSCOPIC STUDIES OF CRYSTALLIZED PIGMENT-PROTEIN COMPLEXES OF *R. PALUSTRIS*

W. Mäntele, K. Steck, A. Becker, T. Wacker, W. Welte, N. Gad'on* and G. Drews*

Institut für Biophysik und Strahlenbiologie, Albertstr. 23
* Institut für Biologie II, Schänzlestr. 1, 78 Freiburg FRG

INTRODUCTION

In bacterial photosynthesis, the processes of light absorption, energy migration, trapping and charge separation have been extensively studied by spectroscopic techniques using membranes and isolated antenna or reaction center (RC) complexes. With the structure of the reaction center available from high-resolution X-ray analysis, electron transport can be "visualized" and a large number of the spectroscopic data can be better understood. A similar progress in the understanding of the processes of energy migration might also be achieved with the crystallization of antenna pigment-protein complexes. Crystallization of different antenna complexes has been reported by several groups. The RC-B875 complex of the purple photosynthetic bacterium *R. palustris*, i.e. the reaction center with the core antenna system which is in close contact to the RC and which is synthesized in a fixed stoichiometry, has recently been crystallized by us [1]. Since this complex contains both photochemically active and inactive bacteriochlorophylls, light absorption, energy migration and charge separation may be studied. The B800-850 complex of the same bacterium has also been crystallized [2,3]. A crystal analysis of both pigment-protein complexes, at a sufficient resolution, should provide structural information that helps to understand the energetic coupling between antennae and reaction center.

Aside from its access to the structure, the crystalline state provides an elegant way to perform spectroscopic studies on oriented pigments. Unlike in uniaxially oriented samples, a three-dimensional orientation of transition moments can be obtained. The absorption strength in three dimensions, represented by the three main axes of an ellipsoid, can thus be determined. Basically, three degrees of order have to be distinguished when conclusions on the arrangement of transition moments are drawn from spectroscopic data obtained on a crystal: (1) the resultant of transition moments in an individual reaction center or antenna molecule; (2) the resultant of transition moments of individual antenna molecules forming a photosynthetic unit or a complete antenna complex; (3) the resultant of the elements of the unit cell given by the order of the crystal. Some crystal forms, for example those involving a 3, 4, or 6-fold symmetry axis, tend to "randomize" an even high order of transition moments. In many cases, however, several crystal forms can be obtained from one pigment-protein complex, one of them usually simple enough to allow an optical analysis. Here, we present spectroscopic data on the pigment arrangement and charge separation of the crystallized RC-B875 complex as well as linear dichroism and polarized fluorescence spectra of the crystallized B800-850 complex from *Rp. palustris.*.

EXPERIMENTAL

A home-built microspectrophotometer was used to record absorbance, linear dichroism, light-induced difference and fluorescence emission spectra. Its design is shown in fig. 1a. A symmetrical optics is used to illuminate the crystal and to enlarge its image, which allows a high measuring light flux for time-resolved measurements. Using two microscope objectives with a long working distance, the crystals can be tilted up to ± 40° with respect to the beam. A silicon diode detector with

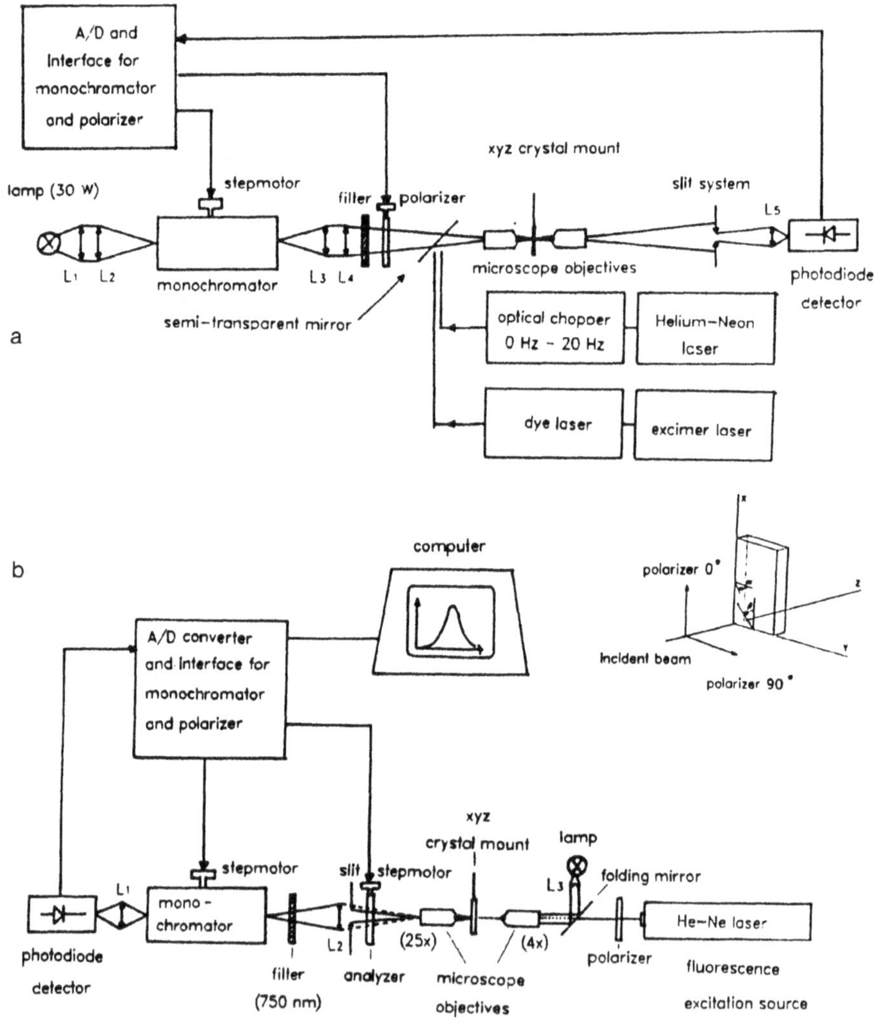

Fig.1. Microspectrophotometer for absorption, linear dichroism and light-induced difference spectra (1a) and fluorescence emission spectra (1b).

peak sensitivity between 700 nm and 900 nm was optimized for the detection of low light levels to record crystal spectra without causing significant photoreactions in the RC-B875 crystal. Light-induced difference spectra were obtained either by using additional bleaching light, excitation by laser flashes or modulated light excitation with light from a yellow He-Ne laser (594 nm). Fluorescence emission spectra recorded by the modified setup as shown in fig. 1b were obtained with excitation at 594 nm. Forward emission was detected by the same detector as used for absorption measurements.

Crystals of the RC-B875 and the B800-850 complex were obtained from the complexes solubilized in LDAO in the presence of polyethylene glycoll. Details on the conditions of crystallization are reported elsewhere [2]. Thin crystalline platelets were directly grown on glass cover slides. A high yield of crystals of all sizes was thus obtained, their largest surface being parallel to the slide. For optical analysis, they were covered with a second slide separated by a spacer of 5 µm to 20 µm thickness.

Absorbance spectra of crystals were obtained by rcording a single beam spectrum (I) through the crystal, a reference spectrum (I_0) next to it and forming log (I_0/I) on the instrument computer. For linear dichroism spectra, I_0 was recorded separately for both directions of polarization ($I_{0,par}$ and $I_{0,perp}$). When the crystal was tilted with respect to the beam, spectra with polarization parallel and perpendicular to the tilt axis were recorded. Using the absorption with the polarization parallel to the tilt axis, a correction factor for the increased pathlength due to tilting was obtained.

RESULTS AND DISCUSSION

The crystallization procedure for the RC-B875 complex yields rectangular crystals with dimensions up to 30 µm * 50 µm. Clearly distinguished long and short axes define three orthogonal morphological axes x, y and z used for the macroscopic orientation of the crystal in the spectroscopic experiment (see insert of fig 1b). A series of spectra shown in fig. 2 (taken from [1]) with the polarizer rotated from the x-axis (0°) to the y-axis (90°) shows a perfect orientation of the Q_x transitions along the x-axis of the crystal, whereas the Q_y transitions are oriented along the y-axis.

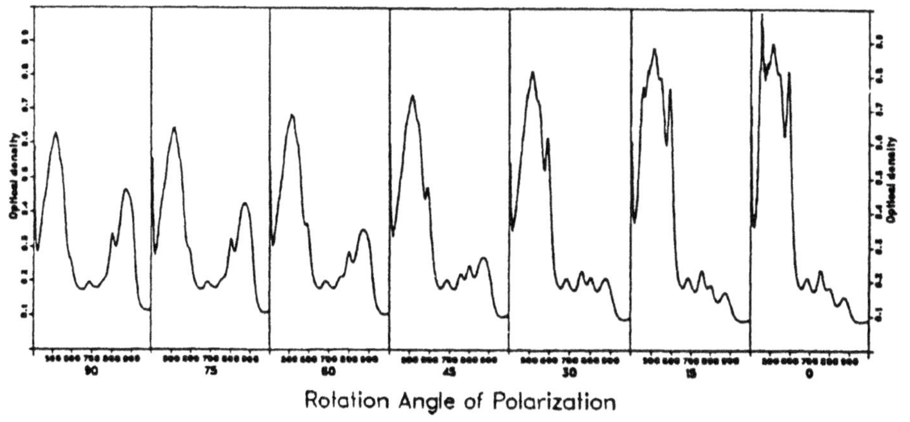

Fig 2. Spectra obtained at different angular positions of the polarizer

The integral absorbance strength of the Q_y transition with respect to the Q_x transition in their respective polarization, however, is smaller than observed for the solubilized complex. Thus, some distribution of the Q_y transitions in the y-z plane can be expected. Tilting series around the x-axis and the y-axis as reported in [1] reveal an almost perfect orientation of the Q_x transitions along the x-edge of the crystal and a predominant orientation of the Q_y transition moments along the y axis, with some components in the z direction. From the magnitude of the Q_y absorption at tilt angles up to ± 30°, a circular degenerate orientation in the y-z plane could be excluded [1]. A preferential orientation of the B875 bacteriochlorophyll Q_y transitions, however, must have implications on the transfer of energy from the antenna to the reaction center.

The series of spectra with varying angle of polarization (fig. 2) not only allows to determine the orientation of the B875 transition moments, but also yields some information on the reaction center pigment orientation. The Q_y transition of bacteriopheophytin in the RC absorbing at 755 nm shows a preferential orientation along the x direction; its direction thus seems to correlate with the direction of the antenna Q_x transitions. The dichroism of the reaction center accessory bacteriochlorophylls is less pronounced: in the y direction, the 800 nm band appears superimposed as a strong band on the antenna peak, but some residual absorption is also observed in the x direction. Nevertheless, the projection of the accessory bacteriochlorophyll absorption seems to be preferentially oriented in the y direction. The distribution in the x-y plane of the 865 nm antenna Q_y absorption, of the 800 nm accessory BChl Q_y absorption, and the 752 nm BPheo Q_y absorption can be represented by the corresponding ellipses as shown in fig 3. Due to the strong absorption of the antenna bacteriochlorophylls at 865 nm, the polarized absorption spectra contain only little information on the orientation of the primary electron donor. The residual absorption at 865 nm in the x direction, although suggestive, cannot be assigned to the primary electron donor absorption.

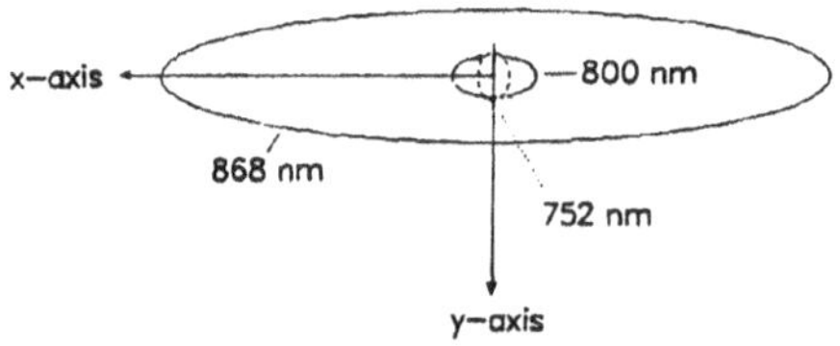

Fig.3: Ellipses representing the absorption strength in the x-y plane

In order to obtain more precise information on the orientation of the pigments involved in primary electron transport, light-induced difference spectra were recorded as described above. The time-resolved absorbance changes and difference spectra thus obtained clearly demonstrate charge separation by the photochemically intact reaction center in the crystal. Figure 4 shows the light-induced absorbance changes induced (a) by additional bleaching light; (b) by modulated light excitation and (c) by a laser flash.

The light-induced difference spectrum obtained with additional bleaching light (fig. 4a) was obtained for measuring light polarized in the x and in the y direction. The intensity of the additional bleaching light at 594 nm, polarized in the x direction, was kept high enough to obtain

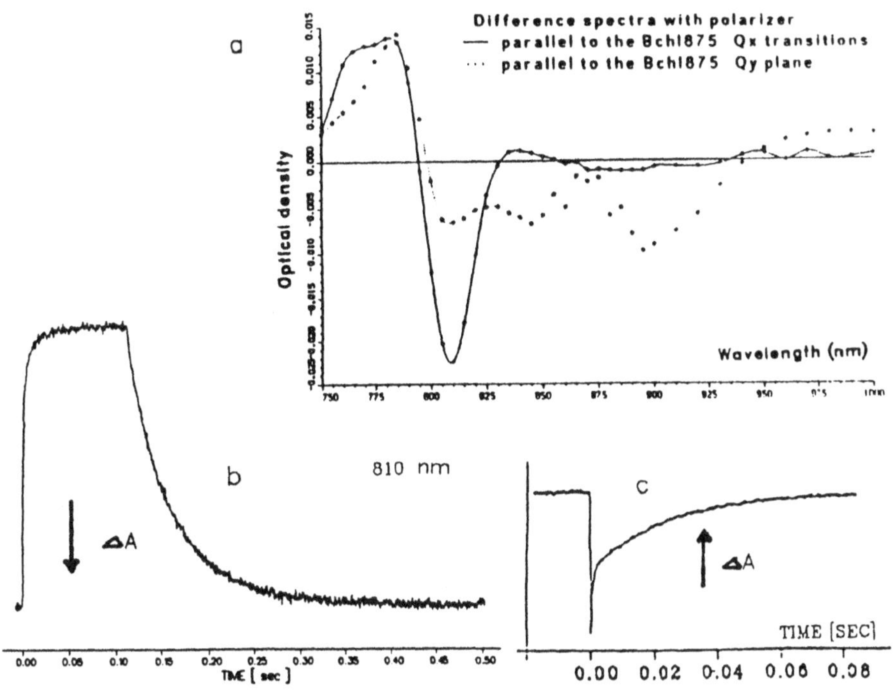

Fig. 4: Absorbance changes obtained with: (a) additional bleaching light, (b) modulated light excitation and (c) laser flash excitation

saturation. The difference spectrum shows the well-known shift of the accessory bacteriochlorophyll and the bacteriopheophytin absorption bands leading to the differential band feature with a zero-crossover at approx. 800 nm. A distinct difference is observed between the two polarization directions in the 750nm - 780 nm spectral region, presumably due to the predominant absorption of the bacteriopheophytin in the x direction (see above). In the primary donor spectral region (820 nm - 900 nm), the high absorption of the antenna pigments strongly reduces the accuracy of the difference spectra. Nevertheless, the differential signal in the x axis is close to zero, whereas a clear absorption decrease is observed in the y axis. The projection of the primary donor transition moment in the x-y plane thus seems to be preferentially oriented along the y direction.

Excitation of the reaction center either by modulated light excitation or by a laser flash (fig. 4b and 4c) has been used to determine the rate constants of the back-reaction from the $P^+Q_A^-$ or $P^+Q_B^-$ state in the crystal. Using modulated light excitation, the exposure time and the intensity of the light was chosen in order to obtain saturating bleaching. The analysis of the decay yields half-times of 20 msec to 30 msec. At pre-crystalline conditions, half-times of 60 msec to 80 msec were observed, with a small proportion of the photoexcited RC reacting with half-times that can be assigned to the $P^+Q_B^- \rightarrow PQ_B$ back reaction. The proportion of reaction centers containing Q_B thus seems to be further reduced upon crystallization. Whether the accelerated decay in the crystal ($t_{½}$ ≈ 25 msec) is due to the crystalline state (one might imagine an influence of the charge-separated state of a reaction center on its well-ordered nearest neighbors) cannot be decided yet. Preliminary electrical measurements of charge separation in RC-B875 crystals using microelectrodes (in cooperation with H.-W. Trissl, University of Osnabrück)

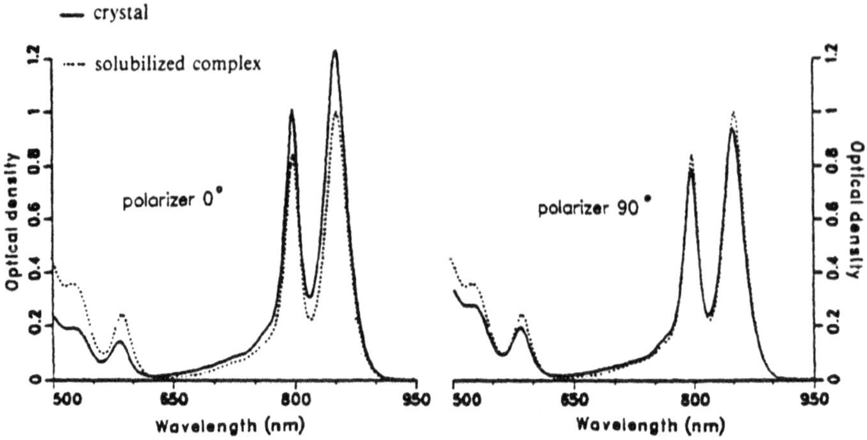

Fig. 5: Absorption spectra of the *Rp. palustris* B800–850 complex in suspension and in the crystalline form

have also indicated half-times of charge recombination of 20 msec to 30 msec. The almost complete lack of a time constant typical for $P^+Q_B^- \rightarrow PQ_B$ charge recombination might as well be explained by assuming partial or complete bleaching of those reaction centers containing Q_B by the measuring light. In order to avoid this, a measuring light intensity corresponding to much less than 1 photon per RC per second would have to be applied, which is in conflict with sensitivity as well as with measuring time. In order to avoid the actinic effect of the measuring light, light -induced absorbance changes were recorded at 950 nm at the far-red slope of the B875 and the RC primary donor absorption band. The kinetic components of the recovery from the photooxidized state, however, were identical to those found at 865 nm and thus favour an interpretation in terms of an almost complete loss of Q_B upon crystallization.

Crystals of the B800–850 light-harvesting complex suitable for spectroscopy were obtained by procedures described in [2,3]. A spectrum of

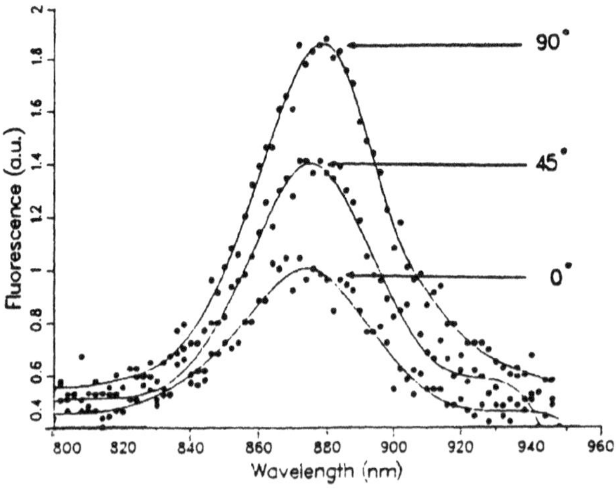

Fig. 6: Emission spectra of a B800–850 crystal at 0°, 45° and 90°

a crystal in two directions of polarisation together with the spectrum in solution is shown in figure 5. A more detailed series of spectra for various rotation and tilting angles is reported in [4]. A distinct dichroism is observed for the Qx as well as for the 800 nm and 850 nm Qy transitions, although less pronounced as in the case of the RC-B875 complex. This can be explained by assuming a crystal symmetry that partially cancels the dichroism of the oriented complex. Preliminary spectra of the crystallized B800-850 complex from *Rs. salexigens*, showing the perfect orientation of the 800nm and 850 nm bands along the y direction (data not shown), indicate a different crystal form. Although this crystal form could not be obtained for the same complex from *Rp. palustris*, these spectra allow to place the Qy transitions perfectly within a plane.

Emission spectra of the *Rp. palustris* B800-850 crystals using 594 nm excitation are shown in fig.6 at three angles of polarisation set for the emitted radiation. The polarisation of the exciting laser light was chosen in the y direction (see fig. 5) in order to obtain optimum excitation. The distinct variation of the intensity of the emitted light reflects the dichroism of the 850 nm absorption band.

CONCLUSIONS

The absorbance spectra, light-induced difference spectra and emission spectra of the crystals clearly show that the crystallized complexes maintain their native spectroscopic and functional properties. A preferential orientation of the core antenna transition moments around the reaction center may point to specific paths of excitation energy to the RC primary donor. Due to the comparatively large distance of the complexes in the crystal, only very small differences in the half-width of absorption bands or in the kinetic parameters are observed. The excellent orientation in the crystalline state can thus be used to study the distribution and relative orientation of the transition moments. However, low-temperature absorption, fluorescence and kinetic data from crystals will be necessary for a detailed study of individual transition moments of the antenna complexes. Together, they will add to the information obtained from uniaxially oriented complexes.

REFERENCES

[1] Wacker, T. et al. (1986) FEBS Lett. 197, 267-273

[2] Wacker, T. et al. (1987) in: Progress in Photosynthesis Research (Biggins, J., ed.) pp. 383-386; M. Nijhoff Publishers

[3] Mäntele, W. et al. (1986) In: Antennas and Reaction Centers from Photosynthetic Bacteria (Michel-Beyerle, M. E., ed.) pp. 88-91; Springer Series in Chem. Phys.

[4] Wacker, T. et al. (1988) in: Proceedings of the II European Conference on the Spectroscopy of Biological Molecules (in press)

PROTEIN-PROSTHETIC GROUP INTERACTIONS IN BACTERIAL REACTION CENTERS

Marc Lutz, Bruno Robert, Qing Zhou,
Jean-Michel Neumann, Wojciech Szponarski,
and Gérard Berger

Département de Biologie, C.E.N. Saclay
91191 Gif-sur-Yvette cedex, France

Specific interactions with their host polypeptides increasingly appear as essential in determining the excitation- and charge-transfer properties of the prosthetic groups within the bacterial reaction center. Detailed information on these interactions can be obtained from resonance Ràman spectroscopy of the reaction center-bound pigments. Indeed, this technique currently permits selective observations of seven out of the eleven prosthetic groups born by wild-type reaction centers of Rhodospirillales. These Raman-observable groups are: i, the carotenoid (Lutz et al 1976, 1987), ii, the primary donor bacteriochlorophylls (Lutz & Robert 1985, Robert & Lutz 1986), iii, the accessory bacteriochlorophylls (Robert & Lutz 1987), iv, the normal acceptor bacteriopheophytin (Lutz 1980) and, v, the second bacteriopheophytin (Lutz 1980). Some recent, structural as well as functional results concerning these groups are outlined in the following.

I Carotenoid conformation

It has been known for a decade that the protein imposes a non all-trans conformation to carotenoids bound to reaction centers of any Rhodospirillale species or strain (Lutz et al. 1976, 1978, Boucher et al. 1977). Up to now, however, the precise assignment of this conformation has remained uncertain and somewhat controversial. We thus reinvestigated the stereoisomerism of the spheroidene molecule bound to reaction centers (RCs) of Rhodobacter sphaeroides (Lutz et al. 1987). A stable cis isomer could be extracted and purified from the reaction centers by working at very low ambient light levels. Resonance Raman (RR) spectroscopy showed that this cis isomer assumed the same configuration as that of the RC-bound molecule. Proton-NMR spectroscopy of the extracted isomer permitted to assign it the 15-15' monocis configuration. Comparisons between RR spectra of the native form and of the 15-15' cis extract showed that, in the RC, 15-15'cis spheroidene is in addition twisted into a non planar conformation . Comparisons of extraction-induced changes in relative intensities of the RR bands of the 760-1060 cm^{-1} regions, which largely correspond to out of plane vibrational modes of the polyene chain, further indicated that the out of plane twist of RC-bound spheroidene should

Figure 1. The C6-C15' section of 15-15'-cis spheroidene, displaying those wavenumbers (cm^{-1}) of Raman-active modes of the 750-1060 cm^{-1} region that can be localized on the molecule on the basis of normal mode calculations (Saito & Tasumi 1983) and of selective deuteriation (Tasumi 1987). Squared wavenumbers refer to modes which are weakened by at least a factor of 2 upon extraction of spheroidene from the reaction center in the dark. These modes localize the out of plane deformation of RC-bound spheroidene to the C8-C12 and/or C8'-C12' regions of the molecule.

predominantly affect C8-C12 and/or C8'-C12' regions of the molecule, rather than the central region (Fig.1). Comparisons between difference electronic absorption spectra of RC-bound spheroidene and of RC-bound methoxyneurosporene showed that the out of plane twisting of both these native forms results in a drastic weakening of their 1C←1A electronic transitions, compared with those of the planar, 15-15' cis forms.

Resonance Raman spectra of other RC-bound carotenoids in their native states show that they must share 15-15' cis isomerisation and out of plane twisting with RC-bound spheroidene. This is in particular the case for: i, methoxyneurosporene and neurosporene present in RCs of the Ga and G1C mutants of Rhb sphaeroides, respectively, ii, spirilloxanthin in RCs of Rsp rubrum and, iii, dihydroneurosporene or dihydrolycopene present in RCs of Rps viridis. Resonance Raman data in particular exclude that the isomer present in Rps viridis RCs might be 13 cis, as suggested from X-ray crystallographic data (Deisenhofer et al, these proceedings). Indeed, 13-cis isomers of C40 carotenoids (Koyama et al. 1983) do not yield the characteristic 1240 cm^{-1} RR band common to 15 cis isomers and to RC-bound carotenoids, including those of Rps viridis. On the other hand, the 13-cis isomers give rise to a conspicuous 1135 cm^{-1} RR band, not observed in RR spectra of RC-bound carotenoids, including those from Rps viridis.

II Interspecific variability of the structure of the primary donor in bacterial reaction centers

We recently developed difference methods permitting selective observations of RR spectra of the primary donor molecules in RCs of Rhodospirillales, in Soret resonance conditions (Robert & Lutz, 1986). In these experiments, variable amounts of states P^+ or P^R are built up in the sample by making use of the actinic effect of the Raman analysis laser beam. As far as no contribution of any of these two states could be detected in RR spectra excited at the top of the Soret transition of the neutral, ground-state pigments, difference spectra obtained e.g. by subtracting RR spectra of bacterial RCs obtained at high irradiance from spectra obtained at low irradiance essentially contain contributions arising from the primary donor alone, at least when carotenoid-containing bacterial strains are considered.

We used these methods in order to compare interactions assumed by the BChls constituting P in different bacterial strains. For example, we recently compared these interactions for the primary donors of Rhb sphaeroides and of Rsp rubrum RCs (Zhou et al. 1987). This work was extended to the primary donors of Rhb capsulatus (Zhou et al. 1987), of Rps viridis (Robert et al. 1987), as well as of Chromatium vinosum and Chromatium tepidum RCs (Zhou et al., in preparation).

Contributions of the primary donor, selectively extracted from the higher frequency regions of RR spectra (1550-1750 cm^{-1}) of RCs by difference spectroscopy are shown in Fig. 2 for Rhb sphaeroides, Rps viridis and Rsp rubrum. In these three spectra, the 1615 cm^{-1} band arises from stretching modes of methine bridges of both BChls constituting the primary donor. This band is known to be sensitive to the coordination state of the central Mg of

BChl, being located at 1600 cm^{-1} when this atom binds two external ligands and around 1615 cm^{-1} when it binds a single external ligand. The position and width of this band in the three difference spectra of Fig. 2 indicate that the Mg of both BChl molecules constituting P in these species are five-coordinated. Refined crystallographic data on RCs from Rps viridis (Michel et al. 1986) and sequence homologies in RC proteins indicate that these ligands are side chains of two histidine residues (His L 173 and His M 200 in Rps viridis).

On the other hand, a limited variability can be observed in the intermolecular liganding of the 2-acetyl and 9-keto, conjugated carbonyl groups of the primary donor BChls. Information on the interaction states of these groups is obtained from the wavenumbers of their stretching modes, which are RR-active at Soret resonance and occur in the 1620-1710 cm^{-1} range (for a review about attributions of the different Raman bands, see Lutz, 1984 or Lutz & Robert, 1987). Typically, 2-acetyl C=O groups of BChl a and of BChl b vibrate at wavenumbers as high as 1665 and 1668 cm^{-1} in nonpolar environments, respectively. These values may shift down to 1625 cm^{-1} when these groups are intermolecularly bound. The stretching mode of the 9-keto group of both BChl a and b is observed at wavenumbers up to 1705 cm^{-1}, when free from interactions in nonpolar environments. This wavenumber may shift down to 1650 cm^{-1} when this group is interacting, e.g. is H-bonded.

RR spectra of the primary donor of Rps viridis RCs yielded four distinct bands in the C=O stretching region, located at 1627, 1633, 1666 and 1683 cm^{-1}. These bands arise from two bound acetyl C=O groups, one bound keto C=O group and one weakly interacting, if at all, keto C=O group, respectively. In RR spectra of the primary donor from Rhb sphaeroides RCs, three different frequencies only can be observed in the same spectral region. These occur at 1633, 1660 and 1683 cm^{-1}. The 1633 cm^{-1} band arises from a H-bonded acetyl C=O group. The 1660 cm^{-1} band arises both from a free acetyl C=O group and from a H-bonded keto C=O group. The 1683 cm^{-1} band arises from a weakly interacting keto C=O. These results largely agree with both X-ray diffraction studies (Michel et al. 1986) and sequence analyses of the RC proteins (Deisenhofer et al 1985) of these both strains. Indeed, these latter data indicate that, in Rps viridis, side chains of His L168 and of Tyr M195 may each form a H-bond with the acetyl C=O groups of each of the primary donor BChls. In Rhb sphaeroides RCs, Tyr M195 is replaced by a phenylalanin,the side chain of which is not able to ligand the acetyl C=O of BChl P_M. We thus conclude that the 1627 cm^{-1} band of Rps viridis arises from the acetyl C=O of BChl P_M, whereas the common 1633 cm^{-1} frequency arises from the His-liganded C=O of BChl P_L in both Rps viridis and Rhb sphaeroides.

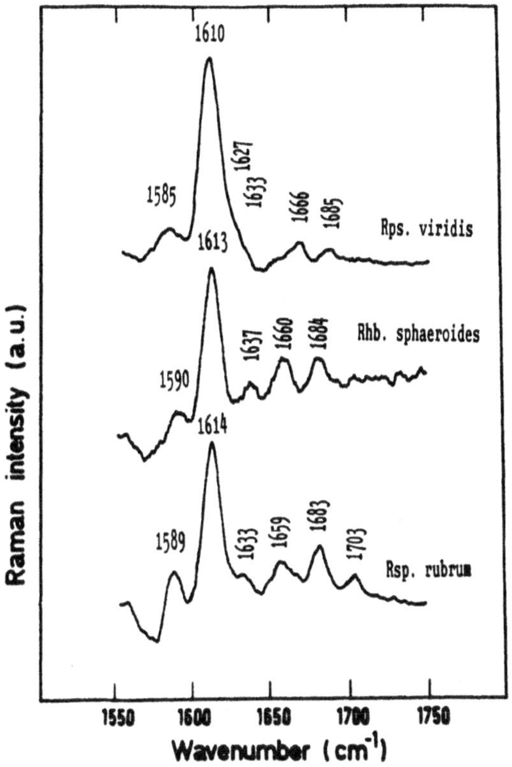

Figure 2. Resonance Raman spectra (1450-1750 cm^{-1} regions) of the neutral, ground states of the primary donors of Rhodopseudomonas viridis, Rhodobacter sphaeroides 2.4.1., and Rhodospirillum rubrum S1. These spectra are differences between RR spectra of untreated reaction centers excited under low and high irradiance conditions, respectively. Excitation wavelength 363.8 nm, sample temperature 20K.

The keto carbonyls of primary donor BChls assume nearly the same stretching frequencies in both Rps viridis and Rhb sphaeroides. However, X-ray diffraction studies and sequence comparisons indicated that a conserved Ile residue (M282 in Rps viridis) is closest neighbour to the keto C=0 group of BChl P_M, and that the nearest neighbour to the keto group of BChl P_L in Rps viridis is the L248 Thr residue, which is replaced by a Met residue in RCs of Rhb sphaeroides. Inasmuch as neither the Ile nor the Met side chains can provide significant H-bonding to a C=O group, the simplest interpretation of our RR data that may be consistent with X-ray and sequence data is that the 1684 cm^{-1} band arises from the keto group of BChl P_M in both Rps viridis and Rhb sphaeroides; the 1684 cm^{-1} wavenumber may actually indicate that these groups are not bonded intermolecularly, but rather are located in a high permittivity local environment (see e.g. Girin and Bakshiev, 1963). The 1660 cm^{-1} wavenumbers should then be ascribed to the BChl P_L keto groups, which might be H-bonded to the peptidic NH groups of the Thr and Met residues of Rps viridis and Rhb sphaeroides, respectively.

The band intensities and frequencies observed in the C=O stretching regions of RR spectra of the primary donor of Rsp rubrum are almost the same as observed for Rhb sphaeroides, but for a weakening of the 1660 cm^{-1} band, a component of which is shifted up to 1700 cm^{-1}. It may thus be predicted from these spectra that the acetyl carbonyl of BChl P_L, still vibrating at 1633 cm^{-1}, is liganded to the same His (L168) side chain as in e.g. Rhb sphaeroides, which hence should be conserved in the L polypeptide of Rsp rubrum. The weakly interacting, or free keto group of BChl P_M might still be in the vicinity of an Ile side chain, although another, chemically equivalent, non bonding side chain may also fit our data. Finally, the keto group of BChl P_L in Rsp rubrum, vibrating at 1700 cm^{-1}, and not around 1660 cm^{-1}, should be free from any interactions with peptidic NH groups, at variance with its state in Rps viridis and in Rhb sphaeroides.

Recent RR results on the primary donor of Rhb capsulatus (Zhou et al. 1986) and on those of the sulfur species Chromatium vinosum and tepidum (Zhou et al, unpublished) indicate that the interactions found to be conserved in Rps viridis, Rps sphaeroides and Rsp rubrum are still conserved in these species. Taken together, RR data on the primary donors of the above-mentioned six species of Rhodospirillales indicate that a limited, but definite interspecific variability occurs in the proteic environments of the primary donor molecules, as manifested by variable molecular interaction states of the keto group of BChl P_M and of the acetyl group of BChl P_L. The variable interaction states of these groups do not appear to affect, to first order, the primary events in the reaction center. On the other hand, the interaction states of the acetyl group of

BChl P_L, interacting with a His residue, and of the keto group of BChl P_M, possibly free but located in a rather polar microenvironment, might constitute important parameters in the structure and function of the bacterial primary donor.

III Protein conformational changes occurring in bacterial RCs during charge separation

We have studied the influence of the primary charge separation on RR spectra of RCs from the carotenoidless mutant (R 26 strain) of Rhb sphaeroides (Robert & Lutz, 1987). Analysis of the C=O stretching region of RR spectra of R 26 RCs indicates that they contain much weaker contributions of the primary donor than those from the carotenoid-containing strains. This was confirmed by the fact that no actinic effect of the laser beam could be detected in RR spectra of R 26 RCs. Indeed, varying the irradiance value at the sample results, in RR spectra of wild-type RCs obtained at Soret resonance of the bacteriochlorins, in a variable bleaching of the primary donor contributions. This in turn results in variations of relative intensities of RR bands arising from the BChl molecules with respect to those of BPheo bands (Robert & Lutz, 1986). This lack of primary donor contributions in RR spectra of R 26 RCs excited at the top of the Soret band (which is common to the six bacteriochlorin pigments) is most probably to be related to the slight variations in structure induced in this band by the presence or absence of a carotenoid in the RC (Lutz et al. 1978). These variations most likely do not concern electronic levels of all six bacteriochlorins equally, and are likely to result in the differences in RR scattering cross sections of the primary donor that are observed between the R 26 and 2.4.1. strains.

Nevertheless, formation of the P^+Q^- state induces two changes in the C=O stretching region of RR spectra of R 26 RCs (Fig. 3). These changes can be simply interpreted as due to a shift of a 1689 cm^{-1} band down to 1675 cm^{-1} and to a weakening of a band at 1660 cm^{-1}. The assignment of these changes to individual bacteriochlorin(s) of the RC has been made as follows. Individual carbonyl stretching frequencies are known for each of the two BPheos of R 26 RCs (Raman experiments at Q_X resonance, Lutz 1980, Agalidis et al. 1984) and for the two primary donor BChls of 2.4.1. RCs (cf section II and Robert and Lutz 1986). None of these individual carbonyl frequencies matches with the 1689 and 1660 cm^{-1} values involved in the experiments on R 26 RCs , which hence can only concern changes in the RR contributions of the accessory bacteriochlorophyll(s). Moreover, the P^+Q^--induced changes are still observed when the accessory BChl B_M is selectively bleached chemically by borohydride treatment of the RCs. It hence can be concluded that these changes essentially concern the accessory molecule B_L, which is located along the normal pathway of the photoelectron.

It is worth noting that these changes appear whether the primary donor is photochemically or chemically oxidized as well. This implies that they are induced by the ionization of P and not by the electron transfer per se. The 14 cm^{-1} downshift of the 1689 cm^{-1} band is not accompanied by any other sizeable change in relative intensity of any other RR band. Hence it is most likely to arise from a change in the local environment of the B_L molecule, rather than from changes in resonance conditions for the Raman scattering of molecule B_L. Such changes might conceivably result e.g. from changes in electronic coupling between B_L and P upon oxidation of the latter, or to electrostatic effects from species P^+ upon the Soret electronic states of B_L.

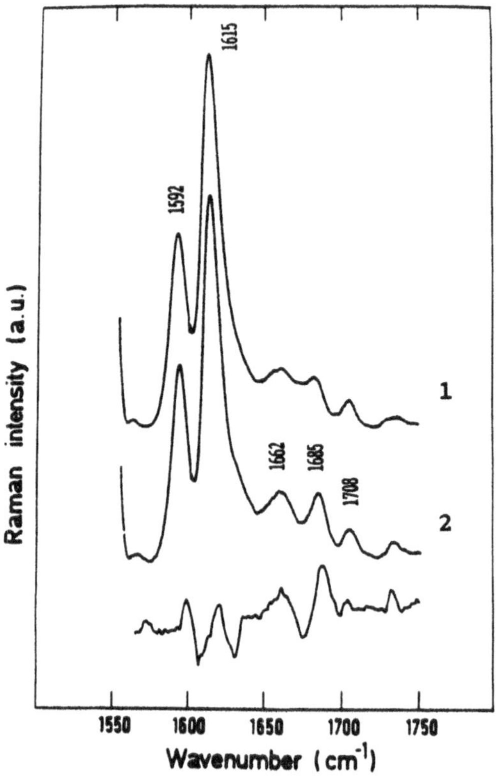

Figure 3. Resonance Raman spectra (1450-1750 cm^{-1} regions) of untreated reaction centers from Rhodobacter sphaeroides R26, excited under low (spectrum 2) and high irradiance conditions (spectrum 1). The lower spectrum is a difference between spectra 2 and 1, normalized on the 1592 and 1615 cm^{-1} bands. Excitation wavelength 363.8 nm, sample temperature 20K.

The 14 cm^{-1} downshift of the 1689 cm^{-1} band corresponds to the creation of a 2.1 kcal H-bond upon the keto carbonyl of B_L, or to an equivalent strengthening of a possibly preexisting bond (cf section II). The weakening of the 1660 cm^{-1} band also may be attributed to a variation in the local environment of the 2 acetyl C=O of the B_L molecule. As far as the observed weakening of the band is not accompanied by any significant frequency shift, this variation may well consist of a steric repulsion from a nearby molecule site.

Two hypotheses can be made about the identity of the partner molecule that binds the 9 keto of the B_L molecule during ionization of P : a first candidate might be a tyrosine residue located near this pigment (residue M 208 in Rps viridis (Michel et al, 1986). However, the hydroxyl group of this sidechain does not appear to be able to bind the C=O grouping in resting reaction centers (Deisenhofer, personal communication); it yet is conceivable that these two groupings could be brought into a more favourable geometry as a consequence of the ionization of P. An alternative hypothesis is that the H-bond on the keto group is establihed with a bound water molecule located near to the B_L molecule. An excellent candidate would actually be the water molecule mentioned by J. Deisenhofer (these proceedings) as being bound, in Rps viridis RCs, to the sidechain of His M200 (itself liganding the Mg atom of BCHl P_M), and located about 3 A from the keto group of BChl B_L. This water molecule should then be brought e.g. closer to this group through a slight proteic motion induced by ionization of P. Because no similar change most probably affects the local environment of the second accessory molecule B_M, the change observed around B_L upon ionization of P may have a functional role in electron transfer within the reaction center. Current resonance Raman and time-resolved Raman experiments should indicate whether the presence of a negative charge on the acceptor Bpheo L is also accompanied by proteic conformational changes around prosthetic groups of R 26 reaction centers.

REFERENCES

Agalidis, I., Lutz, M. and Reiss-Husson, F. (1984) Biochim.Biophys Acta 766, 188
Boucher, F., Van der Rest, M., and Gingras, G. (1977) Biochim. Biophys. Acta 461, 339
Deisenhofer, J., Epp, O., Huber, R. and Michel, H. (1985) Nature 318, 618
Girin, O.P., and Bakshiev, N.G. (1963) Usp. Fiz. Nauk. 79, 235 (Engl Transl. p 106)
Koyama, Y., Takii, T., Saiki, K., and Tsukida, K. (1983) Photobiochem. Photobiophys. 5, 139
Lutz, M., Kléo,J., and Reiss-Husson, F. (1976) Biochem. Biophys. Res. Comm. 69, 711
Lutz, M., Agalidis, I., Hervo, G., Cogdell, R.J., and Reiss-Husson, F. (1978) Biochim. Biophys. Acta 503, 287

Lutz, M., (1980) in : Proc. 7th Int. Conf. on Raman Spectrosc. (Murphy, W.F. ed.) Amsterdam, North-Holland, p. 511
Lutz, M., (1984) Adv. Infrared Raman Spectrosc., 11, 211
Lutz, M., and Robert, B., (1985) in : Antennas and Reaction Centers of Photosynthetic Bacteria (Michel-Beyerle, M.E. ed.) Springer Verl., Berlin, p. 138
Lutz, M., and Robert, B., (1987) in : Biological Applications of Raman Spectrosc. (Spiro, T.G. ed.) Wiley, New York, vol. 3, chap. 9, in the press
Lutz, M., Szponarski, W., Berger, G., Robert, B., and Neumann, J.M. (1987) Biochim. Biophys. Acta , 423, 894
Michel, H., Epp, O., and Deisenhofer, J. (1986) EMBO J. 5-10, 2445
Robert, B., and Lutz, M., (1986) Biochemistry 25, 2303
Robert, B., and Lutz, M., (1987) Biochemistry, submitted
Robert, B., Steiner, R., Zhou, Q., Scheer, H., and Lutz, M., (1987) in : Progr. in Photosynthesis Res., (Biggins, J. ed.) Martinus-Nijhoff, Dordrecht, vol. 1, p. 411
Saito, S., and Tasumi, M., (1983), J. Raman Spectrosc. 14, 310
Tasumi, M., (1987) unpublished results
Zhou, Q., Robert, B., and Lutz, M., (1987) in : Progr. in Photosynthesis Res., (Biggins, J. ed.) Martinus-Nijhoff, Dordrecht, vol. 1, p. 395
Zhou, Q., Robert, B., and Lutz, M. (1987) Biochim. Biophys. Acta 890, 368.

CIRCULAR DICHROISM SPECTROSCOPY OF PHOTOREACTION CENTERS

Ted Mar and Gabriel Gingras

Département de biochimie
Université de Montréal
Montréal, Québec
(Canada) H3C 3J7

Circular dichroism (CD) of photoreaction centers is very sensitive to the interactions felt by the bacteriochlorophyll (Bchl) and bacteriopheophytin (Bph) molecules. In the photoreaction center, these molecules may be involved in exciton interactions (1), in mixed charge transfer-and-exciton interactions (2), in interactions with higher energy transitions (3) and in interactions with the molecular polarizability of the protein (4). CD spectra of Bchl-protein antenna complexes and of the photoreaction center generally have been interpreted in terms of exciton interactions only (1, 5-11). However, since some of the other interactions (2, 12) may significantly contribute to the CD spectra, they may also be studied by CD spectroscopy. This particularly applies to the charge transfer and Bchl-protein interactions which may be important in the mechanism of the primary electron transfer reaction.

I. CIRCULAR DICHROISM SPECTRUM OF THE PHOTOREACTION CENTER

Following the elucidation of the crystal structure of the *Rhodopseudomonas viridis* (13) and *Rhodobacter sphaeroides* (14, 15) photoreaction centers, the geometry and the arrangement of the four Bchl and of the two Bph molecules embedded in the protein have been accurately determined. Using known positions and orientations and assuming only exciton interactions among the Q_y electronic transitions of the six pigments, the lowest lying CD bands have been calculated (9-11). Although the calculated and measured spectra resemble each other, some of their detailed features do not agree. The calculated spectrum is conservative (the sum of its positive and negative rotational strengths is zero) as is expected of any model in which interaction is limited to the exciton coupling of the Q_y transition dipoles (16). However, the experimental CD spectra for many photoreaction centers (17) are not conservative. As an example, Fig 1 shows the CD spectrum of the photoreaction center from *Ectothiorhodospira sp* in which the sum of the positive rotational strengths is greater than that of the negative rotational strengths. This is even more evident when the primary electron donor is oxidized. To account for this experimental lack of conservativeness, other interactions besides the exciton interactions between the chromophores should be considered. There are other features of the CD spectrum that are not accounted for by the simple exciton model such as the energies of the monomer Q_y transitions and the widths of the CD bands. In the calculations quoted above (9-11), their values were assumed to fit the experimental data rather than being obtained ab initio.

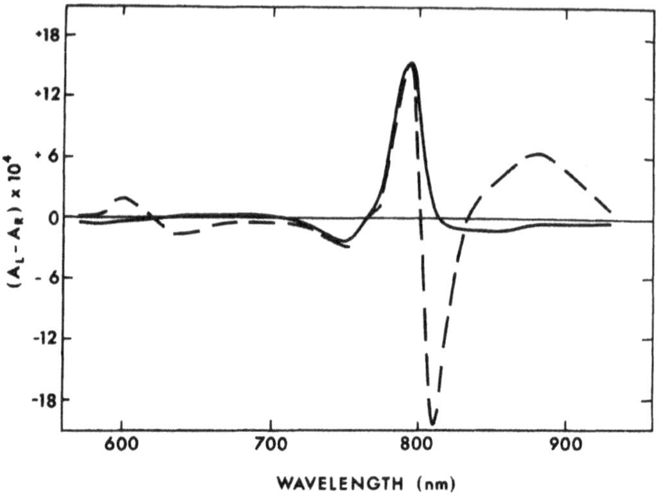

FIG 1. Circular dichroism spectra of reduced (dotted line) and ferricyanide-oxidized (continuous line) photoreaction center preparations from Ectothiorhodospira sp. measured at room temperature.

II. CIRCULAR DICHROISM OF ORIENTED PHOTOREACTION CENTER

More information can be drawn from CD spectroscopy when it is performed on oriented preparations. In the present instance, we had recourse to two methods of orientation bearing either on purified photoreaction center or on the chromatophore membrane where it is embedded. Orientation of purified preparations from *Rhodospirillum rubrum* was achieved by pressing their suspensions in 10% polyacrylamide gel along the direction of the measuring light beam (18). In this case, the pressure compresses the block and stretches the gel equally in all directions normal to the direction of the pressure. Linear dichroism absorption measurements showed that, in the pressed gel, the Q_y transition dipole of P870 is oriented symmetrically in the plane of the gel normal to the direction of the pressure and of the measuring light beam. This implies that pressure is orienting the photoreaction center molecules with their C_2 symmetry axis for the six pigment molecules (13-15) parallel to the direction of the measuring light beam. Because the photoreaction center has a low rotational mobility in the chromatophore membrane (19), orientation of the photoreaction center could also be obtained by air drying chromatophores on a glass plate (20). These chromatophores were first treated with $K_3Fe(CN)_6$ to selectively bleach the antenna Bchl. Air drying chromatophores onto a glass plate causes the photosynthetic membrane to lie flat on the glass slide (20). The C_2 symmetry axis of the photoreaction center being oriented normal to the plane of the membrane (21, 22), it is also normal to the plane of the glass. Fig 2 shows the absorption spectrum of a dried film of $K_3Fe(CN)_6$-treated chromatophores from *Rhodospirillum rubrum* measured with a horizontal measuring light beam linearly polarized parallel and perpendicular to the vertical axis of the glass slide. The plane of the glass is oriented 40° to the direction of the measuring light beam. The results show that the P870 absorption dipole is approximately in the plane of the glass while the P750 and P600 absorption dipoles are approximately normal to the plane. This is in line with the photoreaction center being oriented with its C_2 symmetry axis normal to the plane of the glass.

Fig 3 and Fig 4 show the CD spectra measured in the direction of the C_2 symmetry axis for *Rhodospirillum rubrum* photoreaction centers embedded either in a polyacrylamide gel (Fig 3) or in the chromatophore membrane (Fig 4). In both cases, the CD bands at 870 and 810 nm and at 630 and 600 nm undergo a rotational strength decrease compared with unoriented photoreaction center. Tinoco (16) has shown, in all generality, that the CD excitonic transitions disappear when light is incident along the symmetry axis of the system. If it was due wholly to exciton interaction, the entire CD spectrum should decrease. Figs 3 and 4 show that the positive CD band at 795 nm and the negative CD band at 750 nm do not decrease. This implies that these two CD bands are not due to exciton interactions having a symmetry axis in the direction of measurement. They must be due to other interactions acting asymmetrically about the C_2 axis. Asymmetric Bchl-protein (4) or Bchl-Bph interactions may be suggested. The decrease in rotational strength of the

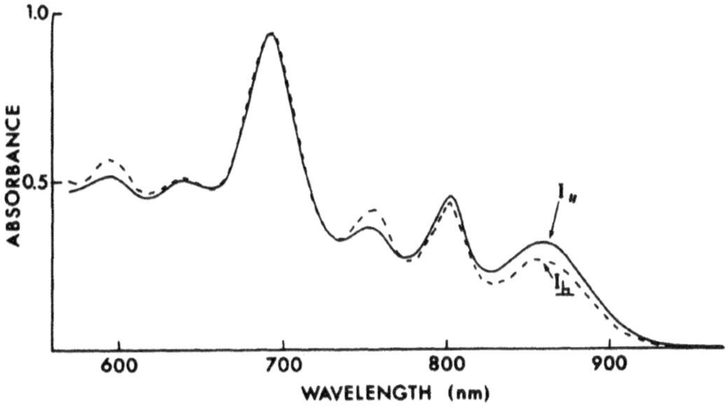

FIG 2. Linear dichroism of a dry film of ferricyanide-treated chromatophores from *Rhodospirillum rubrum*. The vertically ($I_{//}$) and horizontally (I_h) polarized measuring light beam was directed along a horizontal axis of the supporting glass slide. The latter was tilted about a vertical axis so as to form a 40° angle between the plane of the slide and the axis of the measuring beam.

biphasic CD bands at 870 and 810 nm and at 630 and 600 nm can be explained by exciton interactions acting symmetrically about the symmetry axis. However, this decrease could also be due to other types of interactions acting symmetrically that axis. If both the molecular exciton and charge transfer interactions are symmetric about the C_2 symmetry axis, then left and right circularly polarized light incident along that axis will be equally absorbed. There is experimental evidence for charge transfer interactions between the two Bchl molecules of the special pair (23) and for configuration mixing between exciton and charge transfer bands (2). CD spectroscopy of oriented photoreaction center, even if it cannot distinguish between exciton and charge transfer bands, is thus a powerful means of ascertaining which bands are due to interactions acting symmetrically about the C_2 symmetry axis and which bands are due to non symmetric interactions that might occur on only one side of this axis.

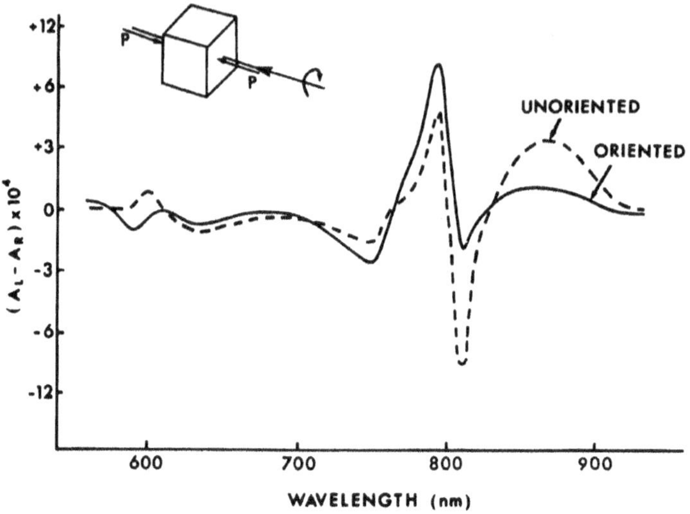

FIG 3. Circular dichroism spectra of oriented (continuous line) and unoriented (dotted line) photoreaction center from *Rhodospirillum rubrum*. The photoreaction center was suspended in 10% polyacrylamide gel and oriented by applying pressure (P) to the gel in the direction of the light beam. The CD spectrum of the unoriented preparation was corrected for optical pathlength so as to normalize the two spectra.

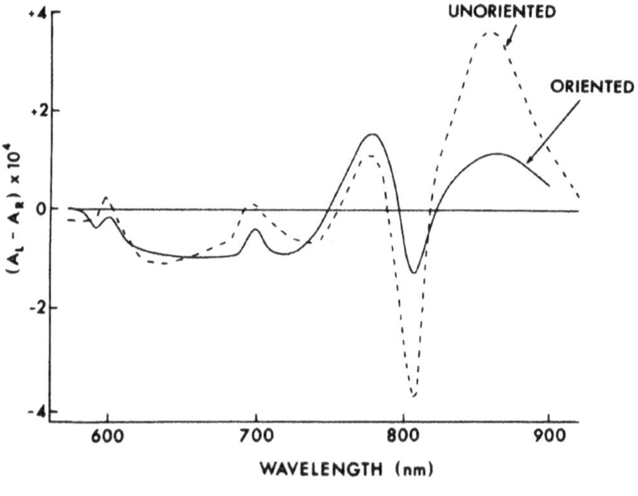

FIG 4. Circular dichroism spectra of oriented (continuous line) and unoriented (dotted line) ferricyanide-treated chromatophores from *Rhodospirillum rubrum*. The chromatophores were oriented by air drying onto a glass plate. The measuring light beam was normal to the plane of the glass plate. The spectrum of the unoriented preparation was measured in a 1 cm pathlength cuvette with a sample having the same A_{870} as that on the glass plate.

III. CD SPECTRA OF PHOTOBLEACHED PHOTOREACTION CENTERS

Selective modification of the chromophores is another approach that can be used to elucidate the CD spectrum of the photoreaction center. Borohydride reduction and subsequent extraction of one of the Bchl monomers (24, 25) resulted in a CD spectrum due to the interaction of the remaining Bchl and Bph molecules (26). Phototrapping of intermediary states (27, 28) by continuous illumination at low potential led to the reduction of one of the Bph molecules with a resulting CD spectrum due to the interaction of four Bchl and one Bph molecules (7). We have extended this technique to the photoreaction center isolated from *Ectothiorhodospira sp.* (29). By using higher light intensities and longer times of illumination, than used previously, we have obtained a complete bleaching of P750 and a large bleaching of P800 (Fig 5) thus isolating the absorption spectrum of the special pair (30). The CD spectrum of this bleached preparation (Fig 6) shows the Q_y transition to be split in two conservative positive and negative components at 852 and 795 nm. The Q_x transition is also split in two conservative negative and positive components at 620 and 598 nm. Although these bands are conservative, they are not symmetric : the 795 and 852 nm bands, on the one hand, and the 598 and 620 nm bands, on the other, have unequal widths. This CD spectrum is simpler than that of the borohydride-treated photoreaction center which showed three Q_y bands for the three remaining Bchl molecules (26). The CD spectrum of the special pair (Fig 6) resembles the theoretical CD spectrum of an exciton dimer (31). Using the geometry of the special pair found by x-ray crystallography (13-15) and the point monopole expansion of Weiss (32), we have calculated the Q_y and Q_x exciton interaction energies to be 268.8 cm^{-1} and 243.3 cm^{-1} respectively. These exciton interaction energies lead to predicted band splittings of 537.6 cm^{-1} and 486.6 cm^{-1} for the Q_y and the Q_x bands respectively (30). The corresponding experimental values (Fig 6) are 841.6 cm-1 for the Q_y and and 593 cm-1 for the Q_x band. The fact that the experimental values are larger than predicted by this calculation could be attributed to configuration mixing of the exciton states with charge transfer states (2) or to an asymmetry of the two Bchl molecules of the special pair (8). Using the asymmetric model (8) with the calculated exciton interaction energies we deduced the theoretical CD spectrum for the special pair (Fig 6). The two Bchl molecules of

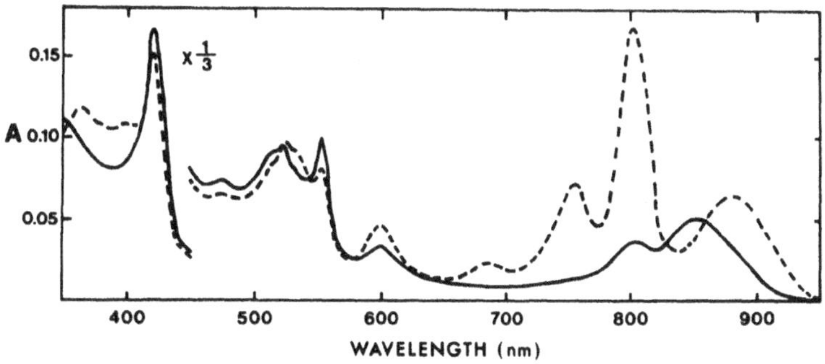

FIG 5. Absorption spectrum of photo-trapped photoreaction center from *Ectothiorhodospira sp.*. The samples were suspended in 100 mM Tris HCl/0.1% (W/V) Triton X-100 in the presence of 8 mg Na ascorbate/ml of suspension. One of the samples was kept in the dark (dotted line) whereas the other received 117.5 min of illumination with white light of 6 x 10^6ergs/cm^2sec under anaerobic conditions and at room temperature (continuous line).

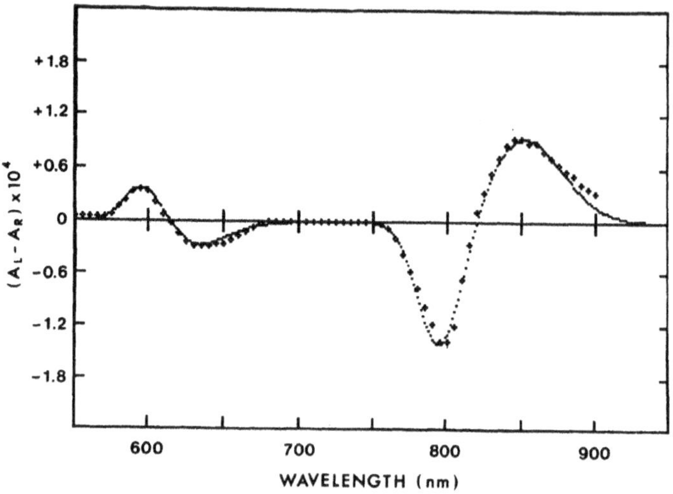

FIG 6. Circular dichroism spectrum of photoreaction center from *Ectothiorhodospira sp.* illuminated in the presence of sodium dithionite. The sample was suspended in 50 mM Tris HCl/0.1%(W/V) Triton X-100 in the presence of 12 mg of Na dithionite/ml of suspension. The sample was illuminated for 44 min at room temperature with white light passing through a heat filter with an intensity of 3.25 x 10^6 ergs/cm^{2}sec under anaeorbic conditions.The experimental spectrum is represented by crosses and the calculated spectrum by a continuous line.

the special pair were calculated to have monomer absorption peaks at 845 and 615.1 nm for one and at 801.1 and 602.5 for the other. This difference in the monomer absorption spectra results in unequal widths for the dimer bands (8). To fit the experimental spectrum, Gaussian shapes with exponential half-widths of 37 and 22 nm were used for the 852 and 620 nm bands and for the 795 and 598 nm bands respectively. The deduced and experimental CD spectra match each other quite well (Fig 6). The shift of the Q_y band of the special pair from 852 nm in the modified photoreaction center to 880 nm in the unmodified photoreaction center may be due to configuration mixing of charge transfer states with exciton states. In this respect, a large change in the dipole moment associated with the Q_y transition of the special pair was recently observed with unmodified photoreaction centers (23). It may be that a configuration mixing of the charge transfer state with the exciton state occurs only when electron transfer can occur. Although the asymmetric exciton model for the special pair does not uniquely fit the experimental data, it certainly is plausible.

IV. CONCLUSION

The CD spectrum of photoreaction center oriented with its C_2 symmetry axis parallel to the direction of the measuring beam shows that the biphasic bands at 870 and 810 nm and at 630 and 600 nm are due to interactions which are symmetric about this axis. Conversely, the CD bands at 795 and 750 nm are due to interactions which are asymmetric with respect to that axis. An extension of the phototrapping technique has allowed CD spectroscopy to be carried out on selectively bleached intermediary states with a reduced number of interacting chromophores. Under these conditions, the CD spectrum of the special pair was isolated and found to consist of two conservative bands.

ACKNOWLEDGEMENTS

This work was supported by a grant from the Natural Science and Engineering Council of Canada.

REFERENCES

1. Sauer, K., Dratz, E.A. & Coyne, L. (1968) Proc. Natl. Acad. Sci. U.S.A. 61, 17-24.
2. Parson,W.W., Sherz,A. & Warshel,A. (1985) Springer Ser. Chem. Phys. 42, 122-130.
3. Sherz, A. & Parson, W.W. (1984) Biochim. Biophys. Acta 766, 666-678.
4. Boxer, S.G. & Wright, K.A. (1979) J. Am. Chem. Soc. 101, 6791-6794.
5. Philipson, K.D. & Sauer, K. (1972) Biochemistry 11, 1880-1885.
6. Pearlstein, R.M. & Hemenger, R.P. (1978) Proc. Natl. Acad. Sci. U.S.A. 75, 4920-4924.
7. Shuvalov, V.A. & Asadov, A.A. (1979) Biochim. Biophys. Acta 545, 296-308.
8. Mar, T. & Gingras, G. (1984) Biochim. Biophys. Acta 764, 283-294.
9. Knapp, E.W., Fischer, S.F., Zinth, W., Sander, M., Kaiser, W., Deisenhofer, J. & Michel, H. (1985) Proc. Natl. Acad. Sci. U.S.A. 82, 8463-8467.
10. Knapp, E.W., Scherer, P.O.J. & Fischer, S.F. (1986) Biochim. Biophys. Acta 852, 295-305.
11. Scherer, P.O.J. & Fischer, S.F. (1987) Biochim. Biophys. Acta 891, 157-164.
12. Bucks, R.R. & Boxer, S.G. (1982) J. Am. Chem. Soc. 104, 340-343.
13. Deisenhofer, J., Epp, O., Miki, K., Huber, R. & Michel, H. (1984) J. Mol. Biol. 180, 385-598.
14. Allen, J.P., Feher, G., Yeates, T.O., Komiya, H. & Rees, D.C. (1987) Proc. Natl. Acad. Sci. U.S.A. 84, 5730-5734.
15. Chang, C.H., Tiede, D., Tang, J., Smith, U., Norris, J. & Schiffer, M. (1986) FEBS Lett. 205, 82-86.
16. Tinoco, I.,Jr. (1962) Advances in Chemical Physics (Prigogine, I.,ed.) Vol IV, pp 113-160, John Wiley, New York.
17. Philipson, K.D. & Sauer, K. (1973) Biochemistry 12, 535-539.
18. Mar, T & Gingras, G. (1984) Biochim. Biophys. Acta 764, 86-92.
19. Mar, T., Picorel, R. & Gingras, G. (1981) Biochim. Biophys. Acta 637, 546-550.
20. Vermeglio, A. & Clayton, R.K. (1976) Biochim. Biophys. Acta 502, 51-60.
21. Paillotin,G., Vermeglio,A., Breton,J. (1979) Biochim. Biophys. Acta 545, 249-264.
22. Yeates, T.O., Komiya, H., Rees, D.C., Allen, J.P. & Feher, G. (1987) Proc. Natl. Acad. Sci. U.S.A. 84, 6438-6442.
23. Lockhart, D.J. & Boxer, S.G. (1987) Biochemistry 26, 664-668.
24. Ditson, S.L., Davis, R.C. & Pearlstein, R.M. (1984) Biochim. Biophys. Acta 766, 623-629.
25. Maroti, P., Kirmaier, C., Wraight, C., Holton, D. & Pearlstein, R.M. (1985) Biochim. Biophys. Acta 810, 132-139.
26. Shuvalov, V.A., Shkuropatov, A.Ya., Kulakova, S,M., Ismailov, M.A. & Shkuroputova, V.A. (1986) Biochim. Biophys. Acta 849, 337-346.
27. Shuvalov, V.A. & Klimov, V.A. (1976) Biochim. Biophys. Acta 440, 587-599.
28. Tiede, D.M., Prince, R.C. & Dutton, P.L. (1976) Biochim. Biophys. Acta 449, 447-467.
29. Lefebvre, S., Picorel, R., Cloutier, Y. & Gingras, G. (1984) Biochemistry 23, 5279-5288.
30. Mar, T., Picorel, R. & Gingras, G., manuscript in preparation.
31. Tinoco, I.,Jr. (1963) Radiat. Res. 20, 133-139.
32. Weiss, C.,Jr. (1972) J. Mol. Spectro. 44, 37-80.

LOW TEMPERATURE LINEAR DICHROISM STUDY OF THE ORIENTATION OF THE PIGMENTS IN REDUCED AND OXIDIZED REACTION CENTERS OF RPS. VIRIDIS AND RB. SPHAEROIDES

J. Breton

Service de Biophysique, Département de Biologie, CEN Saclay
91191 Gif-sur-Yvette Cedex, France

INTRODUCTION

Since the crystal structure of reaction centers (RCs) from purple photosynthetic bacteria has been solved (1-3) one of the important issue in relating this structure to its primary function is to assign the contribution that each of the six individual pigments seen in the model makes to the various absorption bands observed in the absorption spectrum of these RCs. The relevance of this problem, which is tackled both experimentally and theoretically, is twofold : on one hand this assignment is essential in order to interpret unambiguously the results of the femtosecond and picosecond absorbance changes involved in the initial steps of electron transfer. On the other hand if the absorption properties of the RC can be correctly calculated using a non-phenomenological model, the energy levels of the various states which can be calculated by such a theory will give a better description of the mechanisms of the initial step of charge separation (Parson et al., this volume).

Spectroscopy with plane polarized light has proven to be a very efficient tool to investigate several aspects of the organization of the pigments in photosynthetic systems including the bacterial RC (4). For example a linear dichroism (LD) study of Rps. viridis RCs in the intact photosynthetic membrane has led to a prediction of the angular relationships of the monomer bacteriochlorophylls forming the special pair dimer (P) with respect to each other as well as with respect to the membrane plane (5) which is in full agreement with the X-ray data on the same RC (6). Furthermore a comparison of the low temperature absorption (A) and LD spectra of RCs from Rb. sphaeroides and from Rps. viridis showing identical orientations for all the corresponding optical transitions, it could be predicted that the geometrical arrangements of the six pigments in the two RCs are indistinguishable (7) as it is now confirmed by the X-ray results on Rb. sphaeroides RCs (3).

In the present report we compare A and LD spectra for RCs in the reduced (P) and oxidized (P^+) states of the primary donor. These spectra were taken at 10K for RCs from Rps. viridis and RCs from Rb. sphaeroides 241, its R26 mutant and the borohydride-modified RCs from R26 (8). This study was specifically aimed at attempting to assign the molecular origin of the absorption changes ($\triangle$A) observed upon (photo-) oxidation of the special pair P. The data reported here suggest an

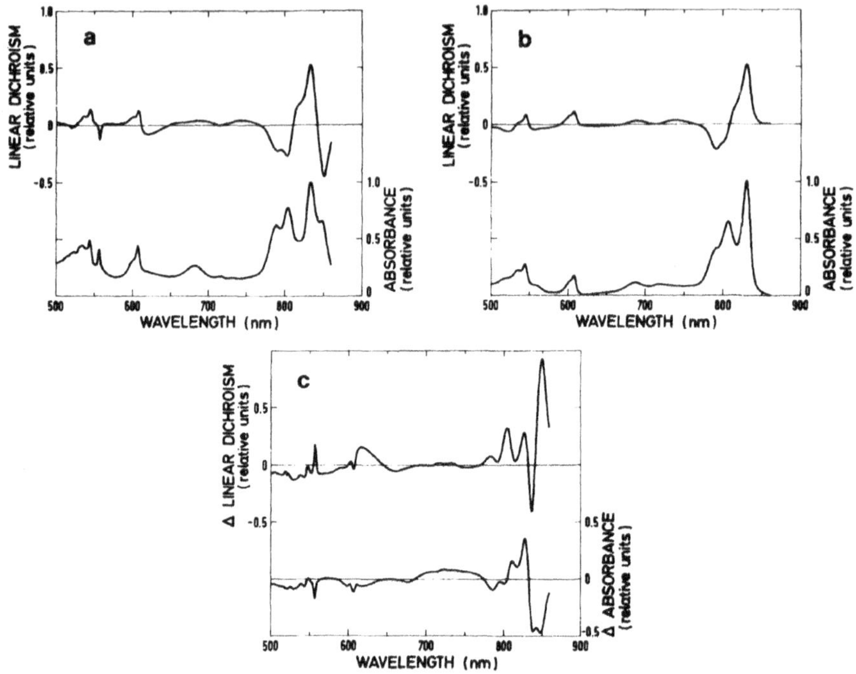

Fig. 1 : Low temperature (10K) absorption and linear dichroism (LD) spectra of reaction centers from Rps. viridis embedded in a squeezed polyacrylamide gel. The primary electron donor is in the reduced (a) or oxidized (b) state. See ref. 7 for experimental details. (c) Calculated oxidized-minus-reduced difference absorption and LD spectra.

assignment for the split sharp bands observed in the region 800 to 815nm for the low temperature ΔA spectra of Rb. sphaeroides RCs (9). They also reveal a small difference in the orientation of one of the accessory bacteriochlorophylls (B_M) when comparing the RCs of Rps. viridis and Rb. sphaeroides.

RESULTS AND DISCUSSION

1. Reaction centers from Rps. viridis

The A and LD spectra of Rps. viridis RCs oriented by uniaxial squeezing of a polyacrylamide gel (7) are shown for the spectral range 500-860nm in Fig. 1a (reduced state) and Fig. 1b (oxidized state). The orientation of the six pigments has been extensively discussed (7) and is summarized together with our assignments in Fig. 2*. It should also

----

*The problem of a convenient nomenclature for the cofactors has been addressed during this Conference. As no consensus was achieved we prefer to keep using a two letter code in which the first letter stands for primary donor (P), monomeric bacteriochlorophyll (B) or bacteriopheophytin (H) while the second letter designates the L or M branch of the polypeptide closest to these cofactors.

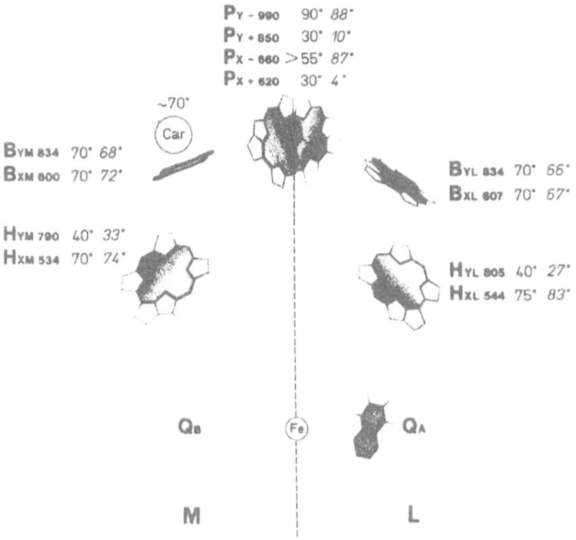

Fig. 2 : Assignment of the optical transitions of the pigments in the reaction center from Rps. viridis. P, B and H refer to the primary donor molecules, the monomeric bacteriochlorophylls and the bacteriopheophytins, respectively. X and Y designate the axis of the optical transitions, while L and M mean the polypeptide branch closest to these cofactors. Next to the peak wavelength (in nm) for each transition, two values for the angle between the transition and the C2 (vertical) axis are given. The first one is estimated from the LD measurement and the second one is calculated, assuming excitonic coupling within P only (7), from the atomic coordinates kindly provided by J. Deisenhofer and H. Michel. Q, quinones ; Car, carotenoid.

be noticed that in the oxidized sample the magnitude of the dichroism of the Q_X transitions of (i) the accessory bacteriochlorophylls (B_L, B_M) absorbing at 600-610nm and (ii) the bacteriopheophytins (H_L, H_M) absorbing at 535-545nm are very similar. The high potential cytochrome C-558 exhibits a negative LD demonstrating an orientation at about 35° from the C2 axis for the 556-nm transition. The orientation of the carotenoid (Car) transition perpendicular (i) to the C2 axis in Rps. viridis and Rb. sphaeroides 241 (7) and (ii) to the P_{Y-} transition at 890nm in Rb. sphaeroides 241 (10), has been used together with the efficient T-T energy transfer from ^3P to Car to locate this molecule. Our initial proposal to further locate the Car on the C2 axis, which was indeed based on weak grounds (7), has turned out to be incorrect as refinements of the structure now indicate an elongated object positionned on top of B_M with the correct orientation to be the Car molecule.

The comparison of the ΔA and ΔLD spectra (Fig. 1c) shows that the S-shaped feature centered at 832nm has the same shape in the two spectra indicating that it can confidently be assigned to the band shift of at least one (and more probably both) B molecule(s). On the other hand the bleaching ΔA signal at 850nm gives rise to a positive ΔLD band peaking also at 850nm. Several authors have interpreted various spectroscopic observations on this band by assigning it to the Q_Y transition of a B molecule (11-13). For this interpretation to be reconciled with the spectra shown in Fig. 1c it would be necessary that upon oxidation of P the 850-nm dipole both shifts to 830nm and changes its orientation from a value close to the C2 axis to a value close to the magic angle (55° from the C2 axis) so that it does not contribute to the ΔLD signal at 830nm. Furthermore the orientation of the 850-nm dipole in this assignment cannot be that of the Q_Y transition of one of the B molecules which are both oriented at 65-70° from the C2 axis (Fig. 2), as calculated from the atomic coordinates of the RC (Deisenhofer and Michel, private communication). This assignment is quite orthogonal to that already discussed in (4,5,7,14) in which we ascribe the 850-nm band to the high-energy exciton component P_{Y+} of P tilted at a small angle to the C2 axis while the low-energy P_{Y-} exciton state is responsible for the 990nm band and is oriented perpendicular to the C2 axis. It should be noted at this stage that several theoretical calculations of the absorption spectra of RCs from Rps. viridis have been recently formulated (15-19). Regarding the 850-nm band, however, these theories still diverge quite radically on its assignment. Some of them (15,16 and Pearlstein, this volume) rather support our model while others consider a quantum mechanical mixture of P_{Y+} with a large contribution from the B_L molecule (17-19).

2. Reaction centers from Rb. sphaeroides 241

The pronounced shoulder at 812nm in the absorption spectrum of the reduced RCs (Fig. 3a, 4a) together with the negative LD signal at 818nm have been already discussed (7). The A and LD spectra of the oxidized sample (Figs. 3b, 4b) shows several new interesting features such as a negative LD band at about 840nm, which corresponds to a very weak absorption band and a shoulder on the long wavelength side of the absorption band (more clearly seen in Fig. 4b) which gives a full width at half maximum (FWHM) of 16nm for the 800-nm absorption band and of only 12nm for the corresponding LD band. This transition around 808nm, which to our knowledge is first reported here, could possibly correspond to the unoxidized bacteriochlorophyll remaining in the P^+ dimer. This seems to us rather unlikely as the orientation of its Q_Y transition should be slightly more perpendicular to the C2 axis than that of the B molecules. This would lead to a larger positive LD contribution at 808nm than at 800nm in contrast to the observed spectra (Fig. 4b). We rather propose the tentative assignment that the 808-nm component in the P^+ spectrum represents one of the two B molecules.

In view of (i) the shoulder at 812nm in the absorption spectrum (Fig. 3a, 4a) which is significantly more pronounced than in the spectra of the RC from Rb. sphaeroides R 26 (Fig. 5a) and of (ii) the location of the Car molecule discussed in the previous section, it is tempting to speculate that part of the 812-nm absorption shoulder corresponds to the Q_Y transition of the B_M molecule red-shifted by the presence of Car and which would shift to about 808nm upon oxidation of P. This assignment is further substanciated by the observation of a fine structure in the 800-nm region of the ΔA spectrum (Fig. 3c) which could be assigned to the superposition of a blue shift centered at 810nm (B_M) giving rise to a positive lobe at 806nm and a negative one at 814nm and of a second blue shift centered at 800nm (B_L) giving rise to a positive lobe at 796nm and a negative one at 804nm. In the corresponding ΔLD spectrum the blue shift of the 800-nm band is

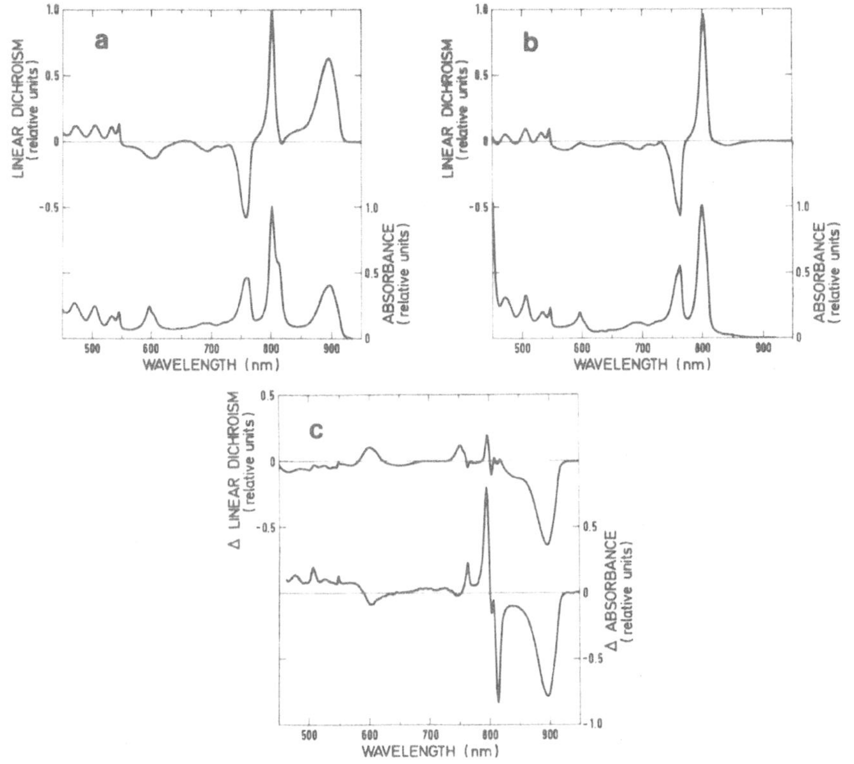

Fig. 3 : Same spectra as in Fig. 1 but for reaction centers from Rb. sphaeroides 241 at 10K. (a) reduced, (b) oxidized, (c) oxidized-minus-reduced.

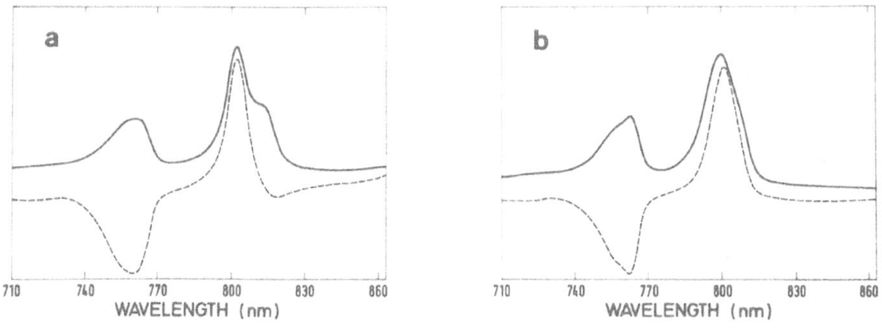

Fig. 4 : Absorption (——) and linear dichroism (----) spectra at 10K for the 800-nm region of reaction centers from Rb. sphaeroides 241 with the primary electron donor in the reduced (a) or oxidized (b) state.

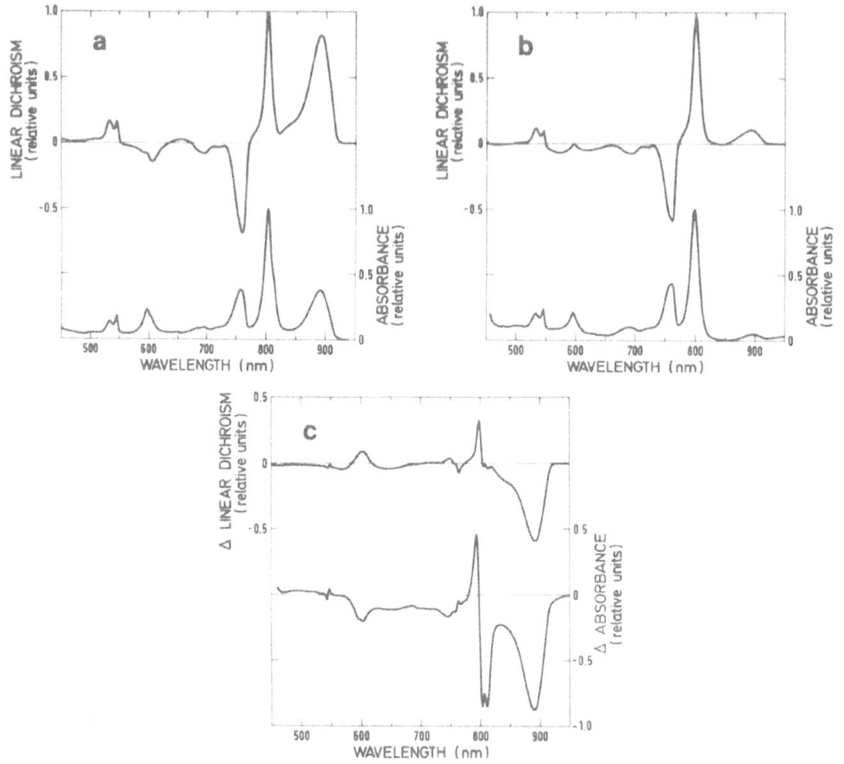

Fig. 5 : Same spectra as in Fig. 1 but for reaction centers from Rb. sphaeroides R26 at 10K. (a) reduced, (b) oxidized, (c) oxidized-minus-reduced.

clearly recognized and has a sign and a magnitude compatible with the orientation of the Q_Y transition of the B_L molecule at about 65-70° from the C2 axis as seen in the structure of Rps. viridis. However the situation is quite different around 810nm where the blue shift of the 810-nm band is seen as a very small △LD signal riding on top of a larger positive △LD band which, as discussed in (7), is assigned to the bleaching of the P_{Y+} exciton component of P responsible for the negative LD signal at 818nm (Fig. 3a, 4a). If this assignment is correct then the small magnitude of the △LD signal attributed to the shift of the 810-nm band indicates that the Q_Y transition of the B_M molecule is close to the magic angle and is thus about 10° closer to the C2 axis than the Q_Y transition of B_L at 800nm. Another interesting observation pertains to the small magnitude of the LD signal for the Q_X transitions of the B molecules relative to those of the H molecules in the LD spectrum of oxidized RCs (Fig. 3b). Compared to the corresponding signals in the LD spectrum of Rps. viridis RCs (Fig. 1b), this indicates that the Q_X transitions of the B molecules are on the average tilted at about 60° from the C2 axis in Rb. sphaeroides 241 compared to a value closer to 70° for Rps. viridis.

3. Reaction centers from Rb. sphaeroides R26

The A and LD spectra for both reduced (Fig. 5a) and oxidized (Fig. 5b) RCs from Rb. sphaeroides R26 are very similar to the corresponding ones from Rb. sphaeroides 241 (Fig. 3). The only noticeable differences are located in the 800-820nm region where the shoulder in the absorption and the dip in the LD of the reduced RCs are less pronounced in the strain R26 than in 241. Although no resolved shoulder is present in the spectrum of oxidized R26 RCs, the FWHM of the 800-nm absorption band (18nm) is larger than that of the corresponding LD band (14nm) indicating that the structure underlying the 800-nm band is essentially the same in the RCs of the two strains. The $\triangle$A and $\triangle$LD spectra (Fig. 5c) also closely resemble the corresponding spectra for the carotenoid-containing RCs. This is taken to indicate that the Q_Y transition of the B_L and B_M molecules have the same geometry in the two RCs and that the main spectral differences are likely related to the slightly larger overlap of their absorption bands in R26 than in 241. In the Q_X region, the dichroism of the B molecules is also identical in the RCs of the two strains. Furthermore the identical contribution of the high energy exciton component of P around 600nm (P_{X+}) to the $\triangle$A and $\triangle$LD spectra for the two strains demonstrates that the geometry of the dimer is the same for these RCs.

4. Borohydride-treated reaction centers from Rb. sphaeroides R26

The A and LD spectra of borohydride-treated RCs from Rb. sphaeroides R26 are shown for a sample cooled either in the dark (Fig. 6a, state P) or in the light (Fig. 6b, state P^+). The particular sample used for these spectra came from a batch of RCs which had a higher than normal 760/800nm ratio indicating some contamination with bacteriopheophytin. Upon $NaBH_4$ treatment the 800nm/860nm ratio at room temperature dropped from 2.2 to 1.5 consistent with the view that one of the two B molecules is removed (8,20,21 ; Scheer et al., this volume). The RCs were used directly after dialysis without further purification. In the reduced sample at 10K (Fig. 6a) the 800-nm absorption band has lost its 810-nm shoulder, as already reported in (21), and the LD spectrum is quite different from that of the unmodified RCs (Fig. 5a) with a large decrease of the amplitude of the 800-nm band relative to the 890-nm band. The LD spectrum still exhibits a dip around 815nm.

A small absorption band at 840nm in the spectrum of the photooxidized RCs, to which is associated a negative LD component, can be detected (Fig. 6b). It corresponds to the bands observed at the same wavelength in the spectra of RCs from Rb. sphaeroides 241 (Fig. 3b). This band can also be seen in the LD/A spectrum of the unmodified RC from Rb. sphaeroides R26 in the oxidized state (data not shown). Thus the 840-nm transition, which corresponds to dipoles oriented close to the symmetry axis, seems an ubiquitous feature in the spectra of all the investigated samples of oxidized RCs from Rb. sphaeroides. Its origin however remains to be determined. Upon photooxidation the amplitude of the 800-nm LD band relative to that of the bacteriopheophytins band at 760nm increases (Fig. 6b), an observation that can also be made in the normal RC and which suggests that the 800-nm band is composed of (at least) two transitions with opposite dichroism. Furthermore the FWHM of this band in the oxidized sample is much closer in the A spectrum (14nm) and LD spectrum (13nm) of the modified RCs than in the corresponding spectra of the native RCs, clearly indicating that the underlying structure of this band becomes simpler after the

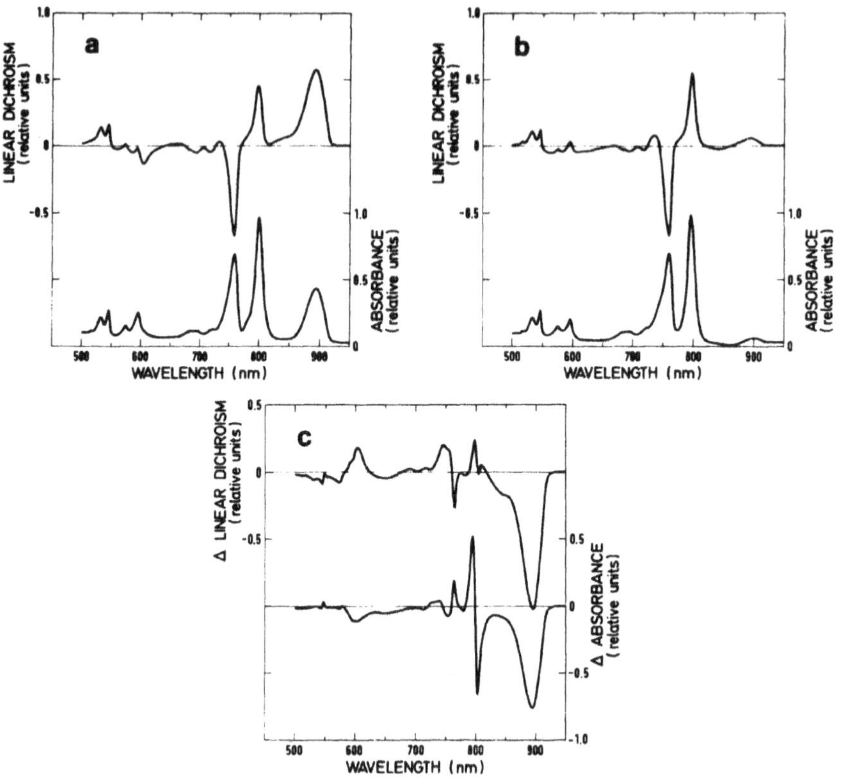

Fig. 6 : Same spectra as in Fig. 1 but for borohydride-treated reaction center from Rb. sphaeroides R26 at 10K. This treatment is though to essentially remove the B_M molecule (see Fig. 2). Furthermore the spectra were obtained for samples cooled either (a) in the dark or (b) under continuous illumination ($\lambda > 700$nm). (c) Oxidized-minus-reduced difference spectra.

borohydride treatment (Figs. 7a,b). These observations are supporting the view that part of the 810-nm shoulder in native RCs belongs to the B_M molecule which is removed upon modification. The $\triangle$A and $\triangle$LD spectra (Figs. 6c, 7c) further strengthen this idea as the long wavelength S-shaped signal centered around 810nm observed in the spectra of the unmodified RCs from Rb. sphaeroides (Fig. 3c, Fig. 5c) is no longer seen after borohydride-treatment (Fig. 7c). It is clear however that the almost symmetrical shift centered at 800nm observed in the $\triangle$A spectrum, and previously described in (21), is strongly distorted in the $\triangle$LD spectrum. This is seen in Fig. 7c where a band shift centered at 800nm in the $\triangle$LD spectrum is seen riding on top of a positive band giving rise to a peak at 809nm. This positive band we assign to the bleaching of the P_{Y+} exciton of P oriented close to the C2 axis. One could argue that this positive $\triangle$LD band can be due to the appearing monomer in P^+. However the existence of a negative LD component in the P spectrum (Fig. 7a, LD/A) makes this hypothesis highly unprobable.

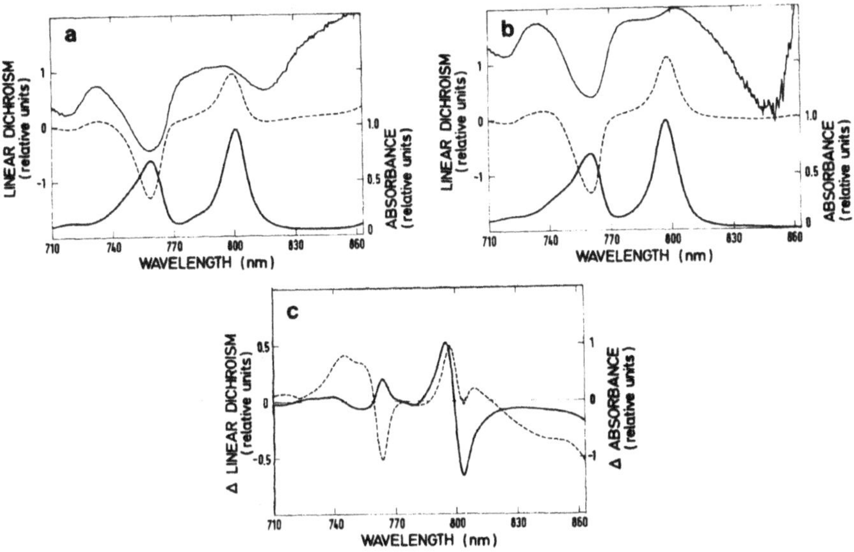

Fig. 7 : Same spectra as in Fig. 6 but the wavelength scale has been amplified in the 800-nm region. In (a) and (b) the upper traces represent the LD/A spectra.

Comparing the $\triangle$A and $\triangle$LD spectra in the 600-nm region for the native (Fig. 5c) and the modified (Fig. 6c) R26 RCs, the orientation of the P_{X+} exciton component of P at a small angle to the C2 axis appears the same before and after modification. Furthermore the A and LD spectra for all the oxidized RCs from Rb. sphaeroides (Figs. 3b,5b,6b) are quite the same in the 600-nm region. This observation is taken to indicate that the B_M molecule contributes very little to the dichroism in the Q_X region and thus that its X direction is oriented close to 55° from the C2 axis.

CONCLUSIONS

The visible and near IR spectral changes ($\triangle$A, $\triangle$LD) associated with the (photo)oxidation of the primary electron donor at 10K in various RC preparations reveal many close similarities in the geometrical organization of the pigments between the two purple bacteria Rps. viridis and Rb. sphaeroides but also suggest that in the RCs of the latter species the two accessory bacteriochlorophylls (B_L,B_M) are in different environments and exhibit slightly different orientations of their X and Y transition moments with respect to the pseudo C2 symmetry axis.

In Rps. viridis the $\triangle$A and $\triangle$LD spectra are interpreted in terms of (i) the bleaching of the high energy exciton component of P (P_{Y+}) absorbing at 850nm and oriented at a small angle from the C2 axis and (ii) the shift from 834nm (state P) to 830nm (state P^+) of the two equivalent B molecules having their Q_Y and Q_X transition moments oriented at about 70° from the C2 axis.

In RCs from the strains 241 and R 26 of Rb. sphaeroides as well as in the borohydride-treated RCs from R26 the geometrical arrangement of the special pair pigments and of the two bacteriopheophytins (H_L, H_M) are spectrally almost indistinguishable from that observed in Rps. viridis. However the new data on the various RCs from Rb. sphaeroides presented here allow to precise the assignment for the shoulder observed around 810nm in the spectra of the reduced RCs from Rb. sphaeroides 241 and R26. This assignment takes into account the ΔA and ΔLD data on the modified RCs from R26 (Figs. 6,7) and the hypothesis that the Car molecule in the RCs from 241 is located in close proximity to B_M. This assignment of the 810-nm shoulder is actually a combination of the two previous proposals that it is due essentially to the Q_Y transition of B_M (9) or to the high-energy exciton component P_{Y+} of the special pair (7,10). With this new combined assignment the sharp features observed in the ΔA spectra are due to the blue-shift of the B molecule(s) and to the bleaching of P_{Y+}. The corresponding ΔLD shows (i) a broad positive band in the 800-820nm spectral range which is assigned to the bleaching of P_{Y+} oriented close to the symmetry axis and (ii) either two band shifts (R26 and 241) or only one band shift (modified R26). The magnitude of the ΔLD due to the 800-nm Q_Y band of B_L indicates an orientation at about 70° from the C2 axis close to that observed (7) and calculated (Fig. 2) for Rps. viridis RCs. On the other hand the very small magnitude of the ΔLD due to the 810-nm Q_Y transition of B_M suggests that this transition is tilted at only 55-60° from this axis. A similar conclusion regarding the orientation of the Q_X transitions of B_M (about 60°) versus B_L (about 70°) is also derived from a comparison of the LD spectra of the oxidized RCs. It will be of interest to see if future refinements of the present Rb. sphaeroides X-ray structure (2,3) will reveal such a slight difference in the geometry and/or environment of the two accessory bacteriochlorophyll molecules.

ACKNOWLEDGEMENTS

I wish to acknowledge the role of Chris Kirmaier and of Hugo Scheer in providing useful details about the borohydride treatment of the RCs and of Sandra Andrianambinintsoa in applying them successfully.

REFERENCES

1. J. Deisenhofer, O. Epp, K. Miki, R. Huber and H. Michel (1984) J. Mol. Biol. 180, 385-398.
2. C. H. Chang, D. Tiede, J. Tang, U. Smith, J. Norris and M. Schiffer (1986) FEBS Lett. 205, 82-86.
3. J. P. Allen, G. Feher, T. O. Yeates, H. Komiya and D. C. Rees (1987) Proc. Natl. Acad. Sci. USA 84, 5730-5734.
4. J. Breton and A. Verméglio (1982) in "Photosynthesis : Energy Conversion by Plants and Bacteria", Vol. I, pp. 153-194 (Govindjee, ed.), Academic Press.
5. G. Paillotin, A. Verméglio and J. Breton (1979) Biochim. Biophys. Acta 545, 249-264.
6. H. Michel, O. Epp and J. Deisenhofer (1986) EMBO J. 5, 2445-2451.
7. J. Breton (1985) Biochim. Biophys. Acta 810, 235-245.
8. P. Maroti, C. Kirmaier, C. Wraight, D. Holten and R. M. Pearlstein (1985) Biochim. Biophys. Acta 810, 132-139.
9. C. Kirmaier, D. Holten and W. W. Parson (1985) Biochim. Biophys. Acta 810, 49-61.
10. A. Verméglio, J. Breton, G. Paillotin and R. Gogdell (1978) Biochim. Biophys. Acta 501, 514-530.

11. V. A. Shuvalov and A. A. Asadov (1979) Biochim. Biophys. Acta 545, 296-308.
12. V. A. Shuvalov and W. W. Parson (1981) Biochim. Biophys. Acta 638, 50-59.
13. H. J. Den Blanken and A. J. Hoff (1982) Biochim. Biophys. Acta 681, 365-374.
14. A. Verméglio and G. Paillotin (1982) Biochim. Biophys. Acta 681, 32-40.
15. J. Eccles, B. Honig and K. Schulten (1988) Biophys. J., in press.
16. Y. Won and R. A. Friesner (1988) J. Phys. Chem, in press.
17. W. W. Parson and A. Warshel (1987) J. Am. Chem. Soc. 109, 6152-6163.
18. E. J. Lous and A. J. Hoff (1987) Proc. Natl. Acad. Sci. USA 84, 6147-6151.
19. E. W. Knapp, P. O. J. Scherer and S. F. Fischer (1986) Biochim. Biophys. Acta 852, 295-305.
20. S. L. Ditson, R. C. Davis and R. M. Pearlstein (1984) Biochim. Biophys. Acta 766, 623-629.
21. D. Holten, C. Kirmaier and L. Levine (1987) In "Progress in Photosynthesis Research", Vol. I (J. Biggins, ed.), Martinus Nijhoff Publishers, Dordrecht, Netherlands, pp. 169-176.

ANISOTROPIC MAGNETIC FIELD EFFECTS OF THE PHOTOSYNTHETIC BACTERIAL REACTION CENTER OF RHODOBACTER SPHAEROIDES R-26, STUDIED BY LINEAR DICHROIC MAGNETO-OPTICAL DIFFERENCE SPECTROSCOPY (LD-MODS) IN THE TEMPERATURE RANGE 1.2 - 310 K

E.J. Lous and A.J. Hoff

Department of Biophysics, Huygens Laboratory
State University of Leiden, The Netherlands

INTRODUCTION

Recently we have introduced magneto-optical difference spectroscopy (MODS) to measure triplet-minus-singlet absorbance difference (T - S) spectra of bacterial photosynthetic reaction centers (RC) over a wide range of temperatures (Hoff et al., 1985; Lous and Hoff, 1986). The MODS technique rests upon the change in yield of the triplet state of the primary donor, ^3P, effected by a magnetic field of small amplitude (a few tens of millitesla). The field is modulated at a few hundred hertz and the resulting modulation in absorbance lock-in detected over a wide range of wavelengths. Since the magnetic field B_0 is a vectorial quantity and the magnetic field effect (MFE) sensitive to the orientation of RC with respect to $\vec{B}_0$ (see below), one expects that it should be possible to perform a linear dichroic (LD)-MODS experiment, which would result in a LD-(T - S) spectrum. Knowledge of the orientational dependence of the MFE should then allow to extract information on e.g. the magnitude and direction of the dipolar interaction of the primary radical pair (RP) P^+I^-, where I is the bacteriopheophytin acceptor. The dipolar interaction between P^+ and I^- plays a significant role in the interpretation of reaction yield detected magnetic resonance (RYDMR) and MFE spectra (Lersch and Michel-Beyerle, 1982; Tang and Norris, 1983; Moehl et al., 1985; Hunter et al., 1987). An independent determination of the dipolar interaction would help in obtaining a reliable value of the isotropic exchange interaction $J(P^+I^-)$, which is of great interest for understanding photoinduced electron transport (Marcus, 1987; Bixon, 1987).

In this communication we report on LD-MODS experiments performed on RC of Rhodobacter (Rb.) sphaeorides R-26. We have obtained (T - S) and LD-(T - S) spectra and we have studied the MFE as a function of B_o over a wide range of temperatures. We find clear evidence that, dependent on the temperature, two processes contribute to the MFE: One is due to the so-called radical pair mechanism (RPM), the other a consequence of the magnetic field-induced quantum mechanical mixing (QMM) of the triplet energy levels of ^3P. The RPM seems to predominate at higher temperatures (50 - 290 K), whereas below 50 K the QMM effect gradually becomes dominant and at T < 10 K is solely responsible for the MFE.

The isotropic (T - S) spectra obtained by MODS are strongly temperature dependent. Since their shape does not depend on the relative contribution of the RPM and QMM mechanisms to the MFE, this dependence is a consequence of (structural) changes in the RC. We present evidence that the spectral changes are well explained taking into account band broadening with increasing temperature (due to the increase in population of higher vibrational states) and in addition a slight conformational change of P resulting in a shift of the maximum of the long-wavelength absorption band from 860 nm at 290 K to 890 nm at 4 K and below. Our results make it unlikely that any drastic change in RC structure occurs upon cooling to cryogenic temperatures (down to 1.2 K).

MECHANISMS RESPONSIBLE FOR THE MAGNETIC FIELD EFFECT

Radical pair mechanism (RPM). In isolated RC the radical pair P^+I^- is formed in a spin-correlated singlet state $^1[P^+I^-]$ with close to 100 % quantum efficiency. Conversion to the triplet state $^3[P^+I^-]$ is effected through local hyperfine (hf) fields. In RC in which the quinone secondary acceptor is reduced or removed, the pair P^+I^- decays by recombination to either the singlet ground or excited state (from $^1[P^+I^-]$ with lumped rate constant k_S) or to the triplet state ^3P (from $^3[P^+I^-]$ with rate constant k_T) (Fig. 1). When an external magnetic field $\vec{B}_o$ of sufficient magnitude is applied, the rate of singlet-triplet conversion is diminished because two of the three triplet energy levels of $^3[P^+I^-]$ are then energetically removed from the singlet $^1[P^+I^-]$ level by the Zeeman interaction. Hence the $^1[P^+I^-]$ state can then only convert into the $m_S = 0$ (T_o) state of $^3[P^+I^-]$ and the yield of ^3P is decreased by up to 45 %. (See for reviews of the RPM - MFE on photosynthetic RC Hoff (1981) and Boxer et al. (1983).)

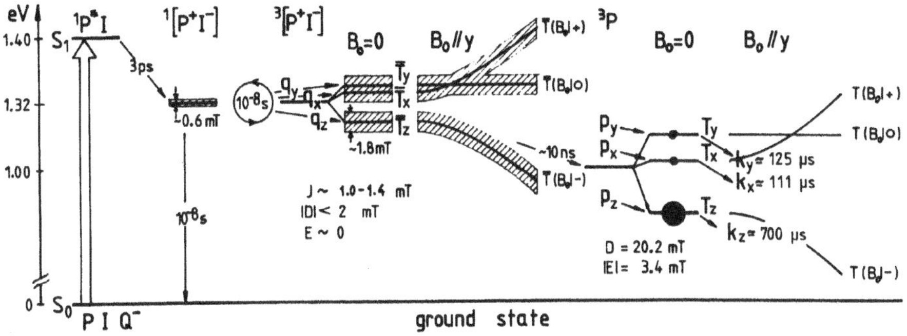

Fig. 1. Energy level scheme of the radical pair and primary donor triplet energy levels.

If all hf fields are taken together into one hf interaction A the triplet yield in the absence of an external field $\vec{B}_0$ is given by (Haberkorn and Michel-Beyerle, 1979)

$$\phi_T(0) = 3A^2k_T(k_S + k_T)/\{(3A^2 + 4\ k_Sk_T)(k_S + k_T)^2 + 16k_Sk_T(J - A/2)^2\} \quad (1)$$

and for large B_0 by

$$\phi_T(B_0 = \infty) = A_T^2k\ (k_S + k_T)/\{(A^2 + 4k_Sk_T)(k_S + k_T)^2 + 16k_Sk_TJ^2\} \quad (2)$$

where J is the exchange interaction (singlet-triplet splitting) in the radical pair P^+I^- (J and A in frequency units), and the g-values of P^+ and I^- are taken to be equal. From these expressions it is seen that for $k_S = 0$, i.e. 100 % triplet yield, there is no RPM-induced MFE. It is thought that this situation prevails at very low temperatures (T < 10 K) where $\phi_T(B_0)$ is reportedly unity (Wraight et al., 1974).

The MFE induced by the RPM is orientation dependent. Firstly, the hf interactions contain anisotropic components (predominantly due to the pyrrole nitrogens of I^-), secondly, the g-values of P^+ and I^- are slightly different and slightly anisotropic, therefore contributing an anisotropic component to the singlet-triplet conversion for $B_0 \neq 0$ (Boxer et al., 1983), and finally the two radicals P^+ and $\dot{I}^-$ experience a magnetic dipole-dipole interaction, which is strongly anisotropic and characterized by the fine structure parameters D and E. The latter parameter (which is a measure of the rhombicity of the dipole-dipole interaction) presumably is small and can be taken zero.

The dipolar interaction can be simply incorporated in expressions (1) and (2). For zero field we note that for E = 0 the energy gap between the singlet RP level and two of the triplet RP levels is given by $J - \frac{1}{3}D$, and for the third triplet RP level by $J + \frac{2}{3}D$. Dividing eq. (1) by 3 and substituting the above energies for J gives the three ϕ_T^i, i = x,y,z. For high field the D parameter depends on the angle θ between the dipolar z axis and the magnetic field $\vec{B}_o$. The effective dipolar interaction is then given by

$$D_{zz} = D(\cos^2\theta - \frac{1}{3}) . \qquad (3)$$

ϕ_{T_o} (no $S - T_\pm$ mixing) is now given by eq. (2) with J substituted by $J - D_{zz}$ (Lersch, 1982; Hore et al., 1986).

Quantum mechanical mixing (QMM). In zero magnetic field the three triplet levels of ^3P are characterized by the wavefunctions $|T_x\rangle = (\beta_1\beta_2 - \alpha_1\alpha_2)/\sqrt{2}$; $|T_y\rangle = i(\beta_1\beta_2 + \alpha_1\alpha_2)/\sqrt{2}$ and $|T_z\rangle = (\alpha_1\beta_2 + \alpha_2\beta_1)/\sqrt{2}$, where α and β are the 'up' and 'down' spin functions of the individual unpaired electrons of ^3P and $\vec{x},\vec{y},\vec{z}$ are now the principal axes of the dipolar tensor of ^3P. When a magnetic field $B_o \gg D_P$ is applied, where D_P is the largest fine structure parameter of ^3P, the spin magnetic moment is quantized along $\vec{B}_o$ resulting in three new energy levels $|T_+\rangle$, $|T_-\rangle$ and $|T_o\rangle$ that are linear combinations of the $|T_i\rangle$:

$$|T_j\rangle = \sum_i c_{ij}|T_i\rangle; \quad j = +,-,0, \quad i = x,y,z. \qquad (4)$$

For example, when $\vec{B}_o$ is parallel to the dipolar z axis,

$$|T_\pm\rangle = \mp(|T_x\rangle \pm i|T_y\rangle)/\sqrt{2}; \qquad |T_o\rangle = |T_z\rangle. \qquad (5)$$

For arbitrary direction of $\vec{B}_o = (B_{ox},B_{oy},B_{oz})$, where B_{oi} is the projection of $\vec{B}_o$ on the i spin axis, the coefficients c_{ij} in eq. (4) are the components of the eigenvectors of the Hamiltonian matrix

$$\begin{bmatrix} X & -ig\beta B_{oz} & ig\beta B_{oy} \\ ig\beta B_{oz} & Y & -ig\beta B_{ox} \\ -ig\beta B_{oy} & ig\beta B_{ox} & Z \end{bmatrix} , \qquad (6)$$

where g is the (isotropic) g-value of ^3P, β the electronic Bohr magneton and X,Y,Z the principal values of the dipolar tensor of ^3P (see e.g. Kottis and Lefebvre, 1963). It will be shown below that, because the decay rates of the individual sublevels of ^3P depend on the composition of these levels in terms of the zero field eigenfunctions $|T_i\rangle$, the QMM induces a MFE at low temperatues even when the quantum yield of ^3P is unity and no MFE is induced by the RPM.

TRIPLET DECAY

The equilibrium triplet concentration under continuous illumination obviously depends on the decay rates of the individual triplet levels. It has been shown recently that the decay of ^3P at higher temperatures proceeds via back reaction to the triplet radical pair state $^3[P^+I^-]$ and subsequent conversion to $^1[P^+I^-]$ and ^1P, with rate constant (Chidsey et al., 1985)

$$k_{RP} = \frac{1}{3} k_S \phi_T e^{-\Delta H/kT} \quad . \tag{7}$$

For the single hf interaction model of Haberkorn and Michel-Beyerle (1979), eq. (7) can easily be derived by first observing that $k_{RP} = k_T \exp(-\Delta H/kT) \cdot \phi_S^!$, where ΔH is the enthalpy gap between $^3[P^+I^-]$ and ^3P, k is Boltzmann's constant, and $\phi_S^!$ the probability of converting $^3[P^+I^-]$ to $^1[P^+I^-]$. One then notes that $\phi_S^!$ is given by eq. (1) by simply interchanging k_S and k_T and introducing a statistical factor $\frac{1}{3}$ to take account of the spin multiplicity. Thus $\phi_S^! = \frac{1}{3} \cdot \frac{k_S}{k_T} \phi_T$ and eq. (7) follows. Since ϕ_T depends on a magnetic field, the MFE on k_{RP} should parallel the MFE on the yield of ^3P. For a field of 50 mT at room temperature Chidsey et al. (1985) report for k_{RP} a factor of 0.54 (calculated from their figure 2 and taking additional decay via intersystem crossing (ISC) into account) and for ^3P a factor of 0.50, a remarkable agreement considering the various assumptions made.

For direct ISC from ^3P to the singlet ground state the triplet sublevel decay rates in the absence of an external field ($B_o = 0$) are given by

$$k_i \propto |\langle T_i|\hat{H}_{so}|S\rangle|^2, \quad i = x,y,z \tag{8}$$

where the operator $\hat{H}_{so}$ represents the spin-orbit interaction governing intersystem crossing. In a magnetic field the triplet levels mix and the corresponding decay rates are given by

$$k_j \propto |\langle T_j|\hat{H}_{so}|S\rangle|^2 \quad \text{or}$$

$$k_j \propto |\langle \sum_i c_{ij} T_i|\hat{H}_{so}|S\rangle|^2, \quad j = +,-,o. \tag{9}$$

Writing out eq. (9) gives

$$k_j = \sum_i c_{ij}^2 k_i + \sum_{i \neq k} \mathrm{Re}\, c_{ij} c_{kj}^{*} \sqrt{k_i k_k}, \quad i,k = x,y,z. \tag{10}$$

If the decay channels x,y,z, are independent, the phases of the triplet wavefunctions are uncorrelated and the cross terms in eq. (10) vanish. The cross terms may be important for coherent decay via x,y,z, which may occur when the triplet axes do not coincide with the normals to the symmetry planes of the molecule (Schadee et al., 1976). We have, however, recently demonstrated that for Rps. viridis the triplet x,y axes of ^3P are lying close to the optical transition moments along the N-N 'symmetry' axes of one of the bacteriochlorophyll macrocycles of P (Lous and Hoff, 1987), so in all probability the cross terms in (10) may be neglected for the decay of ^3P (Hoff, 1979,1982). The k_j are then given by

$$k_j = \sum_i c_{ij}^2 k_i \; . \tag{11}$$

Thus the decay rates k_j for $B_0 \neq 0$ are linear combinations of the zero field k_i's. For randomly oriented samples this has the consequence that when the sublevels are thermally isolated (no spin-lattice relaxation) the decay rate averaged over all orientations in the presence of a magnetic field is different from that in zero field. Consider for example a RC oriented such that $\vec{B}_0 \,//\, \vec{z}$. For sufficiently large B_0, ^3P is populated predominantly or even exclusively in the $m_S = 0$, i.e. here the $|T_z\rangle$ level. Hence, the decay rate of such RC equals k_z. RC that are oriented with the $\vec{x}$ or $\vec{y}$ spin axes parallel to $\vec{B}_0$ have decay rates equal to k_x and k_y, respectively. The probability for RC to be oriented with $\vec{z} \,//\, \vec{B}_0$, however, is much lower than the probability to have $\vec{x}$ or $\vec{y} \,//\, \vec{B}_0$, the distribution function being proportional to sin θ, where θ is the angle between the $\vec{z}$ axis and $\vec{B}_0$. Thus, for $B_0 \gg D_P$, random orientation, thermally isolated triplet levels and population of the $m_S = 0$ sublevel, one has an average decay rate $\bar{k}$ close to $(k_x + k_y)/2$, whereas in zero magnetic field $\bar{k} = \frac{1}{3}(k_x + k_y + k_z)$. Because $k_x, k_y \gg k_z$, the concentration of ^3P in a magnetic field will be less than that in zero field.

The above rationale will be generally applicable whenever a triplet state is populated predominantly in one, or two of the three sublevels, as was demonstrated by Clarke et al. (1977) for zinc chlorophyll b in n-octane. The QMM-magnetic field effect was observed for ^3P of Rb. sphaeroides R-26 by

De Vries and Hoff (1978), who measured at 1.4 K a decrease in the fluorescence of whole cells when a magnetic field was applied. We will show below that the QMM is responsible for the MFE measured on the absorbance of isolated RC at T < ~ 10 K.

MAGNETIC FIELD DEPENDENCE OF THE CONCENTRATION OF ^3P

The relative concentration of each sublevel i of ^3P, $[T_i]$, under conditions of continuous illumination is the result of competition between the rate of population, which is equal to the product of the light flux and the absorption cross section K, times the probability of triplet formation ϕ_T^i, and the rate of decay, which is composed of decay via back reaction to the radical pair state (k_{RP}^i) and decay via ISC to $|S\rangle$ (k_{ISC}^i):

$$[T_i] = \frac{K\phi_T^i[S]}{k_{RP}^i + k_{ISC}^i}, \qquad i = x,y,z. \tag{12}$$

If the quantum efficiency of charge separation is unity and k_T independent of i, ϕ_T^i is given in zero field by eq. (1) modified to take into account the dipolar interaction in the radical pair, and in field B_o by the modified eq. (2). The relative concentration of ^3P in zero field is given by summing the $[T_i]$'s, in field B_o by $[T_o]$. The triplet decay rates k_{RP}^i and k_{ISC}^i in zero field, and k_{RP}^j, k_{ISC}^j in field B_o, depend on the thermalization of the triplet sublevels. We may assume with confidence that during the lifetime of $^3[P^+I^-]$ (about 3 ns), there is no thermalization over the three RP triplet levels. For ^3P, it has been demonstrated with laser flash EPR (Hoff and Proskuryakov, 1985) that the spin-lattice relaxation at 296 K is of the order of 1 µs. Because the transitions $^3P \leftrightarrows {}^3[P^+I^-]$ leave the magnetization, hence the triplet sublevel population invariant, we may assume that during the lifetime of ^3P at 293 K (about 5 µs, Hoff et al., 1977) spin relaxation is efficient enough to thermalize the triplet sublevels. At lower temperatures, this is no longer true and we will present evidence that spin-lattice relaxation becomes inefficient below 50 K. Hence, at the low end of our temperature range we may assume that the spin levels in both ^3P and $^3[P^+I^-]$ are thermally isolated. We may therefore discriminate four limiting cases: Complete thermalization in ^3P (high temperature) with I. $B_o = 0$ and II. $B_o >> |D_P|$ and thermal isolation in ^3P (low temperature) with III. $B_o = 0$ and IV. $B_o >> |D_P|$. The corresponding (relative) ^3P concentrations are given by

$$\text{I (high T, } B_o = 0) \; [^3P]_I = \frac{K \sum_i \phi_T^i(0)}{K \sum_i \phi_T^i(0) + \frac{1}{3} k_S \sum_i \phi_T^i(0)\, e^{-\Delta H/kT} + \frac{1}{3} \sum_i k_i} \quad (13a)$$

$$\text{II (high T, } B_o \neq 0) \; [^3P]_{II} = \frac{K \phi_{T_o}(B_o)}{K \phi_{T_o}(B_o) + \frac{1}{3} k_S \phi_{T_o} e^{-\Delta H/kT} + \frac{1}{3} \sum_j k_j} \quad (13b)$$

$$\text{III (low T, } B_o = 0) \; [^3P]_{III} = \frac{K \sum_i \{\phi_T^i(0)/(k_S \phi_T^i(0)\, e^{-\Delta H/kT} + k_i)\}}{1 + K \sum_i \{\phi_T^i(0)/(k_S \phi_T^i(0)\, e^{-\Delta H/kT} + k_i)\}} \quad (13c)$$

$$\text{IV (low T, } B_o \neq 0) \; [^3P]_{IV} = \frac{K \phi_{T_o}(B_o)}{K \phi_{T_o}(B_o) + k_S \phi_{T_o} e^{-\Delta H/kT} + k_o} \quad (13d)$$

with $i = x,y,z$ and $j = o,+,-$. Expressions (13c,d) simplify somewhat if we assume that at low temperature, say $T < 10$ K, $\phi_T^i(0) \approx \frac{1}{3}$ (Hoff and De Vries, 1978), $\phi_T(B_o) = 1$ (Wraight et al., 1974) and $k_S e^{-\Delta H/kT} \ll k_o, k_i$ (Chidsey et al., 1985). We then obtain

$$\text{III (low T, } B_o = 0) \; [^3P]_{III} = \frac{K \Sigma\, 1/k_i}{3 + K \Sigma\, 1/k_i} \quad (13e)$$

$$\text{IV (low T, } B_o \neq 0) \; [^3P]_{IV} = \frac{K}{K + k_o} \; . \quad (13f)$$

The MFE at high and low temperature is given by

$$R\,(T = 293 \text{ K}) = [^3P]_I/[^3P]_{II} \quad (14a)$$

$$R\,(T < 10 \text{ K}) = [^3P]_{III}/[^3P]_{IV} \; . \quad (14b)$$

ORIENTATION DEPENDENCE OF THE MAGNETIC FIELD EFFECT

As noted above, the MFE depends on the orientation of the RC with respect to $\vec{B}_o$ because of anisotropies in the hf interaction and the g-values of P^+ and I^-, and because the effective dipole interaction D_{ZZ} is proportional to $(\cos^2\theta - \frac{1}{3})$, where θ is the angle between the RP dipolar z axis (which may be taken along the line connecting the centers of P^+ and I^-)

and $\vec{B}_o$. For fields lower than 50 mT the g-anisotropy can be neglected. The anisotropy in hf interaction is mainly due to the I^- pyrrole nitrogens that constitute only a minor part of the total lumped hf values, so that to first order hf anisotropy can also be disregarded. This leaves the dipolar interaction as the main cause of the anisotropic MFE at low fiels and high temperatures. At low temperatures, the MFE on $[^3P]$ is governed by the dependence of k_{ISC} on magnitude and direction of $\vec{B}_o$ in the triplet spin axes coordinate frame of ^3P. The total orientation dependence of $[^3P]$ is therefore given by substituting in eqs. (11) $\phi_T^i(0) = \phi_T^i(\theta)$, $\phi_{T_o}(B_o) = \phi_{T_o}(\theta)$ and $k_j = k_j(\psi,\eta)$ where ψ and η are the polar angles of $\vec{B}_o$ in the $\vec{x},\vec{y},\vec{z}$ spin axes system of ^3P.

The MFE is interrogated by a beam of unpolarized light. Because of its anisotropy, the MFE introduces elliptic polarization that is detected by a photoelastic modulator and an analyzer. For high temperatures (little decay via ISC) the amount of ellipticity for a non-oriented sample follows from (Vermeglio et al., 1978; Meiburg, 1985):

$$\Delta A_{//} = c \int_0^{\pi} Z.G(\theta)|\vec{\mu}_\lambda|^2 \sin\theta \, d\theta \tag{15a}$$

$$\Delta A_{\perp} = c \int_0^{\pi} Y.G(\theta)|\vec{\mu}_\lambda|^2 \sin\theta \, d\theta \tag{15b}$$

$$Z = \tfrac{1}{2} \sin^2\theta \sin^2\alpha + \cos^2\theta \cos^2\alpha \tag{15c}$$

$$Y = \tfrac{1}{4}(\cos^2\theta \sin^2\alpha + \sin^2\alpha) + \tfrac{1}{2} \sin^2\theta \cos^2\alpha \tag{15d}$$

where c is a proportionality constant, $\vec{\mu}_\lambda$ the optical transition moment at a particular wavelength, $G(\theta)$ given by $[^3P(B_o = 0)] - [^3P(B_o)]$, and α the angle between $\vec{\mu}_\lambda$ and $\vec{B}_o$. For low temperatures (predominant decay via ISC) the position of $\vec{\mu}_\lambda$ within the $\vec{x},\vec{y},\vec{z}$ frame has to be specified by two fixed angles β and γ. The projections of $|\vec{\mu}_\lambda|$ on directions // and $\perp \vec{B}_o$ are then given by $|\vec{\mu}_\lambda| R(\eta) R(\psi)(\sin\beta\cos\gamma, \sin\beta\sin\gamma, \cos\beta)^\dagger$ where $R(\eta)$ and $R(\psi)$ are rotation matrices and the dagger sign means transposition.

From the above it follows that at high temperatures LD-MODS gives information on the angle of the optical transition moments with the dipolar $P^+ - I^-$ axis, whereas at low temperature it gives information on the position of the optical transitions in the $\vec{x},\vec{y},\vec{z}$ principal axes system of the dipolar tensor of ^3P.

MATERIALS AND METHODS

The experimental set-up of the LD-MODS spectrophotometer is a combination of the MODS set-up described in Lous and Hoff (1986) and our LD-ADMR set-up (Den Blanken et al., 1984; Lous and Hoff, 1987). A sinusoidal magnetic field is used with modulation frequency f_m = 323 Hz. Since the MFE does not depend on the direction (sign) of the magnetic field, the absorbance is modulated with frequency $2f_m$. This modulation and the simultaneously monitored 100 kHz LD signal are demodulated using fase-sensitive detection yielding a triplet-minus-singlet (T - S) and a LD-(T - S) signal, respectively. These signals are relatively calibrated as in the LD-ADMR experiment (Lous and Hoff, 1987). Two types of cooling were used, a bath cryostat for temperatures below 4 K and a helium-flow system for temperatures above 4 K. The temperature was measured with an absolute accuracy of +0.4 and ±2 K for the bath and the flow cryostat, respectively. The relative accuracy was 0.03 and 0.1 K, respectively.

Reaction centers of the photosynthetic bacterium Rb. sphaeroides R-26 were prepared as described (Den Blanken and Hoff, 1982). The primary electron acceptor Q was reduced by adding 10 mM sodium ascorbate and illuminating the sample with white light for about 1 minute in the bath cryostat at about 230 K, prior to cooling to 4.2 K, or continuously while cooling to the required temperature in the flow cryostat. The samples were diluted to 67 % (v/v) ethyleneglycol and to an optical density of 0.20 at 802 nm.

RESULTS AND DISCUSSION

Amplitude of the magnetic field effect

The MFE expressed as ΔA/A at 860 - 890 nm equals (1 - 32) x 10^{-3} over the temperature range studied. These low values are not surprising, in view of eqs. (12) - (14). If k_{ISC} = 0, one would expect to observe no MFE at all under cw illumination at room temperature, since the magnetic field induced decrease in triplet yield is then exactly compensated by the increase in triplet lifetime. With the values observed for quinone-reduced RC of k_{RP} = 2×10^5 s^{-1} (Hoff et al., 1977), k_{ISC} = 7×10^3 s^{-1} (Hoff, 1976), $\sum_i \phi_T^i$ = 0.15 and ϕ_T = 0.085 (Hoff et al., 1977) one calculates for a (light flux dependent) $^3P^O$concentration of 1 - 5 % a MFE of only ΔA/A = (1.5 - 6) x 10^{-3} at room temperature, well agreeing with our observed value. For quinone-depleted RC with k_{RP} = 2.5×10^4 s^{-1}, $\sum_i \phi_T^i$ = 0.32, ϕ_{T_0} (B = 1000 G) =

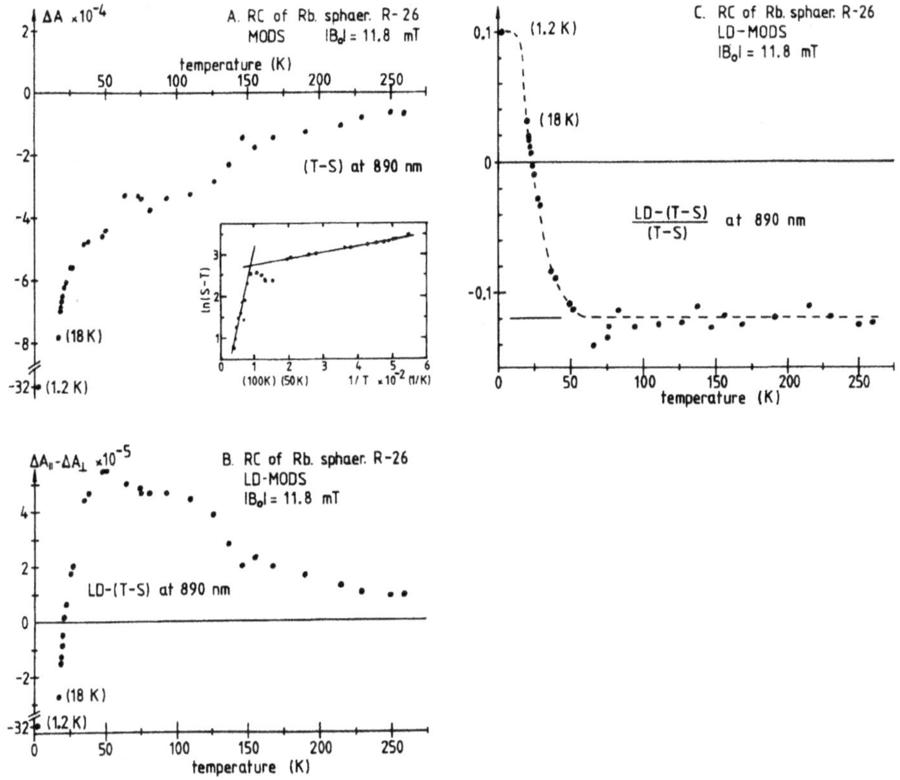

Fig. 2. Simultaneously monitored temperature dependence of (a) (T - S) and (b) LD-(T - S) ΔA signals at 890 nm. In (c) the ratio A (anisotropy) between (b) and (a) is plotted in a reciprocal temperature plot. The resolution of the monochromator was 2.5 nm.

$0.47 \times \sum_i \phi_T^i$ (Chidsey et al., 1984) one has a MFE of $\Delta A/A = (5 - 25) \times 10^{-3}$, again agreeing well with experiment (Moehl et al., 1985).

Temperature dependence of the MFE of P890 (890 nm)

Fig. 2 shows the temperature dependence of the (T - S) (a) and the LD-(T - S) (b) signals as measured in our LD-MODS apparatus at the long wavelength absorbance of the primary donor (890 nm). Decreasing the temperature from 310 K to 18 K the (T - S) intensity increases and seems to continue doing so down to 2.1 K. Between 1.2 and 2.1 K no intensity changes were observed. Note that comparisons between bath cryostat and helium flow MODS intensities can be done only roughly due to possible differences in Q^- concentrations as they were performed with different methods of Q^- accumulation. Effects contributing to the raise in (T - S) intensity as shown in Fig. 2a upon lowering the temperature are:

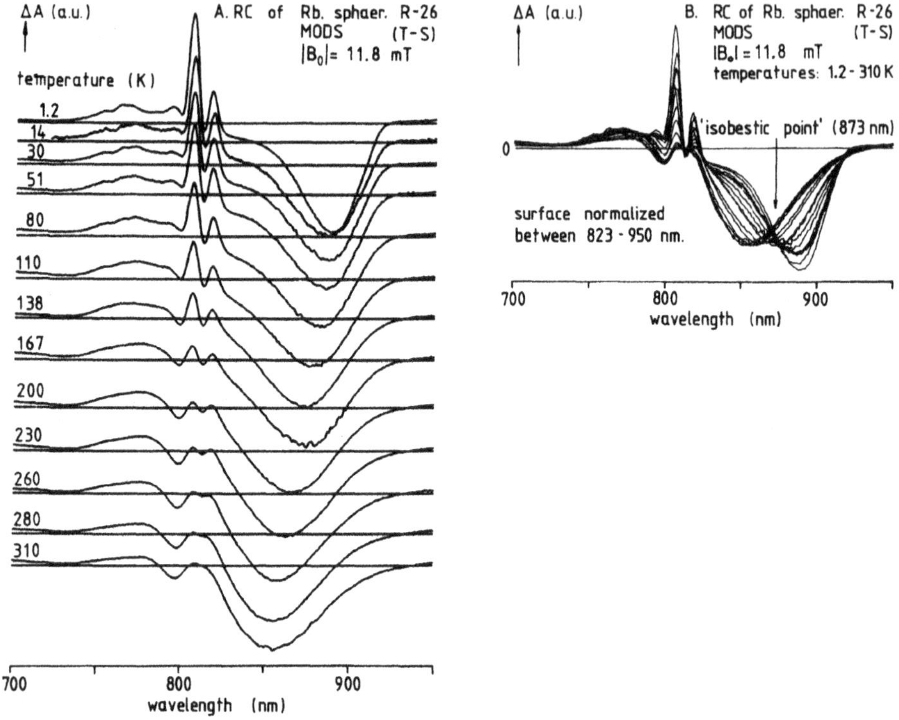

Fig. 3. (a) (T - S) spectra taken at various temperatures and normalized on the surface area of the P band between 823 and 950 nm. Resolution 2.5 nm. (b) Composite of the curves of (a).

1. Increase of the triplet yield.
2. Decrease of the decay rate k_{RP} of the ^3P state (Chidsey et al., 1985; Hunter et al., 1987).
3. Sharpening of the absorption band of P(-) owing to freezing out the vibrational transitions (Scherer et al., 1985).
4. Red shift of the band from 860 to 890 nm.
5. Increase of the spin-lattice relaxation time so that it becomes longer than the ^3P lifetime at 1.2 K (sublevel-averaged lifetime approximately 140 μs, longest sublevel lifetime 0.6 ms (Hoff, 1976)); onset QMM-MFE.

Plotting Fig. 2a in an Arrhenius plot (inset) two approximately straight lines can be drawn between 310 and 50 K and between 50 and 18 K, which reflect the temperature dependence of processes (1) - (4) and process (5), respectively. The decay of ^3P is nearly temperature-independent below 77 K (Chidsey et al., 1985). The breakpoint at 50 K agrees well with the maximum temperature at which microwave transitions in ^3P can be observed by the ADMR technique, which has the same sensitivity as MODS. It presumably reflects the onset of spin-lattice relaxation within the ^3P manifold, which abolishes

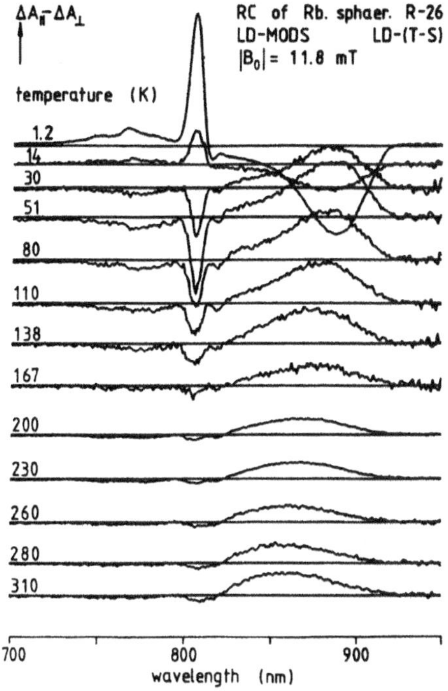

Fig. 4. LD-(T - S) spectra monitored simultaneously with the spectra shown in Fig. 3 and multiplied by the same normalization factors.

the QMM above 50 K. From Fig. 2c it is seen that the QMM has an orientation dependence that is quite different from that of the RPM.

Temperature dependence of the (T - S) and LD-(T - S) spectra

Fig. 3a shows the (T - S) spectra at $|B_0|$ = 11.8 mT and various temperatures. To correct for differences in ^3P concentration we normalized the spectra on the surface of the band between the nearly temperature-independent crossing at 823 and 950 nm. From the presentation of Fig. 3b an 'isosbestic' point at 873 nm can be clearly recognized, the top of the band shifting smoothly without jump as a function of temperature. This may indicate that there are two conformations of the P dimer, one favored at high and one at low temperatures, which have the maximum of the long-wavelength P band at 860 and 890 nm, respectively. The transition between these conformations may be caused by thermal contraction of the protein matrix (Scherer et al., 1985). The spectral region around 810 nm at low and high temperature can be fitted reasonably well with one spectral band structure provided the increasing number with temperature of vibronic states per optical transition is taken into account by varying the width of the various bands (Lous and Hoff, 1986; E.J. Lous, unpublished results). This means that no large changes in the RC structure occur upon varying the temperature.

Simultaneously with the (T - S) spectra in Fig. 3a the LD-(T - S) spectra of Fig. 4 were recorded and scaled with the same normalization factors as in Fig. 4a. The sign change observed betweeen 14 and 30 K corresponds with that observed in Fig. 2b and is exhibited in the whole spectrum except at 819 nm where the sign stays negative.

The temperature dependence of the LD-(T - S) spectra shown in Fig. 4 reflects the change in MFE processes from a mixture of RPM and QMM to pure QMM. As discussed in the theoretical section the axis of preference is for both the RP and the QMM mechanisms the direction of $\vec{B}_o$, but for the RPM one has to average over all orientations of the dipolar RP axis with respect to $\vec{B}_o$ whereas for the QMM one must average over all orientations of the principal axes system of the dipolar tensor of ^3P with respect to $\vec{B}_o$. The different relative contribution of the RP and the QMM mechanism to the MFE at the various temperatures gives rise to the observed changes in sign and magnitude of the LD-MODS signal at 860 - 890 nm and of the LD-(T - S) spectra as such (Fig. 4).

B_o curves at 860 nm

Fig. 5a shows the (T - S) intensity at 890 nm as a function of the intensity of the magnetic field B_o for several temperatures. The curves are normalized at 11.8 mT to bring out more clearly that there are only small differences in the shape of the curves up to 17 mT. The $B_o(½)$ (B_o at half the maximal MFE value) reflects mainly lifetime broadening due to the fastest of the two decay rates of the RP state, k_T, which is nearly independent of temperature (Ogrodnik et al., 1987). Chidsey et al. (1985) have reported that the decay rate of ^3P decreases with B_o, with at room temperature a dependence on B_o that, apart from the sign, is similar in shape to the 289 K curve of Fig. 5a. This reinforces our conclusion that above 50 K the RPM is responsible for the magnetic field effect.

For small B_o, $B_o < 5$ mT, no increase in ^3P triplet yield with increasing B_o is observed. This agrees with observations by Ogrodnik et al. (1987) but contrasts with work reported in Norris and van Brakel (1986). This difference may be attributed to differences in experimental set-up. In

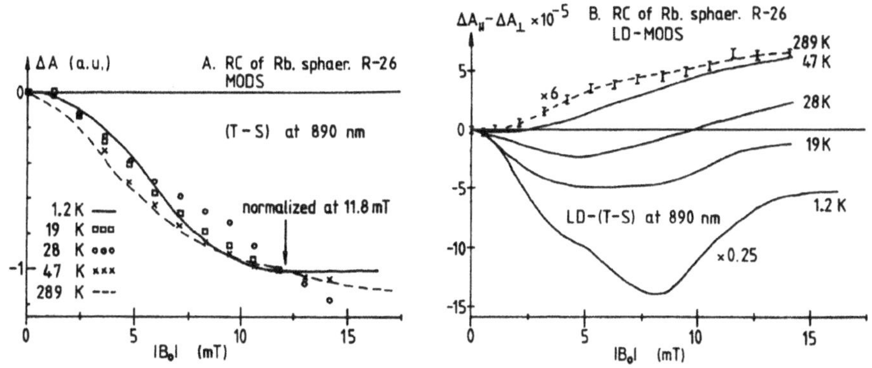

Fig. 5. LD-MODS signals detected at 890 nm as a function of the amplitude of the Zeeman field $B_0(t) = |B_0| \sin \omega t$ at various temperatures. (a) (T - S) signals normalized at an arbitrary point (11.8 mT) to facilitate comparison of the shape of the curves. (b) LD-(T - S) signals shown on an absolute ΔA scale. Note that the 1.2 K trace is in reality 4 times larger and the 289 K trace 6 times smaller then represented. The resolution in (a) and (b) was 2.5 nm.

Fig. 5b we show that at 298 K the B_0 curve of the LD-MODS at 890 nm has a sign opposite to that of the isotropic MODS effect. In the above mentioned work laser excitation has been used, wich due to polarization of the laser beam may lead to sizeable anisotropy in the observed MFE at 298 K. Depending on the configuration of excitation and probe beam such anisotropy may give rise to a postive MFE (increase in ^3P) at low B_0.

Finally we note that the QMM in principle is also operative in the triplet RP state, since its sublevels are presumably isolated at all temperatures. If $\phi_T < 1$ at 1.2 K, then the QMM, possibly in combination with level antricrossing effects, may be responsible for the shape of the B_0 curve around 7.5 mT.

CONCLUDING REMARKS

In this paper we have presented the LD-MODS technique and its application to RC of Rb. sphaeroides R-26. We have given an approximate theory that qualitatively accounts well for the observed temperature and B_0 dependence of the MODS and LD-MODS effects and the corresponding (T - S) spectra. Our results indicate that two mechanisms are responsible for the magnetic field effect. One, the radical pair mechanism, is operative at temperatures above 50 K, the other, the quantum mechanical mixing of the triplet sublevels of ^3P, dominates at $T < 10$ K. For $10 < T < 50$ K both mechanisms contribute.

The (T - S) and LD-(T - S) spectra and their temperature dependence are qualitatively well explained invoking the two above mechanisms. Quantitative fits of the spectra will be presented elsewhere. We note that no appreciable changes in pigment configuration occur upon cooling RC from 298 to 1.2 K.

The B_0 dependence of the MODS and LD-MODS signals at 860 - 890 nm is complex and a quantitative interpretation must await calculation of the radical pair spin dynamics in low and intermediate fields. Our present observations suggest that conflicting results on the sign of the MFE at low values of B_0 as reported in the literature might be due to selection effects resulting from the use of polarized excitation and/or detection beams.

ACKNOWLEDGEMENTS

We thank H.M. Nan for preparing the reaction centers and J. van Egmond and the group of Dr. Th. Wenkenbach for their technical support in constructing the superconducting magnet. We are grateful to Dr. J. Schmidt and his colleagues of the Center for the Study of Excited States of Molecules for using some of their experimental facilities. This work was supported by the Netherlands Foundation for Chemical Research (SON), financed by the Netherlands Organization for the Advancement of Pure Research (ZWO).

REFERENCES

Bixon, M., Jortner, J., Michel-Beyerle, M.E., Ogrodnik, A., and Lersch. W., 1987, On the role of the accessory bacterioclorophyll in reaction centers of photosynthetic bacteria: Intermediate acceptors in the primary electron transfer, Chem. Phys. Lett., in the press.

Boxer, S.G., Chidsey, C.E.D., and Roelofs, M.G., 1983, Magnetic field effects in the solid state: An example from photosynthetic reaction centers. Ann. Rev. Phys. Chem. 34:389.

Chidsey, C.E.D., Kirmaier, C., Holten, D., and Boxer, S.G., 1984, Magnetic field dependence of radical-pair decay kinetics and molecular triplet quantum yield in quinone-depleted reaction centers, Biochim. Biophys. Acta 766:424.

Chidsey, C.E.D., Takiff, L., Goldstein, R.A., and Boxer, S.G., 1985, Effect of magnetic fields on the triplet state lifetime in photosynthetic reaction centers: Evidence for thermal repopulation of the initial radical pair, Proc. Natl. Acad. Sci. USA 82:6850.

Clarke, R.H., Connors, R.E., and Keegan, J., 1977, Magnetic field effect on the low temperature triplet state population of an organic molecule, J. Chem. Phys. 66:358.

De Vries, H.G., and Hoff, A.J., 1978, Magnetic field effect on the fluorescence intensity of Rhodopseudomonas sphaeroides at 1.4 K, Chem. Phys. Lett. 55:395.

Den Blanken, H.J., and Hoff, A.J., 1982, High-resolution optical absorption-difference spectra of the triplet state of the primary donor in isolated reaction centers of the photosynthetic bacteria Rhodopseudomonas sphaeroides R-26 and Rhodopseudomonas viridis measured with optically detected magnetic resonance at 1.2 K, Biochim. Biophys. Acta 681:365.

Den Blanken, H.J., Meiburg, R.F., and Hoff, A.J., 1984, Polarized triplet-minus-singlet absorbance difference spectra measured by absorbance-detected magnetic resonance (ADMR), Chem. Phys. Lett. 105:336.

Haberkorn, R., and Michel-Beyerle, M.E., 1979, On the mechanism of magnetic field effects in bacterial photosynthesis, Biophys. J. 26:489.

Hoff, A.J., 1976, Kinetics of populating and depopulating of the photoinduced triplet state of the photosynthetic bacteria Rhodospirillum rubrum, Rhodopseudomonas sphaeorides (wild type), and its mutant R-26 as measured by ESR in zero-field, Biochim. Biophys. Acta 440:765.

Hoff, A.J., Rademaker, H., van Grondelle, R., and Duysens, L.N.M., 1977, On the magnetic field dependence of the triplet state in reaction centers of photosynthetic bacteria, Biochim. Biophys. Acta 460:547.

Hoff, A.J., and de Vries, H.G., 1978, Electron spin resonance in zero magnetic field of the reaction center triplet of photosynthetic bacteria, Biochim. Biophys. Acta 503:94.

Hoff, A.J., 1979, Application of ESR in photosynthesis, Phys. Reports 54:75.

Hoff, A.J., 1981, Magnetic field effects on photosynthetic reaction, Quart. Rev. Biophys. 14:599.

Hoff, A.J., 1982, ODMR spectroscopy in photosynthesis II, in: 'Triplet state ODMR Spectroscopy', R.H. Clarke, ed., John Wiley & Sons, New York.

Hoff, A.J., and Proskuryakov, I.I., 1985, Triplet EPR spectra of the primary electron donor in bacterial photosynthesis at temperatures between 15 and 296 K, Chem. Phys. Lett. 115:303.

Hoff, A.J., Lous, E.J., Moehl, K.W., and Dijkman, J.A., 1985, Magneto-optical absorbance difference spectroscopy. A new tool for the study of radical recombination reactions. An application to bacterial photosynthesis. Chem. Phys. Lett. 114:39

Hore, P.J., Watson, E.T., Pedersen, J.B., and Hoff, A.J., 1986, Line-shape analysis of polarized electron paramagnetic resonance spectra of the primary reactants of bacterial photosynthesis. Biochim. Biophys. Acta 849:70.

Hunter, D.A., Hoff, A.J., and Hore, P.J., 1987, Theoretical calculations of RYDMR effects in photosynthetic bacteria, Chem. Phys. Lett. 134:6.

Kottis, P., and Lefebvre, R., 1963, Calculation of the electron spin resonance line shape of randomly oriented molecules in a triplet state. I. The $\Delta m = 2$ transition with a constant linewidth. J. Chem. Phys. 39:393.

Lersch, W., 1982, Zur spinselektiven Rekombination von Elektron-Loch-Paaren im bakteriellen Reaktionszentrum. Einfluss von Mikrowellen und anisotropen Wechselwirkungen auf die Spindynamik in äusseren Magnetfeldern, Diplomarbeit, München.

Lersch, W., and Michel-Beyerle, M.E., 1983, Magnetic field effects on the recombination of radical ions in reaction centers of photosynthetic bacteria, Chem. Phys. 78:115.

Lous, E.J., and Hoff, A.J., 1986, Triplet-minus-singlet absorbance difference spectra of reaction centers of Rhodopseudomonas sphaeroides R-26 in the temperature range 24 - 290 K measured by Magneto-Optical Difference Spectroscopy (MODS), Photoynth. Res. 9:89.

Lous, E.J., and Hoff, A.J., 1987, Exciton interactions in reaction centers of the photosynthetic bacterium Rhodopseudomonas viridis probed by optical triplet-minus-singlet polarization spectroscopy at 1.2 K monitored through absorbance-detected magnetic resonance, Proc. Natl. Acad. Sci. USA 84:6147.

Marcus, R.A., 1987, Superexchange versus an intermediate bacteriochlorophyll mechanism in reaction centers of photosynthetic bacteria, Chem. Phys. Lett. 133:471.

Meiburg, R.F., 1985, Orientation of components and vectorial properties of photosynthetic reaction centers, Thesis, University of Leiden, The Netherlands.

Moehl, K.W., Lous, E.J., and Hoff, A.J., 1985, Low-power, low-field RYDMAR of the primary radical pair in photosynthesis, Chem. Phys. Lett. 121: 22.

Norris, J.R., and van Brakel, G., 1986, Energy trapping in photosynthesis and purple bacteria, in: 'Light Emission by Plants and Bacteria', Govindjee, J. Amesz and D.C. Fork, eds., Acad. Press, New York.

Ogrodnik, A., Remy-Richter, N., Michel-Beyerle, M.E., and Ficke, R., 1987, Chem. Phys. Lett., in the press.

Schadee, R.A., Schmidt, J., and van der Waals, J.H., 1976, Intersystem crossing into a superposition of spin states? The system tetramethylpyrazine in durene, Chem. Phys. Lett. 41:435.

Scherer, P.O.J., Fischer, S.F., Hörber, J.K.H., and Michel-Beyerle, M.E., 1985, On the temperature-dependence of the long wavelength fluorescence and absorption of Rhodopseudomonas viridis reaction centers, in: 'Antennas and Reaction Centers of Photosynthetic Bacteria. Structure, Interactions, and Dynamics', M.E. Michel-Beyerle, ed., Springer-Verlag, Berlin.

Tang, J., and Norris, J.R., 1983, Theoretical calculations of microwave effects on the triplet yield in photosynthetic reaction centers, Chem. Phys. Lett. 94:77.

Vermeglio, A., Breton, J., Paillotin, G., and Cogdell, R., 1978, Orientation of chromophores in reaction centers of Rhodopseudomonas sphaeroides: A photoselection study, Biochim. Biophys. Acta 501:514.

Wraight, C.A., Leigh, J.S., Dutton, P.L., and Clayton, R.K., 1974, The triplet state of reaction center bacteriochlorophyll: Determination of a relative quantum yield, Biochim. Biophys. Acta 333:401.

BACTERIAL REACTION CENTERS ARE INTRINSICALLY HETEROGENEOUS

A.J. Hoff

Department of Biophysics, Huygens Laboratory
State University of Leiden, The Netherlands

INTRODUCTION

The reaction center is, by definition, the (aggregate of) protein molecule(s) which houses the components involved in the primary charge separation and charge stabilizing reactions. It can be isolated, purified, crystallized and X-rayed. Its atomic structure is known with a resolution of 2.3 Å (Deisenhofer et al, 1985, and Deisenhofer, J., private communication), the atomic coordinates to within 0.2 Å. Spectral simulations based on the X-ray structure are highly successful (Knapp et al., 1986; Vasmel et al, 1986; Lous and Hoff, 1987a). The conclusion seems warranted that, at least for any particular bacterial species, a unique reaction center structure indeed exists. This notion has led to efforts to calculate from the crystal structure in great detail electronic spin and charge densities, overlap integrals, electron transfer matrix elements etc. (Plato et al., 1986; Warshel and Parson, 1987; Parson and Warshel, 1987). In this note I will show that such ventures may be overly optimistic. A number of experiments bear out that considerable heterogeneity in detailed reaction center structure exists that affects markedly the spectral properties of the reaction center. It is not unlikely that this heterogeneity also affects electron transport properties.

ODMR OF TRIPLET STATES

At cryogenic temperatures the triplet state of the primary donor, ^3P, is generated with high yield (approaching unity) by electron-hole recombination in reaction centers in which the quinone acceptors have been reduced or from which they are removed. The three triplet sublevels are non-degenerate due to the dipole-dipole interaction of the two unpaired electrons. Application of microwaves resonant between two of the three sublevels at temperatures at which spin-lattice relaxation is effectively impeded (about 1.5 K) causes a transfer of population between the two levels, which translates in a change in triplet concentration when the two sublevels have different decay kinetics and equilibrium populations. This change can be optically detected via the phosphorescence (in reaction centers difficult because of the low quantum yield, about 10^{-8} (Boxer, S.G., private communication)), the fluorescence or the singlet ground state absorption. Fig. 1B shows a resonance line obtained by such an optically detected magnetic resonance (ODMR) experiment.

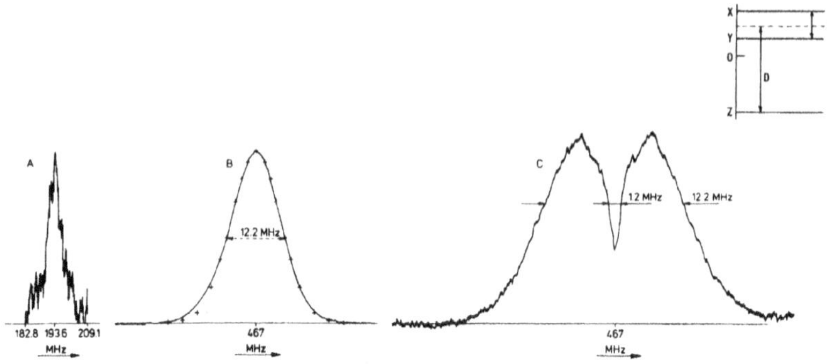

Fig. 1. Fluorescence detected ODMR at 2 K of whole cells cells of Rhodobacter sphaeroides 2.4.1. A) Double resonance line of the $2|E|$ transition, with one microwave oscillator fixed at 467 MHz. B) Single resonance line of the $|D| - |E|$ transition. Crosses represent a Gauss curve. C) Hole burning experiment, with one oscillator set at 467 MHz and one slowly sweeping through the resonance. From Hoff (1976).

For reaction centers two resonances are observed corresponding to the $|D| - |E|$ and the $|D| + |E|$ transitions (see inset of Fig. 1), where D and E are the zero field splitting parameters. The third transition, at a frequency $2|E|/h = 193$ MHz, is only observed in a double resonance experiment, in which one of the other transitions is simultaneously irradiated (Fig. 1A). Note that the $|D| - |E|$ line of Fig. 1B is almost a pure Gaussian, the slight deviation at the low frequency site being caused by the earth magnetic field (that slightly shifts the energy levels). The parameters D and E define the principal values along the principal axes of the dipole-dipole interaction tensor: $X - Y = -2E$, $Z = -\frac{2}{3}D$, $X + Y + Z = 0$. These values are governed by the nature of the triplet carrying molecule itself, and by its surroundings. For example, a large planar aromatic molecule like chlorophyll has a positive D value of $(273 - 290) \times 10^{-4}$ cm, depending somewhat on solvent and side groups and an E parameter that is either positive or negative with a value of $(32 - 52) \times 10^{-4}$ cm^{-1}. When such a molecule forms part of an aggregate of n identical molecules, two limiting cases can be discerned: i) The exchange interaction $|J|$ between the molecules (not to be confused with the singlet-triplet splitting) is much smaller than $|D|$. This is the weak coupling or slow hopping case. The triplet state of the aggregate splits into n triplet states with to first order identical D and E values, but with different directions of the x, y and z dipolar axes with respect to the molecular frame of the aggregate. ii) $|J| >> |D|$, the strong coupling, fast hopping or exciton case. Now one triplet state is observed with values of D and E and directions x, y, z that depend on the geometry of the aggregate (Sternlicht and McConnell, 1961). Moreover, a large $|J|$ is a result of significant orbital overlap of the constituents of the dimer, which will also lead to appreciable resonance integrals, i.e. one would expect in this case the triplet state having a significant charge transfer character. This will lead to reduction of $|D|$.

In addition to the intermolecular couplings, the D and E values are sensitive to the environment. For example, aromatic molecules in a glassy matrix show broad, inhomogeneous ODMR lines of a width of several hundred MHz. Although the ODMR lines of Fig. 1 are much narrower, they are still inhomogeneously broadened. This was demonstrated by microwave hole burning. Samples were irradiated with microwaves of saturating intensity and fixed frequency ($\Delta f < 50$ kHz) within the $|D| - |E|$ resonance, and simultaneously with microwaves whose frequency was scanned over the |D| - |E| line. The result is depicted in Fig. 1C, which shows a hole of width ~ 1.2 MHz. Since the spin-spin relaxation time T_2 under the same conditions is 1.16 ± 0.85 μs (Lous and Hoff, 1987b), corresponding to a homogeneous linewidth $\Gamma = 1/\pi T_2$ = 0.27 MHz and a holewidth of 2Γ = 0.54 MHz, the measured holewidth is to within a factor of two equal to the homogeneous linewidth. The additional broadening is presumably due to spin diffusion, the power broadening being less than ~ 10 kHz, i.e. < Δf. Note that the shortest sublevel lifetime of ^{3}P at 1.2 K exeeds 100 μs, hence the triplet lifetime does not contribute to the homogeneous linewidth.

The hole burned in the |D| - |E| line does not give rise to a hole in the |D| + |E| line, which is decreased in intensity with conservation of lineshape. This means that there are many combinations of |D| and |E| that give one particular (|D| - |E|)/h resonance frequency and a whole range of (|D| + |E|)/h frequencies. This in turn means that both D and E show a (Gaussian) distribution of values. This is confirmed by the width of the 2|E| transition of about 5 MHz, i.e. of comparable Δν/ν as the |D| - |E| line (Fig. 1A). As argued above, the distribution in D and E values may arise from environmental, or from intramolecular interactions.

TRIPLET-MINUS-SINGLET SPECTRA

When ODMR transitions are monitored through the singlet ground state absorbance (Den Blanken et al., 1982), one may scan the monitoring optical wavelength keeping the microwave frequency fixed. The result is the triplet-minus-singlet (T - S) absorbance difference spectrum (Fig. 2). The interpretation of the full T - S spectrum lies outside the scope of the present note, see e.g. Lous and Hoff (1987a). Here I will concentrate on the long wavelength absorbance band of P at 890 nm for Rb. sphaeroides and 1007 nm for Rps. viridis at 1.5 K. From Fig. 2 it is clear that this band in the T - S spectrum is quite asymmetric, with a short wavelength tail extending to 830 nm (900 nm for Rps. viridis). Monitoring the P-band at various micro-

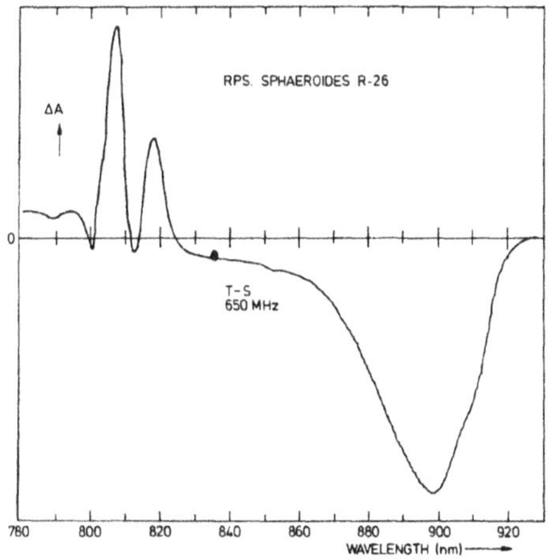

Fig. 2. Triplet-minus-singlet spectrum of reaction centers of Rb. sphaeroides R-26 recorded at 1.5 K with absorbance detected ODMR. The microwave frequency was resonant with the |D| + |E| transition at 650 MHz. From Hoff et al. (1985).

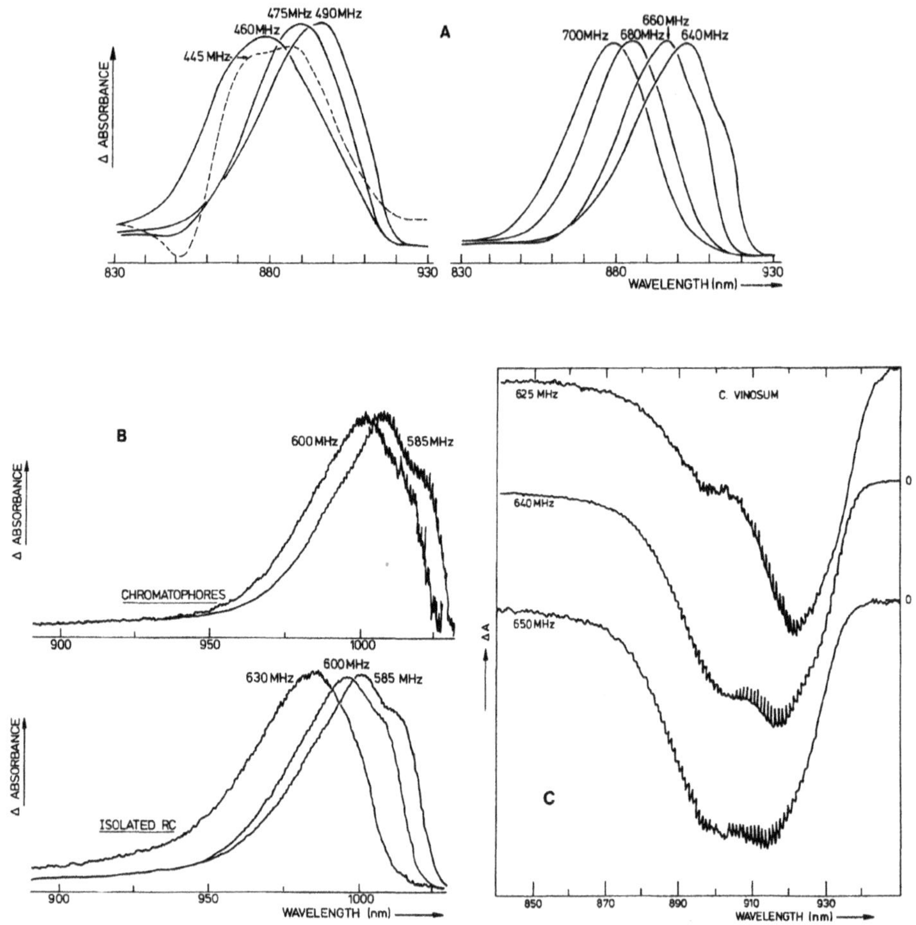

Fig. 3. A) The normalized (inverted) long-wavelength band of the ADMR-monitored triplet-minus-singlet spectrum of chromatophores of *Rb. sphaeroides* R-26 at 1.2 K at various microwave frequencies as indicated. The dashed line is a superposition of the T - S spectra of reaction center and antenna triplet states. B) As in A. for chromatophores and reaction centers of *Rps. viridis*. From Den Blanken and Hoff (1983) and Den Blanken et al. (1983). C) Long-wavelength band of the T - S spectrum of reaction centers of *Chromatium vinosum* at three different microwave frequencies. The spikes are switching artefacts caused by stepping the monochromator. From Hoff et al. (1985).

wave frequencies within one particular, inhomogeneously broadened, ADMR transition (the A stands for absorbance), one observes clear shifts of the maximum absorbance wavelength (Fig. 3). In addition the shape of the absorbance band changes, with a pronounced shoulder coming up at microwave frequencies located in the low frequency tail of the $|D| + |E|$ transition. The variability of the absorbance wavelength and lineshape with microwave frequency is much less pronounced when the 800 nm (830 nm for *Rps. viridis*) is scanned (Fig. 4). Note that the variation of absorbance lineshape with microwave frequency is the same in chromatophores as in reaction centers, i.e. it is not caused by the reaction center isolation procedure.

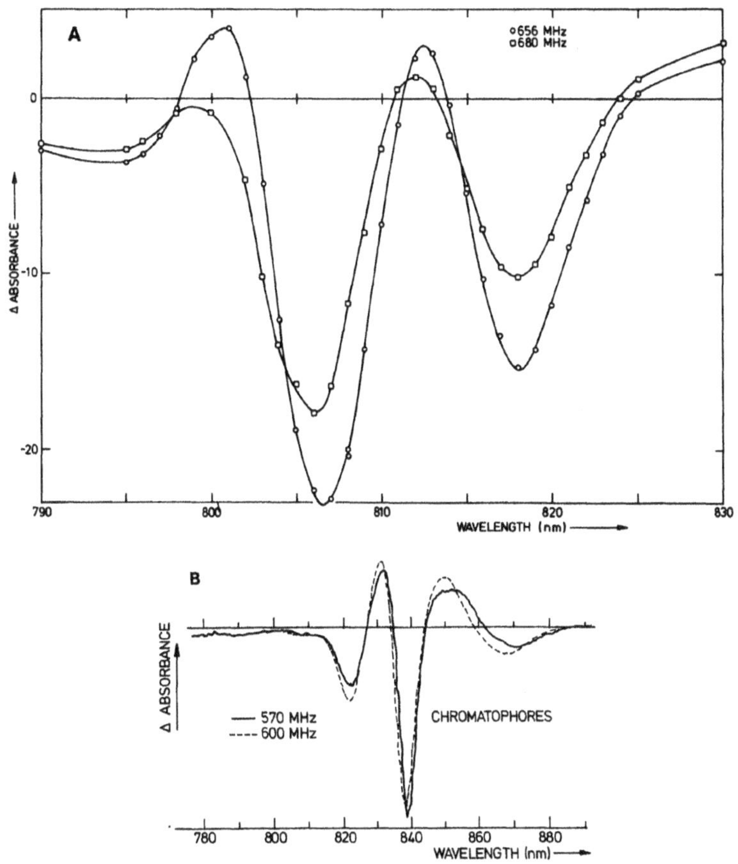

Fig. 4. A) The (inverted) T - S spectrum around 800 nm of reaction centers of Rb. sphaeroides recorded by ADMR at 1.2 K with the microwaves set at two different frequencies as indicated. From Den Blanken et al. (1982b). B) Same for chromatophores of Rps. viridis. From Den Blanken et al. (1983). The spectra of B) but not of A) are normalized.

From the above results it appears that there is a correspondence between the selection of a set of particular values of $|D|$ and $|E|$ (that combined give the sharply defined frequency ($|D| - |E|$)/h) and the selection of a (restricted distribution of) particular P absorbance band(s). This correspondence could be caused by environmental effects, or by a direct relation between the configuration of the P dimer, the principal values of the dipole-dipole interaction tensor,and the lineshape and the wavelength of maximum absorption of the P-band. As argued above, a direct relation would be expected for the strong coupling case. Environmental effects seem to be less likely since the 800 nm region is much less sensitive to microwave selection.

The P-band in the T - S spectrum of Fig. 2 is about 10 % narrower at half height than the P-band in the normal absorbance spectrum at 5 K, the inhomogeneous broadening appararent from the normalized traces of Fig. 3

contributing mainly to the wings of the latter band. Thus the linewidth of the P-band in the T - S spectrum, which is produced with microwaves fixed at a very well defined frequency, reflects either residual inhomogeneity due to the remaining distribution of $|D|$ and $|E|$ values (see above) or the true homogeneous linewidth.

THE HOMOGENEOUS LINEWIDTH OF THE P-BAND

The question what the homogeneous linewidth of the P-band is at 1.5 K can be resolved in two ways: one may perform a time domain experiment with photon echo spectroscopy and determine T_2 ($T_2^{-1} = T_2^{*-1} + (2T_1)^{-1}$ with T_2^* the intrinsic optical dephasing time and T_1 the lifetime of the excited state) or one may perform an optical hole burning experiment. Both routes have been taken (Meech et al., 1985,1986; Boxer et al., 1986a,b) and the result is that the homogeneous linewidth of the P-band of *Rb. sphaeroides* is about 35 nm and of *Rps. viridis* about 50 nm. These values correspond to a 'lifetime' of not more than 25 fs. The interesting question what causes such an extremely short lifetime will be left aside (see e.g. Meech et al., 1986; Boxer et al., 1986b; Friesner, 1987a,c, and Wiersma et al., this volume).

The wavelength at which the optical hole has maximum depth,but not its width or shape, depends markedly on the burn wavelength (Boxer et al., 1986a). This behavior is analogous to the shift with microwave frequency shown in Fig. 3, and reflects additional inhomogeneous broadening of the P-band.

THE STRUCTURE OF THE P-BAND

For certain microwave frequencies the P-band in the T - S spectrum of both *Rps. viridis* and *Rb. sphaeroides* shows a pronounced shoulder at the long wavelength side (Fig. 3). This shoulder is visible in the low temperature absorbance spectrum of reaction centers of *Rps. viridis* (Vermeglio and Paillotin, 1982) but not in that of *Rb. sphaeroides*. The shoulder has been attributed to a charge transfer band (Vermeglio and Paillotin, 1982; Maslov et al., 1983). If this interpretation is correct, then not all reaction centers show this band (since it is absent in some of the T - S spectra) and the purported charge transfer state is unlikely to play a role in electron transport. The assignment of the shoulder (when present) to a charge transfer state, however, is doubtful. It has been demonstrated that the P-band shows uniform polarization over the band, both in the absorbance LD spectrum (Vermeglio and Paillotin, 1982; Breton, 1985) and in the LD-(T - S) spectrum (Hoff et al., 1984; Lous and Hoff, 1987a). Since a charge transfer band will generally have its optical transition moment in a direction different from that of the P-band, its contribution to the oscillator strength of the P-band must be minor. Yet, in at least one reaction center, that of the purple bacterium *Chromatium vinosum*, the 'shoulder' has the bulk of the oscillator strength (Fig. 3C), whereas also for this reaction center the polarization is uniform over the P-band (Hoff et al. 1984). It was therefore concluded that the structure of the P-band is not due to a charge transfer state, but to an inhomogeneity in the configuration of the primary donor dimer.

LINESHAPE SIMULATIONS OF THE P-BAND

There is now a large body of simulations based on exciton theory that show considerable success in the reproduction of the salient features of absorbance (A), ΔA, LD, CD, T - S and LD-(T - S) spectra (see e.g. Knapp et al. 1986). In these simulations the linewidths are generally taken from in vitro spectra and adjusted on a more or less ad hoc basis. For an exciton-

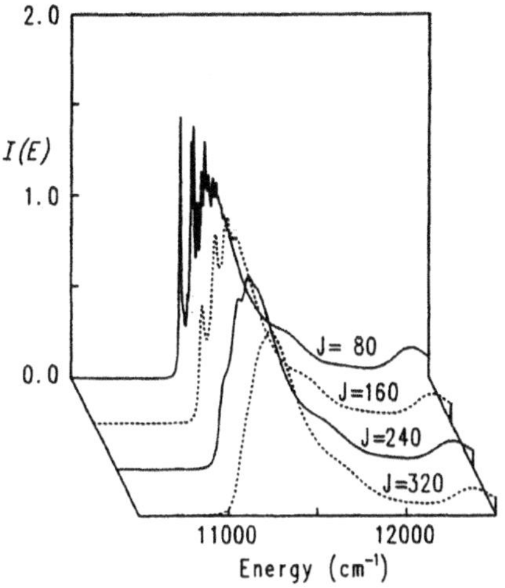

Fig. 5. Calculated homogeneous absorption spectra of Rb. sphaeroides as a function of the coupling strength J between the excited state P^* and a charge transfer state $P^{\pm}$, with energy gap $\Delta E(P^* - P^{\pm}) = 2000\ cm^{-1}$. From Won and Friesner (1987a).

ically coupled BChl dimer one would expect the Q_y band to show to first approximation a width $1/\sqrt{2}$ times smaller than that of the corresponding monomers (Heminger, 1977). The P-band, however, is not narrower, but considerably broader than the Q_y band of BChl in vitro. The first theoretical treatment that seems to have successfully addressed this problem is due to Won and Friesner (1987a-c). They treat two excitonically and vibronically coupled BChl Q_y bands of the primary donor, with vibrational structure modeled after that of Chl a obtained from site-selected Shpol'skii spectra (Platenkamp et al., 1980), that are coupled to a charge transfer state, also with vibrational structure. The complex eigenvalue problem is solved and absorbance spectra calculated as a function of the coupling parameters. Some spectra are shown in Fig. 5. For values of $J = 240 - 320\ cm^{-1}$, and $\Delta = 2000\ cm^{-1}$, where J is the coupling between the excited P^* and the charge transfer state and Δ their energy gap, a P-band is generated that is remarkably similar to the experimental site-selected T - S P-band. It is anticipated that similar calculations for Rps. viridis will be equally successful.

From the theoretical work of Friesner and coworkers (1987a,b) one is tempted to infer that the shoulder in the absorbance and T - S spectra of Rps. viridis, and in the T - S spectra of Rb. sphaerorides and C. vinosum, reflects a severely broadened zero-phonon line. Its presence or absence clearly depends on, amongst others, the coupling strength J and the energy gap Δ. These parameters are likely sensitive to the exact configuration of the primary donor, regardless whether the charge transfer is internal ($P^{\pm}$) or external (P^+B^-). Thus, small variations in this configuration (which are not resolved by X-ray crystallography) that are frozen in by cooling to cryogenic temperatures, are likely responsible for i) the observed lineshape of the P-band, which is a convolution of many spectra as in Fig. 5, ii) the inhomogeneity of the P-band observed both with microwave selection in the T - S spectra and in optical hole spectra and iii) the inhomogeneous lineshape of the ODMR transitions.

Recent theories on electron transport in the bacterial reaction center focus on superexchange and direct exchange mechanisms (Marcus, 1987; Fischer and Scherer, 1987; Bixon et al., 1987; Warshel and Parson, 1987; Parson and Warshel, 1987). Clearly, reaction center heterogeneity, be it static (low temperature) or dynamic (high temperature) has to be taken into account when calculated electron transfer rates are compared to the observed values. Experimentally, the influence of reaction center heterogeneity on electron transfer rates could be demonstrated by careful analysis of the charge separation kinetics at low temperature as a function of the wavelength of narrow-banded laser excitation.

ACKNOWLEDGEMENT

The experimental work in the author's laboratory was carried out with support of the Netherlands Foundation for Chemical Research (SON), financed by the Netherlands Organization for the Advancement of Pure Research (ZWO).

REFERENCES

Bixon, M., Michel-Beyerle, M.E., Ogrodnik, A., and Jortner, J., 1987, Recombination and electronic exchange interactions of radical pairs in photosynthetic reaction centers, Biochim. Biophys. Acta, in the press.

Boxer, S.G., Lockhart, D.J., and Middendorf, T.R., 1986a, Photochemical holeburning in photosynthetic reaction centers, Biochim. Biophys. Acta 123:476.

Boxer, S.G., Middendorf, T.R., and Lockhart, D.J., 1986b, Reversible photochemical holeburning in Rhodopseudomonas viridis reaction centers, FEBS Lett. 200:237.

Breton, J., 1985, Orientation of the chromophores in the reaction center of Rhodopseudomonas viridis. Comparison of low-temperature linear dichroic spectra with a model derived from X-ray crystallography, Biochim. Biophys. Acta 810:235.

Deisenhofer, J., Epp., O., Miki, K., Huber, R., and Michel, H., 1975, Structure of the protein subunits in the photosynthetic reaction centre of Rhodopseudomonas viridis at 3 Å resolution, Nature 318:618.

Den Blanken, H.J., van der Zwet, G.P., and Hoff, A.J., 1982, ESR in zero field of the photoinduced triplet state in isolated reaction centers of Rhodopseudomonas sphaeroides R-26 detected by the singlet ground state absorbance. Chem. Phys. Lett. 85:335.

Den Blanken, H.J., and Hoff, A.J., 1982, High-resolution optical absorption-difference spectra of the triplet state of the primary donor in isolated reaction centers of the photosynthetic bacteria Rhodopseudomonas sphaeroides R-26 and Rhodopseudomonas viridis measured with optically detected magnetic resonance at 1.2 K. Biochim. Biophys. Acta 681:365.

Den Blanken, H.J., and Hoff, A.J., 1983, Resolution enhancement of the triplet-singlet absorbance-difference spectrum and the triplet-ESR spectrum in zero field by the selection of sites. An application to photosynthetic reaction centers, Chem. Phys. Lett. 98:255.

Den Blanken, H.J., Jongenelis, A.P.J.M., and Hoff, A.J., 1983, The triplet state of the primary donor of the photosynthetic bacterium Rhodopseudomonas viridis, Biochim. Biophys. Acta 725:472.

Fischer, S.F., and Scherer, P.O.J., 1987, On the early charge separation and recombination processes in bacterial reaction centers, Chem. Phys. 115:151.

Heminger, R.P., 1977, Optical spectra of molecular aggregates near the strong coupling limit. J. Chem. Phys. 67:262.

Hoff, A.J., 1976, Kinetics of populating and depopulating of the components of the photoinduced triplet state of the photosynthetic bacteria Rhodospirillum rubrum, Rhodopseudomonas sphaeroides (wild type), and its

mutant R-26 as measured by ESR in zero field. Biochim. Biophys. Acta 440:765.
Hoff, A.J., den Blanken, H.J., Vasmel, H., and Meiburg, R.F., 1985, Linear-dichroic triplet-minus-singlet absorbance difference spectra of reaction centers of the photosynthetic bacteria Chromatium vinosum, Rhodopseudomonas sphaeroides R-26 and Rhodospirillum rubrum S1, Biochim. Biophys. Acta 806:389.
Knapp, E.W., Scherer, P.O.J., and Fischer, S.F., 1986, Model studies to low temperature optical transitions of photosynthetic reaction centers, Biochim. Biophys. Acta 852:295.
Lous, E.J., and Hoff, A.J., 1987a, Exciton interactions in reaction centers of the photosynthetic bacterium Rps. viridis probed by optical triplet-minus-singlet polarization spectroscopy at 1.2 K monitored through absorbance-detected magnetic resonance, Proc. Natl. Acad. Sci. USA 84: 6147.
Lous, E.J., and Hoff, A.J., 1987b, Absorbance detected electronspin echo spectroscopy of non-radiative triplet states in zero field. An application to the primary donor of the photosynthetic bacterium Rhodobacter sphaeroides, Chem. Phys. Lett., in the press
Marcus, R.A., 1987, Superexchange versus an intermediate $BChl^-$ mechanism in reaction centers of photosynthetic bacteria, Chem. Phys. Lett. 133:471.
Maslov, V.G., Klevanik, A.V., Ismailov, M.A., and Shuvalov,V.A., 1983, O prirode dlinnovolnovoj polocy poglolshtshenya reaktsionnykh tsentrov Rhodopseudomonas viridis v svyazi s protsessom pervitsnogo razdelenya zaryadov, Doklady AN SSSR 269:1217.
Meech, S.R., Hoff, A.J., and Wiersma, D.A., 1985, Evidence for a very early intermediate in bacterial photosynthesis. A photon-echo and hole-burning study of the primary donor band in Rhodopseudomonas sphaeroides, Chem. Phys. Lett. 121:287.
Meech, S.R., Hoff, A.J., and Wiersma, D.A., 1986, Role of charge transfer states in bacterial photosynthesis, Proc. Natl. Acad. Sci. USA 83: 9464.
Michel, H., Epp, O., and Deisenhofer, J., 1986, Pigment-protein interactions in the photosynthetic reaction centre from Rhodopseudomonas viridis, EMBO J. 5:2445.
Parson, W.W., and Warshel, A., 1987, Spectroscopic properties of photosynthetic reaction centers. II. Application of theory to Rhodopseudomonas viridis, J. Am. Chem. Soc., in the press.
Platenkamp, R.J., den Blanken, H.J., and Hoff, A.J., 1980, Single-site absorption spectroscopy of pheophytin-a and chlorophyll-a in a n-octane matrix, Chem. Phys. Lett. 76:35.
Plato, M., Trankle, E., Lubitz, W., Lendzian, F., and Möbius, K., 1986, Molecular orbital investigation of dimer formations of bacteriochlorophyll. A. Model configurations for the primary donor of photosynthesis. Chem. Phys. 107:185.
Sternlicht, H., and McConnell,H.M., 1961, Paramagnetic excitons in molecular crystals. J. Chem. Phys. 35:1793.
Vasmel, H., Amesz, J., and Hoff, A.J., 1986, Analysis by exciton theory of the optical properties of the reaction center of Chloroflexus aurantiacus, Biochim. Biophys. Acta 852:159
Vermeglio, A., and Paillotin, G., 1982, Structure of Rhodopseudomonas viridis reaction centers. Absorption and photoselection at low temperature, Biochim. Biophys. Acta 681:32.
Warshel, A., and Parson, W.W., 1987, Spectroscopic properties of photosynthetic reaction centers. I. Theory, J. Am. Chem. Soc, in the press.
Won, Y., and Friesner, R.A., 1987a, Simulation of photochemical hole-burning experiments on photosynthetic reaction centers, Proc. Natl. Acad. Sci. USA 84:5511; J. Phys. Chem., in the press
Won, Y., and Friesner, R.A., 1987b, Simulation of optical spectra from the reaction center of Rhodopseudomonas viridis, J. Phys. Chem., in the press.

NOMEN EST OMEN

A note on nomenclature

For a long time photosynthesis research has been marred by a proliferation of symbols, labels and names with no apparent consistency. Especially the nomenclature of the components that function in the primary events is bewildering, with practically every author having his/her own pet photosynthetic alphabet. To a large extent this is due to the many uncertainties that existed with regard to the identity of electron donors and acceptors (a favorite acceptor label being X_n, n = 0, 1, ...). Only recently, with the elucidation of the structure of single crystals of reaction center proteins, has our knowledge become detailed enough to permit the adoption of a rational, unambiguous, easy to use and easy to understand labeling system. At the same time, unfortunately, the problem has been compounded by the discovery of the near-C_2 symmetry of the prosthetic groups of the (bacterial) reaction centers, with one set organized in an active, one set in an apparently non-active electron transport chain. Deisenhofer et al. (1984,1985) have introduced a nomenclature that is precise but not wholly unambiguous, and not easy to use, especially not when orally delivered. It is the purpose of this note to list the requirements of a 'good' nomenclature, focusing on the labels of the prosthetic groups in the bacterial reaction center, and to discuss a few alternatives that were proposed during the workshop.

Labels should be 1) unambiguous, 2) informative, 3) logical, 4) easy to memorize, 5) easily printable, 6) easily pronounceable (and still be unambiguous), 7) a 'common denominator' of existing nomenclature and 8) transparent to newcomers in the field. Above all, they should be <u>short</u>. With all due respect to Deisenhofer et al., their labeling system (Fig. 1A) 1) may give rise to confusion since the quinone acceptor that is in the active 'L' chain, Q_A, is really bound to the non-active 'M' chain and vice versa for the Q_B acceptor, 2) is not easy to memorize or pronounce and 3) is not short. What are the alternatives?

A one-letter code makes it difficult to distinguish between the two chains while keeping the same letter for symmetry related groups, without taking recourse to exotic alphabets, in conflict with points 5) and 6). A two-letter code permits easy discrimination of the two chains, if the letters L or M are avoided for the reason given above. Chain labels A and B have enjoyed already a certain popularity to discriminate the quinone acceptors, and could be used throughout. Subscripting A (active chain) and B (non-active) emphasizes that these letters do not relate to particular pigments.

The next step is labeling the individual components. Q as a generic label for the quinone acceptor is well-accepted and should be retained. The primary donor of all photosystems is generally labeled P for pigment. (This originates from Bessel Kok's label P-700.) There are, however, more pigments in the reaction center and it therefore seems logical to replace it by D for primary donor (dimer), specifying D_A and D_B for the individual monomers on the active (A) and non-active (B) chain.

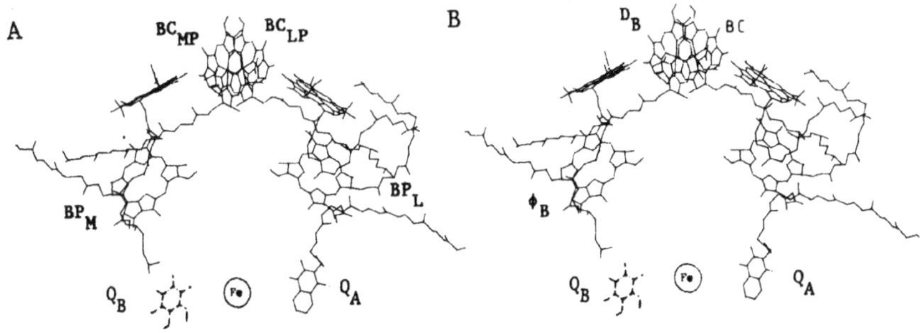

Fig. 1. A) Nomenclature according to Deisenhofer et al. (1985). The figure represents the cofactors of the reaction center of Rps. viridis as depicted by Deisenhofer et al. (1984), with the exception of Q_B that has been added purely as a fancy. B) Proposed nomenclature.

There is as yet no consensus on the labeling of the accessory bacteriochlorophylls and the bacteriopheophytins. Following Deisenhofer et al. the label B is proposed for the accessory bacteriochlorophylls: B_A and B_B. (Note that the term 'voyeur' bacteriochlorophyll is misleading since one of them is likely involved in electron transfer from P to the bacteriopheophytin on the A-chain.) The bacteriopheophytins have sometimes been labeled H, reflecting that they are free base porphyrin derivatives. In the opinion of this note's author, this is somewhat contrived (after all a pheophytin is not a proton) and ambiguous as H also denotes one of the reaction center protein subunits. Furthermore, H denotes N in Russian and is usually not pronounced in French. Since B or P have already been taken one may choose the phonetic Φ (for Greek ph and for free base). This is unambiguous, transparent, would apply also to photosystem II of plants and avoids confusion with ferredoxin nomenclature in photosystem I.

The proposed new nomenclature of electron transport components in the photosystem of purple bacteria is summarized in Fig. 1B. It seems at present too early to extend this labeling system to other photosystems, with the possible exception of photosystem II and that of the Chloroflexaceae. Even for these two systems adjustments are necessary, e.g. C for B in photosystem II and Φ_{B1}, Φ_{B2} for the two bacteriopheophytins on the B-chain of C. aurantiacus. In spite of this, I believe that the nomenclature proposed in Fig. 1b conforms to the requirements listed above and, when generally adopted, should facilitate the communication between workers in photosynthesis, and with their colleagues in other fields of human endeavour.

A.J.H.

Deisenhofer, J., Epp, O., Miki, K., Huber, R., and Michel, H., 1984, X-ray structure analysis of a membrane protein complex, J. Mol. Biol. 130: 385; 1985, Structure of protein subunits in the photosynthetic reaction centre of Rhodopseudomonas viridis at 3 A resolution, Nature 318: 618.

REACTION CENTERS OF PURPLE BACTERIA WITH MODIFIED CHROMOPHORES

H. Scheer, D. Beese, and R. Steiner

Botanisches Institut der Universität
Menzinger Str. 67
D-8000 München 19, FRG

A. Angerhofer

Physikalisches Institut der Universität
Pfaffenwaldring 57
D-7000 Stuttgart 80, FRG

INTRODUCTION

Reaction centers (RC*) of purple bacteria are generally composed of three subunits designated H(igh), M(edium) and L(ow) according to their apparent molecular weigths on SDS PAGE. Four molecules of bacteriochlorophyll (Bchl) are bound to it, together with two bacteriopheophytins (Bphe), two quinones (Q) and one non-heme iron. The crystal structure of RC from the BChl b-containing purple photosynthetic bacterium, Rp. viridis (Deisenhofer et al, 1984) and from the Bchl a-containing Rb. sphaeroides (Chang et al., 1986; Allen et al., 1987) shows a C2 - symmetry axis which divides the reaction center into two very similar sets of pigments interacting mainly with the L and M-subunits, respectively. The reaction center is asymmetric, however, in functional terms. The primary charge separation takes place most probably from the special pair situated on the symmetry axis, via $Bphe_L$ and Q_A situated on the L-(or 'active') branch of the complex, to Q_B on the M or 'inactive' branch (Deisenhofer et al., 1984; Vermeglio and Paillotin, 1982; Zinth et al., 1985).

---

*) Abbreviations: RC = reaction center; Bchl = bacteriochlorophyll; Bphe = bacteriopheophytin, the location of these pigments on the L- or M-branch of the RC is indicated by the respective subscript; P870 = primary donor; Chl = chlorophyll; Q = quinone; subscript indicates the primary (A) or secondary acceptor (B) located on the L- and M-branch, respectively; cd = circular dichroism, ESR = electron spin resonance; SDS-PAGE = sodium dodecylsulfate polyacrylamide gel electrophoresis, LDAO = dimethyldodecylamineoxide, TX-100 = Triton X-100; Rb. = Rhodobacter; Rp. = Rhodopseudomonas, Rs. = Rhodospirillum, Cf. = Chloroflexus, Cr. = Chromatium.

This provides a function for only three of the six chlorophyllous pigments. Much less is known on the function of the remaining two monomeric BChl a molecules absorbing around 800nm ($Bchl_L$, $Bchl_M$), and of the second pheophytin, $Bphe_M$, which shows a well resolved absorption from the 'active' $Bphe_L$ in the Q_X spectral region. $Bchl_L$ is situated between the primary donor, P870, and the "primary" acceptor, $Bphe_L$. It is still unclear, however, if it acts as "pre-primary" acceptor in this process (Shuvalov and Duysens, 1986), or if some other mechanisms like super-exchange, rapid hopping between a close and a distant site or yet another process is operative (Holten, 1986; Ogrodnik et al., 1982; Breton, 1986; Zinth et al., 1985; Knapp et al., 1985; Wasielewski, 1986; Parson, 1982, Michel-Beyerle et al., 1987; Fischer and Scherer, 1987). The participation of the pigments on the M-branch, $Bchl_M$ and $Bphe_M$, is unclear. The latter can also accept an electron (Robert et al., 1985; Michel-Beyerle et al., 1987), but it is questionable if this also takes place under physiological conditions.

In view of the tight packing of the pigments and the highly optimized charge separation, modified RC are of considerable interest to understand the mechanism of charge separation and the function of all pigments present in RC. The possibility to remove one of the six tetrapyrrolic pigments in bacterial reaction centers (Ditson et al., 1984) provided for the first time an experimental tool to modify the tetrapyrroles in this complex. By treatment with sodium borohydride, about 50% of the bacteriochlorophyll (Bchl) absorbing at 800nm can be removed. From kinetic (Maroti et al., 1985; Holten, 1986; Shuvalov and Duysens, 1986; Breton 1986) and spectroscopic evidence (Robert et al., 1986; Scherer and Fischer, 1987), it has been suggested that the 'inactive' $Bchl_M$ located between the primary donor P870 and the bacteriopheophytin $Bphe_M$ on the M-branch, is reduced at the 3-acetyl group and can then be dissociated from the remaining complex (Maroti et al., 1985). A basic requirement for evaluating the results is a thorough characterization of the modified preparations with respect to chemical composition, homogeneity, spectroscopy and kinetics. This is of particular importance because the homogeneity of the sample has been questioned more recently by Shuvalov et al. (1986), who discuss a product mixture in which the $Bphe_M$ is partly reduced as well.

Principally, an exchange of pigments with modified ones would be most useful. Much of the progress in the retinylidene-protein research relies on this technique. However, comparably few such experiments have been reported on (bacterio)chlorophyll proteins, and it is moreover very difficult to verify the exchange. Loach et al. (1975) provided ESR-spectroscopic evidence for an exchange of Bchl in AUT-particles from Rs. rubrum. The line narrowing observed upon incubation of ^1H-RC with ^2H-Bchl was reversed, however, after subsequent washing of the samples (P.Loach, private communication, Beese, 1984). The reincorporation of Chl a' into ether-washed PSI particles led to a ligth induced difference spectrum similar to that of P700, but the bleaching was at shorter wavelengths, and more important it was irreversible (Hiyama et al., 1987). Verification problems are even more severe with antenna complexes (Clayton and Clayton, 1982; Parkes-Loach et al., 1987; Plumley and

Schmidt, 1987, Chadwick et al., 1987) for difficulties in establishing reliable functional criteria in an isolated antenna. These verification problems are mainly related to three factors: pigment adsorption on RC (vide supra) detergent effects on the spectra of pigment-protein complexes (Chadwick et al., 1987), and formation of micellar (bacterio)chlorophyll-detergent complexes which have spectroscopic and chemical properties very similar to the ones of protein complexes with the respective pigments (Gottstein and Scheer, 1983; Scherz and Parson, 1984, 1986, Scheer et al., 1985; Scherz and Rosenbach-Belkin, 1987).

Here, we want to sumarize data characterizing the "$NaBH_4$-removable" Bchl a, its interactions with near-by pigments, and the homogeneity of treated RC from Rb. sphaeroides, as well as the extension of the modification method to other pigment complexes. In addition, results are presented on an improvement of the method of Loach et. al.(1975) by which the 'extra' Bchl molecules $Bchl_L$ and $Bchl_M$ are exchanged with extraneous pigment.

MATERIALS AND METHODS

Preparations

RC from Rb. sphaeroides were prepared by a modification of the method of Feher and Okamura (1978). $Bchl_M$ was removed according to the original method of Ditson et al. (1985) by addition of solid borohydride in the presence of LDAO as detergent. To stop the reaction in kinetic experiments, aliquots were diluted 1:1 with a glucose solution (for spectroscopic studies), or treated with acetone to precipitate the protein (for polypeptide analysis).

Exchange experiments

RC were treated with a 10-20 fold excess of free pigment according to Loach et al (1975). The temperature was raised to 40°C. Purification of RC after modification and detergent

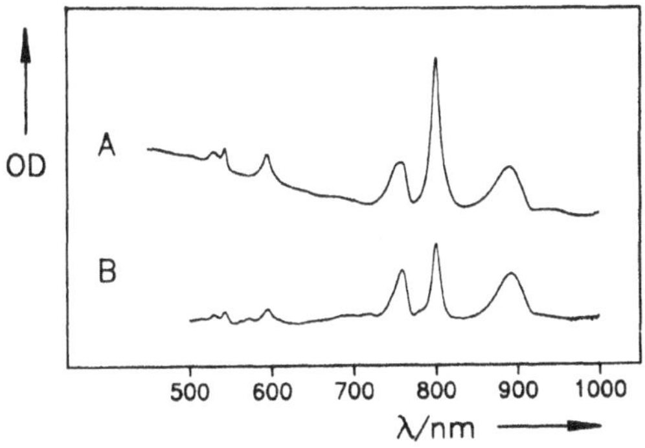

Fig. 1. Absorption spectra at 4 K of RC from Rb. sphaeroides R26 before (A),and after treatment with borohydride and subsequent purification on DEAE cellulose (B).

exchange (if necessary) was done by chromatography on DEAE cellulose (DE 52, Whatman). Chemical oxidation of reaction centers was performed by titration with potassium ferricyanide.

Spectroscopy

Extinction coefficients were determined with respect to the 870nm transition (ε=1.28x10^5 $cm^{-2}M^{-1}$, Clayton and Wang, 1971). Since the intensity of this band remained unchanged during treatment with borohydride or exchange, the same value was used for modified RC. For the determination of molar ellipticities, the reaction was followed in the CD cuvette, and the decrease at the 870nm band then used to calibrate all other bands in purified, modified RC. Low temperature absorption and fluorescence spectra were recorded in the apparatus described by Angerhofer et al. (1986), fluorescence-ODMR and microwave-induced absorption-difference (MIA or ADMR) spectra were measured with the ODMR set-up described by Angerhofer et al. (1985).

RESULTS

Treatment of RC with borohydride from *Rp. sphaeroides*

Upon treatment with borohydride, the 800nm absorption decreases by about 55% as compared to its starting value. Distinct changes occur also in the Q_X spectral region between 500 and 630nm (Fig.1). Typical preparations have an 800/870nm absorption ratio of 1.1 to 1.3. These spectral changes are similar to the ones described by Ditson et al. (1984). The reaction is less complete in the presence of TX-100 instead of LDAO. The low-temperature spectrum shown in Fig.1 reveals some details (Beese et al., 1987) which were not discernible in the room temperature spectra. Two minor bands become apparent at 780 and 572nm, and the shape of the Bphe band at 757nm is different. Instead of consisting of two poorly resolved bands of equal intensity, it now appears to be a main band shifted to longer wavelengths with a short wavelength shoulder.

Since the reaction is accompanied by an increase of the pH to 10-10.5, the effect of high pH alone was tested in an independent experiment. Up to pH 10.5, the spectrum remains

Table 1. Reaction of different bacteriochlorophyll-proteins with potassium borohydride

RC	*Rb. sphaeroides R26*	decrease 804nm
RC	*Rb. sphaeroides 2.4.1*	no effect
RC	*Rp. viridis*	some decrease 820nm
B880	Rs. rubrum	no effect
B800-850	*Rb. sphaeroides*	decrease 800nm
B800-850	*Rp. acidophila* (type I)	decrease 800nm
B800-850	*Rp. acidophila* (type II)	decrease 800nm
B800-840	Cr. vinosum	decrease 800nm
B800-820	Cr. vinosum	decrease 800nm

unchanged. At higher pH, the absorption around 860 nm decreases first, followed by the 800 nm band. Obviously, the high pH obtained under the experimental conditions has no effect on the spectrum. It should be noted, however, that the reaction is accelerated by it.

Treatment of other chlorophyll - proteins and micellar chlorophyll-detergent complexes with borohydride

The reactions of other bacteriochlorophyll - proteins with borohydride are summarized in Table 1. Except for the RC from the wild-type strain Rb. sphaeroides 2.4.1 and from Rp. viridis, which gaved little to no reaction, all other complexes showed a selective and very pronounced reaction of those chromophores absorbing around 800nm, whereas the ones absorbing at longer wavelenghts did not react. This indicates some common property of these chromophores, which sets them apart from the ones absorbing at longer wavelengths. Part of this reactivity may be related to a better accessibility to the hydrophilic borohydride. That the aggregation state of the pigments may be another factor, is indicated by model studies with micellar complexes in the detergent, TX-100. The reaction of monomeric chlorophyll a or b is faster by two orders of magnitude as compared to the respective aggregates. More recently, some experiments were carried out with chlorophyll-protein complexes from green plants showing that they react as well with borohydride in a differential manner (Scheer, Anderson and Porra, unpublished). Another noteworthy and common feature is the regioselectivity of the reaction. Bchl a and the plant pigment, Chl b, bear two carbonyl groups. In solution, e.g. in methanol, the 13^1 C=O group is always much less reactive than 3 C=O or 7^1 C=O, respectively. This is shown in Fig. 2 for the reaction of Chl b. The same regioselectivity has been observed in all experiments, both with the pigment-protein and pigment-detergent complexes.

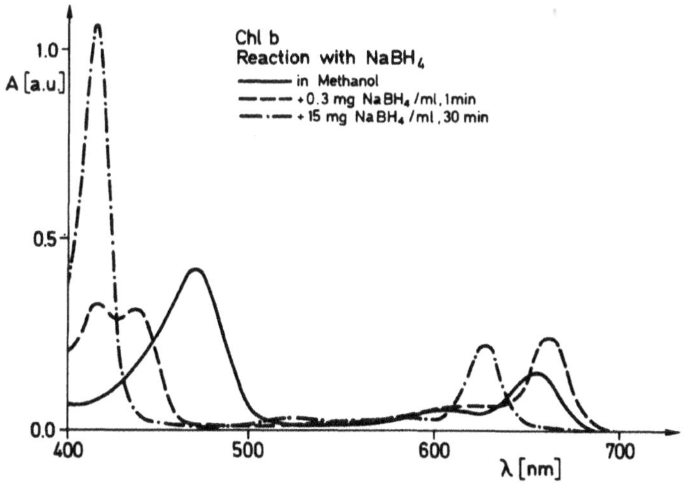

Fig. 2. Reaction of Chl b in methanolic solution with borohydride. See inset for experimental details.

Sample homogeneity: Spectroscopy

The excitation spectra of the P870 fluorescence at low temperatures are similar to the absorption spectra. The emission spectra of modified RC (Beese et al., 1987) have two bands at 917nm (reaction center), and around 776nm (contamination with free (= non-aggregated) pigments). The latter are strongly fluorescent and thus picked up with high sensitivity in the RC, which is only weakly fluorescent itself. The only significant change of the sample after modification with sodium borohydride, is an increase of this free pigment emission by approximately 150%. There is in particular no new band above 800nm. This indicates a homogeneous emission of protein-bound pigments. Homogeneity of the preparation and unchanged hyperfine coupling of the primary donor triplet, is also supported by optically-detected magnetic resonance. The spectra of unmodified and modified RC are identical within the experimental error (Beese et al., 1987) both for absorption and fluorescence detection.

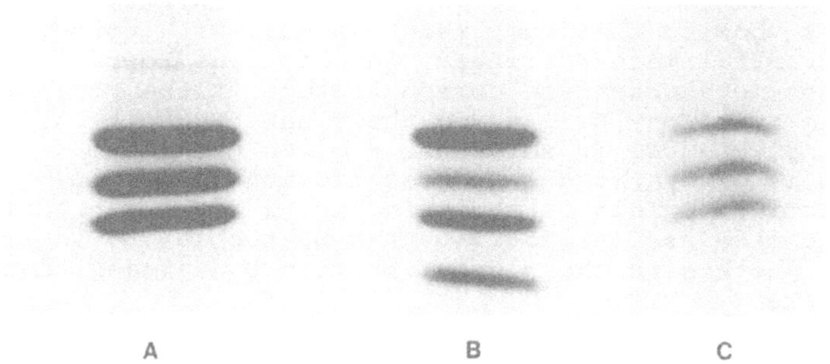

Fig. 3: Polypeptide composition of RC as analyzed by SDS-PAGE. Trace A: Untreated RC. Trace B: RC treated with borohydride. Absorbance ratio 800/870 = 1.3. Trace C: RC incubated for 14 hrs at pH 11.

Sample homogeneity: Polypeptide composition

The RC contains three polypeptides (H,M,L) with apparent molecular weights of 28, 24 and 21 kDa, respectively. An additional peptide (15 kDa) appears upon treatment with borohydride, and the intensity of the M-band is diminished at the same time (Fig. 3), indicating most likely a cleavage of M. Control samples incubated at pH 11 for similar length of time, but without the addition of borohydride, did not show the additional polypeptide. This indicates that only the $NaBH_4$-reduction but not the accompaning pH change leads to cleavage of the M-polypeptide. Cleavage reactions of this reagents have been reported before (Crestfield et al., 1963).

Two conclusions can be drawn from these results: 1) The sample is biochemically heterogeneous in spite of its spectroscopic homogeneity. 2) Cleavage of the M-subunit has only minor effects on stability and functionality of RC. Reversible bleaching has been retained before in membranes after extensive proteolysis (Bachofen and Wiemken, 1984; Steiner et al., 1986; Theiler et al., 1984), but the reaction centers are generally not resistent to the same treatment in solution. This reflects a more specific proteolysis by borohydride.

Coupling of $B800_M$ with P870

The results discussed so far indicate that removal of $Bchl_M$ produces only localized changes and no disturbances of the remaining five tetrapyrrols. Such interactions are revealed, however, by methods which are more sensitive to interactions. As an example, the cd spectra of untreated and

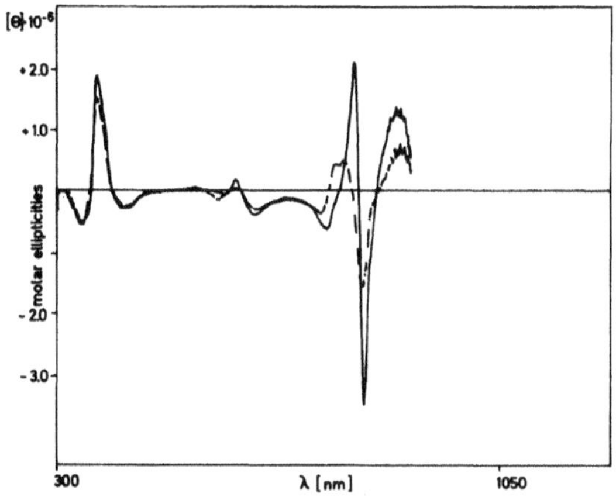

Fig. 4. CD spectra of untreated (——) and borohydride treated RC (---). The treated sample was purified over DEAE cellulose.

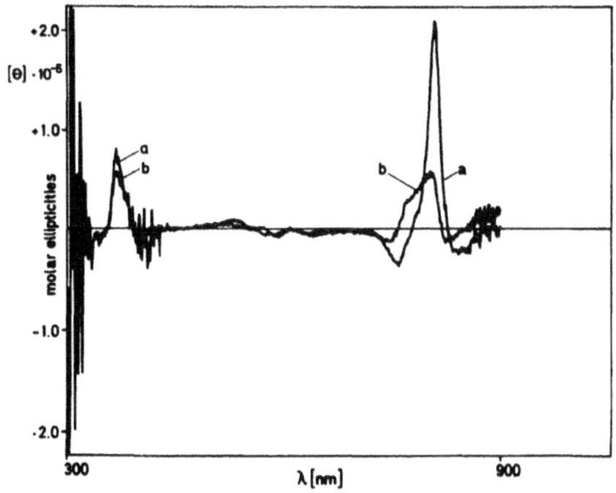

Fig. 5. Circular dichroism spectra of RC oxidized with $Fe(CN)_6^{3-}$ a: Untreated RC. b: Sample after reaction with borohydride and purification.

modified RC are compared in Figs. 4,5. In relaxed (=reduced) RC the changes are no longer confined to the spectral region around 800nm. There is in particular a decrease in optical activity of the 870nm band, indicating some structural change and/or the coupling of $B800_M$ with P870 in unmodified RC. In oxidized RC (Fig.5), the strong band around 800 nm is reduced to 33% in intensity, and the fine structure of the Bphe bands indicates a sign inversion for the one absorbing at longer wavelengths. Distinct changes have also been found in the microwave-induced absorption difference spectra of RC upon modification with borohydride. They could be rationalized by a structural change (Scherer and Fischer, 1987). The Raman resonance spectra show small but distinct shifts as well reflecting interactions of $Bchl_M$ with both neighboring B870 and $Bphe_M$ (Beese et al., 1987).

These examples show that removal of $Bchl_M$ has very pronounced effects on its neighboring pigments, P870 and $Bphe_M$. They involve both minor geometric changes, probably due to a rearrangement in reponse to the hole created, and the elimination of couplings.

Exchange of 'extra bacteriochlorophylls' in RC

Application of the procedure of Loach et al. (1975) to RC from *Rs. rubrum* and *Rb. sphaeroides* did not show any exchange of pigments detectable by labeling with ^{14}C (both ways) or with Bchl bearing a different alcohol. All samples were purified after the incubation by chromatography on DEAE cellulose, to remove any adsorbed pigments. This confirms findings of Loach (private communication), who observed a loss of ESR line-narrowing in ^{2}H-treated RC if they were re-purified after the incubation. Whereas this reversible line-narrowing poses an interesting problem by itself, because it means spin-exchange between P870 and loosely adsorbed Bchl, the results defy the original goal.

When the method was pursued further, it could be modified, however, to produce what looks like a true exchange of the 'extra' Bchl. The main modification is an increase of the incubation temperature to about 40°, which is just at the point where reaction centers begin to unfold. The verification tests involved several different techniques. 1. Labelling of Bchl a with different esterifying alcohol and subsequent analysis by HPLC. 2. Incubation of ^{14}C-RC with cold Bchl a, and vice versa.

Scheme 1. Experimental procedure to establish exchange of 'extra' Bchl in RC.

^{12}C-RC

↓ a) Incubation with excess ^{14}C Bchl, 4°
b) DEAE cellulose chromatography

$^{12/14}C$-RC

↓ a) $NaBH_4$ reduction b) extraction

^{14}C-3-deethyl-3-hydroxyethyl Bchl a | $^{12/14}C$-RC (^{14}C-depleted)

Exchanges of ≤50% of the total Bchl present in RC were observed. In all cases, the purity of the treated RC with respect to free or adsorbed pigment was checked by absorption (intensity ratio of NIR bands) and/or fluorescence spectroscopy (observation of emission bands below 850nm).

Assignment of the exchanged pigment to $Bchl_M$ and $Bchl_L$ was done by two procedures. 1. Exchange of P870 was excluded by the fact that the ESR signal of RC treated with ^2H-Bchl a was unchanged. 2. A positive correlation with the 'extra' Bchl a made use of the aforementioned $NaBH_4$ selectivity to $Bchl_M$ (Scheme 1). The finding of ^{14}C-3-deacetyl-3-hydroxyethyl Bchl a confirms that exchange occurs at least at $Bchl_M$, and the exchange yield then indicates that $Bchl_L$ is acessible as well.

These results indicate that the way first pursued by Loach et al. (1975) could be useful after all. Experiments to explore this reaction further by using structurally modified pigments are in progress.

ACKNOWLEDGEMENTS

This work was supported by the Deutsche Forschungsgemeinschaft (SFB 143 in München, AZ W041/37 in Stuttgart). We are indebted to Dr. Reng (Gesellschaft für iotechnologische Forschung, D-3301 Stöckheim) for mass culture of bacteria. We thank F.Lendzian (Berlin) for measurement of the ESR spectra.

REFERENCES

Allen,J.P, G.Feher, T.O.Yeates, D.C.Rees, J.Deisenhofer, H. Michel und R.Huber, 1986, Structural homology of reaction centers from Rhodopseudomonas sphaeroides and Rhodopseudomonas viridis as detected by x-ray diffraction, Proc.Natl. Acad.Sci.USA, 83:8589

Angerhofer,A., J.U. von Schütz and H.C.Wolf, 1985, Fluorescence-ODMR of light harvesting pigments of photosynthetic bacteria, Z. Naturforsch. 40c:379

Angerhofer, A., R.J. Cogdell and M.F. Hipkins, 1986, A spectral characterization of the light harvesting pigment-protein complexes from Rhodopseudomonas acidophila. Biochim. Biophys. Acta 848:333

Beese,D., Diplomarbeit, Universität München, 1984

Beese,D., R.Steiner, H.Scheer, A.Angerhofer, B.Robert, and M.Lutz, 1987, Chemically modified photosynthetic bacterial reaction centers: Circular dichroism, Raman resonance,low temperature absorption, fluorescence and ODMR spectra and polypeptide composition of borohydride treated reaction centers from Rhodobacter sphaeroides R26, Photochem. Photobiol., in press

Breton, J., J.Deprez, B.Tavitian, and E.Nabedryk, 1986, Spectroscopy, structure and dynamics in the reaction center of Rhodopseudomonas viridis, in: "Progress in Photosynthesis Research," J. Biggins, ed., M.Nijhoff, Dordrecht, Vol.I., p.387

Chadwick,B.W., C.Zang, R.J.Cogdell, and H.A.Frank, 1987, The effects of lithium dodecyl sulfate and sodium borohydride on the absorption spectrum of the B800-850 light harvesting complex from Rhodopseudomonas acidophila 7750, Biochim. Biophys. Acta, in press; see also contribution to this book

Chang, C.-H., D. Tiede, J.Tang, V. Smith, J. Norris, and M.Schiffer, 1986, Structure of Rhodopseudomonas sphaeroides R-26 reaction centers, FEBS Lett., 205:82

Clayton, R.K. and R.T.Wang, 1971, Photochemical Reaction Centers from Rhodopseudomonas sphaeroides., Methods Enzymol, 23:696

Clayton, R.K. and B.J. Clayton, 1981, B850 pigment-protein complex of Rhodopseudomonas sphaeroides: extinction coefficients, circular dichroism, and the reversible binding of bacteriochlorophyll, Proc. Natl. Acad. Sci. USA 78:5583

Crestfield, A.M., S. Moore and W.H. Stein, 1963, The preparation and enzymatic hydrolysis of reduced and S-Carboxymethylated proteins, J. Biol. Chem. 238:622

Deisenhofer, J., O. Epp, K. Miki, R. Huber and H. Michel, 1984, X-ray structure analysis of a membrane protein complex. Electron density map at 3 Å resolution and a model of the chromophores of the photosynthetic reaction center from Rhodopseudomonas viridis, J. Mol. Biol., 180:385

Ditson, S.L., R.C. Davis and R.M. Pearlstein, 1984, Relative enrichment of P-870 in photosynthetic reaction centers treated with sodium Borohydride, Biochim. Biophys. Acta, 766:623

Fischer, S.F. and P.O.J. Scherer, 1987, On the early charge separation and recombination processes in bacterial reaction centers. Chem. Phys., 115:151; see also contribution to this book

Gottstein,J. and H.Scheer; 1983, Long-wavelength absorbing forms of bacteriochlorophyll a in solutions of Triton X-100, Proc. Natl. Acad. Sci. USA, 80:2231

Holten, D., C.Kirmaier, and D.Levine, 1986, Picosecond studies of the kinetics and mechanisms of electron trasfer in bacterial reaction centers, in: "Progress in Photosynthesis Research," J.Biggins, ed., M.Nijhoff, Dordrecht, Vol.I, p.169

Hiyama,T., T.Watanabe, M.Kobayashi, and M.Nakazato, 1972, Interaction of chlorophyll a' with the 65kDa subunit protein of photosystem I reaction center, FEBS Lett., 214:97

Loach,P.A., M.Kung, and B.Hales, 1975, Characterization of the phototrap in photosynthetic bacteria, Ann. N.Y. Acad. Sci., 244:297

Maroti, P., C. Kirmaier, C. Wraight, D. Holten and R.M. Pearlstein, 1985, Photochemistry and electron transfer in borohydride-treated photosynthetic reaction centers, Biochim. Biophys. Acta, 810:132

Michel-Beyerle, M.E., M.Plato, J.Deisenhofer, H.Michel, M.Bixon, and J.Jortner, 1987, Unidirectionality of charge separation in reaction centers of photosynthetic bacteria. Biochim. Biophys. Acta, submitted for publication, see also contibution to this book

Ogrodnik, A., H.W. Krüger, H. Orthuber, R. Haberkorn, M.E. Michel-Beyerle and H. Scheer, 1982, Recombination dynamics in bacterial photosynthetic reaction centers. Biophys. J., 39:91

Parkes-Loach, P., J. Riccobono, and P.Loach, 1987, Preparation of subunits from the light-harvesting complex of Rhodospirillum rubrum, in: "Progress in Photosynthesis Research," J.White, ed., M.Nijhoff, Dordrecht, Vol.II, p.25

Parson, W.W., 1982, Photosynthetic bacterial reaction centers: Interactions among the bacteriochlorophylls and bacteriopheophytins, Ann. Rev. Biophys. Bioeng., 11:57

Plumley,F.G. and G.W.Schmidt, 1987, Reconstitution of chlorophyll a/b light harvesting complexes - Xanthophyll-dependent assembly and energy transfer, Proc. Natl. Acad. Sci. USA, 84:146

Robert, B., M. Lutz and D.M. Tiede, 1985, Selective photochemical reduction of either of the two bacteriopheophytins in reaction centers of Rhodopseudomonas sphaeroides R-26. FEBS Lett., 183:326

Robert, B., R. Steiner, Q. Zhou, H. Scheer and M. Lutz, 1986, Structures of antenna complexes and reaction centers from bacteriochlorophyll b-containing bacteria: Resonance raman studies, in: Progress in Photosynthesis Research, J. Biggins, ed., M.Nijhoff, Dordrecht, p.I.411

Scheer,H., B.Paulke, J.Gottstein, 1985, Long-wavelength absorbing forms of bacteriochlorophyll a: II. Structural requirements for formation in Triton X-100 micelles and in aqueous methanol and acetone, in: "Optical Properties and Structure of Tetrapyrroles", G.Blauer, H.Sund, eds., de Gruyter, Berlin-New York, 1985, S.587

Scherz, A. and W.W.Parson, 1984, Oligomers of bacteriochlorophyll and bacteriopheophytin with spectroscopic properties resembling those found in photosynthetic bacteria, Biochim.-Biophys.Acta., 766:653

Scherz, A. and W.W.Parson, 1986, Interactions of the bacteriochlorophylls in antenna bacteriochlorophyll-protein complexes of photosynthetic bacteria, Photosynthesis Res., 9:21; see also contribution to this book

Shuvalov, V.A., and L.N.M. Duysnes, 1986, Primary electron transfer reactions in modified reaction centers from Rhodopseudomonas sphaeroides. Proc. Natl. Acad. Sci. USA, 83:1690

Shuvalov, V.A., A.Ya. Shkuropatov, S.M. Kulakova, M.A. Ismailov and V.A. Shkuropatova, 1986, Photoreactions of bacteriopheophytins and bacteriochlorophylls in reaction centers of Rhodopseudomonas sphaeroides and Chloroflexus aurantiacus. Biochim. Biophys. Acta, 849:337

Steiner, R., B. Kalumenos and H. Scheer, 1986, The photosynthetic apparatus of Ectothiorhodospira halochloris 3. Effect of proteolytic digestion on the photoactivity, Z.Naturforsch., 41c:873

Theiler, R., F. Suter, H. Zuber and R.J. Cogdell, 1984, A comparison of the primary structures of the two B 800-850-apoproteins from wild-type Rhodopseudomonas sphaeroides strain 2.4.1. and a carotenoidless mutant strain R 26.1, FEBS Lett., 175:231

Vermeglio, A. and G. Paillotin, 1982, Structure of Rhodopseudomonas viridis reaction centers, absorption and photoselection at low-temperature, Biochim.Biophys. Acta, 681:32

Wasielewski, M.R., 1986, Ultrafast electron and energy trasfer in reaction center and antenna proteins from photosynthetic bacteria, presented at the VIIth Int. Congr. Photosynthesis, Providence, 1986

Wiemken,V. and R.Bachofen, 1984, Probing the smallest functional unit of the reaction center of Rhodospirillum rubrum G9 with proteinases, FEBS Lett., 166:155

Zinth,W., M.C.Nuss, M.A.Franz, W.Kaiser, and H.Michel, 1983, in: Antennas and reaction centers of photosynthetic bacteria, M.E.Michel-Beyerle, ed., Springer, Berlin, p.286

QUANTITATIVE ANALYSIS OF GENETICALLY ALTERED REACTION CENTERS

USING AN *IN VITRO* CYTOCHROME OXIDATION ASSAY

Edward J. Bylina, Raffael Jovine, and Douglas C. Youvan

Department of Applied Biological Sciences
Massachusetts Institute of Technology
Cambridge, MA 02139

INTRODUCTION

In the analysis of complex phenomena such as the biogenesis or photochemical function of reaction center proteins, the approach of a geneticist is to generate and characterize a large number of mutants affecting specific steps in the process. In some analyses, spontaneous mutants which possess selectable characteristics may be informative. For example, the role of specific amino acid residues in herbicide and quinone binding can be deduced from the analysis of spontaneous herbicide resistant mutants. However, the analysis of spontaneous mutations as a general approach to study structure-function relationships has serious limitations. Many mutations have phenotypes which possess unselectable characteristics. With the determination of the structure of the reaction center[1], this class of mutations can be generated by replacing important amino acid residues adjacent to chromophores throughout the protein by site-directed mutagenesis.

We have recently reported the construction of herbicide resistant reaction centers[2] and other directed mutations that affect the optical and electronic properties[3] of reaction centers from *R. capsulatus*. In one sense, all of these genetically modified reaction centers are new proteins that require a myriad of spectroscopic experiments for full biophysical characterization. As a prerequisite to more sophisticated studies, one of the first measurements which one would like to perform is a quantitative enzymatic assay that is generally applicable to all modified reaction centers.

Analyses of reaction centers are in many ways more advanced than other enzymes in that single steps may be investigated in exquisite detail through many different types of spectroscopy[4]. Time-resolved absorption spectroscopy is one of the most imformative. Transient absorption changes in different time domains, ranging from 4 picoseconds for bacteriopheophytin (Bphe) reduction to 100 microseconds for Q_b reduction, may be used to deduce the path of the photoelectron. Thus, spectroscopists have the means for analyzing the effects of mutations in exact detail. However, such spectroscopy is complex and time consuming; we must look to simpler techniques for purposes of screening large numbers of mutant proteins.

An *in vitro* assay has already been developed[5] which tests reaction center function from cytochrome oxidation to quinone reduction. In the presence of excess cytochrome, excess quinone, continuous saturating light, and catalytic levels of reaction centers, the assay yields (V_{max}) values of approximately 200 cytochromes oxidized per reaction center per second[6]. The low turnover of cytochromes in this assay suggests that there is a rate limiting step in reaction center photochemistry with a turnover time of 5 milliseconds (reciprocal of V_{max}). Hence, the putative rate-limiting step is 50 times slower than the Q_a to Q_b electron transfer time, the slowest characterized electron transfer step. One possibility is that the rate limiting step is quinone release from the Q_b pocket.

Theoretically, one would not expect a cytochrome oxidation assay carried out in the presence of saturating light to detect subtle differences among genetically modified reaction centers, since multiple photons may be absorbed by defective complexes while the assay is idling at the slowest step. For example, modified reaction centers with decreased quantum yields for Bphe reduction could absorb multiple photons (cycling through the 2 nanosecond back-reaction) before this partially blocked step would become competitive with the 5 millisecond kinetic barrier for complete turnover.

A simple but effective improvement can be made in the cytochrome oxidation assay by using subsaturating light intensity, such that the absorption of photons by any given reaction center is separated in time by greater than tne reciprocal of V_{max} (5 msec). We have developed and employed such an assay using subsaturating flashes of light to characterize a diverse set of reaction center mutations. While no single assay is perfect, we believe that our assay provides a rapid means of comparing large numbers of modified reaction centers affected at various steps in their photochemistry.

METHODOLOGY

Preparation of Reaction Centers. Reaction centers from *R. capsulatus* were prepared as previously described[7], with the following (miniprep) modifications. This method uses a one-step column purification on DEAE to prepare reaction centers from a background (U43) lacking LHII[8]. For the work described herein, we have scaled down the procedure from 8 l to eight 1 l preparations starting from RCV Plus media[9] grown in 2 l Erlenmyer flasks (anaerobic respiration) shaking at 150 RPMs in the dark. This procedure ensures that photosynthetically impaired strains are not overgrown by photosynthetically competent (spontaneous) revertants. As prepared, reaction centers are spectroscopically devoid of all light harvesting chlorophyll and typically yield A_{800}/A_{280} ratios of 1.5 to 2 (except in some cases where the mutation has modified the environment of the voyeur bacteriochlorophylls).

Time-Resolved Optical Absorption Spectrophotometer. We have constructed a single beam absorption spectrophotometer which utilizes a McPherson EU-701-50 tungsten light source and two 0.35 meter McPherson monochromators arranged linearly on both side of an Oriel 78100 sample compartment. Light is detected by a Hamamatsu R928 (extended red) PMT which is connected to a MACADIOS (G.W. Instruments, Cambridge, MA.) analog-digital converter. A diagram of this instrument is presented elsewhere[10]. A 495 nm long pass filter is placed in the measuring beam to attenuate 2nd order light. Ten microsecond (width at half height) light pulses are delivered by a programmable xenon flash lamp constructed by the University of Pennsylvania Biomedical Instrumentation Group. The flash is filtered through a 680 nm long pass filter and a 1 OD neutral density filter (optional) placed between the lamp and the fiber optic. The MACADIOS and flashlamp are programmed to collect 26,000 points over 5 seconds, with 1 second of data capture prior to eight flashes spaced 64 milliseconds apart (low power setting).

Flash intensities are normalized by measuring the scattered light from the side of the cuvette with an MFOD72 photodiode which has a broad response at 825 nm. The signal from the photodiode is used to charge a 0.1 µF capacitor through a 330 kΩ resistor. Oscilloscope connects are made to both sides of the capacitor so that the 10 microsecond flashes are visible on 5 second triggered sweeps (33 msec time constant). Total flash intensity may be estimated by the steady-state voltage or, more precisely, by adding together all (eight) rapid deflections.

Cytochrome Oxidation Assay with Subsaturating Light. Cytochrome oxidation is detected at 550 nm with the entrance slit set at 200 microns and the exit slit at 2000 microns. Assays were performed in 1 ml of Buffer B [10 mM potassium phosphate buffer (pH 7.35) with 0.05% LDAO (Fluka)], 2 mM UQ_0 (2,3-dimethoxy-5-methyl-1,4-benzoquinone, Sigma #D-9150), 0.2 mg / ml of reduced cytochrome *c* (horse heart, Sigma Type IV #C-7752), and 0.1 ODV (OD_{802} x ml) of reaction centers. The UQ_0 was dissolved to 100 mM in Buffer B (heated to 55 C for a few minutes) and stored at -80 C. Ten ml of the stock cytochrome solution (10 mg/ml) was reduced by bubbling hydrogen gas (generated by one half of a # 70304 BBL GasPak $NaBH_4$ tablet in 10 ml of water) through the cytochrome solution in the presence of catalyst (five # 70303 BBL tablets). The reduction was stopped after no further increase in A_{550} was observed; aliquots were frozen at -80 C.

Cytochrome Oxidation Assay with Saturating Light. These measurements were performed as described above except that a 250 W projector tungsten halogen lamp (Sylvania model ENH) was mounted directly to the side port of the Oriel sample compartment. Light was filtered through a 570 nm long pass filter and one cm of water. Incident light at the sample was ca. 500 mW/cm^2 but sample heating was less than 2 degrees C per five second experiment. The lamp circuit was interfaced to a TTL signal from the MACADIOS via a relay which closed one second after data collection began. Initial velocities were calculated based on the first 1.7 seconds of the reaction.

RESULTS

Cytochrome Oxidation Assay with Saturating Light. Wild-type reaction centers from strain U43(pU2922) were analyzed for their ability to oxidize cytochrome in the presence of increasing concentrations (15 to 500 micromolar) of UQ_0. Assay conditions were very similar to those previously described for *R. sphaeroides* [5]. As in *R. sphaeroides*, we have confirmed that reducing the light intensity from 500 mW/cm^2 to 150 mW/cm^2 or the cytochrome concentration from 0.2 mg/ml to 0.1 mg/ml has negligible effects on the velocity of the reaction. We conclude that that the assay conditions are saturating with respect to light and cytochrome for wild-type, although this may not be the case for certain modified reaction centers.

The *R. capsulatus* turnover data were analyzed (not shown) using a double reciprocral plot of (initial velocity)$^{-1}$ vs. (UQ_0 concentration)$^{-1}$. K_m for UQ_0 is 90 micromolar (µM) and V_{max} is 270 cytochromes oxidized per reaction center per second, very similar to the published values[6] for *R. sphaeroides* (90 µM and 200 sec^{-1}). The values from the double reciprocal plot were used to generate the solid line in Fig. 1 which show good agreement with the experimental data.

We concluded from the saturating light experiments that the quinone and cytochrome *c* concentrations described for the *R. sphaeroides* assays are also saturating for *R. capsulatus*. For both species, the maximum turnover rate (270 cytochromes oxidized per reaction center per second) is again too low to be used for comparing reaction centers. A 4 millisecond turnover time for wild-type will undoubtly obscure many impaired steps in mutant reaction centers, since multiple photons may be absorbed during this time.

Light Intensity Response Curve for the Cytochrome Oxidation Assay. In the next series of experiments with wild-type *R. capsulatus* reaction centers, we investigated the effect of reducing the actinic light intensity. Figure 2 shows a light intensity reponse curve for the number of cytochromes oxidized per reaction center per flash (averaged over 8 flashes). A relative light intensity measurement of 10 (photodiode measurement in mV) is well within the linear portion of the response curve. This corresponds to an oxidation of 0.25 cytochromes per reaction center per flash. With a train of eight flashes (64 millisecond separation), each reaction center will absorb an average of two photons with an average time of 256 milliseconds between photons. Under these

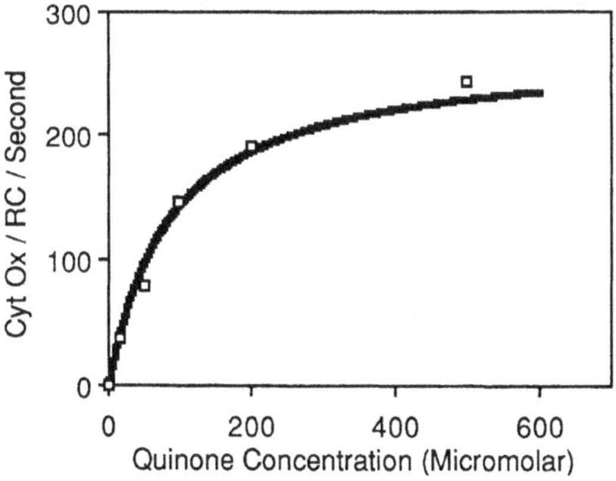

Figure 1. Dependence of cytochrome oxidation on concentration of added quinone.

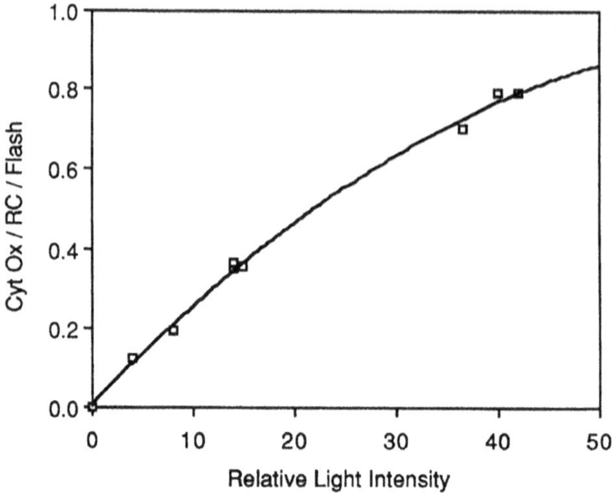

Figure 2. Dependence of cytochrome oxidation on light intensity.

conditions, we avoid the 4 to 5 millisecond slow step in the saturating light assay for two reasons: First, the time between absorption of photons is fifty times the slow step; second, most reaction centers absorb three photons or less, so the most probable cause of the slow step (quinol release) is avoided.

The effect of herbicide addition on flash induced cytochrome oxidation is shown in Fig. 3. In the presence of 0.2 mg/ml ametryne (2-ethylamino-4-isopropylamino-6-methylthio-s-triazine), the total cytochrome oxidation after eight flashes is reduced approximately 50%. The increase in cytochrome oxidation over the course of the eight flashes is no longer linear. The same result is found when UQ_0 is removed from the assay mixture. The plot shown in Fig. 3 is representative of a reaction center deficient in electron transfer from Q_a to Q_b, since purified reaction centers lack Q_b, and ametryne acts as an inhibitor of Q_b function.

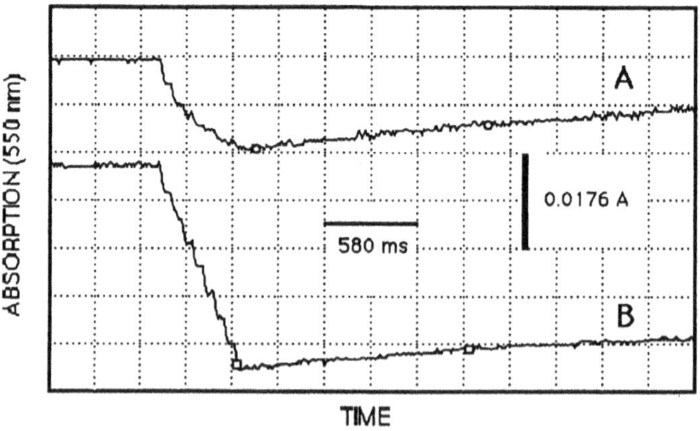

Figure 3. Cytochrome oxidation assay of 0.2 ODV wild-type reaction centers (A) with, and (B) without 0.2 mg/ml ametryne.

In vitro Assays on Mutant Reaction Centers: Comparison with Photosynthetic Growth. We have recently completed the construction and preliminary characterization of a diverse set of directed mutations in the reaction center[3]. One particular example from this group of reaction center mutations is EL104L (nomenclature: glutamic acid residue, L subunit, residue number 104, changed to leucine). Data from the crystal structure[1] and resonance Raman experiments[11] suggest that EL104 hydrogen bonds to the C-9 keto group on ring V of the L subunit bacteriopheophytin (Bphe) and could be important in directing electrons down the L, rather than M, branch of the symmetrically arranged prosthetic groups within the reaction center. Cytochrome oxidation assays using subsaturating light intensities show that this mutant reaction center is impaired approximately 30% relative to wild-type (Fig. 4; data presented are not normalized for variations in flash intensity).

A range of amino acid substitutions have been placed at structurally important residues in the vicinity of all of the prosthetic groups in the reaction center [special pair bacteriochlorophylls (Bchls), accessory Bchls, BPhes, quinones, and iron][3]. Table 1 summarizes the effect that a representative number of these mutations have on photosynthetic growth and activity (units defined below) in the cytochrome oxidation assay. Using these data, it is possible to compare the *in vitro* and *in vivo* properties of mutant reaction centers. In specific cases we observe that photosynthetically defective strains yield reaction centers with wild-type cytochrome oxidation values (such as YL162K).

DISCUSSION

Defining "Unit" Activity for Reaction Centers. In this communication, we define 100 units of reaction center activity as the absorption change ($\Delta A_{550} = 0.0179$) that is associated with the oxidation of excess (0.2 mg/ml) horse heart cytochrome *c* by 0.1 ODV (A_{802} x ml) of wild-type *R. capsulatus* reaction centers after eight 25% saturating flashes (delivered at 64 msec intervals; 10 microsecond duration) in the presence of saturating UQ_0 (2 mM). This assay is more sensitive to changes in the quantum yield of fast charge-separation steps than the saturating light intensity assay, since it avoids an undefined 5 msec kinetic barrier. As demonstrated in figure three, this assay can also identify reaction centers with normal Q_a activity that are blocked in Q_b function. The assay has been effective in providing an overall scale (values ranging from <2% to 102% relative to wild-type) for assessing a diverse set of mutations affecting all seven of the reaction center prosthetic groups.

Disparity Between Photosynthetic Growth Assays and *in vitro* Assays. We can attribute many of the differences observed in overall reaction center activity between *in vitro* and *in vivo* assays to two principle causes: 1) membrane pleiotropic effects, and 2) purification/stability artifacts.

Our *in vitro* assay requires that the purified reaction centers be capable of reducing UQ_0 and oxidizing cytochrome *c*. Since we are using subsaturating light intensities, the activity that is measured should follow the overall quantum yield for Q_a reduction. With the absorption of multiple

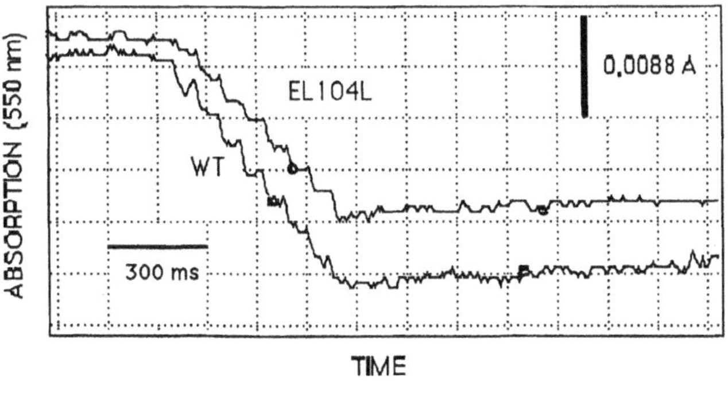

Figure 4. Cytochrome oxidation assay of reaction centers from wild-type and from EL104L.

Table 1. *In vivo* and *in vitro* activity of reaction center mutations.

MUTATION	COMMENT / LOCATION	PS GROWTH	UNIT ACTIVITY
Wild-type	BamHI Site at IL229	+	100
YL162V	Heme - Special Pair	+	102
YL162K	Heme - Special Pair	-	97
HL173Q	Special Pair Bchl	+	21
HL153T	Voyeur Bchl	+	31
EL104L	Bacteriopheophytin	+	70
WM250V	Q_a	-	<2

Mutant nomenclature is discussed in the text. The single letter amino acid code is used. See reference 3 for characterization of mutants listed, with the exception of mutations at L162 (see ref. 12). The photosynthetic growth assay is described elsewhere[3].

photons, electron conduction to Q_b and quinone turnover are expected to be important parameters. In light limiting conditions, *in vivo* photsynthetic growth assays are much more complicated, in that they are subject to additional parameters which may influence the overall yield of ATP regenerated per photon absorbed by the bacterium.

Although the mutations we have studied are localized to the reaction center, deleterious structural affects may propagate to associated complexes. For example, some mutations may cause global changes in the structure of the reaction center and have a dramatic affect on the structure and function of the closely associated LHI antenna complex. Reaction center mutations may also affect interactions with the cytochrome bc_1 complex- again lowering the overall efficiency for converting light to chemical energy. Such interactions can be invoked to explain the category of reaction center mutations that result in modified proteins which function better in the *in vitro* assay than in the *in vivo* assay. Conversely, some categories of mutated reaction centers may not survive detergent solubilization and chromatography steps that are required for protein isolation. Reaction centers that function poorly *in vitro* (such as HL173Q and HL153T), but which have little effect on photosynthetic growth assays may be indicative of this situation. In these cases, milder solubilization conditions (new detergents) and/or reconstitution procedures may be required for accurate *in vitro* measurements.

ACKNOWLEDGEMENTS

This work was supported by the National Science Foundation (DMB-8609614) and the United States Department of Agriculture (8700295).

REFERENCES

1. J. Deisenhofer, O. Epp, K. Miki, R. Huber, and H. Michel, Nature 318, 618-624 (1985).; H. Michel, O. Epp, and J. Deisenhofer, EMBO J. 5, 2445-2451 (1986).
2 E.J. Bylina and D.C. Youvan, Z. Naturforsch. 42c, 769-774 (1987).
3. E.J. Bylina and D.C Youvan, in preparation.
4. for reviews, see M.E. Michel-Beyerle, Ed., Antennas and Reaction Centers of Photosynthetic Bacteria- Structure, Interaction and Dynamics, Springer, Berlin, 1985.; W.W. Parson and B. Ke, inPhotosynthesis (Govindjee, Ed.), Vol. I, 331-385, Academic Press, New York, 1982.
5. M.Y. Okamura, R.J. Debus, D. Kleinfeld, and G. Feher, in Function of Quinones in Energy Conserving Systems (B.L. Trumpower, Ed.), 299-317, Academic Press, New York, 1982.
6. M.L. Paddock, J.C. Williams, S.H. Rongey, E.C. Abresch, G. Feher, and M.Y. Okamura, in Progress in Photosynthesis Research (J. Biggins, Ed.), Vol III, 775-778, Martinus Nijhoff, Dordrecht 1987.
7. R.C. Prince and D.C. Youvan, Biochim. Biophys. Acta 890, 286-291 (1987).
8. D.C Youvan, S. Ismail, and E.J. Bylina, Gene 38, 19-30 (1985).
9. H.-C. Yen and B. Marrs, Arch. Biochem. and Biophys. 181, 411-418 (1977).
10. D.C. Youvan, E.D. Lickerman, and M.M. Yang, Photonics Spectra 21,109-110 (1987).
11. D.F. Bocian, N.J. Boldt, B.W. Chadwick, and H.A. Frank, FEBS Lett. 214, 92-96 (1987).
12. R. Jovine and D.C. Youvan, in preparation.

PROPERTIES OF REACTION CENTERS FROM THE GREEN PHOTOSYNTHETIC BACTERIUM *Chloroflexus aurantiacus*

Robert E. Blankenship, Jeffrey T. Trost, and L. J. Mancino*

Department of Chemistry
Arizona State University
Tempe, AZ 85287-1604 USA

*Current Address:
Department of Plant Science
University of Arizona
Tucson, AZ 85721 USA

INTRODUCTION

The photosynthetic reaction center is an elegant machine that transduces light energy into redox energy. Purple photosynthetic bacteria have been instrumental in advancing our understanding of how this transduction is accomplished, because it has been possible to purify and characterize their reaction centers in terms of chemical composition, spectral, kinetic and thermodynamic properties[1,2]. The recent determination of the three dimensional structures of two purple bacterial reaction centers has brought this study to a stage where the function of the complexes can be analyzed in detail with respect to their structure[3-7].

To better understand the interplay of structure and function, it is useful to examine many examples of systems with the same general organization. If a certain aspect is not found to be conserved in different systems, such as a particular amino acid in a given position, then it is presumably not of critical importance to the function of the complex. Those structural features that are conserved over the widest range of evolutionary distance are likely to be in some manner critical to the function of the complex.

A "family tree" of photosynthetic bacteria is shown in Figure 1. The evolutionary relatedness of different groups of bacteria has been extensively analyzed by Woese and colleagues by determining catalogues and more recently sequences of 16S ribosomal RNA[8]. This molecule is found in all organisms, is relatively constant in function and is complex enough to provide information on organisms that span a very wide evolutionary range. When the photosynthetic bacteria are analyzed using this method, the purple bacteria are found to cluster together in a single large group or phylum, which also contains a number of nonphotosynthetic bacteria. The purple bacteria are subdivided into four divisions. Most of the widely studied organisms, including *Rhodobacter sphaeroides*, *Rhodobacter capsulatus* and *Rhodopseudomonas viridis* are in the alpha division, with the purple sulfur bacteria, including *Chromatium vinosum*, in the gamma division.

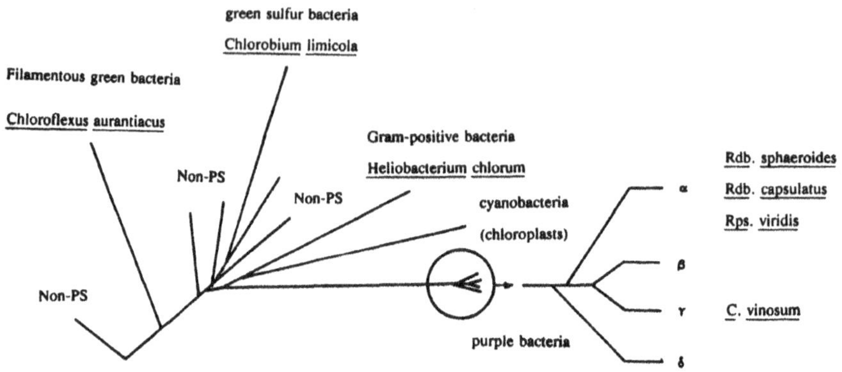

Figure 1. Phylogenetic tree of photosynthetic bacteria adapted from Ref. 8.

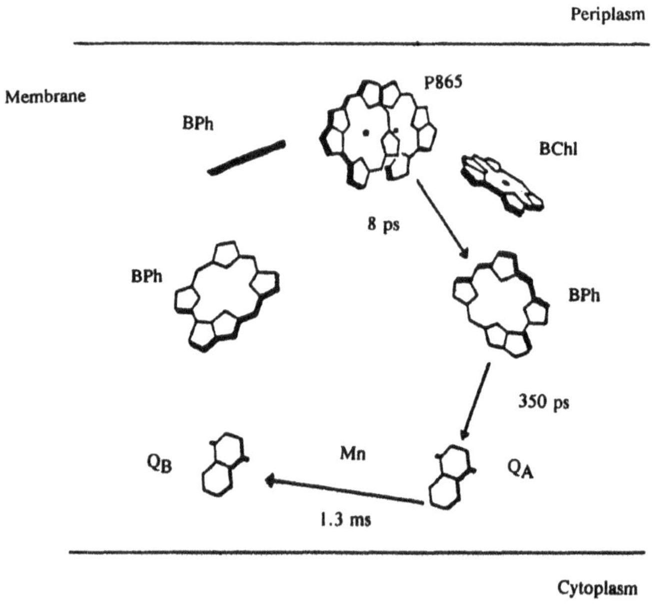

Figure 2. Electron transfer rates and likely cofactor arrangement in reaction centers from Chloroflexus aurantiacus.

To gain an understanding of the range of possible structures that function as photosynthetic reaction centers, we have examined the properties of the filamentous green bacteria, a group of bacteria that are only distantly related to the purple bacteria, yet contain a reaction center that is generally similar to that found in the purple bacteria in terms of chemical composition and reaction pathway. The filamentous green bacteria represent one of the earliest branchings of the true bacteria, and appear to occupy a critical place in the evolutionary development of photosynthesis[8]. The thermophilic organism Chloroflexus aurantiacus is the only member of this group that has been studied extensively.

Previous reports from ours[9-16] and other[17-23] laboratories have characterized the chemical composition, kinetic and spectral properties of the reaction centers from Chloroflexus. Briefly, it has been found that the reaction centers contain six tetrapyrrole pigments, similar to the situation in purple bacteria. However, three bacteriochlorophyll a (BChl) and three bacteriopheophytin a (BPh) are found instead of the pattern of four BChl and two BPh found in the purple bacteria. The simplest explanation for this difference is that the monomer BChl on the inactive or M branch of the reaction center is BPh in Chloroflexus. This assignment is consistent with dichroism measurements, and would appear to represent a relatively minor variation on the general structural theme of the purple bacterial reaction center, especially since a variety of evidence indicates that the M side monomeric BChl can be removed from purple reaction centers without any significant alterations in kinetic behavior. In addition to the tetrapyrrole pigments, the Chloroflexus reaction center contains menaquinones as both Q_A and Q_B, and a manganese atom instead of an iron. The proposed structural arrangement of pigments and cofactors in the Chloroflexus reaction center is shown in Figure 2.

The kinetic behavior of Chloroflexus reaction centers is remarkably similar to that of Rhodobacter sphaeroides, and is also shown in Figure 2. Interestingly, the rates of all the reactions that have been measured in Chloroflexus are somewhat slower than the corresponding reaction in sphaeroides. For example, the $P^* \rightarrow P^+BPh^-$ reaction time is about 8 ps in Chloroflexus[15] and about 3-4 ps in sphaeroides[2].

The protein composition of the Chloroflexus reaction center is significantly different from that found in all the purple bacteria that have been studied. The reaction center appears to contain only two peptides[18,16,23] instead of the three or four found in all purple bacteria. The two peptides are very difficult to separate using SDS-PAGE. This has slowed the progress in understanding the protein composition of the reaction center complex. Recently, Shiozawa et al.[23] have presented N-terminal amino acid sequence data for the two peptides, which they indicate to be identical for the first 18 amino acids. The sequence determined so far shows no apparent similarity with the L and M peptides of any of the purple bacteria.

MATERIALS AND METHODS

Reaction centers of Chloroflexus aurantiacus were isolated using a modification of the method of Pierson and Thornber[17]. Briefly, 200 g cells were sonicated in 20 mM Tris pH 8.0, 1 mM phenylmethylsulfonylfluoride (PMSF) and membranes isolated by centrifugation at 200,000 xg. Membranes (A865=16) were treated with 1.5% lauryldimethylamineoxide (LDAO), 100 mM NaCl, 20 mM Tris, pH 8.0 for 1 hour at 37 C, and centrifuged for 90 min at 200,000 xg. The supernatant liquid, containing the crude reaction centers, was dialyzed overnight against 50 mM Tris, pH 9.0, 0.1% LDAO and applied to a 8x30 cm DEAE Sephacel column. The column was washed extensively with 50 mM Tris, pH 9.0, 0.1% LDAO, until the green color due to the solubilized antenna pigment BChl c was removed, and the reaction centers eluted by increasing the NaCl concentration to 50 mM. Additional purification was achieved by repeating the DEAE ion exchange treatment on a smaller column. Final purification was achieved using a 0.9x25 cm SOTA Chromatography DEAE analog HPLC column, with the reaction centers eluted with an NaCl gradient of 0-500 mM, in 50 mM Tris, pH 9.0, 0.1% LDAO. The reaction centers were then dialyzed into 20 mM Tris, pH 8.0, 0.1% LDAO and concentrated with an Amicon ultrafilter using a YM-30 membrane. This procedure yielded reaction centers with a 280nm/812nm absorbance ratio of 1.4-1.5.

The reaction center peptides were separated using a reverse-phase HPLC method using an ISCO 2300 gradient HPLC system. A 100μl sample containing up to 1 mg total protein was applied to a 0.45x25 cm Whatman Partisil 10-ODS column that had been equilibrated with 60% formic acid, 40% water. The peptides were eluted using a linear gradient running from 0 to 30% isopropanol in 60% formic acid at a flow rate of 2 ml/min over 30 minutes. The peptides were detected by their absorbance at 280 nm using an ISCO V4 absorbance monitor.

Amino acid analysis was performed by the Biotechnology Instrumentation Facility at U. Calif. Riverside.

RESULTS AND DISCUSSION

The UV/VIS absorption spectrum of reaction centers from *Chloroflexus* are shown in Figure 3, along with a sample of reaction centers from *Rhodobacter sphaeroides* R-26. The spectra have been normalized at 865 nm (the measured extinction coefficients of the two complexes are essentially the same at this wavelength[13,24]). The spectra are very similar in the 865 nm region, but differ substantially in the B800 and BPh regions, due to the differing content of BChl and BPh. Despite the fact that *Chloroflexus aurantiacus* contains large quantities of carotenoids, the reaction center as purified contains no carotenoid, as judged by the absence of the characteristic carotenoid peaks in the 400-500 nm region. It is not entirely clear if the reaction center contains carotenoid *in vivo* and it is lost during the isolation procedure or if it lacks it at all times.

If a rapid electron donor is added to a sample of *Chloroflexus* reaction centers, along with a quinone such as ubiquinone or menaquinone, absorbance oscillations at 450 nm are observed (Figure 4). These oscillations are characteristic of the action of the two electron gate that functions in purple bacterial reaction centers and Photosystem II of oxygen evolving organisms[25]. If o-phenanthroline, an inhibitor of the $Q_A \rightarrow Q_B$ reaction is added, the oscillations are abolished. This experiment establishes that reaction centers from *Chloroflexus*, isolated as described above, can carry out electron transfer all the way to Q_B.

The protein subunit composition of the *Chloroflexus* reaction center is shown by the HPLC separation shown in Figure 5. Two peptides were resolved, both of which are extremely hydrophobic. The first peptide eluted, referred to as P1, runs on SDS PAGE with an apparent molecular weight of 26 kDa, while the second peptide eluted runs with a slightly higher apparent molecular weight (data not shown). In similar experiments using reaction centers isolated from *Rhodobacter sphaeroides*, three peptides were observed (data not shown). These peptides were found to be the H, M and L peptides (in order of increasing retention time, corresponding to increasing hydrophobicity) by running them on SDS-PAGE. This order is as expected based on the amino acid composition of the three peptides[1].

The amino acid compositions of the two peptides are shown in Table I, along with the composition of the active complex (minor discrepancies between the average of the composition of the two individual peptides and the complete reaction center are probably due to the presence of minor amounts of other proteins in the complete reaction center sample). The amino acid compositions of the two peptides are similar, although there are some differences. P1 contains 33% and P2 contains 32% polar or charged residues. Table II compares the amino acid composition of the *Chloroflexus* peptides

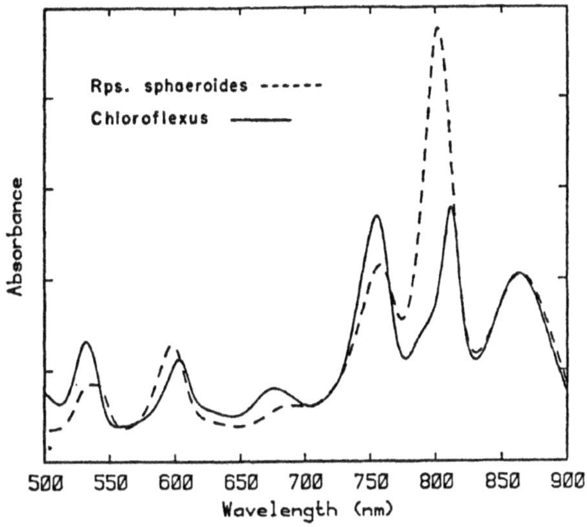

Figure 3. Absorption spectra of isolated reaction centers from Chloroflexus aurantiacus (solid line) and Rhodobacter sphaeroides (dashed line).

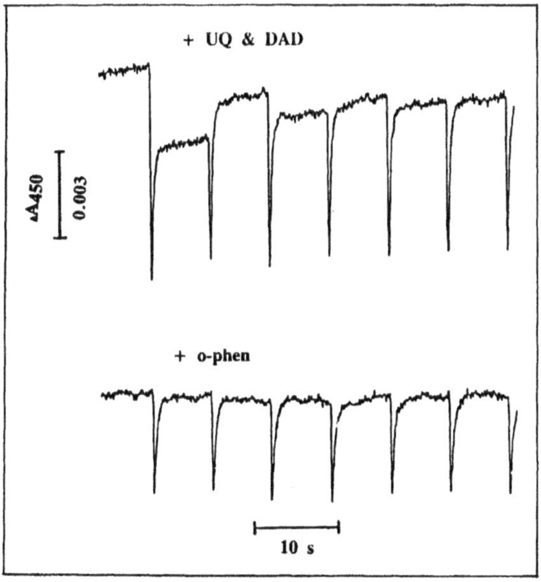

Figure 4. Quinone oscillations in isolated reaction centers from Chloroflexus aurantiacus. UQ, ubiquinone; DAD, diaminodurene; o-phen, o-phenanthroline.

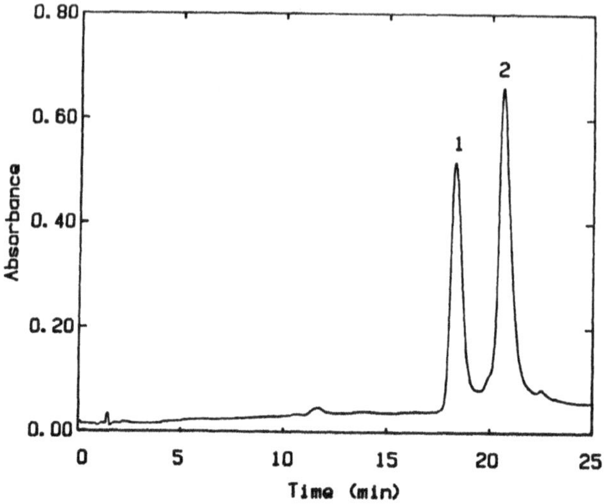

Figure 5. Reverse-phase HPLC separation of two peptides from the isolated reaction center of *Chloroflexus aurantiacus*. Peak 1 is due to protein P1 and peak 2 due to protein P2.

Table I Amino Acid Composition of Reaction Centers From *Chloroflexus*

Amino acid	Complete RC	P1 peptide	P2 peptide
Asx	7.4	7.1	7.4
Thr	4.6	3.8	5.2
Ser	5.7	6.0	4.8
Glx	6.8	6.4	6.3
Gly	10.1	12.4	10.4
Ala	9.3	7.5	9.7
Val	4.7	6.0	4.1
Met	2.3	2.3	3.0
Ile	7.2	8.3	7.4
Leu	10.2	7.8	11.5
Tyr	4.9	4.9	5.6
Phe	9.9	11.3	9.3
Lys	3.0	2.3	3.0
His	2.3	2.3	2.2
Arg	5.3	4.9	3.3
Pro	6.3	6.8	6.7
Cys	NA	NA	NA
Trp	NA	NA	NA

Values are in mole% NA= not analyzed

Table II Compositional Divergence Index

	Cfx	P1	P2	D1D2	L	M	H	LM	LH	MH	LMH
Cfx	--										
P1	xxx	--									
P2	xxx	0.062	--								
D1D2	0.050	xxx	xxx	--							
L	xxx	0.075	0.045	xxx	--						
M	xxx	0.085	0.060	xxx	0.047	--					
H	xxx	0.111	0.092	xxx	0.098	0.086	--				
LM	0.055	xxx	xxx	0.052	xxx	xxx	xxx	--			
LH	0.056	xxx	xxx	0.064	xxx	xxx	xxx	0.041	--		
MH	0.062	xxx	xxx	0.058	xxx	xxx	xxx	0.045	0.025	--	
LMH	0.051	xxx	xxx	0.051	xxx	xxx	xxx	xxx	xxx	xxx	--

$$d = \left(\sum_{i=1}^{N} (X_{ai} - X_{bi})^2 \right)^{1/2}$$

X_{ai} = mole fraction of ith amino acid in protein a

d	Relatedness of proteins
0.01-0.03	Identical proteins (error due to analysis)
0.03-0.05	Closely related proteins
0.05-0.06	Distantly related proteins
0.06-0.10	Unrelated similar proteins (e.g. different membrane proteins)
>0.10	Dissimilar proteins

D1D2 composition taken from Ref. 26.
Rhodobacter sphaeroides composition data for L, M and H taken from Ref. 1.

with the three peptides from *Rhodobacter sphaeroides* and the PS II reaction center proteins D1 and D2[26,27]. The proteins are compared using the compositional divergence, a simple quantitative measure of how similar are the amino acid compositions of two proteins[28,1]. This measure, while far cruder than either a sequence or three dimensional structure, gives an overall indication of the similarity of two proteins, and does correctly show the relationship of L and M, as well as the similarity of LM and D1D2 (Table II).

The results indicate that the P1 and P2 peptides are both more closely related to the L peptide of *Rhodobacter sphaeroides* than they are to the M or H peptides. The correspondence of the P2 peptide to L is the closest. Neither P1 nor P2 are very closely related to the H peptide.

The results of the protein analysis are consistent with the earlier suggestions[18,16,23] that *Chloroflexus* contains only two peptides in the active reaction center complex. In purple bacteria it is possible to dissociate H from LM, with the LM complex retaining photochemical activity[1]. However, the presence of the H peptide is required for $Q_A \rightarrow Q_B$ activity[29]. *Chloroflexus* is apparently able to carry out the $Q_A \rightarrow Q_B$ reaction without an analog to the H peptide.

A possible explanation for this difference lies in the very different membrane organization in the two types of organisms. In the purple bacteria the H peptide protrudes into the cytoplasmic region of the cell. In *Chloroflexus* the very large chlorosome antenna complex almost certainly occupies the same relative position[16].

In conclusion, all available evidence is consistent with the general picture that the overall pattern of electron flow in *Chloroflexus* and the purple bacterial reaction centers is the same, despite an enormous evolutionary distance between the two classes of organisms. The observed substantial differences in pigment composition and protein complement between *Chloroflexus* and any of the purple bacteria, coupled with this remarkable similarity of function, provides an opportunity to make a much more exacting comparison of structure to function than is available using the structures of purple bacteria that are presently available.

ACKNOWLEDGEMENTS

Supported by grants from the Competitive Research Grants Program of the U.S. Department of Agriculture (84-CRCR-1-1523) and the Herman Frasch Foundation.

REFERENCES

1. Okamura, M. Y., Feher, G. and Nelson, N. (1982) in: "Energy Conversion by Plants and Bacteria", Vol. I, Govindjee, ed., Academic Press, pp. 195-272.
2. Kirmaier, C. and Holten, D. (1987) Photosyn. Res. 13, 225-260.
3. Deisenhofer, J., Epp, O., Miki, K., Huber, R. and Michel, H. (1985) Nature 318, 618-624.
4. Michel, H., Epp, O. and Deisenhofer, J., (1986) EMBO J. 5, 2445-2451.
5. Chang, C.-H., Tiede, D., Tang, J., Smith, U., Norris, J. and Schiffer, M. (1986) FEBS Lett. 205, 82-86.
6. Allen, J. P., Feher, G., Yeates, T. O., Komiya, H. and Rees, D. C. (1987) Proc. Nat'l Acad. Sci USA 84, 5730-5734.
7. Yeates, T. O., Komiga, N., Rees, D. C., Allen, J. P. and Feher, G. (1987) Proc. Nat'l Acad. Sci USA 84, 6438-6442.
8. Woese, C. R. (1987) Microbiol. Rev. 51, 221-271.
9. Bruce, B. D., Fuller, R. C. and Blankenship, R. E. (1982) Proc. Nat'l Acad. Sci. USA 79, 6532-6536.
10. Blankenship, R. E., Feick, R., Bruce, B. D., Kirmaier, C., Holten, D. and Fuller, R. C. (1983) J. Cell. Biochem. 22, 251-266.

11. Kirmaier, C., Holten, D., Mancino, L. J. and Blankenship, R. E. (1984) Biochim. Biophys. Acta 765, 138-146.
12. Kirmaier, C., Blankenship, R. E. and Holten, D. (1986) Biochim. Biophys. Acta 850, 275-285.
13. Blankenship, R. E., Mancino, L. J., Feick, R., Fuller, R. C., Machnicki, J., Frank, H. A., Kirmaier, C. and Holten, D. (1984) in "Advances in Photosynthetic Research", C. Sybesma, ed., Vol. I, pps. 203-206.
14. Mancino, L. J., Hansen, P. L., Stark, R. E. and Blankenship, R. E. (1985) Biophys. J. 47, 2a.
15. Becker, M., Middendorf, D., Woodbury, N. W., Parson, W. W. and Blankenship, R. E. (1986) in "Ultrafast Phenomena", G. R. Fleming and A. E. Siegman, eds., Springer Verlag, Berlin, pps. 374-378.
16. Blankenship, R. E. and Fuller, R. C. (1986) in "Photosynthesis III: Encyclopedia of Plant Physiology, New Series, Vol. 19, Springer-Verlag, Berlin, pps. 390-399.
17. Pierson, B. K. and Thornber, J. P. (1983) Proc. Nat'l. Acad. Sci. USA 80, 80-84.
18. Pierson, B. K., Thornber, J. P. and Seftor, R. E. B. (1983) Biochim. Biophys. Acta 723, 322-326.
19. Vasmel, H. and Amesz, J. (1983) Biochim. Biophys. Acta 714, 118-122.
20. Vasmel, H., Meiburg, R. F., Kramer, H. J. M., DeVos, L. J. and Amesz, J. (1983) Biochim. Biophys. Acta 724, 333-339.
21. Parot, P., Delmas, N., Garcia, D. and Vermeglio, A. (1985) Biochim. Biophys. Acta 809, 137-140.
22. Shuvalov, V. A., Shkuropatov, A. Ya., Kulakova, S. M., Ismailov, M. A. and Shkuropatova, V. A. (1986) Biochim. Biophys. Acta 849, 337-346.
23. Shiozawa, J., Lottspeich, F. and Feick, R. (1987) Eur. J. Biochem. 167, 595-600.
24. Straley, S. C., Parson, W. W., Mauzerall, D. C. and Clayton, R. K. (1973) Biochim. Biophys. Acta 305, 597-609.
25. Crofts, A. R. and Wraight, C. A. (1983) Biochim. Biophys. Acta 726, 149-185.
26. Trebst, A. and Depka, B. (1985) in "Antennas and Reaction Centers of Photosynthetic Bacteria--Structure, Interactions and Dynamics" M. E. Michel-Beyerle, ed. Springer-Verlag, Berlin, pps. 216-224.
27. Nanba, O. and Satoh, K. (1987) Proc. Nat. Acad. Sci. USA 84, 109-112.
28. Harris, C. E. and Teller, D. C. (1973) J. Theor. Biol., 38, 347-362.
29. Debus. R. J., Feher, G. and Okamura, M. Y. (1985) Biochemistry 24, 2488-2500.

STRUCTURAL AND FUNCTIONAL PROPERTIES OF THE REACTION CENTER OF GREEN BACTERIA AND HELIOBACTERIA

J. Amesz

Department of Biophysics, Huygens Laboratory
University of Leiden, The Netherlands

INTRODUCTION

The most thoroughly studied reaction centers are undoubtedly those of purple bacteria, and in particular those of the non-sulfur purple bacteria Rhodopseudomonas viridis and Rhodobacter sphaeroides, which form the main subject of these proceedings. However, bacterial photosynthesis is not confined to purple bacteria alone: the green bacteria and the recently discovered heliobacteria (Gest and Favinger, 1983; Beer-Romero and Gest, 1987) form equally interesting groups of photosynthetic organisms.

There are various lines of evidence to suggest that there exist basically two types of photosynthetic reaction centers. The first group comprises the reaction center of purple bacteria and of the green non-sulfur (green filamentous) bacterium Chloroflexus aurantiacus, and recent experiments by Satoh and coworkers (Nanba and Satoh, 1987; Danielius et al., 1987) strongly support the notion that the reaction center of photosystem II also belongs to this group. The reaction center of green sulfur bacteria appears to belong to a second category, which may include the photosystem I reaction center and that of heliobacteria. In this communication we shall discuss some recent results concerning the functional and structural properties of the reaction center of green bacteria and heliobacteria. First we shall briefly discuss the reaction center of Chloroflexus. The second part of this contribution will concern similarities and possible differences between the reaction centers of green sulfur bacteria and heliobacteria.

GREEN NON-SULFUR BACTERIA

There is little doubt that the reaction center of Chloroflexus aurantiacus in most aspects strongly resembles that of purple bacteria. However, the pigment composition is not identical; the reaction center does not contain carotenoid, and instead of four bacteriochlorophylls (BChls) and two bacteriopheophytins (BPhs) it contains three BChls and three BPhs (Blankenship et al., 1984). A polypeptide equivalent to the H-subunit of purple non-sulfur bacteria appears to be missing. Flash spectroscopic measurements indicate a very similar mechanism of primary and secondary electron transport in the reaction center, with two menaquinone molecules serving as secondary electron acceptors (Kirmaier et al., 1983, 1984; Vasmel and Amesz, 1983).

The reaction center of <u>Chloroflexus</u> has not been crystallized as yet, and consequently X-ray data on its structure are not available. However, the optical properties of the reaction center have been extensively studied (Vasmel et al., 1983a), and these have been used to obtain information on its structure. As a matter of fact both the absorption spectrum of the isolated reaction center (see Fig. 1) and the difference spectrum of the oxidation of the primary electron donor P-865, first measured by Pierson and Castenholz (1974), already suggested a structural similarity to the reaction center of purple bacteria. Vasmel et al. (1986) have recently analyzed the low-temperature absorption, linear dichroism, circular dichroism and fluorescence polarization spectra of the reaction center of <u>C. aurantiacus</u> by means of exciton theory, employing the so-called point-dipole approximation. In this calculation the coordinates were used that have been obtained by Deisenhofer et al. (1984) by X-ray analysis for the chromophores in the reaction center of Rps. viridis. It was assumed that the 'extra' BPh a re-

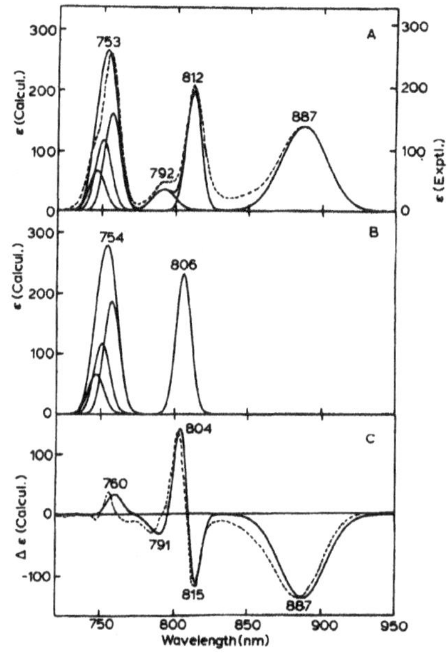

Fig. 1. Calculated (——) and experimental (-----) absorption spectra (77 K) of reaction centers of <u>C. aurantiacus</u>. The third Bph <u>a</u> is assumed to be located in the 'M-chain'. Extinction coefficients in $mM^{-1}cm^{-1}$. 'Gaussian dressing' was applied to the calculated stick spectra. (A) Reaction centers with reduced P-865; (B) with oxidized P-865. (C) Oxidized-minus-reduced difference spectra; the experimental spectrum was obtained with a different batch of reaction centers.

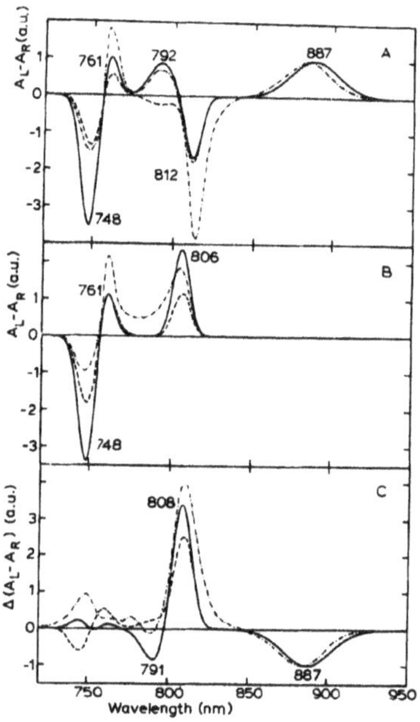

Fig. 2. Circular dichroism spectra. Experimental spectra at 77 K (-.-.-), and calculated spectra for two arrangements of the third Bph <u>a</u>: in the 'M-chain' (——) and in the 'L-chain' (-----). The spec were normalized at 887 nm; the calculated rotational strength at this wavelength was 30 % too low in both cases. Further details as for Fig. 1.

placed one of the 'accessory' BChls, either in the so-called M- or in the L-chain. Considerable red shifts had to be assumed for the monomeric (unperturbed) transitions, especially of those molecules that constitute P-865 (see also Knapp et al., 1985).

As Fig. 1A shows, a quite satisfactory simulation of the absorption spectrum was obtained in this way. The spectrum shown applies to a configuration where the accessory BChl a in the M-chain was replaced by BPh a, but practically the same result was obtained upon replacement of the BChl a in the L-chain. For the spectrum of Fig. 1B it was assumed that the absorption of P-865 completely vanished upon oxidation. The calculated wavelength of the remaining BChl a transition was at 810.5 nm; to obtain a fit with the experimental spectrum an additional 4.5 nm blue shift by the positive charge on P-865$^+$ had to be assumed. A reasonable simulation was also obtained of the circular dichroism spectra (Fig. 2); in this case a somewhat better fit was obtained if the third Bph a was located in the M-chain. We thus conclude that the arrangement of the chromaphores in the reaction center of C. aurantiacus is probably very similar to that in purple bacteria, with the replacement of one accessory BChl a, presumably that in the M-chain, by BPh a.

An interesting result of the calculation is that it explains the origin of the weak absorption band at 792 nm as being essentially the high energy transition of the P-865 dimer. It was earlier thought that this explanation was unlikely because of the rather small angle between the transitions at 792 and 887 nm and because both bands have the same sign in the circular dichroism spectrum (Vasmel et al., 1983a). However, our calculations showed that although the contributions by the other pigments (primarily the accessory BChl a molecule) to the transition are small, they are sufficient to cause a considerable deviation from 90°. In first approximation, the absorption spectrum of the reaction center of C. aurantiacus may thus be described in a simple way, with two BChl a dimer transitions at 792 and 887 nm and a 'monomeric' transition at 812 nm. The two BPhs in the same chain showed a clear exciton splitting, but interaction with the BPh in the other chain is very small. A similar calculation for Rb. sphaeroides showed that in purple bacteria the presence of a second accessory BChl complicates matters considerably. With the exception of the low-energy P-870 transition all BChl a bands now have a strongly mixed character and the 'high energy' transition carries almost no dipole strength, in agreement with experimental observations. In this respect it is of interest to note that the low-temperature absorption spectrum of so-called 'modified' reaction centers of Rb. sphaeroides shows a band near 790 nm of comparable amplitude as that of C. aurantiacus (Scherer and Fischer, 1987; Ganago et al., 1988). In these reaction centers one of the accessory BChls has been converted to BPh and partially removed by treatment with borohydride.

GREEN SULFUR BACTERIA AND HELIOBACTERIA

There is various evidence now that the mechanism of electron transport in the reaction centers of green sulfur bacteria is quite different from that in purple bacteria and in Chloroflexus. The first step that can be observed by flash spectroscopy is the transfer of an electron from the primary electron donor, P-840, to a pigment absorbing at 670 nm (Nuijs et al., 1985a; Shuvalov et al., 1986). Fig. 3 shows the difference spectrum of the reduction of the primary acceptor. Analysis of the pigment composition of membranes of the green sulfur bacterium Prosthecochloris aestuarii has shown that this pigment can only be BChl c (or a closely related pigment): the membranes were found to contain fairly large amounts of this BChl c-like pigment, but only traces of Bph c (or Bph a). The pigment is more lipophilic than the BChl c of the chlorosomes, mainly because it is esterified with phytol instead of farnesol (Braumann et al., 1986).

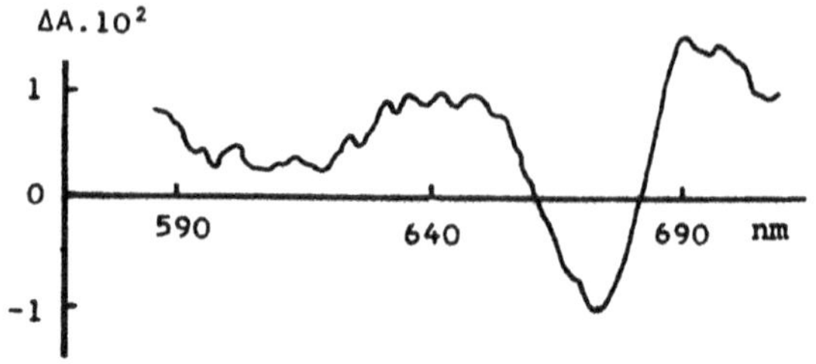

Fig. 3. Absorption difference spectrum of membranes of P. aestuarii, obtained upon excitation with 33 ps-flashes at 850 ns, and measured at 250 ps after the flash. The spectrum was corrected for contributions by P-840 oxidation by subtracting the spectrum measured at 3 ns.

The charge separation takes place within 10 ps. The bleaching at 670 nm reverses again in about 500 - 600 ps (Nuijs et al., 1985a;, Shuvalov et al., 1986), presumably by reduction of a second electron acceptor. This acceptor has not been identified as yet by flash spectroscopy, but there is indirect evidence, based on photoaccumulation experiments, that it is an iron-sulfur center (Swarthoff et al., 1981a). Recent evidence suggests that a quinone may also be involved as early electron acceptor (Nitschke et al., 1987).

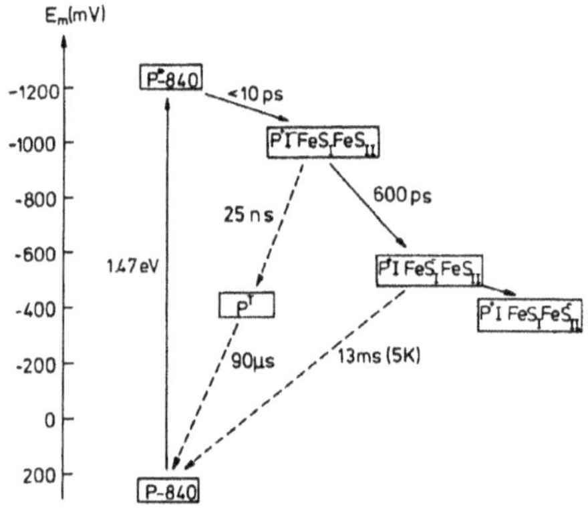

Fig. 4. Energy scheme of electron transfer in the reaction center of green sulfur bacteria. The excited primary donor P-840 is indicated by P^*. The main electron pathways are indicated by solid arrows; back reactions by broken arrows. P^T, triplet state of P-840; I, primary electron acceptor (BChl c); FeS, iron-sulfur center. Unless otherwise indicated, the time constants apply to room temperature.

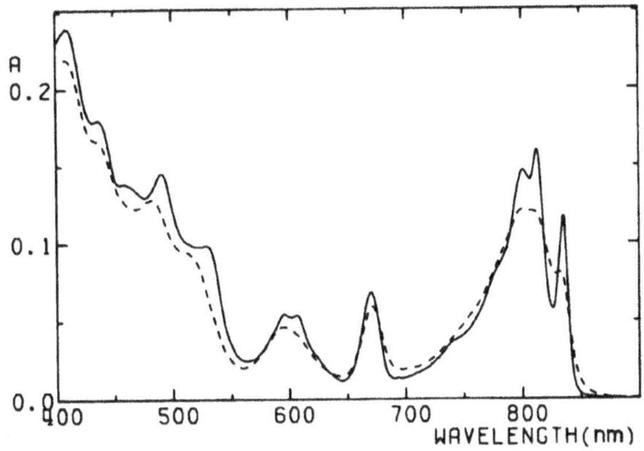

Fig. 5. Absorption spectra of the core complex of *P. aestuarii* at room temperature (-----) and at 77 K (———).

Studies of the formation of the triplet of P-840 indicate that the midpoint potential of the acceptor is well below -500 mV (Smit and Amesz, 1988). Fig. 4 gives a scheme of electron transport for green sulfur bacteria based on these and other observations.

The recently discovered bacterium *Heliobacterium chlorum* (Gest and Favinger, 1983) contains BChl *g* as its major pigment (Brockmann and Lipinski, 1983). The primary electron donor, P-798 is probably BChl *g* too (Fuller et al., 1985; Prince et al., 1985). The oxidation of P-798 is accompanied by a bleaching at 670 nm, with a difference spectrum very similar to that of *P. aestuarii*, and as in *P. aestuarii*, the bleaching reverses in about 500 ps (Nuijs et al., 1985b). So these data suggest that with respect to their primary charge separation in heliobacteria and green sulfur bacteria resemble each other. It is not clear, however, to what extent this similarity applies also to secondary electron transport. Photoaccumulation of a reduced iron-sulfur center has been observed in membranes of *H. chlorum*

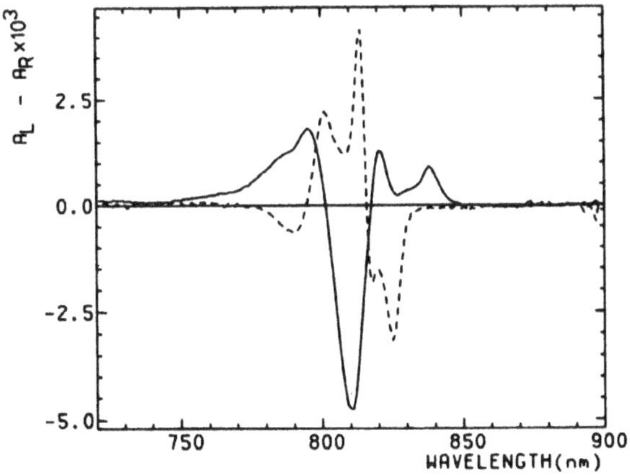

Fig. 6. Circular dichroism spectrum of the core complex, measured at 77 K (———). For comparison the spectrum of the 'soluble' BChl *a* protein is also given (-----).

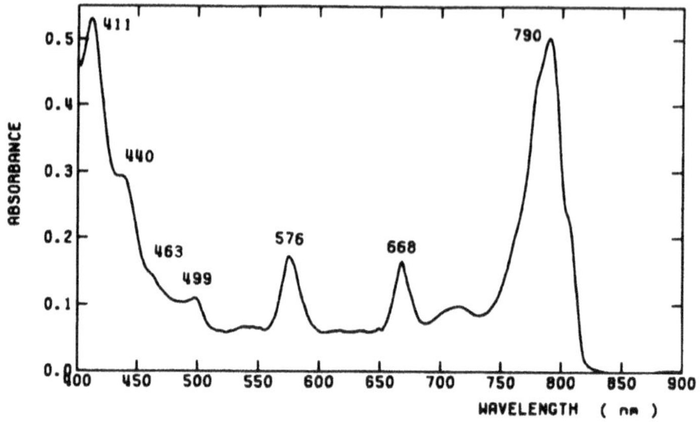

Fig. 7. Absorption spectrum of membranes of H. chlorum, measured at 4 K.

(Brok et al., 1986; Smit et al., 1987) and, as in green sulfur bacteria, there is some evidence for the photoreduction of a quinone (Brok et al., 1986). However, experiments of Smit et al. (1987) indicate that formation of the triplet of P-798 occurs already at a relatively high redox potential and is not dependent on the redox state of the iron sulfur center.

The question now arises if the observed similarity in primary electron transport reflects a similar structure for the reaction center of green sulfur bacteria and heliobacteria. A comparison of the structural properties of the two types of reaction centers at present can only be based on optical measurements, and is complicated by the different spectroscopic and chemical properties of the pigments involved. Nevertheless, the available data do not seem to indicate a structural similarity.

There are no indications that the reaction center of green sulfur bacteria exists as a separate entity, as in purple bacteria or Chloroflexus. The smallest complex obtained so far by detergent solubilization of the membrane is the so-called core complex (Vasmel et al., 1983b; Hurt and Hauska, 1984), which contains about 20 BChl a and 15 BChl c molecules. The core complex accounts for only about 20 - 25 % of the BChl a associated with the membrane; most of the BChl a is contained in the well-known 'soluble' BChl a protein (Olson, 1980). The absorption and circular dichroism spectra of the core complex are shown in Figs. 5 and 6. An analysis of these spectra showed the presence of five Q_y transitions in the region 780 - 840 nm (Vasmel et al., 1983b), which suggested that the core complex may contain subunits of five interacting BChl a molecules. BChl c transitions are seen near 670 nm.

The membrane of H. chlorum contains approximately 30 BChls g per reaction center (Nuijs et al., 1985b). Pigment-protein complexes have not been isolated yet from the membrane, but the relatively small size of the photosynthetic unit suggests that the optical properties of the membrane might be compared with those of the core complex of P. aestuarii. Fig. 7 shows the absorption spectrum. At 4 K three Q_y transitions of BChl g can discerned at 808, 793 and 778 nm. As in P. aestuarii transitions can also be seen near 670 nm. However, in bacteria grown under highly anaerobic conditions the amount of the 670 nm pigment is much smaller than in P. aestuarii (Gest, 1988). The circular dichroism spectrum of H. chlorum (Fig. 8) is relatively simple and shows only one, approximately conservative band centered at 793 nm, indicating that significant interaction occurs only between BChls g_{793} (van Dorssen et al., 1985).

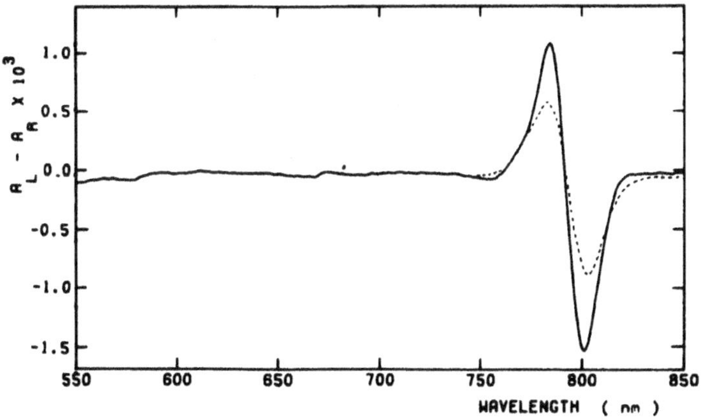

Fig. 8. Circular dichroism spectra of membranes of H. chlorum, measured at 77 K (———) and at 293 K (-----).

Some additional information may be obtained by comparison of the difference spectra of the oxidation of the primary electron donors. In the near-infrared region both spectra show band shifts of BChl in addition to the bleaching of the primary donor, but the spectrum of P-840$^+$ formation is clearly more complicated (Swarthoff et al., 1981b). However, since the pigments involved in the band shifts are different, it may be more useful to compare the region around 670 nm. Here, the P-840 spectrum shows red shifts of at least two different spectral transitions of the BChl c-like pigment (Swarthoff et al., 1981b). These involve mainly BChl c molecules other than the primary electron acceptor (Shuvalov et al., 1986); at room temperature the absorbance changes are approximately of the same magnitude as those caused by the reduction of the primary electron acceptor; at 80 K they are even larger (Swarthoff et al., 1981b). A red shift near 670 nm is also seen in the spectrum of P-798, but here the absorbance changes are much smaller (Nuijs et al., 1985b). The low-temperature difference spectrum gives no evidence for the involvement of more than one transition (Fig. 9) and in this case only the acceptor may be involved.

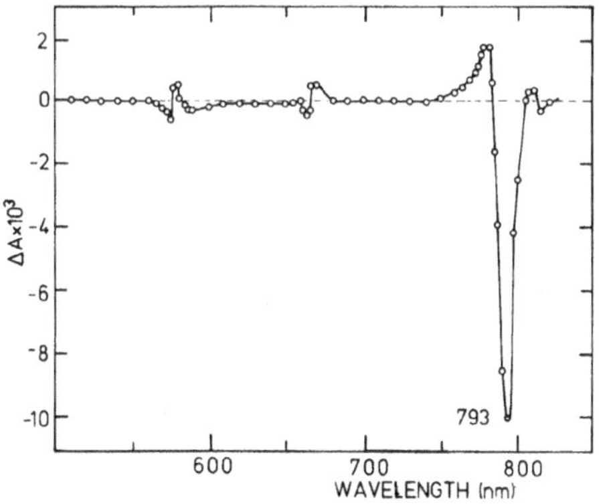

Fig. 9. Absorption difference spectrum of P-798 oxidation at 4 K, obtained by illumination with 10 μs flashes (R.J. van Dorssen, unpublished).

Pending further evidence we conclude that, although the primary charge separation in heliobacteria resembles that in green sulfur bacteria, both the reaction center and the antenna core appear to show significant structural differences in the two groups of bacteria.

ACKNOWLEDGEMENTS

Research supported by the Netherlands Foundation for de Advancement of Pure Research (ZWO) through the Netherlands Foundations for Chemical Research (SON) and for Biophysics.

REFERENCES

Beer-Romero, P., and Gest, H., 1987, Heliobacillus mobilis, a peritrichously flagellated anoxyphototroph containing bacteriochlorophyll g, FEMS Microbiol. Lett., 41: 109.

Blankenship, R.E., Mancino, L.J., Feick, R., Fuller, R.C., Machnicki, J., Frank, H.A., Kirmaier, C., and Holten, D., 1984, Primary photochemistry and pigment composition of reaction centers isolated from the green photosynthetic bacterium Chloroflexus aurantiacus, in: "Advances in Photosynthesis Research", C. Sybesma, ed., Vol. I, p. 203, M. Nijhoff-Dr. W. Junk, The Hague.

Braumann, T., Vasmel, H., Grimme, L.H., and Amesz, J., 1986, Pigment composition of the photosynthetic membrane and reaction center of the green bacterium Prosthecochloris aestuarii, Biochim. Biophys. Acta, 848: 83.

Brockmann, H., and Lipinski, A., 1983, A new bacteriochlorophyll from Heliobacterium chlorum, Arch. Microbiol., 136: 17.

Brok, M., Vasmel, H., Horikx, J.T.G., and Hoff, A.J., 1986, Electron transport components of Heliobacterium chlorum investigated by EPR spectroscopy at 9 and 35 GHz, FEBS Lett., 194: 322.

Danielius, R.V., Satoh, K., van Kan, P.J.M., Plijter, J.J., Nuijs, A.M., and van Gorkom, H.J., 1987, The primary reaction of photosystem II in the D_1-D_2-cyt b_{559} complex, FEBS Lett., 213: 241.

Deisenhofer, J., Epp, O., Miki, K., Huber, R., and Michel, H., 1984, Structure analysis of a membrane protein complex. Electron density map at 3 Å resolution and a model of the chromophores of the photosynthetic reaction center from Rhodopseudomonas viridis, J. Mol. Biol., 180: 385.

Fuller, R.C., Sprague, S.G., Gest, H., and Blankenship, R.E., 1985, Unique photosynthetic reaction center from Heliobacterium chlorum, FEBS. Lett., 182: 345.

Ganago, A.O., Gubanov, V.S., Klevanik, A.V., Melkozernov, A.N., Shkuropatov, A.Ya., and Shuvalov, V.A., 1988, Comparative study of the spectral and kinetic properties of electron transfer in purple and green photosynthetic bacteria, in: "Proc. EMBO Workshop on Green Photosynthetic Bacteria", J.M. Olson, ed., Plenum Press, New York, in the press.

Gest, H., 1988, Physiological and biochemical characteristics of heliobacteria, in : "Proc. EMBO Workshop on Green Photosynthetic Bacteria", J.M. Olson, ed., Plenum Press, New York, in the press

Gest, H., and Favinger, J.L., 1983, Heliobacterium chlorum, an anoxygenic brownish-green photosynthetic bacterium containing a 'new' form of bacteriochlorophyll, Arch. Microbiol., 136: 11.

Hurt, E.C., and Hauska, G., 1984, Purification of membrane-bound cytochromes and a photoactive P840 protein complex of the green sulfur bacterium Chlorobium limicola f. thiosulfatophilum, FEBS Lett., 168: 149.

Kirmaier, C., Holten,. D. Feick, R., and Blankenship, R.E., 1983, Picosecond measurements of the primary photochemical events in reaction centers isolated from the facultative green photosynthetic bacterium Chloroflexus aurantiacus. Comparison with the purple bacterium Rhodopseudomonas sphaeroides, FEBS Lett., 158: 73.
Kirmaier, C., Holten, D., Mancino, L.J., and Blankenship,. R.E., 1984, Picosecond photodichroism studies on reaction centers from the green photosynthetic bacterium Chloroflexus aurantiacus, Biochim. Biophys. Acta, 765: 138.
Knapp, E.W., Fischer, S.F., Zinth, W., Sanda, M., Kaiser, W., Deisenhofer, J., and Michel, H., 1985, Analysis of optical spectra from single crystals of Rhodopseudomonas viridis reaction centers, Proc. Natl. Acad. Sci. USA, 82: 8463.
Nanba, D., and Satoh, K., 1987, Isolation of a photosystem II reaction center consisting of D-1 and D-2 polypeptides and cytochrome b-559, Proc. Natl. Acad. Sci. USA, 84: 109.
Nitschke, W., Feiler, U., Lockau, W., and Hauska, G., 1987, The photosystem of the green sulfur bacterium Chlorobium limicola contains two early electron acceptors similar to photosystem I, FEBS Lett., 218: 283.
Nuijs, A.M., Vasmel, H., Joppe, H.L.P., Duysens, L.N.M., and Amesz, J., 1985a, Excited states and primary charge separation in the pigment system of the green photosynthetic bacterium Prosthecochloris aestuarii as studied by picosecond absorbance difference spectroscopy, Biochim. Biophys. Acta, 807: 24.
Nuijs, A.M., van Dorssen, R.J., Duysens, L.N.M., and Amesz, J., 1985b, Excited states and primary photochemical reactions in the photosynthetic bacterium Heliobacterium chlorum, Proc. Natl. Acad. USA, 82: 6865.
Olson, J.M., 1980, Chlorophyll organization in green photosynthetic bacteria, Biochim. Biophys. Acta, 594: 33.
Pierson, B.K., and Castenholz, R.W., 1974, Studies of pigments and growth in Chloroflexus aurantiacus, a phototrophic filamentous bacterium, Arch. Mikrobiol., 100: 283.
Prince, R.C., Gest, H., and Blankenship, R.E., 1985, Thermodynamic properties of the photochemical reaction center of Heliobacterium chlorum, Biochim. Biophys. Acta, 810: 377.
Scherer, P.O.J., and Fischer, S.F., 1987, Application of exciton theory to optical spectra of sodium borohydride treated reaction centres from Rhodobacter sphaeroides R26, Chem. Phys. Lett., 137: 32.
Shuvalov, V.A., Amesz, J., and Duysens, L.N.M., 1986, Picosecond spectroscopy of isolated membranes of the photosynthetic green sulfur bacterium Prosthecochloris aestuarii upon selective excitation of the primary electron donor, Biochim. Biophys. Acta, 851: 1.
Smit, H.W.J., and Amesz, J. 1988, Electron transfer in the reaction center of green sulfur bacteria and Heliobacterium chlorum, in: "Proc. EMBO Workshop on Green Photosynthetic Bacteria", J.M. Olson, ed., Plenum Press, New York, in the press.
Smit, H.W.J., Amesz, J., and van der Hoeven, M.F.R., 1987, Electron transport and triplet formation in membranes of the photosynthetic bacterium Heliobacterium chlorum, Biochim. Biophys. Acta, in the press.
Swarthoff, T., Gast, P., Hoff, A.J., and Amesz, J., 1981a, An optical and ESR investigation on the acceptor side of the reaction center of the green photosynthetic bacterium Prosthecochloris aestuarii, FEBS Lett., 130: 93.
Swarthoff, T., Amesz, J., Kramer, H.J.M., and Amesz, J., 1981b, The reaction center and antenna pigments of green photosynthetic bacteria, Israel J. Chem., 21: 332.
van Dorssen, R.J., Vasmel, J., and Amesz, J., 1985, Antenna organization and energy transfer in membranes of Heliobacterium chlorum, Biochim. Biophys. Acta, 809: 199.

Vasmel, H., and Amesz, J., 1983, Photoreduction of menaquinone in the reaction center of the green photosynthetic bacterium *Chloroflexus aurantiacus*, *Biochim. Biophys. Acta*, 724: 118.

Vasmel, H., Meiburg, R.F., Kramer, H.J.M., de Vos, L.J., and Amesz, J., 1983a, Optical properties of the photosynthetic reaction center of *Chloroflexus aurantiacus* at low temperature, *Biochim. Biophys. Acta*, 724: 333.

Vasmel, H., Swarthoff, T., Kramer, H.J.M., and Amesz, J., 1983b, Isolation and properties of a pigment-protein complex associated with the reaction center of the green photosynthetic sulfur bacterium *Prosthecochloris aestuari*, *Biochim. Biophys. Acta*, 725: 361.

Vasmel, H., Amesz, J., and Hoff, A.J., 1986, Analysis by exciton theory of the optical properties of the reaction center of *Chloroflexus aurantiacus*, *Biochim. Biophys. Acta*, 852: 159.

Molecular Dynamics Simulation of the Primary Processes in the Photosynthetic Reaction Center of *Rhodopseudomonas viridis*

H.Treutlein *, K.Schulten *, J.Deisenhofer +, H.Michel +,
A.Brünger †, and M.Karplus †

* Physik-Department, Technische Universität München
D-8046 Garching, FRG
+ Max-Planck-Institut für Biochemie, D-8033 Martinsried, FRG
† Department of Chemistry, Harvard University, Cambridge
MA 02138, USA

Abstract

We have carried out a computer simulation of the photosynthetic reaction center of *Rhodopseudomonas viridis* based on the available molecular structure[1,2]. Our simulation employed the **CHARMM** program[3] in conjunction with the socalled **stochastic boundary method**[4]. This method allowed us to study a functionally important segment of the photosynthetic reaction center with 3634 atoms, including the prosthetic groups involved in the primary electron transfer processes. Electron transfer has been modeled by re-charging the respective chromophores assuming charge distributions based on quantumchemical (MNDO) calculations. We discuss to which extent the protein matrix and chromophore arrangement control the relevant electron transfer steps.

1. Introduction

The main open question regarding the photosynthetic reaction center concerns the high efficency with which an electron and a hole are separated by light absorbtion in the system and their recombination is being prevented. Three factors can contribute to this efficiency.

A first factor is the fast rate of the forward transfer of the electron after light excitation of the special pair. In order to understand this rate one needs to know which material properties of the reaction center are actually making the electron transfer possible.

A second factor contributing to the efficiency can be the electrostatic potential inside the reaction center which may favour the separation of electron and hole. In a second article[5] we consider in how far the electrostatic energy in the reaction center contributes in this respect.

A third factor contributing to the efficiency of the reaction center function lies in the fact that the time scales of back-transfer of the electron to the hole are much longer than the time scales for the forward transfers. In the following we will investigate control of electron transfer rates through mechanical motion of the chromophores and their surrounding protein matrix which can either establish or hinder electrical contact. Since the time scale of the process under consideration, i.e. photoinduced electron transfer, is close to the time scale accessible to computer simulations of protein dynamics, the reaction center is an excellent candidate for a molecular dynamics study. In order to obtain an answer to the issues just raised we ask in our computer simulations the following key questions:

— What is the contribution of inherent atomic mobilities to electron transfer?

— How does the protein structure rearrange during electron transfer?

— What are the effects of 'site-directed mutagenesis' on the reaction center?

The last question also provides an important possibility for a comparison between experiments and theory.

2. METHOD

The simulations we describe in this article were based on the X-ray structure of the reaction center of *Rhodopseudomonas viridis* at 3 Å resolution[1,2]. The calculations involved the **CHARMM** program as described by B. R. Brooks et al.[3]. All charge distributions of the neutral and ionized chromophores have been calculated by means of the all valence electron **MOPAC** quantum chemistry program[6].

The whole reaction center with more than 12000 atoms is actually too large to be handled within reasonable computer time by the simulation program. Therefore, we made use of the feature of **CHARMM** which by means of the **method of stochastic boundaries** allows to select a central part of the reaction center for the simulation. The selected part contained all chromophores except three heme units of the cytochrome. Since the selection has an important influence on the results, inducing actually some artifacts, we will briefly explain this method in the following.

The protein is divided into three disjunct sets of atoms. The first set, the *reaction region*, contains that region of the protein which one wants to describe in detail by the molecular dynamics method. For atoms in this region the deterministic Newtonian equations are integrated. Surrounding the reaction region is the *buffer region*. This region involves a shell of atoms which are described by Langevin molecular dynamics. For this purpose noise and friction is added to the Newtonian equations, these terms keeping the respective atoms at the proper temperature in order to prevent a cooling down of the atoms in the reaction region. In addition, a harmonic restoring force is added to the Newtonian equations which keeps the atoms in the buffer region on the average at their initial positions. The third region, the *reservoir region*, contains all other atoms. These atoms are completely neglected in the molecular dynamics description. Our choice of the three regions of the reaction center led us to include 3634 out of about 12000 atoms of the reaction center.

Figure 1 displays the chromophore structure inside the simulated protein segment (reaction and buffer region) together with their abreviated names. Bonds between atoms inside the buffer region are marked by thick solid lines. Phytol chains of the non-functional branch chromophores (BCMP, which is part of the *special pair* dimer, BCMA and BPM) are partially located inside the buffer region, which implies that some of their atoms are constrained to their initial positions. This induces dynamical artifacts, that must be considered when trying to compare the dynamical properties between the functional (BCLP, the other cromophore of the *special pair*, BCLA, BPL, QA and the iron-ion FE1) and the non-functional chromophores. (Of the functional chromophores only a few atoms of the quinone QA lie inside the buffer region.

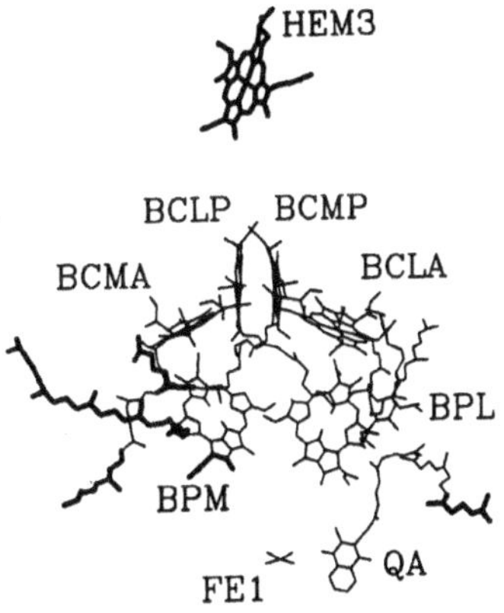

Fig. 1. Structure of chromophores in the simulated protein-segment together with their abbreviated names. Bonds between atoms inside the buffer region are denoted by thick solid lines.

The following calculations were carried out:

— We started the simulation, after minimizing the atomic coordinates provided by Deisenhofer and Michel according to **CHARMM**'s energy function, with randomly chosen velocities conforming to physiological temperatures. The minimization and first simulation run allowed the reaction center segment to equilibrate for 20 ps to a state with relaxed sterical interactions.

— The second run, in the following referred to as run A, simulated for 20 ps the dynamics of the reaction center before the primary electron transfer. For this purpose the charge densities of the special pair bacteriochlorophylls and of the bacteriopheophytin in the functional branch was that of the neutral chromophores as determined by an **MNDO**[6] calculation. The motion resulting from run A had been analyzed.

— We then transferred an electron charge from the *special pair* to the bacteriopheophytin (BPL) by removing one electron from the highest occupied orbital of the *special pair* bacteriochlorophyll distant from the functional branch and depositing this electron into the lowest unoccupied orbital of the functional bacteriopheopytin. The corresponding charge densities of the chromophores were communicated to the **CHARMM** program.

— We then carried out a third simulation, in the following referred to as run B, which monitored the dynamics of the reaction center, thus perturbed, for 20 ps. The resulting motion was analyzed and compared with the behaviour before the electron transfer.

— Finally, we simulated an iron-depleted reaction center for 20 ps. We removed the iron-ion FE1 from our segment and investigated the resulting structural modifications.

3. Results

Comparison of Average Structure from X-ray Data and Resulting from Molecular Dynamics Simulation

In order to test the molecular dynamics simulation in the light of available observations, we analyzed first structural changes occuring during the simulation. For this purpose we determined the average structure of run A and compared it to the X-ray data. The result is shown in Fig. 2. The mean differences in atomic positions between the molecular dynamics structure and the X-ray data is 1.05 Å. Most of the structure has remained remarkably stable during the dynamics before electron transfer (run A). However, the functional pheophytin group (BPL) and the phytol chains of the functional chromophores show larger structural changes indicating a higher degree of flexibility of these moieties than the other chromophores. Another structural deviation observed, namely that of the accessory bacteriochlorophyll of the non-functional branch (BCMA), might be explained by the fact that in our simulation we left out a carotenoid, which has been observed recently to lie close to BCMA[7].

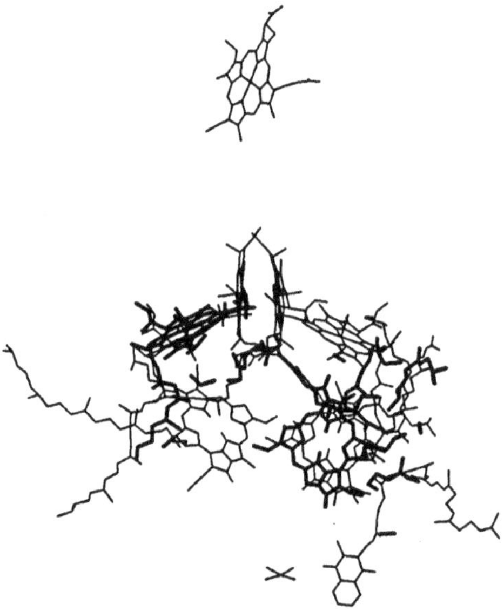

Fig. 2. Comparison of dynamics and X-ray data. The average chromophore structure of run A is represented by thin lines. The X-ray structure is drawn by thick lines for those atoms the positions of which differ by more than 1.2 Å between the two structures.

Thermal Motion

Thermal motion of the protein atoms induces fluctuations of the atomic positions $\vec{r}_j$ around their equilibrium values $< \vec{r}_j >$. The simpliest measure of these fluctuations is given by the root mean square deviations $\sigma_j = \sqrt{< (\vec{r}_j - < \vec{r}_j >)^2 >}$. σ_j, which can

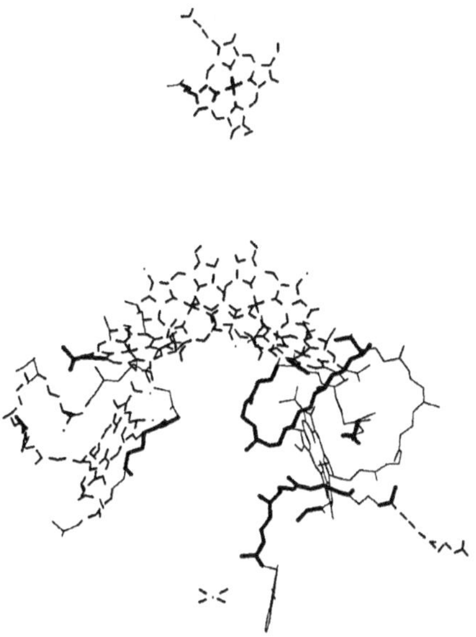

Fig. 3. Flexibility of chromophore atoms during run A. Dashed bonds are drawn between atoms whose root mean square deviations of the mean atomic position σ is less than 0.4 Å, very flexible atoms (σ > 0.7 Å) are connected by thick lines.

differ for different locations in the protein, provides a measure of the local flexibility of a protein. We have evaluated σ_j from run A separately for all atoms in the simulated reaction center segment.

The root mean square deviations of the chromophores during run A are illustrated in Fig. 3 : Dashed bonds are drawn between atoms that exhibit only a small degree of flexibility during run A, thick bonds denote atoms exhibiting a high degree of flexibility. The special pair ring structure turns out to be the most rigid part of the chromophore branches ($\sigma < 0.3$ Å) . This rigidity might be important for the function of the *special pair* . The pheophytin ring and the phytol chains are found to be most flexible. The pheophytine BPL appears to be located in a small "pocket" inside the protein, which accounts for its mobility. The high flexibility might contribute to the different rates of forward- and backward electron transfer by allowing BPL to be shifted after primary electron transfer into a position unfavourable for the back-transfer.

As a second test for our calculations we compared the σ -values of all atoms with the root mean square values calculated from the temperature-factors (Debye-Waller-factors) of the X-ray analysis. The molecular dynamics σ -values are obtained as an average over a 20 ps time periode for a single protein. The X-ray temperature-factors involve an average over 10^{20} molecules during several hours of observation time. The two sets of values, therefore, are strictly not equivalent. In fact, we would expect that the molecular dynamics σ -values are slightly smaller than the ones resulting from X-ray analysis. However, a comparison of the two sets of values may indicate how realistic the atomic fluctuations simulated by **CHARMM** actually are. For this purpose we compare in Figs. 4a,b the two sets of σ -values. Figure 4a shows the

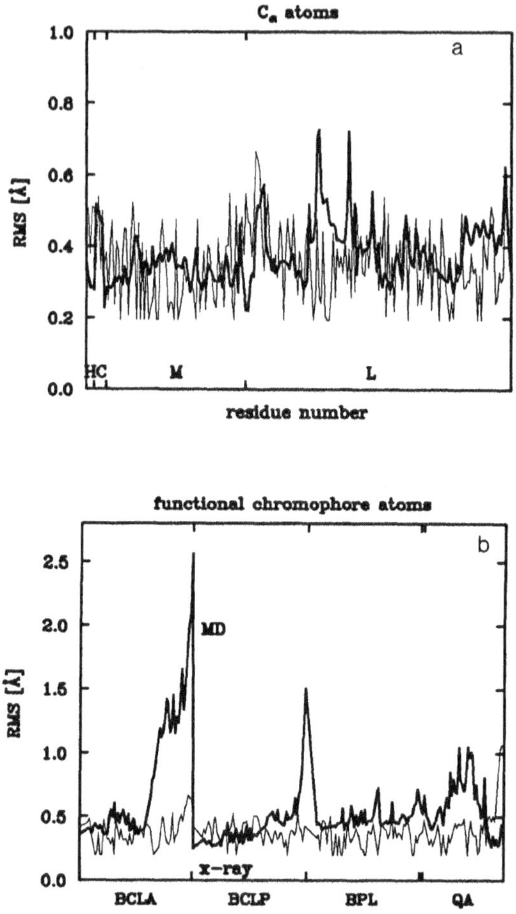

Fig. 4. Comparison of the root mean square deviations of the atomic positions as described by molecular dynamics simulation (run A) and as resulting from X-ray temperature-factors. Figure 4a compares the deviations for all C_α atoms in the simulated protein segment, Fig. 4b compares the deviations for the atoms of some of the chromophores. Thick lines represent simulation data, thin lines data from X-ray analysis.

values for the C_α -atoms, Fig. 4b the values for chromophore atoms. The simulated fluctuations of the C_α -atoms as well as of the atoms belonging to the chromophore rings on the average compare rather well with the results from the X-ray analysis. Large differences can be observed, however, for atoms belonging to the phytol chains of BCLA and BCLP as well as to the chain of QA. The reason for these deviations might be due to faults in the simulation but it might also be due to false interpretation of the X-ray data[8].

Correlated Fluctuations

Another interesting measure of dynamical properties of the reaction center is furnished by the covariance C_{ik} between fluctuations of pairs of atoms i and k. This quantity is defined as

$$C_{ik} = \frac{c_{ik}}{\sqrt{c_{ii}c_{kk}}}$$

$$c_{ik} = < (\vec{r}_i - < \vec{r}_i >) \cdot (\vec{r}_k - < \vec{r}_k >) >$$

Values of the covariance near +1 or −1 indicate that atomic motions are tightly coupled in phase (+1) or out of phase (−1), values around zero indicate a loose coupling. C_{ik} can differentiate between domains of the protein for which thermal fluctuations do not alter very much interatomic (inter-chromophore) distances ($C_{ik} \approx 1$) from domains where thermal fluctuations lead to strong variations of these distances ($C_{ik} \approx 0$).

Figure 5 provides results on the covariances between the *special pair* atoms (Fig. 5a) and between the atoms belonging to BCLP and BPL (Fig. 5b). Figure 5a illustrates that there exists a high degree of covariance for the motions of the two bacteriochlorophyll rings. This implies that the rings form a sandwich complex and move in phase. In contrast, the motions of the two special pair phytol chains are rather uncorrelated with the motion of the rings and are also uncorrelated with each other. Figure 5b shows on the other hand that the motion of the phytol chain of one of the *special pair* chlorophylls (BCLP) is strongly coupled in phase to the pheophytine (BPL) ring. This dynamic coupling, essentially an intermolecular attraction, could provide the interaction needed for the photo-induced primary electron transfer between the special pair and BPL in case the transfer involves the BCLP phytol chain. This attraction would prevent any thermal motion from impeding the fast (3 ps) primary electron transfer.

We have also investigated the covariance between motions of the remaining chromophores. The chromophores consecutive along the electron transfer route BCMP, (BCLA ?), BPL, QA are coupled pairwise in phase, i.e. BCMP to BCLA, BCLA to BPL .... This implies that thermal fluctuations do not affect very much the relative distances between these chromophores. This could indicate that the chromophore arrangement is optimized for the electron (forward) transfer and that structural disturbances due to the inherent thermal mobility are kept at a minimum. This feature obviously can have important implications for the mechanism of primary ellectron transfer: edge to edge couplings between the chromophores, which without this feature may be unreliable due to thermal motions, might have been tuned to rather precise values in the reaction center to provide a most effective forward electron transfer, i.e. along the lines suggested by Plato, Fischer and others in this workshop.

Response to Electron Transfer

We want to investigate now in how far structural and dynamical properties of the reaction center change when an electron charge is suddenly moved from the *special pair* to the functional bacteriopheophytine. For this purpose we have compared the motion of the reaction center monitored during run A before the transfer with the motion monitored during run B after the electron transfer. The simulations showed that the electron transfer disturbs the protein structure.

In Fig. 6 we present the average structure of run B. The mean structural difference to the average structure of run A is found to be 0.32 Å. For atoms whose position has changed by more than 0.4 Å the avrrage structure of run A is also displayed in Fig. 6. The largest structural differences occur for the pheophytine BPL and the (phytol) chains. After the electron is transferred, BPL is shifted towards the *special*

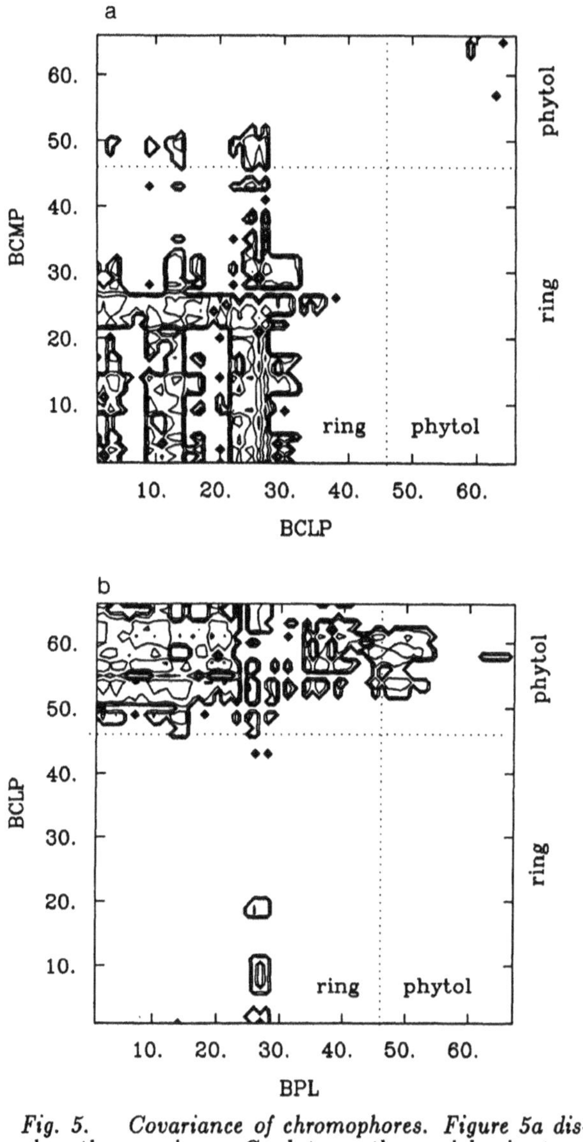

Fig. 5. Covariance of chromophores. Figure 5a displays the covariances C_{ik} between the special pair atoms and Fig. 5b the covariances between the atoms of BCLP and BPL. The axes present the atomic labels i and k. All correlations C_{ik} are positive. The blank area denotes pairs of atoms i, k with correlations smaller than 0.3. *Figure 5 shows further the contour lines which separate regions with larger correlations; the contour lines correspond to increments of* 0.1.

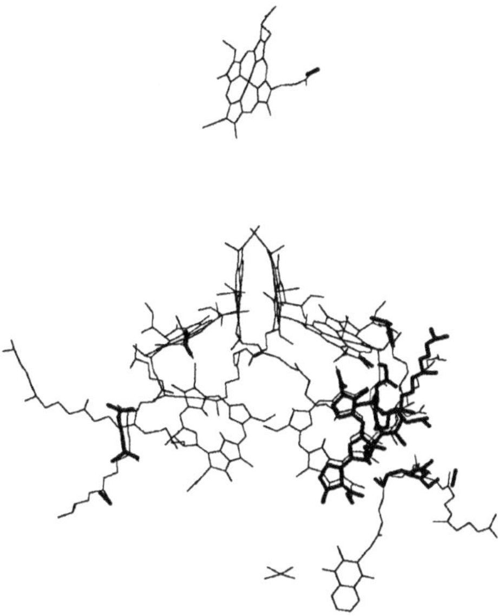

Fig. 6. Structural differences induced by primary electron transfer. The average structure of run B is represented by thin lines. The average structure of run A is overlayed by thick lines for those atoms the positions of which differ between the two structures by more than 0.4 Å.

pair . This motion is induced by the additional Coulomb interaction between the (after the electron transfer) positively charged special pair and the negatively charged pheophytine. As already mentioned above this might hinder the backtransfer of the electron by altering the alignment of the chromophores BCMP, BCLA, BPL or the tight interaction of BPL with the BCLP phytol chain. We will discuss this issue in more detail in a second contribution to this workshop, concerning electrostatic properties of the reaction center[5]. The degree of mobility (flexibility as measured by σ_j) does not alter very much after electron transfer except that BPL becomes somewhat less flexible. This decrease in flexibility might be due to BPL being shifted into a cleft inside the reaction center. The existence of a cleft-like pocket is indicated by the shape of BPL as well as by the fact that our simulations also showed significant differences for the BPL location compared to the location observed through X-ray scattering. It is possible that this pocket inside the L subunit furnishes some degree of control during the primary electron transfer by altering the chromophore – chromophore interactions through the resulting mobility and also allowing for fast and effective dielectric relaxation (see also Ref. 5).

Iron Depleted Reaction Center

In order to elucidate how the amino acid, Fe, and chromophore composition of the reaction center controls structure, dynamics and function one may modify the reaction center. For an unequivocal interpretation of such experiments the modifications need to be very specific. Currently it is attempted to modify the photosynthetic reaction center of *Rps. viridis* by genetic engeneering methods (site directed mutagenesis). We expect that molecular dynamics simulations will play an important role in this respect in suggesting amino acid replacements as well as in interpreting the effects of

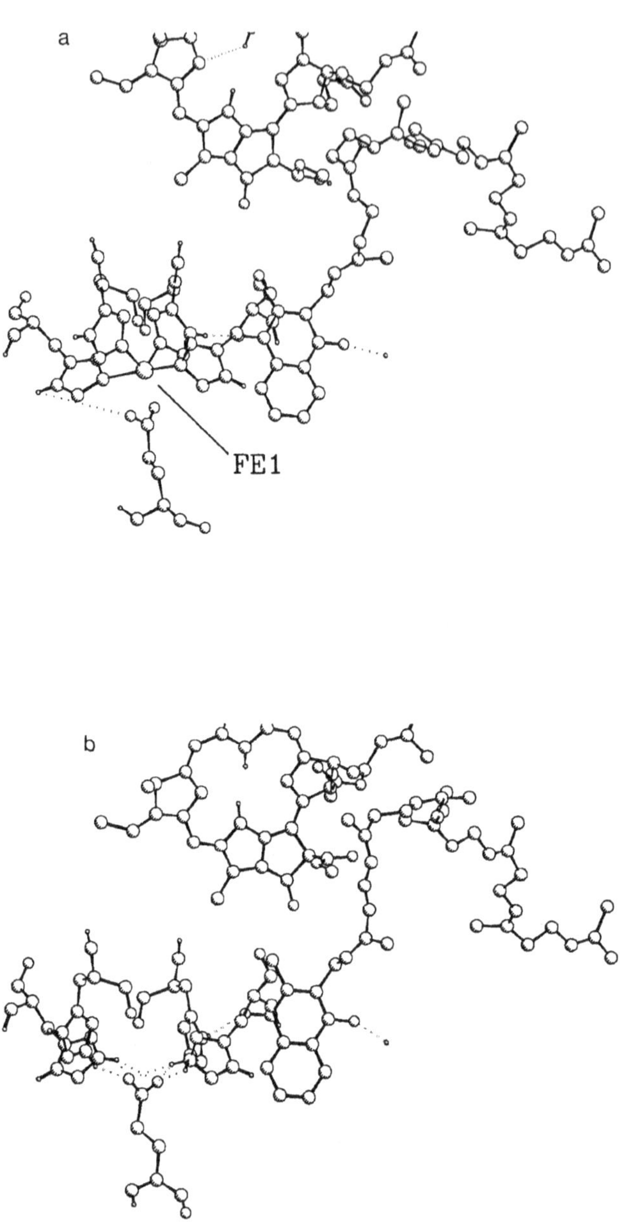

Fig. 7. Structural differences induced by iron depletion. The site of iron FE1 for the native (X-ray) structure is shown in Fig. 7a. Part of the tetrapyrol-ring of BPL is seen near the top and the menaquinone QA with its phytol chain is seen on the right side of the diagram. The iron-ion is bonded to four histidine groups shown on the left side as well as to a glutamate (M232) seen near the bottom. Hydrogen bonds are represented by dashed lines. Fig. 7b displays the same region for the iron depleted reaction center.

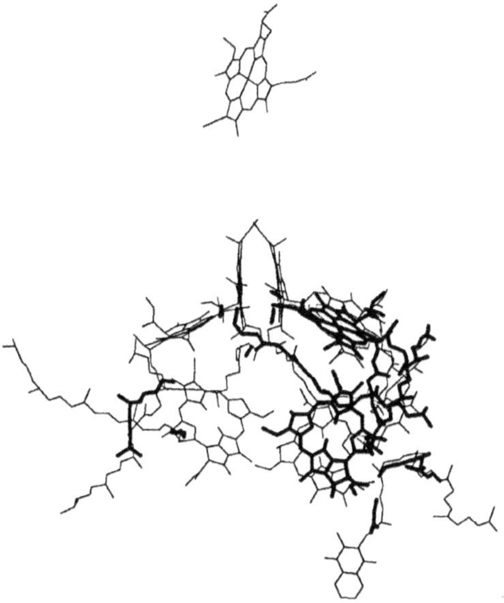

Fig. 8. Structural differences between the iron-depleted and the native reaction center. Thin lines represent the average structure resulting from a simulation of the iron depleted reaction center, thick lines represent the native structure (average structure from run A) for atoms which differ in their positions between the two structures by more than 0.7 Å.

such replacements. As a first attempt in this direction we consider in this contribution the effect of iron depletion on the structure of the photosynthetic reaction center.

Iron-depleted reaction centers have been prepared and examined in the case of the *Rps. sphaeroides* reaction center[9,10]. The main effects observed in this case after iron depletion was a decrease in the yield of the electron transferred to the quinone QA from 100% to 47% and a 20-fold increased lifetime of the negatively charged BPL. The native behaviour could be restored by reconstitution with several metal ions. These results are somehow surprising since the iron-ion is located far away from BPL. The authors[9,10] suggest mainly three possible explanations: (i) structural rearrangements, (ii) alteration of the electrostatic potential and (iii) change of vibronic couplings. In order to understand the effect of iron depletion we simulated the photosynthetic reaction center of *Rps. viridis* without the iron FE1. Because of the close structural similarities between the reaction centers of *Rps. sphaeroides* and *Rps. viridis*[11] we assume that iron depletion may have similar effects on the latter and, hence, our simulation might contribute to an explanation of the effect of iron depletion on the former reaction center.

The simulation has been carried out in the following way. First, we removed the iron FE1 from the simulated segment of the reaction center. Second, we minimized the energy by relaxing the atomic coordinates. Third, we carried out a 30 ps simulation to allow further relaxation of the altered protein. After this we performed finally a 21 ps simulation which was then analyzed.

In Fig. 7 we present the structural differences near the location of FE1. Figure 7a displays the native structure with the iron present. The iron is shown bonded to four histidine groups. Hydrogen bonds are represented by dashed lines. Also a glutamate

(M232) can be seen near the iron as well as, further away, the menaquinone (QA) and part of the bacteriopheophytine (BPL). The same region is shown for the the iron-depleted reaction center in Fig. 7b. The important change which occurs at the site of FE1 is that glutamate M232 substitutes for the missing iron-ion forming hydrogen bonds to the four histidines. As a result the local structures at the FE1 site in the native and iron-depleted protein are very similar. However, a further analysis showed that compared to the native structure the two protein subunits L and M in the iron-depleted reaction center are shifted slightly with respect to each other. This shift changes the arrangement of chromophores in the reaction center as shown in Fig. 8.

The average chromophore arrangement resulting from the simulation of the iron-depleted reaction center is compared in Fig. 8 to the average structure which resulted from run A describing the native reaction center (see above). The mean difference of the atomic positions in the two structures is 0.79 Å. One notices in Fig. 8 that the orientations of BCLA and BPL are very much disturbed. It appears that the iron is needed as a glue between the L and M subunits of the photosynthetic reaction center which also keeps the chromophores in their proper arrangement. The fact that iron depletion has a profound effect on the reaction center quantum yield[9,10] in connection with our findings hints again at the importance of the proper relative arrangement of the chromophores for an effective primary electron transfer process in the reaction center.

ACKNOWLEDGEMENTS

The authors like to thank Zan Schulten for advice and help. This work has been supported by the Deutsche Forschungsgemeinschaft (SFB 143-C1) and by the National Center of Supercomputer Applications in Urbana, IL.

REFERENCES

1. J. Deisenhofer, O. Epp, K. Miki, R. Huber, H. Michel, X-ray Structure Analysis of a Membrane Protein Complex J Molec Biol 180:385 (1984)
2. J. Deisenhofer, O. Epp, K. Miki, R. Huber, H. Michel, Structure of the protein subunits in the photosynthetic reaction center of Rhodopseudomonas viridis at 3 Å resolution, Nature 318:618 (1985)
3. B. R. Brooks et al. , CHARMM: A Program for Macromolecular Energy Minimization, and Dynamics Calculations, J Comp Chem 4:187 (1983)
4. C. L. Brooks, A. Brünger, M. Karplus, Active Site Dynamic s in Protein Molecules: A Stochastic Boundary Molecular-Dynamics Approach, Biopolymers 24:843 (1985)
5. H. Treutlein et al., Electrostatic control of electron transfer in the photosynthetic reaction center of Rps. viridis, contribution to this workshop
6. Dewar Group, University of Texas at Austin, Austin, TEXAS 78712
7. J. Deisenhofer's contribution to this workshop
8. J. Kuriyan et al., Effect of anisotropy and anharmonicity on protein crystallographic refinement, J Molec Biol 190:227 (1986)
9. C. Kirmaier et al., Primary photochemistry of iron-depleted an d zinc-reconstituted reaction centers from Rps. sphaeroides, PNAS 83:6407 (1986)
10. R. J. Debus, G. Feher, M. Y. Okamura, Iron-depleted React ion Centers from Rps. sph. R-26.1: Characterization and Reconstitution with Fe^{2+}, Mn^{2+}, Co^{2+}, Ni^{2+}, Cu^{2+} and Zn^{2+}, Biochemistry 25:2276 (1986)
11. J. P. Allen et al., Strucural homology of reaction centers f rom Rps. sph. and Rps. v. as determined by X-ray diffraction, PNAS 83:8589 (1986)

THE STARK EFFECT IN PHOTOSYNTHETIC REACTION CENTERS FROM *RHODOBACTER SPHAEROIDES* R-26, *RHODOPSEUDOMONAS VIRIDIS* AND THE D_1D_2 COMPLEX OF PHOTOSYSTEM II FROM SPINACH

M. Lösche, G. Feher, and M. Y. Okamura

University of California, San Diego
La Jolla, California 92093

I. INTRODUCTION

At the predecessor of this conference in Feldafing (1985) great interest in the Stark effect in photosynthetic reaction centers (RCs) was expressed [1]. At that time only an abstract [2] had been published showing that the bacteriochlorophyll dimer ($BChl_2$; the primary donor) had a large Stark effect compared to the other (accessory) chromophores. In the intervening period, several groups have contributed to this topic [3-5]. In this report we present detailed spectra and quantitative data on RCs from several organisms, and compare the results with those obtained by other groups and with theoretical predictions. Since much of the work has already been published, we shall endeavor to minimize duplication and focus on newer developments, qualitative and quantitative discussions, and prospects for the future.

Stark spectroscopy is the measurement of the change in absorption of a molecule due to an electric field, which may be either applied externally to a sample or may arise from charges associated with the structure, e.g. from ion gradients across a membrane. In the latter case one usually refers to the absorbance changes as "electrochromic shifts". The Stark effect provides information about the change in the charge distribution of the ground and excited states of the chromophore. Since changes in electronic charge distribution in the excited state of the RC are likely to be important in the primary electron transfer events in photosynthesis, Stark spectroscopy is expected to provide insights into the primary charge separation processes. We report here the results of Stark spectroscopy obtained on unoriented samples of RCs (in polyvinyl alcohol films at T=77K) from two purple bacteria, *Rhodobacter sphaeroides* R-26 and *Rhodopseudomonas viridis*, and from the D_1D_2 complex of photosystem II from spinach.

II. THEORY

The effect of an applied electric field $\vec{F}$ on the absorption spectrum of a chromophore [6-8] (without reorientation of the molecules in the field) can be due to several types of interactions between the field and the charge distribution on the molecule, as schematically represented in Fig. 1. In the upper panel the change in the energy levels of the molecules due to the electric field is shown; the top of the lower panel displays the absorption spectra A of a single transition without field (solid line) and with an externally applied field (dashed line). The resulting difference spectrum, ΔA, is shown at the bottom of the lower panel.

First consider a change in energy due to the interaction of the electric field $\vec{F}$ with the *dipole moment* $\vec{\mu}$, i.e. $\Delta E = -\vec{\mu}\cdot\vec{F}$. In oriented samples (left column in Fig. 1), the difference between the dipole moments in the ground and excited states leads to the linear relation,

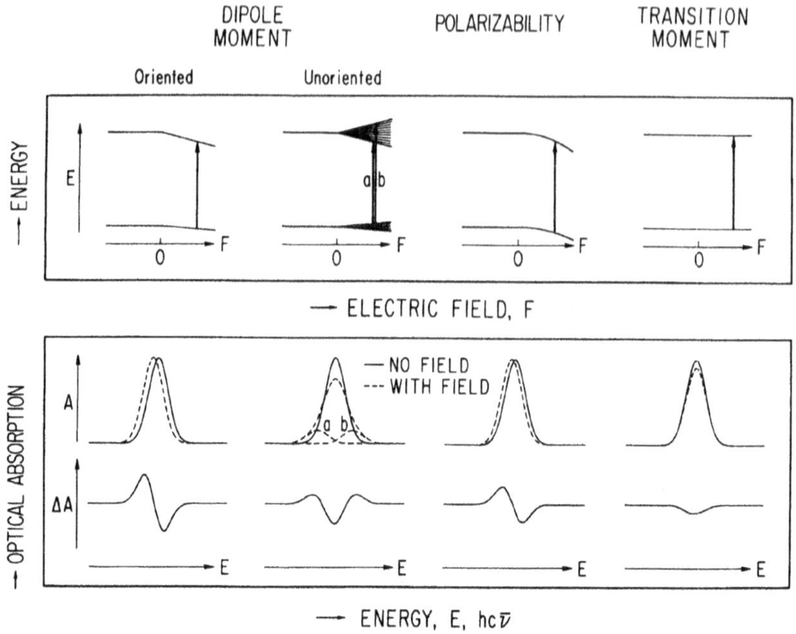

Fig. 1. *Schematic representation of the different effects of the electric field,* $\vec{F}$, *on the energy levels (upper panel) and the concomitant changes in the optical absorption (lower panel) of chromophores. Solid lines represent the absorption, A, in the absence and dashed lines in the presence of a field. The observed difference spectra,* ΔA, *are shown at the bottom of the lower panel. The following interactions are illustrated:*

i) Interaction with permanent dipole moments $\vec{\mu}$, *for oriented and unoriented chromophores. The shift in energy depends on the difference in dipole moments,* $\Delta\vec{\mu}$, *of the ground and excited states. For* oriented *chromophores the observed difference spectrum is linear in F and its shape is proportional to the derivative of the absorption. For unoriented samples, there is a continuous distribution of shifts depending on the relative orientations of* $\Delta\vec{\mu}$ *and* $\vec{F}$. *Two limiting cases* ($\Delta\vec{\mu}$ *and* $\vec{F}$ *parallel* [**a**] *and antiparallel* [**b**]) *are shown to lead to shifts of the wings of the absorption. Since upward and downward shifts in energy are equally likely, the difference spectrum is second order in field (i.e.* $\Delta A \propto F^2$) *and the absorption, A, is broadened resulting in a second derivative shape of* ΔA.

ii) Interaction with induced dipole moments, $\vec{\mu}_{\text{induced}} = \alpha\vec{F}$ *where* α *is the polarizability. The change in the transition energy is given by* $\Delta E = \int \vec{\mu}_{\text{induced}} \cdot d\vec{F} = (1/2)\alpha F^2$. *Since the induced dipole moment is in the direction of the applied field, a uniform shift in the absorption, proportional to the first derivative, is observed.*

iii) Interaction with the transition moments μ_{tr}. *In this case the energy of the transition does not change and the shape of the difference spectrum is the same as the absorption. The effect is proportional to* F^2.

$\Delta E_{\text{trans}} = -(\vec{\mu}_{\text{excited}} - \vec{\mu}_{\text{ground}}) \cdot \vec{F} = -\Delta\vec{\mu} \cdot \vec{F}$. The absorption line is uniformly shifted to lower energy and since we are dealing with a small absolute change of the absorption, the difference spectrum ΔA resembles the first derivative of A with respect to energy.

In unoriented samples (the case we will be dealing with), the dependence of ΔE_{trans} on the angle between $\Delta\vec{\mu}$ and $\vec{F}$ leads to a continuous spread in splittings of the energy levels in the field (second column). Accordingly, downward shifts in energy (indicated by **a** in Fig. 1) are as likely as upward shifts (**b**) (**a** and **b** represent limiting cases, corresponding to the maximal shifts with

$\Delta\bar{\mu}$ and $\vec{F}$ parallel and antiparallel, respectively). The effect of the field on the manifold of transitions leads to the situation indicated by the dashed line in the respective absorption spectrum. As a result, the original transition is broadened and ΔA has the shape of the second derivative. Although the shift of each individual energy level depends linearly on $\vec{F}$, cancellation of oppositely directed shifts of the same magnitude leads to a second order Stark effect, i.e. ΔA is proportional to $(\Delta\bar{\mu}\cdot\vec{F})^2$.

The second mechanism contributing to the Stark effect is the interaction of the electric field with the *induced dipole moment* of the molecule (Fig. 1, third column), $\bar{\mu}_{induced} = \alpha\vec{F}$, where α is the polarizability. The change in energy due to this interaction is $\Delta E = -\int \bar{\mu}_{induced} d\vec{F} = -(1/2)\,\alpha F^2$; it leads to a change in energy for the transition $\Delta E_{trans} = -(1/2)\,\Delta\alpha F^2$ where $\Delta\alpha$ is the difference in polarizabilities between the excited and ground states. The induced dipoles and the electric field point in the same direction, regardless of the relative orientation of the chromophores in the sample. This leads to the same energy shift in both oriented and unoriented samples. The uniform shift of the absorption spectrum A results in a difference spectrum ΔA proportional to the first derivative. For oriented samples, however, the polarizability effect is usually much smaller than the interaction of $\vec{F}$ with $\Delta\bar{\mu}$.

A third contribution is the interaction of the electric field with the *transition moment* $\bar{\mu}_{tr}$. Since the energy of the transition does not change, the Stark spectrum has the shape of the absorption spectrum.

For the quantitative evaluation of the Stark effect due to the $(\Delta\bar{\mu}\cdot\vec{F})^2$ term obtained from unoriented samples we use the relation given in [4, Eq. (2)] and explicitly derived in [7]:

$$\Delta A = \frac{(\Delta\mu)^2 F^2}{10\,h^2c^2}\,\bar{\nu}\,\frac{\partial^2(A/\bar{\nu})}{\partial\bar{\nu}^2} \times (3\cos^2\delta\,\cos^2\chi + 2 - \cos^2\delta - \cos^2\chi) \tag{1}$$

where h is Planck's constant, c the velocity of light, $\bar{\nu}$ the wavenumber (in cm^{-1}), χ the angle between the applied electric field $\vec{F}$ and the electric field vector, $\vec{E}$, of the probing light, and δ the angle between $\Delta\bar{\mu}$ and $\bar{\mu}_{tr}$. Equation 1 contains two parameters that characterize the system, δ and $\Delta\bar{\mu}$. They have been determined experimentally from two independent measurements [4].

(i) The angle δ was obtained by measuring the dependence of ΔA on $\cos^2\chi$ at a fixed wavelength using polarized light and rotating the sample around an axis perpendicular to the measuring beam.

(ii) The change in dipole moment, $\Delta\bar{\mu}$, was determined from the Stark spectra normalized to the second derivative of the absorption spectra. In these experiments the probing light was unpolarized with its $\vec{E}$ vector parallel to the plane of the sample ($\chi = 90°$).

III. THE EFFECTIVE FIELD PROBLEM

A serious problem in determining the values for $\Delta\bar{\mu}$ and $\Delta\alpha$ arises from the uncertainty in our knowledge of the local electric field, $\vec{F}_{local}$, at the molecule. The local field is in general different from the electric field applied to the sample. We express this difference by a correction factor, f; i.e. $\vec{F}_{local} = f\cdot\vec{F}_{applied}$. The difference arises from additional fields due to charges and dipoles in the sample. To calculate all the contributions to the local field is, in general, a difficult problem, which has been solved only for very simple systems. One usually resorts, therefore, to approximations involving correction factors containing the macroscopic dielectric constants of the media in which the chromophore is embedded. Two correction factors are commonly used, the Lorentz and the spherical cavity corrections [9]. In addition, a correction must be made for the change of the macroscopic electric field due to the difference between the dielectric constants of the protein and the surrounding polymer matrix in which the protein is embedded. The respective correction factors for the Lorentz and spherical cavity approximations are [4]:

$$f_L = \frac{3\epsilon_{PVA}}{2\epsilon_{PVA} + \epsilon_P}\left(\frac{\epsilon_P + 2}{3}\right) \tag{2a}$$

$$f_S = \frac{3\epsilon_{PVA}}{2\epsilon_{PVA} + \epsilon_P}\left(\frac{3\epsilon_P}{2\epsilon_P + 1}\right) \tag{2b}$$

where ϵ_{PVA} and ϵ_P are the dielectric constants of PVA and the protein, respectively. The first term corrects for the macroscopic field within the protein and the second term corrects for the local field at the chromophore. Although these approximations are rather crude, the local field corrections given by Eqs. 2a and 2b do not differ greatly from each other if the dielectric constant ϵ_P is low (for $\epsilon_P = 1$ they are, of course, identical). For high values of ϵ_P there is a significant difference between the two approximations. Thus, for Stark measurements made at low temperature (T=77K), where contributions to the dielectric constant due to motion of dipoles are frozen out, the uncertainty in estimating the local field becomes less pronounced, although it is still considerably larger than the statistical error of the measurements. Whereas ϵ_P cannot be measured directly, ϵ_{PVA} was determined to be:

$$\epsilon_{PVA} = 10.2 \pm 0.6 \text{ (T=295K) and } \epsilon_{PVA} = 3.5 \pm 0.2 \text{ (T=77K)}$$

The conversion factor, f, obtained at 77K with the above value of ϵ_{PVA} and an estimated value of $\epsilon_P \cong 3$ ranged from 1.3 to 1.7 (Eqs. 2a, 2b).

IV. SAMPLE PREPARATION, THE STARK SPECTROMETER AND DATA ANALYSIS

Samples were prepared by mixing 500 μl of protein solution (optical absorbance of $A_{670}^{1cm} = 10$ for the PSII complex and A_{800}^{1cm}; $A_{835}^{1cm} = 25$ for RCs from *Rb. sphaeroides* and *Rp. viridis*, respectively) into a 10% (w/v) polyvinyl alcohol (PVA) solution, from which ionic impurities were removed [4]. The mixture was dried on a plexiglass support at 20°C for *Rb. sphaeroides* RCs or at 4°C for PSII and *Rp. viridis* RCs. The dry, clear transparent film, typically had a thickness of ~ 200 μm; its exact dimension was determined mechanically by two independent methods [4]. Transparent glass electrodes (500 Å In-SnO over 1000 Å SiO_2 on float glass obtained from Delta Technologies, Stillwater, MN) were cut to a 2.5 cm × 4 cm size and etched to leave conducting areas of ~ 12 mm × 12 mm square with a narrow conducting strip for contact. The sample was glued between two opposing transparent glass electrodes with a cyanoacrylate compound (Fig. 2). The capacitance of the sample was measured in a capacitance bridge. From the dielectric constant, the sample thickness was determined and checked against the mechanical thickness measurements. The sample was placed in a liquid Nitrogen Dewar with optical windows, where it could be rotated around a vertical axis, thereby varying the angle, χ.

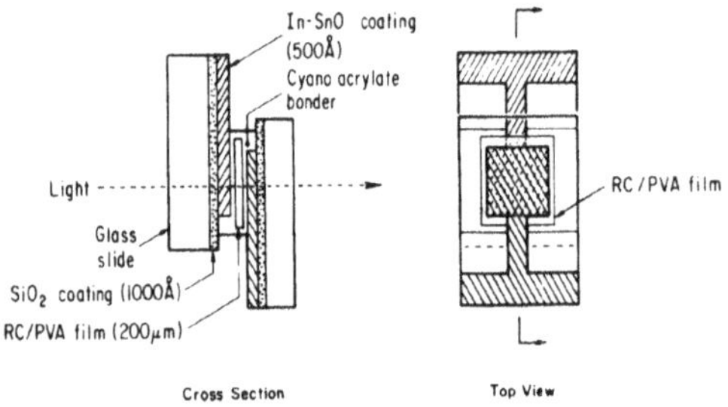

Fig. 2. *Schematic drawing of the Stark effect sample (not to scale). A RC/PVA film is glued between two glass slides coated with conducting, transparent In-SnO films on top of a* SiO_2 *coating (for chemical insulation). The RC/PVA film is slightly larger than the area of the electrodes ~ 12 mm × 12 mm. Contacts are made via narrow conducting strips leading to the opposite ends of the sample.*

The Stark spectrometer (Fig. 3) consisted of a home made absorption spectrometer (resolution: 3.3 nm), an oscillator (f=1 kHz), a high voltage amplifier [10,4] and detector. Both the dc light transmission and the second harmonic signal (at 2kHz) were amplified, simultaneously digitized, and transferred to a microcomputer for the computation of the absorbance change ΔA, the absorption A and its first and second derivatives [4].

The data were analyzed in terms of the interaction of the applied field $\vec{F}$ with the permanent dipole moments $\Delta\vec{\mu}$ of the chromophores, neglecting the contributions from the changes in polarizability and transition moment. From Eqs. (1) with $\chi = 90°$ one obtains for the dipole moment:

$$\Delta\mu_{app} = \frac{hc}{F_{rms}} \left[\frac{10}{1 + \sin^2\delta} \; \frac{\sqrt{2}\,\Delta A}{\bar{\nu}\,\frac{\partial^2(A/\bar{\nu})}{\partial\bar{\nu}^2}} \right]^{1/2} \tag{3}$$

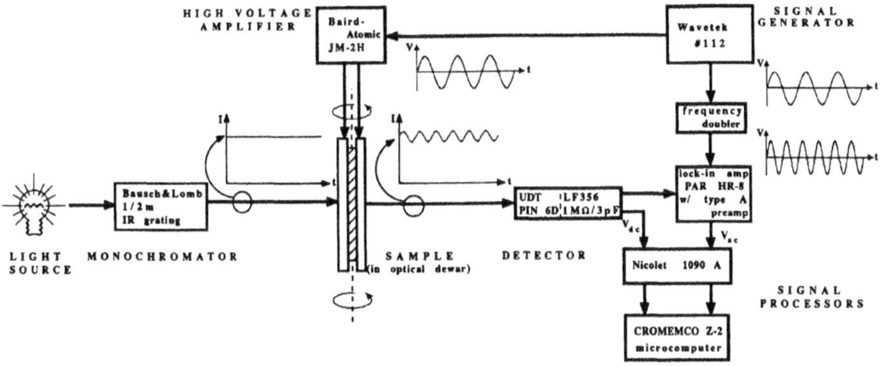

Fig. 3. *Block diagram of the Stark spectrometer. The sample is inside a liquid Nitrogen Dewar (not shown) where it can be rotated around a vertical axis. The high voltage (f=1 kHz) is applied across the sample film. Changes in the probing light intensity, I, are lock-in detected at the second harmonic of the oscillator frequency. Both the ac and dc signals are digitized and transferred to a microcomputer for processing.*

where the apparent dipole moment, $\Delta\mu_{app}$, does not take into account the local field correction. Signals detected at the fundamental frequency of the applied field were smaller than 1% of the second harmonic signal (data not shown), which justifies the assumption of randomly oriented chromophores. Furthermore, the shape of the Stark spectra is dominated by the spectral shape of the second derivative of the absorption (see below). For well resolved transitions, both ΔA and the second derivative of the absorption were measured at the wavelength at which the first derivative (data not shown) vanished. For poorly resolved transitions, estimates of $\Delta\mu_{app}$ were obtained by measuring ΔA and the second derivative of the absorption between their respective minima and maxima or, as in the case of the D_1D_2 complex, at the position of the minimum of the second derivative of the absorption.

V. RESULTS

A. Bacterial Reaction Centers

Figure 4 shows the results obtained from bacterial RCs at T=77K. The absorption, second derivative, and Stark spectra are displayed in the near infrared region for RCs from *Rb. sphaeroides* (Fig. 4a-c) and *Rp. viridis* (Fig. 4d-f). The Stark spectrum obtained from *Rb. sphaeroides* (Fig. 4c) displays five resolved peaks (at λ = 877 nm, 817 nm, 802 nm, 761 nm and 754 nm), which arise from the Q_y transitions of the chromophores. Table 1 summarizes values of $\Delta\mu_{app}$ and δ for these transitions evaluated as described above. In addition, we resolved two peaks at shorter wavelengths (λ = 697 nm and 683 nm) with a $\Delta\mu_{app}$ of ~ 5 Debye. For RCs from *Rb. sphaeroides*, the minima of the Stark spectra correspond to the minima of the second derivative of the absorption spectrum ($\Delta\lambda < \pm$ 1 nm). This indicates that the dominant interaction of the applied field with the molecules is due to the permanent dipole term. The overlap of transitions in the 800 nm and 760 nm regions causes uncertainty in the determination of $\Delta\mu_{app}$ and δ.

The Stark spectrum of RCs from *Rp. viridis* showed 7 resolved peaks in the near infrared region. Six of the peaks (at λ = 989 nm, 849 nm, 833 nm, 818 nm, 803 nm and 787 nm) correspond to the Q_y transitions of the six pigment molecules. The broad Stark signal around λ = 865 nm ($\bar{\nu}$ = 11,560 cm^{-1}) had no match in the absorption spectrum and may represent either a vibronic band of the dimer or a charge transfer state. The Stark spectrum, generally, had the shape of the second derivative of the absorption (minima of both spectra agreed within $\pm$ 1 nm). A significant exception was observed for the band having its peak at λ = 833 nm. The minimum of the Stark spectrum was shifted by ~ 3 nm to longer wavelength with respect to the

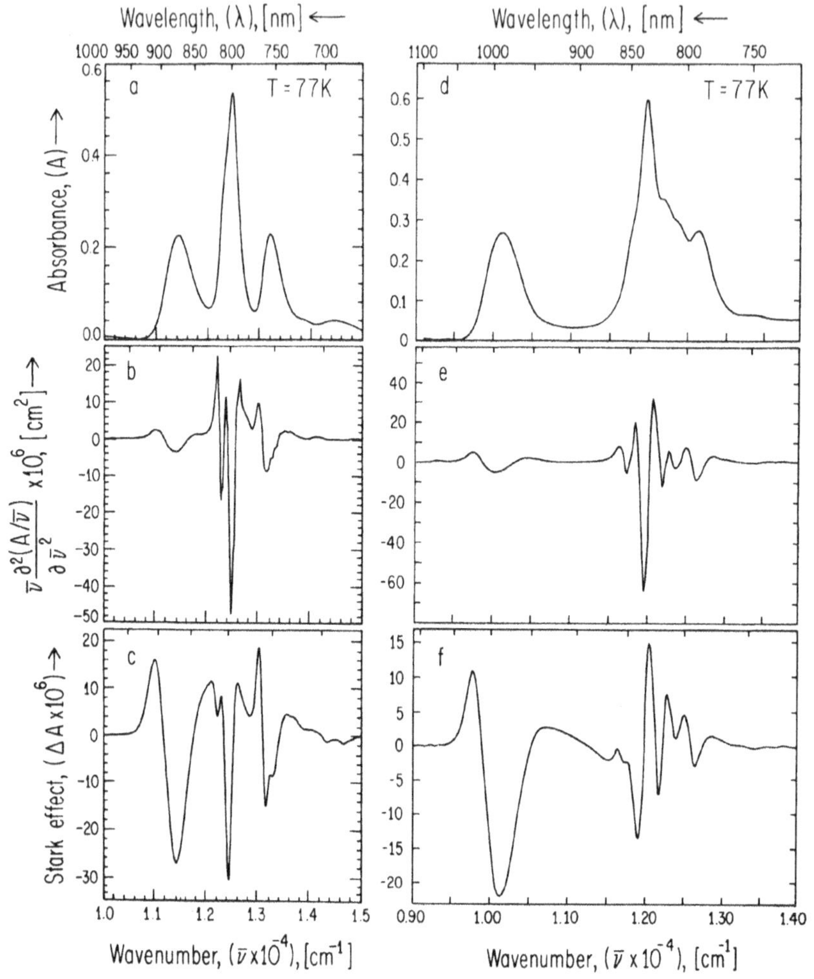

Fig. 4. *Absorption spectra (top panels), second derivatives (middle) and Stark spectra (bottom) of the* Q_y *region of RCs from* Rb. sphaeroides *R-26 (left;* $F_{rms} = 7.14 \times 10^4$ *V/cm) and* Rp. viridis *(right;* $F_{rms} = 4.5 \times 10^4$ *V/cm) at T=77K.*

minimum of the second derivative of the absorption, indicating a contribution from the first derivative (i.e. from the polarizability term). Similarly, a deviation was observed for the broad donor transition at $\lambda = 989$ nm; the Stark spectrum was shifted by ~ 3 nm to shorter wavelength with respect to the second derivative spectrum. The values for $\Delta\mu_{app}$ and δ are shown in the lower half of Table 1. As in the case for *Rb. sphaeroides*, the most striking result is the large value of $\Delta\mu_{app}$ for the long wavelength donor band as compared to the values for the accessory chromophores.

In all our analyses we have neglected the contributions of the first derivative to the Stark spectrum. For those cases, in which a mismatch between the Stark spectrum and the second derivative of the absorption was observed (in particular the 833 nm band in *Rp. viridis*), we evaluated $\Delta\mu$ and δ at the wavelengths at which the first derivative was zero. At $\lambda = 833$nm there is, unfortunately, a strong wavelength dependence of the result causing a considerable error in the values of $\Delta\mu_{app}$ and δ.

A comparison of the lineshapes of the Stark and second derivative spectra of the dimer transition (shown in Fig. 5 for RCs from *Rb. sphaeroides*) shows that although *positions* of the extrema agree, *shapes* do not. The mismatch is much more pronounced on the high energy side of the transition than on the low energy side (the Stark signal is broader than the second derivative of the absorption). This observation cannot be understood from the simple analysis given above, since the contribution from the first derivative is obviously small (the minima of both spectra agree) and the other interactions considered can only account for effects that are symmetric in λ about the peak position. The same effect was observed in RCs from *Rp. viridis*, although in this case the situation was complicated by an additional peak shift. A qualitative, speculative, explanation of the deviation described above was proposed in [4]. Further experimental and theoretical work is needed to account for the effect quantitatively.

B. Photosystem II Reaction Centers

The D_1D_2 complex of PSII from spinach was prepared by the method of Satoh and Nanba [11]. These RCs are believed to be analogous to the bacterial RCs. They contain ~ 4 chlorophylls, 2 pheophytins, and 4 polypeptides comprising the D_1 and D_2 subunits and 2 subunits of cytochrome b_{559}. The D_1D_2 complex is active with respect to pheophytin photoreduction [11] and exhibits a spin polarized triplet state that is characteristic of the bacterial RC [12]. The Stark spectrum (Fig. 6c) shows two prominent peaks at $\lambda = 680$ nm and 670 nm [13]. Both of them are observed in the second derivative of the absorption spectrum (Fig. 6b). In addition there is a small

Table 1. Magnitude and direction of the apparent dipole moments, $\Delta\mu_{app}$, of the chromophores in reaction centers of Rb. sphaeroides *R-26 and* Rp. viridis *(T=77K)*

Bact. species	λ [nm]	δ [deg]	$\Delta\mu_{app}$ [Debye]*
Rb. sphaeroides	877	38 ± 1	6.5 ± 0.2
	817	N.D.	≈ 1
	802	23 ± 2	2.1 ± 0.1
	761	8 ± 4	3.5 ± 0.2
	754	≈ 5	≈ 3.5
Rp. viridis	989	40 ± 2	8.2 ± 0.8
	849	N.D.	N.D.
	833	~ 50	1.8 ± 0.6
	818	14 ± 6	3.4 ± 0.3
	803	N.D.	N.D.
	787	0 ± 5	2.7 ± 0.3

N.D. Not determined

**Errors represent SD (5 independent runs). The dipole moment $\Delta\mu = f^{-1}\Delta\mu_{app}$ where f at 77K is estimated to be 1.3-1.7 (Eqs. 2a and 2b).*

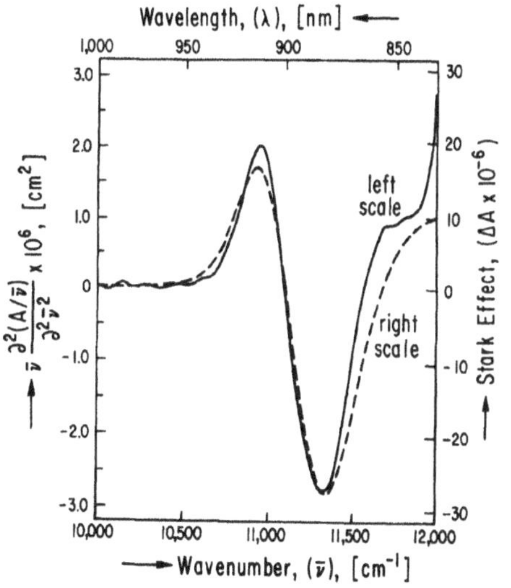

Fig. 5. *Comparison of the second derivative (solid line) with the Stark spectra (dashed line) of the* $BChl_2$ *donor absorption region of* Rb. sphaeroides *R-26 at T=77K. The data are the same as in Fig. 4b,c; amplitudes were normalized at the minimum of the signals.*

band at $\lambda \sim 630$ mm which seems to be absent in the absorption spectrum. The minima of the Stark spectrum were shifted to longer wavelengths with respect to the minima of the second derivatives of the absorption (~ 1.7 nm for the 680 nm band and $\leqslant 1.0$ nm for the 670 nm band). The Stark effect was evaluated at the position of the respective minima of the second derivative spectra. This procedure reduced significantly the calculated value of $\Delta\mu_{app}$ for the 680 nm band. The magnitudes of $\Delta\mu_{app}$ and (δ) for the 680 nm and 670 nm bands obtained by the above procedure were 0.8 Debye (40°) and 1.4 Debye (20°), respectively. In contrast to the results for bacterial RCs the low value of $\Delta\mu_{app}$ of the 680 nm band, which contains the primary donor species P_{680} is smaller than that for the peak due to the accessory pigment at $\lambda = 670$ nm. Although the data presented are preliminary in nature, they strongly suggests a principally different organization of the donor entity, despite the close relation of the amino acid sequences of the D_1D_2 polypeptides and the L and M subunits of bacterial RCs [14].

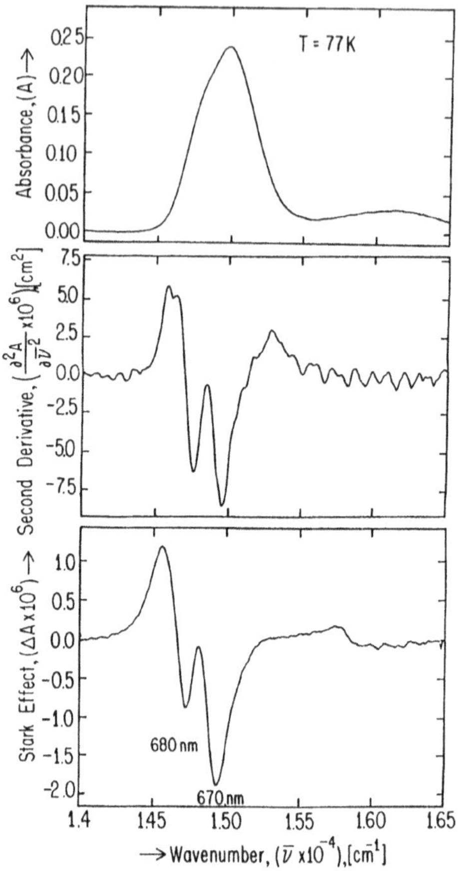

Fig. 6. *Absorption spectrum (top panel), second derivative (middle) and Stark spectrum* ($F_{rms} = 6.93 \times 10^4$ *V/cm) of the* Q_y *region of the* D_1D_2 *complex from spinach at T=77K. The 680 nm band contains the donor absorption and the 670 nm band is ascribed to accessory pigments. The* D_1D_2 *complex was prepared by the method of Satoh and Nanba [11].*

VI. SUMMARY AND DISCUSSION

The major feature of the Stark spectra of the bacterial RCs is that the long wavelength band due to the primary donor has a significantly larger $\Delta\mu$ than that of the accessory BChl. This has been explained by the mixing of charge transfer states into the excited state of the dimer. In contrast, the 680 nm band containing the primary donor, P_{680}, in RCs from PSII does not have a large $\Delta\mu$.

In addition to the major peaks we observed several smaller bands at $\lambda = 697$ and 683 nm in RCs from *Rb. sphaeroides* and at $\lambda = 865$ nm in RCs from *Rp. viridis* and a peak at $\lambda = 630$ nm in PSII RC. These have been assigned to vibronic side bands although the role of charge transfer states cannot be ruled out. Finally the shape of the long wavelength transition deviates significantly from the profile of the second derivative of the absorption spectrum; besides the possible explanation given in [4], this effect may be due to transitions to charge transfer states. Since charge transfer states are believed to play an important role in the Stark effect in bacterial reaction centers, we proceed to discuss them in more detail.

*The absolute value of ΔA is only 65% of the value at the minimum.

A. Charge Transfer States

The charge transfer (CT) states of the RC that may contribute to the large $\Delta\mu$ include: $D^+D^-B\phi$, $DD^+B^-\phi$, $DDB^+\phi^-$, (where DD = $BChl_2$ donor; B = monomeric BChl, ϕ = BPhe). These states have been proposed to be intermediates in the electron transfer process [1,15-22] and are, therefore, of considerable interest. In principle it should be possible to extract information about charge transfer states from Stark effect data. The Stark spectrum may either exhibit new bands due to direct transition to the CT states or may enhance the bands of nearby excited states as has been observed for the long wavelengths donor band in bacterial RCs.

B. Mixing of Charge Transfer Character into the Excited State.

One indication of a CT state is the increase in the dipole moment due to the mixing of CT character into the excited state [23]. If the excited state of the dimer is coupled to a CT state, its wavefunction can be expressed as a linear combination of the wavefunction of the dimer and the CT state:

$$\Psi = \Psi_{DD^*} + c\,\Psi_{CT} \tag{4}$$

The mixing coefficient, c, is given by perturbation theory in the form

$$c = \frac{T}{(E_{DD^*} - E_{CT})} \tag{5}$$

where T is the matrix element connecting DD* with CT and E_{DD^*} and E_{CT} are the energies of the dimer excited state and CT state, respectively. The expectation value of the dipole moment of the excited state can be estimated roughly in terms of the admixture of dipole moments from the DD* and CT states

$$\vec{\mu} = \langle\Psi|\vec{\mu}|\Psi\rangle \cong \vec{\mu}_{DD^*} + c^2\vec{\mu}_{CT} = \vec{\mu}_{DD^*} + \left[\frac{T}{E_{DD^*} - E_{CT}}\right]^2 \vec{\mu}_{CT} \tag{6}$$

where we have neglected the cross terms involving Ψ_{DD^*} and Ψ_{CT}; $\vec{\mu}_{CT}$ is the dipole moment of the charge transfer state. In view of the terms that were neglected in Eq. 6 this estimate represents a lower limit to the dipole moment due to contribution by the CT state. Since the dipole moments of the charge transfer states are very large (for instance, μ_{CT} of the state D^-D^+ with a distance between Mg atoms of 7 Å [24,25] has a value of 35 Debye), the dipole moment of the excited state may be considerably increased due to intermolecular charge transfer. The amount of dipole moment mixed into the excited state depends on the relative contributions of $D_A^+D_B^-$ and $D_A^-D_B^+$, where the subscripts A and B refer to the two BChls of the dimer. If both of these states contribute equally to the excited state there is no increase in the net dipole moment.

The charge transfer states $D_A^+D_B^-$ and $D_A^-D_B^+$, in which charge is transferred within the special pair have been investigated in detail by Parson and Warshel [26]. The matrix element connecting the low energy dimer state to the CT states are expected to be large and appreciable charge transfer character in the excited state would be expected for reasonable values for the energy gap between the CT and the dimer state, if the complex is sufficiently asymmetric. Parson and Warshel [26] have estimated the energy gap between the low energy dimer state and the lowest CT state to be 3600 cm^{-1} in RCs from *Rp. viridis*. With an energy difference between the two CT states of 2000 cm^{-1} they obtained a dipole moment in the excited state of 2.8 Debye and a value of $\delta = 32°$. M. Plato calculated the dipole moment of the dimer in RCs of *Rp. viridis* using the INDO/S (intermediate neglect of differential overlap/spectroscopy) method. He obtained a value of $\Delta\mu$ = 5.3 Debye and $\delta = 30°$ [27].

The apparent lack of a large $\Delta\mu$ in the excited state of P_{680}, the primary donor in PSII, is in contrast with the results in bacterial reaction centers. Possible explanations of this low $\Delta\mu$ include: i) the primary donor is a monomer, ii) the donor is a symmetric dimer (see later discussion) and, iii) the energy of the charge transfer state of the dimer is very high and does not mix significantly into the excited state wavefunction. This possibility is in accord with the high potential (necessary for water oxidation, +1.1 eV [28]) needed to oxidize P_{680}.

C. The Role of Charge Transfer States in Electron Transfer

Charge transfer states have been proposed to be involved in electron transfer from the BChl dimer to Bphe [22,29]. For example the charge transfer state $DD^+B^-\phi$ could serve as an intermediate between the excited state of the dimer and the final charge separated state $DD^+B\phi^-$ in the following electron transfer scheme:

$$\underset{1}{DD^*B\phi} \rightarrow \underset{2}{DD^+B^-\phi} \rightarrow \underset{3}{DD^+B\phi^-}$$

The proposed *intermediate CT state* has not been observed by picosecond spectroscopy [30-32]. This could be accounted for by a rate of decay of the intermediate state that is much faster than the rate of formation. Furthermore, the energy of the putative intermediate state should be close to or lower than the energy of the excited dimer state to account for the near temperature independence of the electron transfer rate.

An alternate mechanism, a *superexchange*, has been postulated to explain the high electron transfer rate [15,16,18,19,30]. In this mechanism the CT state (2) is mixed into the excited state (1) by the matrix element T_{12}. The electron transfer matrix element then contains a superexchange contribution proportional to the mixing coefficient given by

$$T_s = cT_{23} = \frac{T_{12}T_{23}}{E_1 - E_2} \tag{7}$$

where the states 1 and 2 have energies E_1 and E_2 and T_{12} and T_{23} are the matrix elements connecting the respective states. The conditions that give rise to a large superexchange (i.e. $E_1 - E_2 \cong T_{12}$) can be accompanied by a significant mixing of charge transfer character into the excited state. Using the values for the matrix element T_{12} (17 cm^{-1}) and the upper limit for the energy gap (60 cm^{-1}) estimated by Marcus [15] to be required by the superexchange mechanism and the distance between the dimer and accessory BChl of 14 Å obtained from the crystal structure, the contribution to the dipole moment of the excited state from Eq. (6) is 5.4 Debye. Although the magnitude of this contribution may be reduced by the presence of charge transfer states with the opposite moment as well as by a vibrational broadening of the energy gap $E_1 - E_2$, Stark effect measurements should be sensitive to the presence of near resonant CT states involved in the superexchange mechanism.

In the alternative case in which the CT state 2 is present as a real intermediate state between state 1 and 3, the requirement for energy matching is not as severe. The observed temperature independent rate can be explained if $E_2 - E_1$ were of the order of 200 cm^{-1} ($\simeq$kT at room temperature). Assuming the same matrix element T_{12} of 17 cm^{-1} yields a contribution to the dipole moment of only 0.5 Debye.

Another CT state that has been suggested to play a role in the electron transfer process is $DDB^+\phi^-$. Parson and Warshel [21] and Fischer and Scherer [20] have suggested that this state may serve as an intermediate in the electron transfer pathway. Coupling of this state to the excited singlet state of the dimer should also produce an increase in the dipole moment.

D. Which of the Charge Transfer States are Responsible for the Observed Dipole Moment?

In order to determine the contribution of the various CT states to the dipole moment of the excited state, a detailed comparison must be made between experiment and theory. The experimentally determined value of $\Delta\mu$, corrected for the local field (Eqs. 2a, 2b with $\epsilon_P = 3$ and $\epsilon_{PVA} = 3.5$), of the long wavelength dimer band in *Rp. viridis* at 77 K ranged from **4.8-6.3** Debye; the uncertainty being due to the local field correction. Assuming that the dipole of the ground state is small, we can compare the experimentally determined $\Delta\mu$ to the value of **5.3** Debye calculated by M. Plato [27] for the contribution of the CT state $D^+D^-B\phi$ to the dipole moment. The good agreement between experiment and theory suggests that contributions due to other CT states such as $DD^+B^-\phi$ may be small or absent. Although this seems to argue against a superexchange mechanism, the uncertainty in the theory and the local field correction does not permit us to definitely rule out superexchange.

In addition to the dipole moment, there is another quantity, the angle δ between $\Delta\bar{\mu}$ and the transition moment, that, in principle, can be used to identify CT states. For *Rp. viridis* at 77K, δ was determined to be 40±2° (see Table 1). This value is to be compared to the theoretically calculated one of 30° for the $D^+D^-B\phi$ state [27]. The agreement between experiment and theory is only fair, precluding a definitive assignment. Lockhart and Boxer [3] have attempted to rule out the charge transfer contribution due to the $DD^+B^-\phi$ state on the basis of the measured value of δ by arguing that it deviated from the angle between the transition moment and the vector connecting the dimer to the BChl monomer (45°). However, the value for the angle initially obtained was in error and the corrected angle (40°) is close to the value expected for the CT transfer to the BChl monomer.

Another test for the nature of the involvement of CT bands in the Stark effect is the magnitude of the polarizability. This term is expected to be significant for the mixing of a CT state with closely lying excited states as was pointed out by Scherer and Fischer [17] in their analysis of the Stark effect in RCs from *Rb. sphaeroides*. They argue that the absence of a first derivative character in the low energy dimer band is evidence against the coupling to a nearby CT state. This would also argue against a superexchange mechanism.

E. **Direct Observation of Charge Transfer Bands**

In addition to the effect of increasing the dipole moments of nearby states the presence of a CT state may be determined by observing the direct transition from the ground state to the CT state. The modulation of the transition moment with the electric field should give rise to a Stark effect which is proportional to the absorption spectrum of the charge transfer state. This should provide a method for detecting weak transitions in the presence of strong absorption bands [17]. The Stark spectra show several peaks that have not been assigned in the absorption spectra. These include bands at $\lambda = 697$ nm and 683 nm in *Rb. sphaeroides* and at 865 nm in *Rp. viridis* and $\lambda = 630$ nm in PSII RCs. Although it is possible that these bands are vibronic side bands, charge transfer bands cannot be ruled out. Inspection of the Stark spectra do not reveal the band predicted by Parson and Warshel [26] at $\lambda \cong 715$ nm in *Rp. viridis*. The 775 nm band seen in the Stark effect spectra of *Rb. sphaeroides* assigned by Scherer and Fischer [17] to a CT band seems to be due to a positive peak in the second derivative of the absorption spectrum (see Fig. 4).

An important region of the Stark spectrum to look for CT bands is near the long wavelength dimer transition. There is no distinct new band seen in our spectra in this region. This is in contrast to the negative absorption dip seen in this region by Braun et al. [5]. The origin of this discrepancy is not understood although it should be pointed out that an offset in the baseline can easily arise from pickup of a component from the high voltage modulation in the detector. A way that unresolved CT bands may manifest themselves is via a deviation of the Stark spectrum from the second derivative of the absorption spectrum. Such deviations were observed in the dimer region of the Stark spectra of both bacterial species (see Fig. 5). The absence of prominent CT bands is not too surprising since they may be weak and broad. More detailed analyses of the spectra are required before further conclusions can be drawn.

F. **Prospects for the Future**

Stark spectroscopy is an important technique for studying excited states that play a role in photochemical reactions. It is particularly sensitive to charge transfer states, which have been postulated to be involved in electron transfer reactions in photosynthesis. However, before the full potential of the technique can be utilized, several problems need to be solved.

A serious problem is the uncertainty associated with the *local field correction*, which prevents one at present from obtaining an accurate value of $\Delta\mu$. As was mentioned in Section III, there are two types of corrections: One is due to the difference in the dielectric constant of the protein and the matrix in which the protein is embedded. This correction can be investigated experimentally by embedding the protein in matrices having different, known, dielectric constants or by simultaneously measuring the temperature dependence of the dielectric constant of the matrix and the Stark effect. The second correction involves the polarizability of the protein, which affects the local field at the chromophore. Its value can, in principle, be calculated theoretically from the known structure of the RC. In practice this is, however, a very difficult task. Studies of model compounds may help to elucidate this problem. One could, for instance, compare $\Delta\mu$ of the bacteriochlorophyll monomer in a known matrix (e.g. polystyrene) with that obtained in the protein. This approach, unfortunately, cannot be applied to the dimer, which cannot be extracted from the RC in its native configuration.

Related to the above problem is the *variability in the Stark effect* measured on different (seemingly identical) samples. This variability exceeds several times the standard deviation of the measurement. We believe that this is due to the variability in the local field. Specifically, we found that the state of hydration affected the dielectric constant of PVA, which in turn affected the local field (see Eq. 2a,b). To overcome this problem it would be desirable to find a convenient marker-chromophore that can be embedded together with the RC in the matrix and to which the Stark spectra would be normalized. This would also greatly facilitate comparisons of results obtained in different laboratories.

To characterize the Stark spectra quantitatively, we are currently engaged in a phenomenological *simulation of the spectra*. The approach is to first deconvolute the optical spectra into their individual components. The Stark spectra are then fitted with a linear combination and varying coefficient of the zeroth, first and second derivatives of the optical spectrum. From the value of the coefficients the contribution of each type of interaction (see Fig. 1) is determined. This procedure should also help to uncover contributions from direct transitions to CT states.

We next discuss the Stark effect in samples that have been prepared in specific ways. We start with *oriented samples*; ideally single crystals of RCs, although lesser ordered structures (e.g. stretched PVA films, dried layers of chromatophores, squeezed gels, etc.) may also offer advantages over randomly oriented samples. The main result of the orientation is that the Stark effect due to the dipolar term will be first order in the applied electric field. It consequently will be much larger than that observed in unoriented samples and will be detected at the fundamental of the modulation frequency rather than the second harmonic. The Stark effect due to the polarizability term will remain quadratic in F and will not admix into the dipolar term. Thus, oriented samples will provide a better separation of the different contributions to the Stark effect. An additional advantage of working with a single crystal whose structure has been solved [24,25] is that the angle between $\Delta\bar{\mu}$ and the molecular axis can be determined directly. This would eliminate the uncertainty associated with the orientation of the transition moment.

As pointed out in Section VIB, the Stark effect is very sensitive to the energy difference between the CT state and the excited singlet state of the dimer, $E_{DD^{*}} - E_{CT}$ (Eq. 6). An *internal electric field* created by charges inside the RC will shift the energies of the CT states and thereby modify the Stark effect. Changes in the charge distribution may be accomplished by varying the redox potential (e.g. creating Q_A^- and $Q_B^=$), pH, site specific mutations of charged residues or by cross-illumination. From a calculation of the change in local field and the concomitant change in the Stark effect, values of $E_{DD^{*}} - E_{CT}$ may be deduced.

Another handle on the energies of the CT states and their possible involvement in electron transfer would be the determination of the *change in transfer kinetics* in the presence of an electric field. Experiments along these lines have so far only been performed on the charge recombination kinetics between Q_A^- and D^+ [33-36], in which CT states are not believed to play a role. However, it would be instructive to measure the effect of an electric field on the forward electron transfer rate between D^+ and ϕ. The possible involvement of CT states in this process has been discussed in section VIC.

Finally, we wish to stress again the importance of *theoretical studies* to understand the Stark spectrum in terms of the different CT states. In conclusion, we believe that Stark spectroscopy has a promising future in elucidating the primary photochemistry and electron transfer mechanisms in RCs.

ACKNOWLEDGEMENT

We thank Roger A. Isaacson for important technical help, Ed Abresch for the preparation of the RCs, K. Satoh for help with the PSII RCs, D. Fredkin for helpful discussions, and M. Plato for making his theoretical results available to us. This work was supported by grants from the National Science Foundation (DMB 85-18922 and DMB 87-04920) and from the National Institutes of Health (GM13191) and a research scholarship to M. L. from the Deutsche Forschungsgemeinschaft (LO 352/1).

REFERENCES

1. J. Jortner and M. E. Michel-Beyerle, Some aspects of energy transfer in antennas and electron transfer in reaction centers of photosynthetic bacteria, *in*: "Antennas and Reaction Centers of Photosynthetic Bacteria," M. E. Michel-Beyerle, ed. Springer, Berlin, pp. 345-366 (1985).
2. D. deLeeuw, M. Malley, G. Buttermann, M. Y. Okamura, and G. Feher, The Stark effect in reaction centers from *Rhodopseudomonas sphaeroides*, *Biophys. J.* (Abstr.), 111a (1982).

3. D. J. Lockhart and S. G. Boxer, Magnitude and direction of the change in dipole moment associated with excitation of the primary electron donor in *Rhodopseudomonas sphaeroides* reaction centers, *Biochemistry* **26**:664-668 (1987). (Correction in *Biochemistry* **26**:2958 (1987)).
4. M. Lösche, G. Feher and M. Y. Okamura, The Stark effect in reaction centers from *Rhodobacter sphaeroides* R-26 and *Rhodopseudomonas viridis*, *Proc. Natl. Acad. Sci. USA* **84**:7537-7541 (1987).
5. H. P. Braun, M. E. Michel-Beyerle, J. Breton, S. Buchanan and H. Michel, Electric field effect on absorption spectra of reaction centers of *Rb. sphaeroides* and *Rps. viridis*, *FEBS Lett.* **221**:221-225 (1987).
6. W. Liptay, Dipole moments and polarizabilities of molecules in excited electronic states, *in* "Excited States", E. C. Lim, ed., Academic Press, Inc., New York, pp. 129-229 (1974).
7. R. Reich and S. Schmidt, Über den Einfluss elektrischer Felder auf das Absorptionsspektrum von Farbstoffmolekülen in Lipidschichten. I. Theorie, *Ber. Bunsenges. Phys. Chem.* **76**:589-598 (1972).
8. R. Reich, Intrinsic probes of charge separation, *in* "Light-Induced Charge Separation in Biology and Chemistry", H. Gerischer and J. J. Katz, eds., Verlag Chemie, Weinheim (West Germany), pp. 361-387 (1979).
9. C. J. F. Böttcher, "Theory of Electric Polarization", Elsevier, New York, pp. 159-204 (1973).
10. M. Malley, G. Feher and D. Mauzerall, The Stark effect in porphyrins, *J. Mol. Spectrosc.* **25**:544-548 (1968).
11. O. Nanba and K. Satoh, Isolation of a photosystem II reaction center consisting of D-1 and D-2 polypeptides and cytochrome b-559, *Proc. Natl. Acad. Sci. USA* **84**:109-112 (1987).
12. M. Y. Okamura, K. Satoh, R. A. Isaacson, and G. Feher, Evidence of the primary charge separation in the D_1 D_2 complex of photosystem II from spinach: EPR of the triplet state, *in*: "Progress in Photosynthesis Research," Vol. I, pp. 379-381, J. Biggins, ed., Martinus Nijhoff, Boston (1986).
13. M. Lösche, K. Satoh, G. Feher, and M.Y. Okamura, Stark effect in PSII from spinach, *Biophys. J.* (Abstract), February 1988, in press.
14. J. C. Williams, L. A. Steiner, and G. Feher, Primary Structure of the Reaction Center From *Rhodopseudomonas sphaeroides*, *Proteins* **1**:312-325 (1986).
15. R. A. Marcus, Superexchange versus an intermediate $BChl^-$ mechanism in reaction centers of photosynthetic bacteria, *Chem. Phys. Lett.* **113**:471-477 (1987).
16. M. E. Michel-Beyerle, M. Plato, J. Deisenhofer, H. Michel, M. Bixon and J. Jortner, Unidirectionality of charge separation in reaction centers of photosynthetic bacteria, *Biochim. Biophys. Acta*, in press.
17. P. O. J. Scherer and S. F. Fischer, On the Stark effect for bacterial photosynthetic reaction centers, *Chem. Phys. Lett.* **131**:153-159 (1986).
18. A. Ogrodnik, N. Remy-Richter, M. E. Michel-Beyerle and R. Feick, Observation of activationless recombination in reaction centers of *R. sphaeroides*. A new key to the primary electron-transfer mechanism, *Chem. Phys. Letters* **135**:576-581 (1987).
19. R. A. Marcus, Electron transfer and the bacterial photosynthetic reaction centers; in these Proceedings.
20. S. F. Fischer and P. O. J. Scherer, Analysis of different mechanisms for the initial charge separation within the reaction center *Rps. viridis*; in these Proceedings.
21. W. W. Parson and A. Warshel, Calculations of electronic interactions in reaction centers; in these Proceedings.
22. W. W. Parson, N. W. T. Woodbury, M. Becker, C. Kirmaier and D. Holten, Kinetics and mechanisms of initial electron-transfer reactions in *Rhodopseudomonas sphaeroides* reaction centers, *in*: "Antennas and Reaction Centers of Photosynthetic Bacteria," M. E. Michel-Beyerle, ed. Springer, Berlin, pp. 278-285
23. R. S. Mulliken and W. B. Person, "Molecular Complexes", Wiley-Interscience, New York, pp. 9-22 (1969).

24. J. P. Allen, G. Feher, T. O. Yeates, H. Komiya and D. C. Rees, Structure of the reaction center from *Rhodobacter sphaeroides* R-26: The cofactors, *Proc. Natl. Acad. Sci. USA* **84**:5730-5734 (1987).

25. H. Michel, O. Epp and J. Deisenhofer, Pigment-protein interactions in the photosynthetic reaction centers from *Rhodopseudomonas viridis*, *EMBO J.* **5**:2445-2451 (1986).

26. W. W. Parson and A. Warshel, Spectroscopic properties of photosynthetic reaction centers. II. Application of the theory to *Rhodopseudomonas viridis*, *J. Am. Chem. Soc.* **109**:6152-6163 (1987).

27. M. Plato, personal communication.

28. P. Jursinic and Govindjee, Temperature dependence of delayed light emission in the 6 to 340 microsecond range after a single flash in chloroplasts, *Photochem. Photobiol.* **26**:617-628 (1977).

29. A. V. Shuvalov, A. V. Klevanik, A. V. Sharkov, Ju. A. Matveetz, and P. G. Krukov, Picosecond detection of BChl-800 as an intermediate electron carrier between selectively excited P_{870} and bacteriopheophytin in *Rhodospirillum rubrum* reaction centers, *FEBS Lett.* **91**:135-139

30. N. W. Woodbury, M. Becker, D. Middendorf and W. W. Parson, Picosecond kinetics of the initial photochemical electron-transfer reaction in bacterial photosynthetic reaction centers, *Biochemistry* **24**:7516-7521 (1985).

31. J. L. Martin, J. Breton, A. J. Hoff, A. Migus and A. Antonetti, Femtosecond spectroscopy of electron transfer in the reaction center of the photosynthetic bacterium *Rhodopseudomonas sphaeroides* R-26: Direct electron transfer from the dimeric bacteriochlorophyll primary donor to the bacteriopheophytin acceptor with a time constant of 2.8 ± 0.2 psec, *Proc. Natl. Acad. Sci. USA* **83**:957-961 (1986).

32. J. Breton, J. L. Martin, A. Migus, A. Antonetti and A. Orszag, Femtosecond spectroscopy of excitation energy transfer and initial charge separation in the reaction center of the photosynthetic bacterium *Rhodopseudomonas viridis*, *Proc. Natl. Acad. Sci. USA* **83**:5121-5125 (1986).

33. A. Gopher, Y. Blatt, M. Schönfeld, M. Y. Okamura, G. Feher, and M. Montal, The effect of an applied field on the charge recombination kinetics in reaction centers reconstituted in planar lipid bilayers, *Biophys. J.* **48**:311-320 (1985).

34. Z. D. Popovic, G. J. Kovacs, P. S. Vincent, G. Alegria, and P. L. Dutton, Electric field dependence of recombination kinetics in reaction centers of photosynthetic bacteria, *Chem. Phys.* **110**:227-237 (1986).

35. T. Arno, A. Gopher, M. Y. Okamura, and G. Feher, Dependence of the recombination rate $D^+Q_A^- \rightarrow DQ_A$ on the electric field applied to reaction centers from *Rb. sphaeroides* R-26 incorporated into a planar lipid bilayer, *Biophys. J.* (Abstract), February 1988, in press.

36. G. Feher, T. Arno, and M. Y. Okamura, The effect of an electric field on the charge recombination rate of $D^+Q_A^- \rightarrow DQ_A$ in reaction centers from *Rhodobacter sphaeroides* R-26; in these Proceedings.

THE NATURE OF EXCITED STATES AND INTERMEDIATES IN BACTERIAL PHOTOSYNTHESIS

Steven G. Boxer, Richard A. Goldstein, David J. Lockhart, Thomas R. Middendorf, and Larry Takiff

Department of Chemistry
Stanford University
Stanford, California 94305

INTRODUCTION

In bacterial reaction centers (RCs) charge separation is initiated by excitation of the special pair primary electron donor (P) to its first excited singlet state (^{1}P). Within a few ps an electron moves from P to I, the intermediate electron acceptor (Woodbury et al.,1985; Breton et al., 1986; Martin et al., 1986). As described in the paper by Flemming, Martin, and Breton in this volume, the rate of this initial reaction increases as the temperature is lowered. The electron transfers from $I^{\cdot}$ to Q_A within about 200ps at room temperature and somewhat faster at lower temperature (Kirmaier and Holten, 1987). For most of the experiments described in this paper we have used quinone-depleted RCs when the species is *Rhodobacter sphaeroides* and $Q^{\overline{\cdot}}$-RCs when the species is *Rhodopseudomonas viridis*. In this case the fate of the initial charge-separated radical pair state, $^1(P^{\dot{+}}I^{\dot{-}})$, is more complex, as illustrated in Figure 1. $^1(P^{\dot{+}}I^{\dot{-}})$ can decay either by charge recombination to ^{1}PI or ground state PI, or the spin multiplicity of the radical pair can evolve to $^3(P^{\dot{+}}I^{\dot{-}})$. $^3(P^{\dot{+}}I^{\dot{-}})$ can decay by charge recombination to ^{3}PI or the spin multiplicity can continue evolving back to $^1(P^{\dot{+}}I^{\dot{-}})$. ^{3}PI can decay by intersystem crossing or reform $^3(P^{\dot{+}}I^{\dot{-}})$. Interconversion between the singlet and triplet radical pair states is characterized by a parameter ω, whose value depends on the magnetic properties of the radicals $P^{\dot{+}}$ and $I^{\dot{-}}$ and on the externally applied magnetic field strength (Boxer et al., 1983).

Several topics from our current research were discussed at the NATO Workshop: photochemical holeburning of the primary electron donor; Stark effect spectroscopy at 77K on the absorption spectrum of both *Rb. sphaeroides* and *R. viridis* RCs; Stark effect spectroscopy on the fluorescence spectrum of *Rb. sphaeroides* RCs; the phosphorescence spectrum of the triplet states in both *Rb. sphaeroides* and *R. viridis* RCs and for pure monomeric bacteriochlorophyll (BChl) and bacteriopheophytin (BPheo); and the energetics of the initial charge separation step based on an analysis of the decay of the triplet state in very high magnetic fields.

PHOTOCHEMICAL HOLEBURNING SPECTROSCOPY

Several years ago we suggested that the unusually large linewidth of the Q_y absorption band of the special pair might be due to homogeneous broadening (Boxer, 1983). By exciting with a narrow bandwidth laser at various frequencies within the special pair absorption band, it is possible

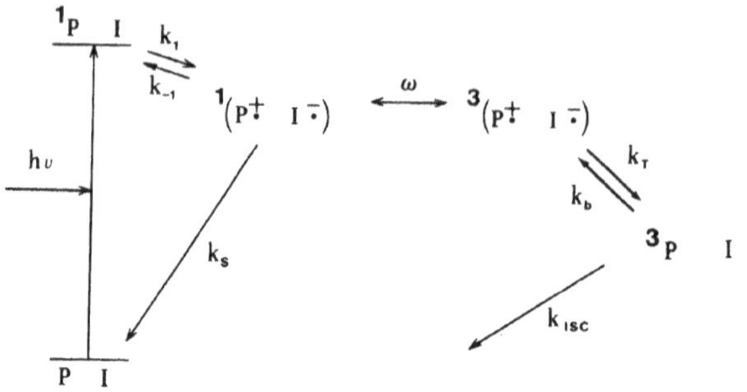

Figure 1. Reaction scheme used to analyze reaction dynamics in RCs in which electron transfer to the quinone has been blocked.

to measure the apparent homogeneous linewidth by measuring the lineshape of the transient bleaching of this band (the photochemical holeburning spectrum). In contrast to the very narrow holes which are typically observed for chromophores in glasses (Thijssen et al., 1985) or proteins (Boxer et al., 1987a), we observed very broad holes (~400cm^{-1}) for both *Rb. sphaeroides* (Boxer et al., 1986a) and *R. viridis* (Boxer et al., 1986b) RCs. Similar, though not quantitatively identical, results were obtained by Wiersma and co-workers who also performed accumulated photon echo measurements (Meech et al., 1985; 1986).

Two classes of interpretations have emerged to explain this unusual result. The first postulates an ultra-fast decay of the initially excited state. Wiersma and co-workers proposed a 25fs decay time based on the holewidth, and suggested that the decay involves formation of a charge transfer state of the special pair from the initially excited, essentially neutral ^{1}P state (Meech et al., 1986). We proposed the same mechanism, but suggested a lifetime on the order of 200fs, arguing that underlying vibronic structure must be considered (Boxer et al., 1986a,b); this value is also consistent with the photon echo measurements. We also proposed the alternative hypothesis that the broad hole was the result of a substantial difference in the equilibrium nuclear configuration between the ground and the excited state (Boxer et al., 1986a,b, 1987b). Such a difference could be the result of movement of the macrocycles or the surrounding medium following excitation or could indicate the existence of a highly dipolar initial excited state. The latter possibility seemed attractive because of the qualitative Stark effect data presented as an abstract by DeLeeuv et al. (1982). More elaborate treatments which are related to our proposal have been presented by Hayes and Small (1986) and Won and Friesner (1987; see also the Friesner paper in this volume).

STARK EFFECT ON THE ABSORPTION SPECTRUM OF BACTERIAL RCs

Stimulated by the holeburning results and the preliminary report of DeLeeuv et al. (1982), we made a quantitative analysis of the Stark effect (electromodulation) spectrum of bacterial RCs. The Stark effect provides quantitative information on the difference in permanent dipole moment between the ground and excited state ($|\Delta\mu_A|$), and on the angle ζ between $\Delta\mu$ and the transition dipole moment of the transition being probed. Details of the methodology can be found in Lockhart and Boxer (1987a). An example of an RC spectrum at 77K is shown in Fig. 2 (see also Boxer et al., 1987b; Lockhart and Boxer, 1987b). The change in absorbance for an immobilized, randomly oriented ensemble of molecules in an electric field is (approximately) proportional to the square of the electric field felt by the chromophores (F_{int}), the square of $|\Delta\mu_A|$, and the second derivative of the absorption spectrum. It is evident from the spectrum in Fig. 2 that $|\Delta\mu_A|$ for the special pair absorption band is substantially greater than that for the monomeric BChl or BPheo bands. Quantitative results are presented in Table 1. A complication in the analysis of these data is the local field correction, f, which accounts for the difference between the applied field F_{ext} and F_{int}: $F_{int} = f \cdot F_{ext}$. The value of f can be estimated from standard theories and is related to the local dielectric properties of the medium around each chromophore (a potentially very important quantity). Since the value of f is not known precisely, the data in Table 1 are presented as the observed value of $|\Delta\mu_A|$ assuming $F_{ext}=F_{int}$, with f as a correction (f is likely to be about 1.2). It is seen that both $|\Delta\mu_A|$ and ζ are identical, within experimental error, for the special pair of both species and that these values are substantially different from the values for the monomers. These observations have led us to suggest that the initial ^1P excited state is substantially more dipolar than the ground state, possibly due to mixing with a charge-transfer state involving charge-separation within the special pair. This mixing would likely cause the angle between $\Delta\mu$ and the transition dipole moment for the Q_y transition to be substantially less than 90° as observed (90° would be the expected angle between $\Delta\mu$ and the Q_y transition moment if the dimer states had perfect C_2-symmetry because the transition moment is believed to be approximately perpendicular to the RC C_2-axis and any charge asymmetry would have to be along the C_2-axis). Although physical pictures of this type are reasonable, detailed calculations which successfully predict the observed values of $|\Delta\mu|$ and ζ for both the monomers and the special pair are likely to provide the greatest insight. The contribution by Plato and co-workers in this volume describes such calculations.

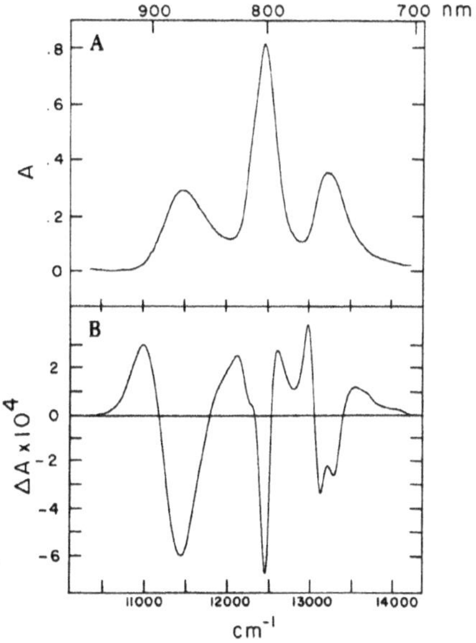

Figure 2. Absorption (A) and Stark effect (B) spectra for *Rb. sphaeroides* RCs in a PVA film in the Q_y region at 77K ($F_{ext} = 2.59 \times 10^5$V/cm).

Table 1. Experimental values for the magnitude of the change in dipole moment, $|\Delta\mu|$, and the angle, ζ, between $\Delta\mu$ and the transition dipole moment for the Q_y transition of the special pair in reaction centers and for pure bacteriochlorophylls.

	T (°K)	$\lvert\Delta\mu\rvert$ (D) [a]	ζ
R. viridis special pair	298	$(10.5\pm0.7)/f$	$36.7\pm2°$
	77	$(6.5\pm0.4)/f$	$37.1\pm2°$
Rb. sphaeroides specialpair	298	$(9.6\pm0.7)/f$	$39.5\pm2°$
	77	$(7.0\pm0.5)/f$	$38.1\pm2°$
BChl b (6-coordinate)	77	$(2.9\pm0.17)/f$	$22.1\pm2°$
BChl a (6-coordinate)	77	$(2.4\pm0.14)/f$	$11.6\pm2°$
BPheo b	77	$(2.6\pm0.15)/f$	$23.8\pm2°$
BPheo a	77	$(2.6\pm0.15)/f$	$9.5\pm2°$

[a] The value of the local field correction f may be different for each chromophore in its particular environment and at different temperatures.

STARK EFFECT ON THE FLUORESCENCE SPECTRUM OF *Rb. SPHAEROIDES* RCs

The fluorescence quantum yield is very small for RCs because of the high rate of forward electron transfer. At low temperatures, where all of the fluorescence is prompt (the back reaction is a thermally activated process), the quantum yield in *Rb. sphaeroides* RCs is approximately that predicted using the recently measured charge-separation rate and a reasonable estimate of several ns for the fluorescence decay rate constant. We have measured the effect of an electric field on this fluorescence using the same approach as described above for the Stark effect on absorption (Lockhart and Boxer, 1987c). In order to test the apparatus and properly adjust the phase for detection (necessary to determine the absolute sign of the change), we first examined BPheo a and BChl a. Each exhibited primarily a second derivative lineshape; a quantitative analysis demonstrated that $|\Delta\mu_F|$ (the difference dipole between the ground and emitting state) for BPheoa was about 30% smaller than $|\Delta\mu_A|$ (Lockhart and Boxer, 1987c).

The effect of an electric field on the RC fluorescence is shown in Fig. 3 and is seen to be completely different from that seen in absorption. The overall fluorescence *increases* in the electric field; the band shifts slightly to the red and there is little evidence for a second derivative component. We can estimate an upper limit for $|\Delta\mu_F|$ as follows. A high-quality fluorescence spectrum was obtained. This spectrum was fit using a sum of Gaussian components and the first and second derivatives of the fit were obtained (the Gaussian components are not taken to have physical significance; because of weak background fluorescence on the blue side of the band, only the red 2/3 of the band was used in the fit). The experimental Stark effect spectrum was fit to a linear combination of the zeroth, first and second derivatives of the emission lineshape (Fig. 4). It is possible to obtain an excellent fit without any contribution from the second derivative. We have taken a conservative approach in this fitting procedure and also show the deviation between the best fit and the observed spectrum for various fixed fractions of the second derivative component. The fractional contribution of the second derivative is directly related to the value of $|\Delta\mu_F|$ as shown. It is evident that $|\Delta\mu_F|$ can not be greater than $|\Delta\mu_A|$, and is likely smaller. There is one cautionary note in the interpretation of this data. The fluorescence quantum yield is very low, so it is possible that some fluorescence comes from RCs which are impaired such that electron transfer from ^1P to I is slower (the emission must be from an intact special pair because it is too far to the red to be associated with monomeric BChla).

This result can rule out the possibility that the state which is populated initially upon excitation decays in less than hundreds of fs into

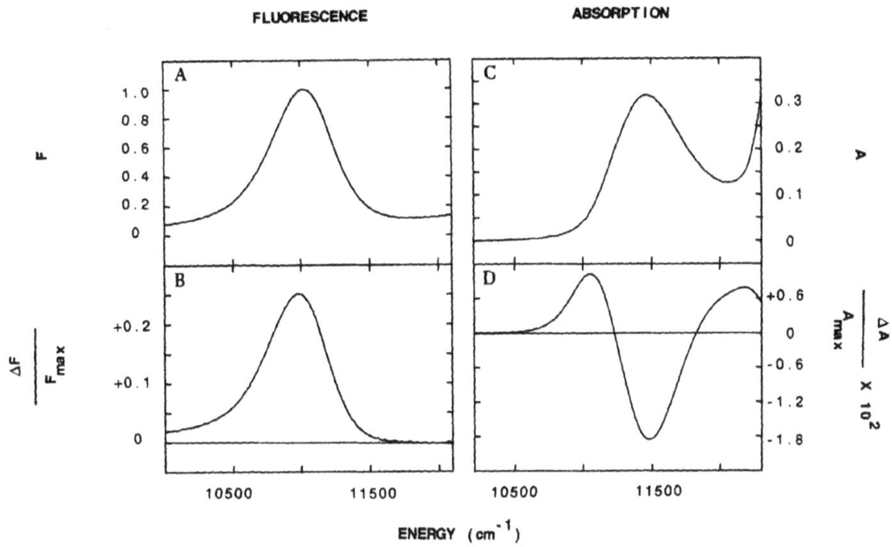

Figure 3. (A) Fluorescence spectrum of *Rb. sphaeroides* RCs in zero applied field at 77K; (B) the effect of an electric field on the fluorescence spectrum at 77K (F_{ext} = 8.9x10^5V/cm); (C) absorption spectrum of the Q_y transition of the special pair in zero applied field at 77K; (D) the effect of an electric field on the absorption spectrum at 77K (F_{ext} = 8.9x10^5V/cm). The same sample was used for all of the above spectra; similar results were obtained using samples that were five times less concentrated.

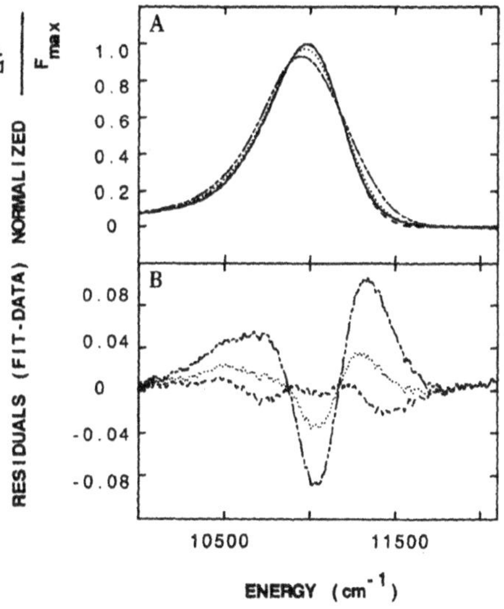

Figure 4. (A) The observed change in fluorescence in the presence of an electric field (——), the best fit to a sum of zeroth, first, and second derivative components of the fluorescence spectrum (---), and best fits using a fixed second derivative component of the fluorescence corresponding to $|\Delta\mu_F| \leq 4.7D/f$ (· · ·) and $|\Delta\mu_F| \leq 8.7D/f$ (— ·· —). The indicated values of $|\Delta\mu_F|$ are the maximum consistent with the magnitude of the second derivative contribution; (B) residuals (fit minus data) of the fits in (A).

a state with greater charge transfer character. Although the best-fit to the data suggests that $|\Delta\mu_P|<|\Delta\mu_A|$, a conservative interpretation is that the value is at most the same, but likely smaller. $|\Delta\mu_P|$ could be smaller than $|\Delta\mu_A|$ for several reasons. (i) There could be a reduction in the mixing with charge-transfer states as the excited state relaxes. (ii) The macrocycles may move more closely together as in excimer formation, or they may move parallel to each other in a sliding motion decreasing the Mg-to-Mg separation. Either motion could reduce the degree of charge separation (changes of 0.2Å correspond to changes of roughly 1D) or change the nature of the state as in (i).

The overall increase in the fluorescence intensity is ascribed to a net reduction in the the rate of $P^{+}\cdot I^{-}\cdot$ formation due to a change in the energy difference between the ^1P and $P^{+}\cdot I^{-}\cdot$ states in the presence of a field. The energy of the $P^{+}\cdot I^{-}\cdot$ state is expected to be very sensitive to an electric field since the dipole moment of this state is about 80D. Assuming that the fluorescence increase is due solely to a change in the rate of the initial electron transfer step, in an applied field of 8.9×10^5V/cm the average lifetime of the ^1P state for the isotropic sample increases by about 0.2ps.

PHOSPHORESCENCE SPECTRA OF RCs AND PURE BACTERIOCHLOROPHYLLS

The ^3P state is formed in high quantum yield upon photoexcitation of RCs at low temperature when Q_A is removed or reduced (Fig. 1). The absolute energy of this state has not been determined previously because it is difficult to sensitively detect emission in the 1100-1500nm wavelength range. Recent advances in liquid nitrogen cooled solid-state Ge-photodiode detectors make it possible to observe this region of the emission spectrum (our detector was generously provided by the ADC Corp., Fresno, CA).

The luminescence spectrum of Q-depleted *Rb. sphaeroides* R-26 RCs at 20K in the infrared is shown in Fig. 5 (Takiff and Boxer, 1988a). No signal was observed for Q-containing RCs (where the ^3P state is not formed); the same signal was observed for Q-reduced RCs in a glycerol/water glass. The luminescence decay rate was found to be identical to that measured for ^3P by transient absorption spectroscopy. At 80K the magnetic field effect on the emission is identical with that measured for ^3P by absorption. Taken together, these data demonstrate that the 1318nm emission band in Fig. 5 is due to phosphorescence from the triplet state of the primary electron donor. The emission maximum does not change appreciably with temperature; however, the emission linewidth does increase from $240cm^{-1}$ at 20K to $580cm^{-1}$ at 280K.

These measurements have been extended to *R. viridis* RCs (Figure 6) and to pure BChl*a*, BChl*b* and BPheo*a* (Takiff and Boxer, 1988a,b). The results are summarized in Table 2. The actual energy of the triplet state could be determined if the lineshape were known for the $S_0 \rightarrow T_1$ absorption band; there is no information on this band and it is unlikely that such a weak absorption can be detected. It is unlikely that the true triplet state energy is more than $\sim200cm^{-1}$ above the emission maxima given in Table 2. Assuming that the Stokes shifts are similar for the first singlet and triplet states (certainly true to within a few hundred cm^{-1}), the data in Table 2 provide information on the singlet-triplet splitting: it is seen that the S-T splitting for the special pair of both species is comparable and that both are about 30% smaller than those of their respective monomers. A reduction in the S-T splitting may result from a degree of delocalization of the excited state of the special pair due to partial charge transfer (Nagakura, 1975), which is consistent with the reduction in the zero-field-splittings in the ^3P state relative to monomeric BChls (Levanon and Norris, 1978). The observation that the phosphorescence linewidth of the special pair is close to that of the monomer in an organic glass, but only about half that of the absorbance and fluorescence of the special pairs, suggests a stronger interaction of the singlet excited state

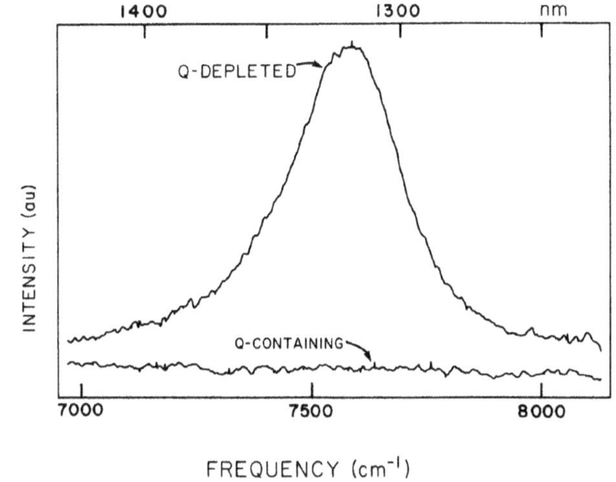

Figure 5. Luminescence spectrum of quinone-depleted *Rb. sphaeroides* RCs in a PVA film at 20K compared with that of a comparably concentrated quinone-containing RC film taken under identical conditions.

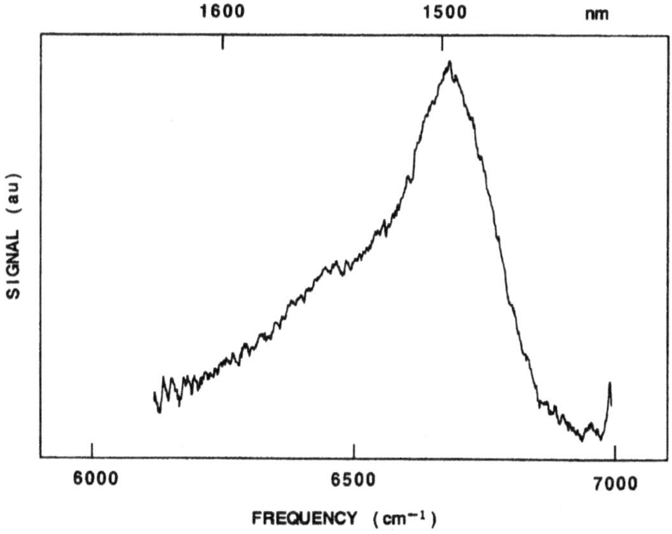

Figure 6. Luminescence spectrum of quinone-reduced *R. viridis* RCs in a deuterated glycerol/buffer glass at 77K.

than the triplet state with the protein matrix. The similarity in the S-T splitting for the two species runs counter to a current hypothesis which suggests that the triplet state in *Rb. sphaeroides* RCs is delocalized while that in *R. viridis* is localized (Norris et al., 1987).

Table 2. Absorption, fluorescence and phosphorescence maxima for reaction centers and pure bacteriochlorophylls.

Chromophore	Absorption λ_{max}	Fluorescence λ_{max}	Phosphorescence λ_{max}	λ_{max}(Fl)- λ_{max}(Phos)[a]
Rb. sphaeroides[b]	868nm $11520cm^{-1}$	919nm $10880cm^{-1}$	1318nm $7590cm^{-1}$	$3290cm^{-1}$
BChl a (6-coord)[c]	777nm $12870cm^{-1}$	785nm $12740cm^{-1}$	1226nm $8157cm^{-1}$	$4583cm^{-1}$
R. viridis[d]	1000nm $10000cm^{-1}$	1019nm $9810cm^{-1}$	1497nm $6680cm^{-1}$	$3130cm^{-1}$
BChl b (6-coord)[e]	818nm $12320cm^{-1}$	821nm $12180cm^{-1}$	1255nm $7970cm^{-1}$	$4210cm^{-1}$
BPheo a[f]	752nm $13300cm^{-1}$	759nm $13180cm^{-1}$	1097nm $9110cm^{-1}$	$4070cm^{-1}$

[a] Singlet-triplet splitting if corrected for possible difference in Stokes shift between absorption and fluorescence vs. triplet absorption (not measured) and phosphorescence; [b] PVA film, 20K; [c] 2-MeTHF/pyridine glass, 77K; [d] D_2O/D-exchanged glycerol glass, 77K; [e] Toluene/pyridine glass, 77K; [f] 2-MeTHF glass, 77K.

ENERGETICS OF INITIAL CHARGE SEPARATION IN *Rb. SPHAEROIDES* RCS

Several years ago we discovered that the ^3P decay rate depends on an applied magnetic field in Q-depleted *Rb. sphaeroides* RCs at room temperature (Chidsey et al., 1985). We proposed that ^3P could decay either by magnetic field independent intersystem crossing with rate k_{isc}, or by activated formation of the $^3(P^{+}I^{-})$ state, singlet-triplet (S-T) evolution to $^1(P^{+}I^{-})$ and decay by k_S (see Fig. 1). Consistent with this hypothesis, the observed ^3P decay rate, k_{obs}, decreases and becomes independent of magnetic field strength as the temperature is lowered. Based on a detailed theoretical analysis (Chidsey et al.,1985) we showed that k_{obs} is given by the following expression:

$$k_{obs} = k_{isc} + (1/3)k_S\Phi_{^3PI}\exp(-\Delta G^\circ/kT),$$

where ΔG° is for $^3(P^{+}I^{-})\rightarrow{}^3P$, $\Phi_{^3PI}$ is the quantum yield of ^3P, and k is the Boltzmann constant. For the derivation of this expression we assumed that nuclear and electron spin polarization were not factors and that anisotropic magnetic interactions could be ignored. We noted, however, that the data deviated from the linear relationship between the ^3P decay rate and ^3P quantum yield predicted by this expression. We suggested that this discrepancy might be the result of nuclear spin polarization in the ^3P state, a notion which we have discussed in a detailed theoretical paper (Goldstein and Boxer, 1987). Nuclear spin polarization effects are expected to be most significant at relatively low magnetic fields (several thousand gauss or less), and will become negligible at much higher fields where the Δg-effect dominates S-T mixing in the radical pair. The magnetic field and temperature dependence of k_{obs} can provide a measure of the energy difference between the $^3(P^{+}I^{-})$ and ^3P states, a critical piece of information which, when combined with the phosphorescence measurement, makes it possible to obtain the driving force of the initial charge separation step, $^1P\rightarrow{}^1(P^{+}I^{-})$. For this reason we have studied k_{obs} at very high magnetic field strengths (Goldstein et al., 1988a). We note that any analysis the energetics of charge separation based on the equation above

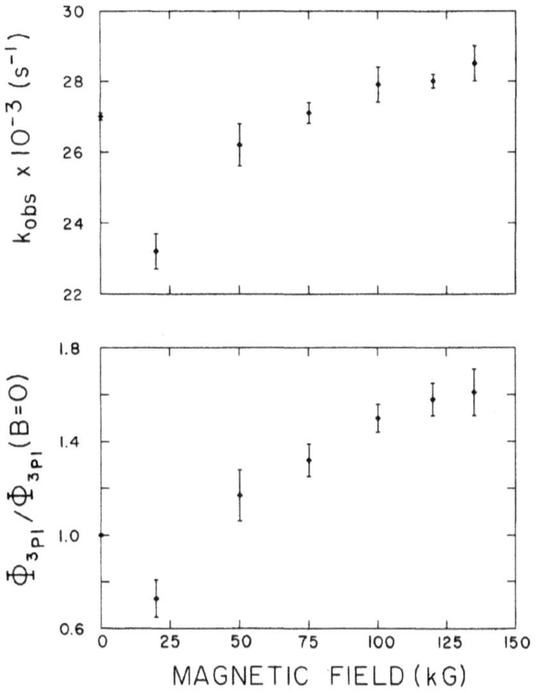

Figure 7. Dependence of (A) k_{obs} and (B) $\Phi_{^3PI}/\Phi_{^3PI}(B=0)$ on magnetic field between 0 and 135 kG at 15°C for quinone-depleted RCs in aqueous buffer. k_{obs} is the observed rate of ^3PI decay, $\Phi_{^3PI}$ is the quantum yield of ^3PI formation, and $\Phi_{^3PI}(B=0)$ is the ^3PI quantum yield at zero applied field.

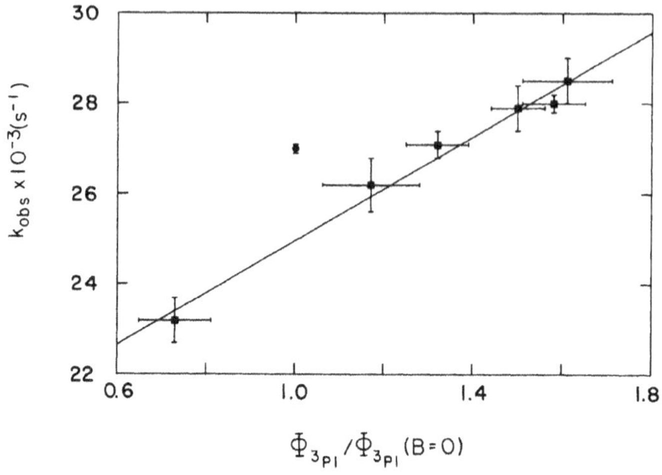

Figure 8. k_{obs} plotted as a function of relative triplet yield at 15°C from 0 to 135 kG. The sample was in aqueous buffer. The data from 20 kG to 135 kG (■) is fit to a straight line with intercept $19.17 \times 10^3 s^{-1}$ and slope $5.80 \times 10^3 s^{-1}$; the data point at 0 G (●) deviates significantly from this line.

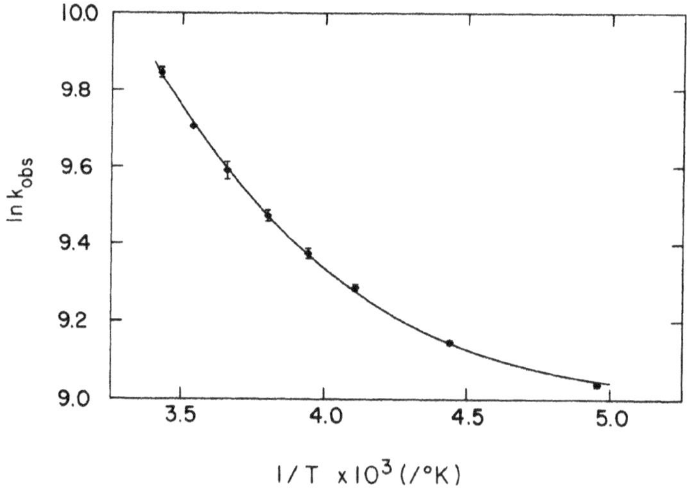

Figure 9. Temperature dependence of k_{obs} at 135 kG for RCs in viscous buffer. Solid line is a fit to an Arrhenius expression plus constant with a low-temperature asymptote of $8.05 \pm .2 \times 10^3 s^{-1}$, a slope of $1440 \pm 40 cm^{-1}$, and a pre-exponential of $13. \pm 3. \times 10^6 s^{-1}$.

and data at zero or low magnetic field is likely to be incorrect due to nuclear spin polarization effect.

The dependence of k_{obs} and the ^{3}P quantum yield (Goldstein et al., 1988b) on magnetic field up to 135kG are shown in Fig. 7. It is seen that that the field effects are parallel at high field, but deviate at low field. k_{obs} is plotted as a function of the relative ^{3}P yield in Fig. 8. It is seen that a good linear fit is obtained for all data obtained from 20 to 135kG, but data obtained at zero field deviates significantly from the fit. Fig. 9 presents the temperature dependence of k_{obs}, which is fit to an Arrhenius expression plus a constant. The high field data in Fig. 8 are taken under conditions where nuclear polarization is not a factor. From the slope of the plot in Fig. 8 and the expression above, we obtain $\Delta G°=1360\pm40cm^{-1}$. The room temperature point in Fig. 9 at 135kG is likewise free from complications due to nuclear spin polarization: from this point and from the low temperature asymptotic value of k_{obs} we obtain $\Delta G°=1350\pm50cm^{-1}$. From the temperature dependence of k_{obs} in Fig. 9 we obtain $\Delta H°=1442\pm42cm^{-1}$. These three values are obtained entirely independently. We conclude that $\Delta G°$ for $^{3}(P^{+}\cdot I^{-}\cdot)\rightarrow{}^{3}P$ is $1360\pm40cm^{-1}$ and that $\Delta G°\sim\Delta H°$, indicating that entropy changes are very small.

The free energy change for the $^{1}PI\rightarrow P^{+}\cdot I^{-}\cdot$ reaction can now be estimated as follows. We take the observed phosphoresence maximum (Table 1) and add a small Stokes shift to give a ^{3}P energy of $7730cm^{-1}$ above the ground state. Taking into account the spin multiplicity of the ^{3}P state we place the ^{3}P free energy $7500cm^{-1}$ above the ground state. Combined with value for the free energy difference between the ^{3}P and $^{3}(P^{+}\cdot I^{-}\cdot)$ states above, we obtain a free energy of $8860cm^{-1}$ for $^{3}(P^{+}\cdot I^{-}\cdot)$, and, accounting for the degeneracy of $^{3}(P^{+}\cdot I^{-}\cdot)$, of $9090cm^{-1}$ for the $^{1}(P^{+}\cdot I^{-}\cdot)$ state. The energy of the ^{1}PI state is known to be about $11,220cm^{-1}$ above the ground state; thus we obtain a free energy difference of $2130cm^{-1}$ (0.264eV) for the initial $^{1}PI\rightarrow{}^{1}(P^{+}\cdot I^{-}\cdot)$ electron transfer step. We note that these conclusions are quite different from those reached by Woodbury and Parson (1984).

Acknowledgements Various portions of this work are supported by the National Science Foundation and the Gas Research Institute. S.G.B. is the recipient of a Presidential Young Investigator Award.

REFERENCES

Boxer, S.G., Chidsey, C.E.D., and Roelofs, M.G., 1983, Ann. Rev. Phys. Chem. 34: 389.
Boxer, S.G., 1983, Biochim. Biophys. Acta Rev. Bioenerg. 726:265.
Boxer, S.G., Middendorf, T.R., and Lockhart, D.J., 1986a, Chem. Phys. Lett. 123:476.
Boxer, S.G., Lockhart, D.J., and Middendorf, T.R., 1986b, FEBS Lett. 200:237.
Boxer, S.G., Gottfried, D., Lockhart, D.J., and Middendorf, T.R., 1987a, J. Chem. Phys., 86:2439.
Boxer, S.G., Lockhart, D.J. and Middendorf, T.R., 1987b, Springer Proc. Phys., 20:80.
Breton, J., Martin, J.-L., Migus, A., Antonetti, A. and Orszag, A., 1986, Proc. Natl. Acad. Sci. U.S.A., 83:5121.
Chidsey, C.E.D., Takiff, L., Goldstein, R., and Boxer, S.G., 1985, Proc. Natl. Acad. Sci. U.S.A. 82:6850.
DeLeeuv, D., Malley, M., Butterman, G., Okamura, M.Y., and Feher, G., 1982, Biophys. J., 37:111a (abstract).
Goldstein, R.A. and Boxer, S.G., 1987, Biophys. J., 51:937.
Goldstein, R.A., Takiff, L. and Boxer, S.G., 1988a, Biochim. Biophys. Acta, submitted.
Goldstein, R.A., Takiff, L. and Boxer, S.G., 1988b, Biochim. Biophys. Acta, submitted.

Hayes, J.M. and Small, G.J., 1986, J. Phys. Chem., 90:4928.
Kirmaier, C. and Holten, D., 1987, Photosyn. Res., 13:225.
Levanon, H. and Norris, J.R., 1978, Chem. Rev., 78:185.
Lockhart, D.J. and Boxer, S.G., 1987a, Biochem. 26:664.
Lockhart, D.J. and Boxer, S.G., 1987b, Proc. Natl. Acad. Sci, in press.
Lockhart, D.J. and Boxer, S.G., 1987c, Chem. Phys. Lett., in press.
Martin, J.-L., Breton, J., Hoff, A.J., Migus, A. and Antonetti, A., 1986, Proc. Natl. Acad. Sci. U.S.A., 83:957.
Meech, S.R., Hoff, A.J., and Wiersma, D.A., 1985, Chem. Phys. Lett., 121:287.
Meech, S.R., Hoff, A.J., and Wiersma, D.A., 1986, Proc. Natl. Acad. Sci. U.S.A., 83:9464.
Nagakura, S., 1975, in "Excited States," E. Liu, ed., 2:321, Academic Press, New York.
Norris, J.R., Liu C.P., and Budil, D.E., 1987, J. Chem. Soc. Faraday Trans., 83:12.
Takiff, L. and Boxer, S.G., 1988a, Biochim. Biophys. Acta, in press.
Takiff, L. and Boxer, S.G., 1988b, J. Am. Chem. Soc., submitted.
Thijssen, H.P.H., van den Berg, R., and Völker, S., 1985, Chem. Phys. Lett. 120:503.
Won, Y. and Friesner, R.A., 1987, Proc. Natl. Acad. Sci. U.S.A., 84:5511.
Woodbury, N.W. and Parson, W.W., 1984, Biochim. Biophys. Acta, 767:345.

ON THE ENERGETICS OF THE STATES $^1P^*$, $^3P^*$ and P^+H^- IN REACTION CENTERS OF *Rb. SPHAEROIDES*

A. Ogrodnik, M. Volk and M.E. Michel-Beyerle

Institut für Physikalische und Theoretische Chemie
Technische Universität München
Lichtenbergstr. 4, 8046 Garching (FRG)

ABSTRACT

Magnetic field dependent recombination measurements together with magnetic field dependent triplet lifetimes [Chidsey et al. (1985), Proc. Natl. Acad. Sci USA **82**, 6850] allowed the determination of the free energy change $\Delta G(P^+H^- - {}^3P^*) = 0.15$-$0.16$ eV between 185K-290K. This value being (almost) temperature independent indicates $\Delta G(P^+H^- - {}^3P^*) \simeq \Delta H(P^+H^- - {}^3P^*)$ and is consistent with $\Delta G({}^1P^* - P^+H^-)$ and $\Delta H({}^1P^* - {}^3P^*)$ from previous delayed fluorescence and phosporescence data implying $\Delta G \simeq \Delta H$ for all combinations of these states.

I. INTRODUCTION

The primary electron transfer process in reaction centers (RCs) of *Rb. sphaeroides* proceeds from the bacteriochlorophyll dimer in its excited singlet state ($^1P^*$) to bacteriopheophytin (H) forming the radical pair (RP) state P^+BH^- with a rate $k_1 = 3.6\ 10^{11}\ s^{-1}$ [1] at room temperature. Femto- and picosecond time-resolved spectroscopy is excluding a kinetic involvement [2-4] of the accessory bacteriochlorophyll (B) located between P and H as revealed in the X-ray structure analysis [5-7]. A complementary experimental key to the mechanism of the primary charge separation is the magnetic field dependent recombination dynamics of P^+BH^- [8-10] leading to either the triplet state ($^3P^*$) or the ground state (PBH). For any mechanistic interpretation [11-15] of the forward and the reverse electron transfer rates the free energy difference ΔG between $^1P^*$ and P^+BH^- and between P^+BH^- and $^3P^*$ is essential.

Recombination occurs when electron transfer between P^+BH^- and ubiquinone is blocked, e.g. by extraction of the quinone. In this case the RP initially formed in its singlet state $^1(P^+BH^-)$ undergoes singlet-triplet mixing under the influence of hyperfine interaction (HFI), yielding the triplet-phased RP $^3(P^+BH^-)$. Recombination of $^3(P^+BH^-)$ into $^3P^*$ with the rate k_T will now compete with recombination of $^1(P^+BH^-)$ leading to the ground state (PBH) with the overall singlet rate k_S which represents both, the direct process and the indirect route via $^1P^*$. As indicated in Fig.1, the energy splitting between the singlet and the triplet RP states impedes HFI induced singlet-triplet mixing. This allows to manipulate the singlet and triplet recombination yields via Zeeman interaction in an external magnetic field (H).

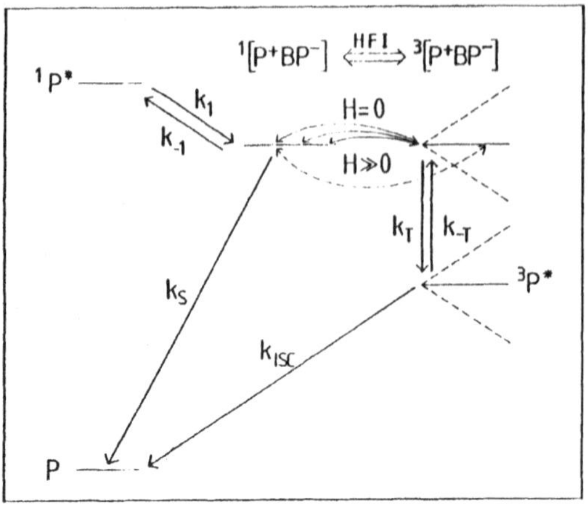

Figure 1: Reaction scheme for charge separation and recombination of singlet and triplet excited states connected via the radical pair mechanism.

The observation of a pronounced temperature dependence [16,17] together with the magnetic field dependence of the thermally activated contributions to the observed $^3P^*$ decay rates [18] points to an additional decay path of $^3P^*$. This pathway involves temperature activated back-transfer of the electron to form $^3[P^+BH^-]$ with the rate k_{-T}, HFI induced formation of $^1(P^+BH^-)$ and final decay to PBH with the rate k_S.

With the knowledge of the parameters k_S, k_T and the triplet recombination yield and their temperature dependence for quinone-depleted RCs of *Rb.sphaeroides*, the rate k_{-T} will be evaluated using the temperature and magnetic field dependent $^3P^*$ lifetimes reported by Chidsey et al. [18]. The ratio k_T/k_{-T} allows the determination of the free energy difference $\Delta G(RP\text{-}^3P^*)$. The result will be related to $\Delta G(^1P^*\text{-}RP)$ and $\Delta G(^1P^*\text{-}^3P^*)$ as derived from delayed fluorescence [19-21] and to the recent phosphorescence data [22] yielding $\Delta H(^1P^*\text{-}^3P^*)$ of quinone-depleted RCs of *Rb.sphaeroides*.

II. ANALYSIS OF MAGNETIC FIELD DEPENDENT TRIPLET LIFETIMES

Based on a suitable treatment [23] of the stochastic Liouville equation the observed overall decay rate of $^3P^*$ is according to [18] given by

$$k_{OBS}(H) = k_{ISC} + \frac{1}{3}\frac{k_S}{k_T}\Phi_T(H)\,k_{-T} \tag{1}$$

with k_{ISC} denoting the rate of intersystem crossing $^3P \rightarrow P$.

In order to eliminate k_{ISC} from Eq.(1), we express the rate k_{-T} for the charge separation from $^3P^*$ by the parameters k_{OBS} and Φ_T in different magnetic fields, H_1 and H_2 according to

$$k_{-T} = \frac{k_{OBS}(H_1) - k_{OBS}(H_2)}{\Phi_T(H_1) - \Phi_T(H_2)}\,\frac{3k_T}{k_S} \tag{2}$$

For the evaluation of k_{-T} the experimental derivation of temperature dependent values for $\Phi_T(H_1)$ and k_S will be presented in the following. The rate $k_T = 6\ 10^8\ s^{-1}$ has been shown to be temperature independent in the range 295-90 K [10].

III. TEMPERATURE AND MAGNETIC FIELD DEPENDENT RECOMBINATION MEASUREMENTS

EXPERIMENTAL DETAILS

Pump and probe experiments were performed with two dye lasers (pulse width 1.4 ns) being pumped by a N_2-laser. The concentration c_0 of quinone-depleted RCs of *Rb.sphaeroides,R26* (50 Vol.% glycerol) was adjusted to give an optical density of 0.5 at the detection wavelengths (pathlength d=2 mm). The samples were excited at 600 nm with an energy density of 0.5 mJ/cm² corresponding to an excitation yield of Φ_{EXC} = 30 % excited RCs/pulse. Time-dependent absorption changes ΔA_P were detected in the Q_y transition of P around 880nm. Measurements were performed at zero magnetic field and at the saturating field strength of 700G in the temperature range 295 K to 90 K. The contribution ΔA_Q to the measured absorption change due to residual quinone was tested in situ by monitoring the decay of ΔA_P resulting from recombination of the state P^+BHQ^-, and was accounted for. This test on quinone was achieved with a pulse sequence of one excitation pulse followed by 5 probing pulses at the repetition rate (15 Hz) of the laser. On the timescale of the RP lifetime the excitation pulse can be regarded as a delta-function, ensuring that no nuclear spin polarization can be built up due to multiple turnovers during excitation [24]. Since the average turnover-rate is smaller than 2 s^{-1} the nuclear spin relaxation time in the RC ground state has to be considerably slower while concomitantly it has to be of the order of the triplet lifetime in the $^3P^*$ state to build up a steady state nuclear spin polarization. Preliminary checks show no changes of the experimental data, when the repetition rate of excitation was changed to 0.1 Hz, indicating either a very slow nuclear spin relaxation in the ground state or the mere fact that no spin polarization is accumulated.

THE RADICAL PAIR LIFETIME AND TRIPLET YIELDS

ΔA_P consists of a bleaching due to the contributions of both, P^+BH^- and of $^3P^*$. The contribution from P^+BH^- is exclusive at time t=0 and decays completely with the RP lifetime τ. At long times the signal is completely originating from bleaching of the P absorption in the presence of $^3P^*$. The triplet yield is expected to evolve approximately as the decay product of a exponentially decaying parent. Deviations due to the quantum mechanical nature of singlet triplet mixing are expected to be smaller than the accuracy of measurement, since $k_T >> k_S$ is valid throughout the temperature range, as we will see later. Thus we fit our data to the relation

$$\Delta A_P(t) = d\, c_0\, \epsilon_{880}\, \Phi_{EXC} \left[e^{-t/\tau_{RP}} + \Phi_T(1 - e^{-t/\tau_{RP}})\right] - \Delta A_Q \qquad (3)$$

The RP lifetimes τ_{RP} are listed in Table I(A), the triplet yields in Table I(B). For comparison, at room temperature lifetimes of 13 ns (0G) and 17 ns (1 kG) have been reported [25,26]; also the room temperature triplet yield is in good agreement with previous values [25,26].

THE RECOMBINATION RATES k_S AND k_T

In order to analyse the magnetic field dependent $^3P^*$ lifetimes $1/k_{obs}$ we need knowledge on k_S and k_T. From analysis of the lifetime broadening of the RP state as reflected in the half width at half maximum (HWHM) of the magnetic field dependence of the triplet yield of room temperature data [10] we know that at least one of the recombination rates is $\simeq$ (3-6) 10^8 s^{-1} . Since the inverse RP lifetimes are considerably smaller than this value, a bottleneck has to be effective in the RP decay channel. Apparantly, the hyperfine induced singlet triplet-mixing constitutes this bottleneck and identifies the triplet recombination rate k_T to be responsible for the lifetime broadening. Since a magnetic field reduces HFI induced singlet-triplet mixing, the RP lifetime is expected to increase or decrease in case the triplet channel is faster or slower than the singlet channel, respectively. An increase of the RP lifetime at 1kG was indeed observed (Table I,A) corroborating $k_T > k_S$.

Low temperature measurements of the magnetic field dependence of the triplet yield as reflected in the absorption changes 44 ns after excitation at 543nm showed almost a constant width of the singlet triplet resonance as a function of temperature [10] indicating k_T to be temperature independent. We conducted improved measurements of the magnetic field dependence of the triplet yield utilizing absorption changes $\Delta A_P(H,t{=}92ns)$ of the Q_y absorption band of P at considerably longer delay times (not shown), confirming the results obtained in [10].

As shown by Haberkorn et al. [23] the RP lifetime can be calculated from

$$\tau = \frac{\Phi_S}{k_S} + \frac{\Phi_T}{k_T} \qquad (4)$$

Eq.(4) is the basis for the determination of k_S. Because of the S-T-mixing being the bottleneck in $^3P^*$ formation, the triplet yield is smaller with respect to the recombination rate k_T than the singlet yield with respect to k_S. Therefore, the RP-lifetime is closely connected to the decay rate k_S in the singlet decay channel and the contribution of the second term in Eq.(4) is small. The rate k_S calculated from Eq.(4) and the data given in Table I(A) and (B), are listed in Table I(C) using $\Phi_S + \Phi_T = 1$. The rates obtained from the measurements at high and low field are fairly consistent. The rate k_S obtained from Eq.(4) is in good agreement with the value $k_S = 5.6\ 10^7$ [26] obtained from a more rigorous analysis, involving the explicit numerical solution of the stochastic Liouville Equation and its fit to a complete set of magnetic field and time dependent data at room temperature. The surprisingly small temperature dependence of τ is due to the small temperature variation of k_S decreasing by a factor of $\simeq 4$ between 295 K and 90 K. Such weak temperature dependence can be expected for electron transfer reactions in the inverted regime involving free energy changes larger than the reorganization energy [27].

IV. KINETICS AND ENERGETICS ASSOCIATED WITH $[P^+BH^-]$ and $^3P^*$

THE CHARGE SEPARATION RATE k_{-T}

Inverse lifetimes of the triplet state $^3P^*$ at magnetic fields of 0 G and 1 kG were extracted from Fig.3 in [18] and are listed in Table I(D) for various temperatures. Table I(E) shows the charge separation rate k_{-T} calculated from the magnetic field modulation of the $^3P^*$ lifetime according to Eq.(2). Triplet yields obtained at 700 G were put together with triplet lifetimes determined at 1 kG. This certainly is justified, because the triplet yield is expected to be constant in this magnetic field range, since Zeeman splitting exceeds by far the hyperfine interaction and additional singlet triplet mixing due to differences in g-value of the to radicals are not important yet. At 90 K k_{-T} cannot be evaluated, since the change of k_{OBS} with magnetic field is to small.

In order to test the consistency of the obtained data, we calculate the intersystem crossing rate k_{ISC} from Eq.(1). As shown in Table I(F,a), k_{ISC} is almost constant with temperature, the low temperature value coinciding with the value given in [18].

As pointed out in [24] Eq.(1) is only valid exactly if nuclear spin lattice relaxation time is shorter than the $^3P^*$ lifetime. In order to test the implication of nuclear spin polarization on the observed triplet decay rate we follow the approach given in [24] (see FOOTNOTE). Eq.(19) in [24] at the two field values 0G and 700G provides two equations for k_{OBS}, thus giving the possibility to obtain k_{-T} and k_{ISC} by a least square fit with a Marquardt-algorithm from the experimental data. For the triplet yield as a function of the hyperfine field we utilized Eqs.(7) and (8) of [23]. As shown in Tab. I(E,b) the rate k_{-T} obtained by accounting for nuclear spin polarization is only slightly smaller than the one obtained after equilibration. In fact, the influence of nuclear spin polarization is smaller than the uncertainty of measurement. The accompanying intersystem crossing rates in Tab. I(F,b) change only slightly as well.

CHANGE OF FREE ENERGY

Table I(G,a) and (G,b) lists the change in free energy $\Delta G(RP\text{-}^3P^*)$ as a function of temperature calculated from

$$k_{-T} = k_T \exp(-\Delta G/kT) \tag{5}$$

without and with accounting for nuclear spin polarization, respectively. The values are the same in both cases within the error of measurement. Note that these values do not depend on k_T, since it cancells in Eqs.(2) and (5). In Table I(E,b) we explicitly calculated k_{-T} to be twice as large as indicated in Table I(E,b) if we double the value of k_T. The small reduction of $\Delta G(RP\text{-}^3P^*)$ with temperature (295K-185K) by an amount of 0.017 ± 0.015 eV (0.018 ± 0.015 eV in the absence of nuclear spin polarization) is almost within the order of uncertainty.

Evaluation of the energy difference $\Delta H(RP\text{-}^3P^*)$ from the temperature dependence of k_{OBS} [18] did not account for temperature dependent changes of the parameters involved in Eq.(1). We recalculate ΔH (and the entropy change ΔS) from the new experimental data. Assuming ΔH and ΔS to be temperature independent a linear regression to the temperature dependent data gives an estimate of $\Delta H(RP\text{-}^3P^*) = 0.12 \pm 0.05$ eV in agreement with ref.[18]. The corresponding value of the entropy change is $\Delta S(RP\text{-}^3P^*) = -(1.6\ 10^{-4} \pm 1.5\ 10^{-4})$ eV/K, the sign indicating an entropy reduction upon charge separation.

V. COMPARISON WITH DATA FROM DELAYED FLUORESCENCE AND PHOSPHORESCENCE MEASUREMENTS

From the temperature dependence of the μs delayed fluorescence $\Delta H(^1P^*\text{-}^3P^*) = 0.4$ eV was deduced [21]. Phosphorescence of the $^3P^*$ state has recently been detected at 1317 nm independently of temperature [22]. Together with the fluorescence peaks at 913 nm and 925 nm for 295 K and 90 K respectively [29], estimates of $\Delta H(^1P^*\text{-}^3P^*) \simeq 0.42$ and 0.40 eV can be made, independent of temperature.

Time-resolved fluorescence measurements have been performed on quinone-depleted RCs with sub-ns [20] and ns [19] time resolution yielding free energy differences of $\Delta G(^1P^*\text{-}RP) = 0.148$ eV and 0.26 eV, respectively. Though two different kinetic models have been invoked in the analysis [29,30], this does not account for the discrepancy [31]. The difference in data rather originates from the mere fact that the smaller value pertains to a non-relaxed RP state, while the larger one refers to a RP state with a lifetime in the 10 ns range which gives rise to magnetic field dependent recombination dynamics. However, independent of the two models used, $\Delta G(^1P^*\text{-}RP_{relaxed})$ can be extracted from the ratio of the quantum yields of the fastest (prompt) and the slowest fluorescence components [31]. Such an analysis of data taken from [20] yields $\Delta G(^1P^*\text{-}RP) = 0.245$ eV in good agreement with [19].

Together with $\Delta G(RP\text{-}^3P^*) = 0.153\text{-}0.166$ eV we get $\Delta G(^1P^*\text{-}^3P^*) = 0.40\text{-}0.42$ eV. Comparing this value with the phosphorescence data [22], we recognize that $\Delta G(^1P^*\text{-}^3P^*) \simeq \Delta H(^1P^*\text{-}^3P^*)$, indicating that $^1P^*$ and $^3P^*$ are similar in entropy. Since $\Delta G(RP\text{-}^3P^*)$ has only little entropy contribution as well, also the entropy change between $^1P^*$ and RP has to be small, implying $\Delta G \simeq \Delta H$ for all combinations of states $^1P^*$,RP and $^3P^*$ in quinone depleted RCs.

ACKNOWLEDGEMENTS

We thank our collaborators R. Letterer, U. Eberl and Dr. W. Lersch as well as Dr. E.W. Knapp for their involvement in the experimental work and fruitful discussions of the results. Financial support from the Deutsche Forschungsgemeinschaft (Sonderforschungsbereich 143) is gratefully acknowledged.

TABLE I

T [K]	290	270	250	230	185	90
A: RADICAL PAIR LIFETIMES						
$\tau_{RP}(0G)$ [ns] ±1	13.0	13.7	15.0	16.4	18.1	21.2
$\tau_{RP}(700G)$ [ns] ±1	15.8	16.9	18.1	21.3	23.2	34.3
B: TRIPLET YIELDS						
$\Phi_T(0G)$ ±.02	.30	.31	.34	.39	.45	.71
$\Phi_T(700G)$ ±.02	.19	.20	.24	.27	.36	.52
C: SINGLET RECOMBINATION RATES (total)						
$k_{S(0G)}$ [$10^7 s^{-1}$]±.6	5.8	5.4	4.8	4.0	3.3	1.5
$k_{S(700G)}$ [$10^7 s^{-1}$]±.6	5.3	4.9	4.4	3.6	2.9	1.5
D: TRIPLET DECAY RATES						
$k_{OBS}(0G)$ [10^4 s^{-1}]	1.96	1.48	1.24	1.08	0.86	0.73
$k_{OBS}(1kG)$ [10^4 s^{-1}]	1.50	1.28	1.09	0.99	0.84	0.72
E: TRIPLET CHARGE SEPARATION RATE						
a: k_{-T} [10^5 s-1] ±30%	6.6	3.2	2.8	1.7	0.6	
b: k_{-T} [10^5 s-1] ±30%	3.9	1.8	1.5	1.0	0.3	
F: INTERSYSTEM CROSSING RATE						
a: k_{ISC} [10^3 s^{-1}] ±30%	7.2	9.1	7.6	7.9	7.6	
b: k_{ISC} [10^3 s^{-1}] ±30%	9.2	10.3	8.9	8.7	8.1	
G: FREE ENTHALPY DIFFERENCE $\Delta G(RP\text{-}^3P^*)$						
a: ΔG [eV] ±.008	.153	.159	.150	.148	.135	
b: ΔG [eV] ±.008	.166	.172	.164	.158	.149	

TABLE I: Temperature dependent radical pair lifetimes, triplet yields, electron transfer rates involved in recombination dynamics, intersystem crossing rate $^3P^* \rightarrow P$ and free energy difference $\Delta G(RP\text{-}^3P^*)$.

A: Radical pair lifetime obtained from Eq.(3)

B: Triplet yields obtained from transient absorption measurements at 880 nm as described.

C: Singlet recombination rates k_S calculated according to Eq.(5)

D: Triplet decay rates from [18].

E: Triplet charge separation rate k_{-T},

a: calculated from Eq.(2)

b: calculated allowing for nuclear spin polarization.

F: Intersystem crossing rate; a: and b: as in E.

G: Free enthalpy difference $\Delta G(RP\text{-}^3P^*)$ from **E** according to Eq.(5); a: and b: as in **E**.

FOOTNOTE

According to Eq.(3) of [24] A_i corresponds to half of the peak-to-peak width of the hyperfine broadened derivative EPR line. Thus, we took half the values of Table I in [24]: A_P = 4.75 G and A_{RP} = 6.5 G. Setting k_{-1} = 0, the expression for the triplet yield from ref.[23,28] differs from the one of Goldstein et al. [24] by a factor of 4 in the first term of Eq.(15) of [25]. Finally k_b in Eq.(1) ref. [24] must be equal to $k_S \exp(-\Delta H/kT)$ as shown in [18].

REFERENCES

[1] Martin, J.-L., Breton, J., Hoff, A.J., Migus, A. and Antonetti, A. (1986), Proc. Nat. Acad. Sci. USA **83**, 957-961

[2] Woodbury, N.W., Becker, M., Middendorf, D. and Parson, W.W. (1985), Biochemistry **24**, 7516-7521

[3] Breton,J., Martin, J.-L., Petrich J., Migus, A. and Antonetti, A. (1986), FEBS Letters **209**, 37-43

[4] Breton, J., Fleming, G. and Martin, J.L., this volume

]5] Allen, J.P., Feher, G., Yeates, T.O., Rees, D.C., Deisenhofer, J., Michel, H. and Huber, R. (1986), Proc. Natl. Acad. Sci. USA **83**, 8586-8593

[6] Chang, C.H., Tiede, D., Tang, J., Smith, U., Norris, J.R. and Schiffer, M. (1986), FEBS Letters **205**, 82-86

[7] Allen, J.P., Feher, G., Yeates, T.O., Komiya, H. and Rees, D.C. (1987), Proc. Nat. Acad. Sci. USA **84**, 5730-5734

[8] Hoff, A.J. (1981), Quart. Rev. Biophys. **14**, 599-665

[9] Hoff, A.J. (1986), Photochem. Photobiol. **43**, 727-745

[10] Ogrodnik, A., Remy-Richter, N., Michel-Beyerle, M.E. and Feick, R. (1987), Chem. Phys. Lett. **135**, 576-581

[11] Marcus, R. A. (1987), Chem. Phys. Lett. **133**, 471-477

[12] Fischer, S.F. and Scherer, P. O. J. (1987), Chem. Phys. **115**, 151-158

[13] Bixon, M., Jortner, J., Michel-Beyerle, M.E., Ogrodnik, A. and Lersch W. (1987), Chem. Phys. Lett. **140**, 626-630

[14] Bixon, M., Michel-Beyerle, M.E. Ogrodnik, A., and Jortner, J. (1987), (submitted)

[15] Parson, W.W. and Warshel, J (1987), J. Am. Chem. Soc.

[16] Parson, W.W., Clayton, R.K. and Cogdell R.J. (1975), Biochim. Biophys. Acta **387**, 265-278

[17] Shuvalov, V.A. and Parson, W.W. (1981), Biochim. Biophys. Acta **638**, 50-59

[18] Chidsey, E.D., Takiff, L., Goldstein, R.A., and Boxer, S.G. (1985), Proc. Nat. Acad. Sci. USA **82**, 6850-6854

[19] Hörber, J.K.H., Göbel, W., Ogrodnik, A., Michel-Beyerle, M.E. and Cogdell, R.J. (1986), FEBS Letters **198**, 273-278

[20] Woodbury, N.W.,Parson, W.W., Gunner, M.R., Prince, R.C. and Dutton, P.L. (1986) Biochim. Biophys. Acta **851**, 6-22

[21] Shuvalov, V.A. and Parson, W.W. (1981) Proc. Nat. Acad. Sci. USA 78, 957-961.

[22] Takiff, L. and Boxer, S. (1987) Photochem. Photobiol. 45, Supplement 61S.

[23] Haberkorn, R. and Michel-Beyerle, M.E. (1979) Biophys. J. 26, 489-498.

[24] Goldstein, R.A. and Boxer, S.G. (1987), Biophys. J. **51**, 937-946

[25] Chidsey, C.E.D., Kirmaier, C., Holten, D. and Boxer, S.G. (1984) Biophys. Acta 424-437.

[26] Ogrodnik, A., Krüger, H.W., Orthuber, H., Haberkorn, R., Michel-Beyerle, M.E. and Scheer H. (1982) Biophys. J. 39, 91-99.

[27] Marcus, R. and Sutin, N. (1985),Biophys. Biochem. Acta **811** 265-322.

[28] Lersch, W. (1982), Diplomarbeit TU München

[29] Woodbury, N.W.T., Parson, W.W. (1984), Biophys. Biochem. Acta **767**, 345-361

[30] Hörber, J.K.H., Göbel, W., Ogrodnik, A., Michel-Beyerle, M.E. and Knapp, E.W. (1985) in: *Antennas and Reaction Centers of Photosynthetic Bacteria - Structure, Interactions and Dynamics* (Michel-Beyerle, M.E. ed.) p.292, Springer Verlag, Berlin.

[31] Ogrodnik, A. and Michel-Beyerle, M.E. , submitted for publication

THE POSSIBLE EXISTENCE OF A CHARGE TRANSFER STATE WHICH PRECEEDS THE FORMATION OF $(BChl)_2^+$ BPh^- IN RHODOBACTER SPHAEROIDES REACTION CENTERS

P. Leslie Dutton, Guillermo Alegria and M. R. Gunner

Department of Biochemistry and Biophysics
University of Pennslyvania
Philadelphia, Pa. 19104

INTRODUCTION

The nature of the first electron transfer step in the photosynthetic reaction center protein is far from certain. Several investigators have considered a monomeric BChl to be important in promoting forward electron transfer from $(BChl)_2^*$ to BPh. The center of the BChl is positioned in the X-ray crystal structure 0.25nm from the center of the $(BChl)_2$, measured in a direction parallel with the z-axis of the protein (4,5). The monomer is displaced out of the direct line joining $(BChl)_2$ and BPh centers but nevertheless it remains an obvious candidate to be on the electron transfer reaction pathway. However, careful searches in the picosecond time domain for absorbance changes that may be associated with transient redox changes on the BChl have failed to demonstrate its involvment in the sequence over a wide temperature range (1-3 although see ref. 6). Instead, spectroscopic investigations with picosecond and subpicosecond resolution have revealed that the loss of the excited singlet state of the special pair of bacteriochlorophylls, $(BChl)_2^*$, coincides with the appearance of the reduced bacteriopheophytin (BPh^-). Thus, the formation of the state $(BChl)_2^+$ BPh^-, which is positioned some 0.2ev below the $(BChl)_2^*$ state appears to occur in a single step with a rate of approximately $3x10^{11}$ s^{-1} (1-3). This separates charge across the approximately 1.1 nm between the centers of the $(BChl)_2$ and BPh, again measured along the line parallel to the z-axis of the protein (4,5). Because of the closly matched kinetics of $(BChl)_2^*$ decay and BPh^- appearance, the involvement of BChl as a conventional redox carrier is cryptic and in doubt. However,it is acknowledged in these studies that, for technical reasons, levels of $BChl^+$ or $BChl^-$ must achieve 15% of the total reaction center population to be detected with any certainty. Thus, there are several viable models (see refs 1-3,6,7-10 for discussion) that can explain these early steps in photosynthesis leading to formation of $(BChl)_2^+$ BPh^-. These include:

1. The BChl plays no part in electron transfer from $(BChl)_2$ to BPh.

2. The BChl serves to increase the electron coupling between $(BChl)_2$ and BPh by the mechanism of superexchange.

3. The BChl is a <u>bona fide</u> redox agent that accepts an electron from $(BChl)_2^*$ to form $(BChl)_2^+$ $BChl^-$ which is followed by electron donation to BPh.

4. The BChl is a <u>bona fide</u> redox agent, but the reaction sequence is that singlet energy transfer from $(BChl)_2^*$ to BChl first induces an electron transfer from BChl to BPh to form $BChl^+BPh^-$. The cation $BChl^+$ so formed then moves to the $(BChl)_2$.

The absence of detected BChl redox chemistry is readily explained by possibilities 1 and 2. However, in mechanisms 3 or 4 the observations place several interesting constaints on possible reaction kinetics. In particular the rate of the primary interaction of $(BChl)_2^*$ with BChl is required to be substantially slower than the ensuing step, otherwise a build-up of a BChl redox intermediate would be detected. In addition, an initial, significant uphill step from $(BChl)_2^*$ to BChl, adding a thermodymanic limitation for the detection of BChl redox changes, is ruled out due to the lack of temperature sensitivity of the reactions between $(BChl)_2$ and BPh.

Thus, the availible data on the nature of early photochemical events in reaction centers in solution can accomodate a number of possible mechanisms for formation of $(BChl)_2^+$ BPh^-. Therefore, additional information is necessary that is derived from experiments that can significantly perturb the reaction(s) leading to $(BChl)_2^+$ BPh^-. One opportunity comes from the fact that the earliest steps of electron transfer in the reaction center separate charge along the z-axis of the protein (4,5) and that this can be modified by application of electric fields across the reaction centers along the same axis (11,18-22). The data for this report were obtained from our experiments on the effect of an applied electric field, E, on the quantum yield, Φ, of $(BChl)_2^+$ Q_A^- in monolayer films of <u>Rb sphaeroides</u> reaction centers placed between planar electrodes (11). In that study we made the observation that Φ decreased as the E opposing charge separation was increased. It was suggested then that this was due to a failure of the system to separate charge at a point before $(BChl)_2^+BPh^-$ formation. This conclusion, however, was contingent on the absence of any significant field induced effects occurring <u>after</u> the formation of the $(BChl)_2^+$ BPh^- state that would prevent the formation of the measured product, $(BChl)_2^+$ Q_A^-. Since that time, we have described the $-\Delta G°$ dependence of the rate of the transition from $(BChl)_2^+$ BPh^- to $(BChl)_2^+$ Q_A^- (13,14). The results of that study lend support to our original assumption. Also a useful X-ray crystal structure for the <u>Rb sphaeroides</u> reaction center has been presented providing a more precise measure of the transmembrane distances between components, necessary for the analysis. This has spurred us to examine in more quantitative terms the effect of applied field on the value of Φ for $(BChl)_2^+$ Q_A^- formation.

Our effort to explore possible sources of loss in Φ as a function of oppposing electric field is providing:

a) A continuing promise that there is a field sensitive step between $(BChl)_2^*$ and $(BChl)_2^+ BPh^-$ which could involve the monomer BChl,

b) A clear framework which provides a simple set of predictions that should disprove or prove the involvement of BChl as a conventional redox carrier of the reaction center.

c) An opportunity, if indeed BChl is proven to be a redox intermediate, that electric field application to monolayer films could be used to stabilize the intermediary state, and even trap it indefinitely.

RESULTS AND DISCUSSION

a) The Initial Approach to the Problem

The simplest treatment of the problem, shown in figure 1, was presented at the meeting in Cadarache (see also ref 14). It explores the applied electric field, E, dependency of the rate calculated of electron transfer, k_{obs}, from $(BChl)_2^*$ to BPh. k_{obs} was calculated from equation 1 (figure 1) using a) the observed Φ at each value of applied field (11), and b) k_d, the summed rates for all the routes of decay of the $(BChl)_2^*$ other than forward electron transfer. The value of k_d is not known but it is considered to be about 10^9 s^{-1} (see 23); it is also reasonable to consider that it is independent of E. Our calculations show that the actual value of k_d does not change the shape of the k_{obs}/E relationship; it simply controls its location on the ordinate. We took k_d to be $3x10^9$ s^{-1} so that at E = 0 when Φ = .98 k_{obs} is $3x10^{11}$ s^{-1}, which is close to that observed experimentally at ambient temperatures.

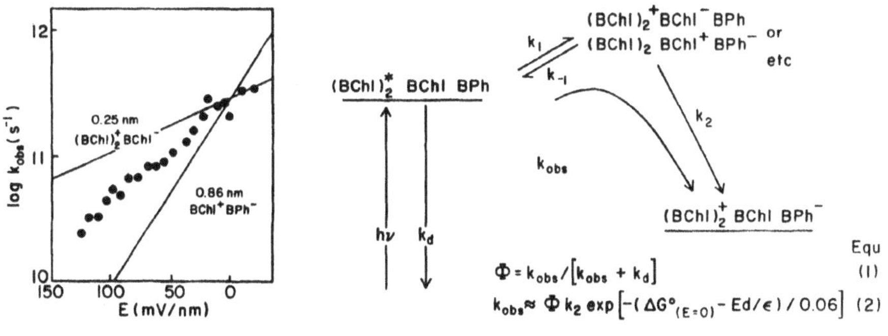

Figure 1. Simple Analysis and Model of the Effect of Applied Electric Field on the Quantum Yield of $(BChl)_2^+ Q_A^-$ Formation

It is clear in figure 1 that k_{obs} derived in this simple way is remarkably sensitive to E and displays a straight line for the relationship between log k_{obs} and E. A rationale for this dependence is provided from the simple reaction scheme shown on the right of figure 1 in which k_{obs} is governed by the presence of an intermediate charge separation state of the kind we have mentioned above in the models 3 or 4.

If for now we focus on the possibilities that a conventional redox intermediate exists as in models 3 and 4, and that the electric field effects the system solely by changing the $-\Delta G°$ values between the singlet state and the intermediate energy levels, then the dependency of k_{obs} on E will be governed by equation 2 in figure 1. The distance for charge separation, d, is measured along the z-axis parallel to the field direction. Note that d is different in models 3 and 4; for $(BChl)_2^+$ $BChl^-$ it is 0.25nm and for $BChl^+$ BPh^- it is 0.86nm (4,5). ϵ is the effective dielectric constant of the intervening medium. In the experimental situation the applied voltage acts across the capacitive cell containing the reaction center layer (11). The values of the dielectric constant throughout this system are unknown. Hence a constant value of 3 has been assumed and used to calculate the strength of the electric field applied.

The left panel of figure 1 shows that the data falls between the line derived from model 3 in which $(BChl)_2^+$ $BChl^-$ is an intermediate state and from model 4 in which $BChl^+$ BPh^- is an intermediate state. However, in view of the uncertainty in the value of the effective dielectric constant, any detailed theoretical assignments of the fits to the data are premature. Never-the-less while this initial treatment is likely to be an oversimplification, the results provide a demonstration that the application of electric fields to monolayer films of reaction centers is a viable way to open up an investigation of this refractile region of electron transfer in the reaction center.

We can now consider the problem at the next level of complexity. Figure 2 describes a more complete description of the kinetic arrangement that includes BChl in the role of intermediary between $(BChl)_2$ and BPh. The scheme is discussed with the following restrictions derived from experiment:

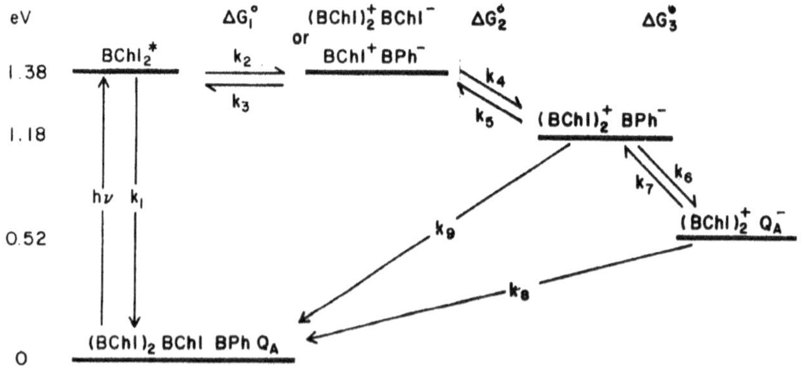

Figure 2. Kinetic Scheme of Electron Transfer Reaction in Reaction Centers.

a) The quantum yield for formation of $(BChl)_2^+ Q_A^-$ is approximately 98% at zero applied field (16,17).

b) The observed rate for formation of $(BChl)_2^+$ BPh^- at zero applied field is approximately $3x10^{11}$ s^{-1} (1-3).

c) The maximum transient concentration of reduced or oxidized BChl is equal to or less than 15% at zero applied field.

d) The observed rate of formation of $(BChl)^+$ BPh^- is temperature insensitive at zero applied field.

e) The $-\Delta G°$ between $(BChl)_2^*$ and $(BChl)_2^+$ BPh^- is 0.2eV at zero applied field; this value is not certain (see ref 12) and it can have significant effect on the relationship between Φ and E. We will discuss this later.

There is a limited range of kinetic and thermodynamic values that satisfy these constraints. The zero field arrangment chosen is:

a) $\Delta G°_1$ = 0eV, $\Delta G°_2$ = -0.2eV

b) k_1 (k_d in figure 1) = $3x10^9$ s^{-1}, k_2 = $3x10^{11}$, k_3 = $3x10^{11}$, k_4 = $2x10^{12}$, k_5 = $7x10^8$ s^{-1}.

c) $\Delta G°_3$ = -0.66eV and k_6 = $4x10^9$ s^{-1}; k_7 is therefore expected to be $4x10^{-2}$ s^{-1}: which is much slower than k_8 at 10 s^{-1} . These reactions are not particularly important to the considerations presented.

d) k_9 = $7x10^7$ s^{-1}. This is assumed to be independent of E, a point that still needs investigation.

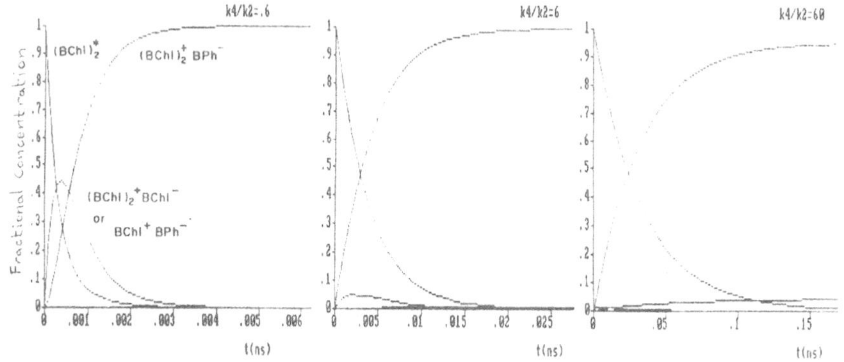

Figure 3. The effect of a possible field induced change in the ratio of k_4/k_2 on the transient fractional concentration of the states involved in forming $(BChl)_2^+$ BPh^-.

Investigation of the system of linear differential equations of the scheme in figure 2 have been confined to considering BChl as a conventional electron carrier as in models 3 and 4, and that the only effect of electric fields is to change the $-\Delta G°$ values between the different redox states. The influence of field on a mechanism involving superexchange (model 2) has not been pursued here. The details of the analysis and the determination of the "best fit" to the experimental data are subject to the same uncertainties as the simple approach.

Figure 3 (center) shows an example of computed concentrations of possible states involved in forming $(BChl)_2^+ BPh^-$ at E = 0 with the above arrangement of kinetics and free energies at E = 0. The time scale shown does not include the decay of $(BChl)_2^+ BPh^-$ to form $(BChl)_2^+ Q_A^-$ or the return to ground state governed by k_9. The traces conform to the experimental findings; the decay of $(BChl)_2^*$ matches the rise of BPh^- and the intermediate rises to no more than 8%. The values of the free energies of the states at E = 0 are displayed in the center panel of figure 4. To the left and the right are examples of what happens to these levels upon application of positive or negative applied fields with the assumptions of uniform dielectric constant and the distances provided in the x-ray structure.

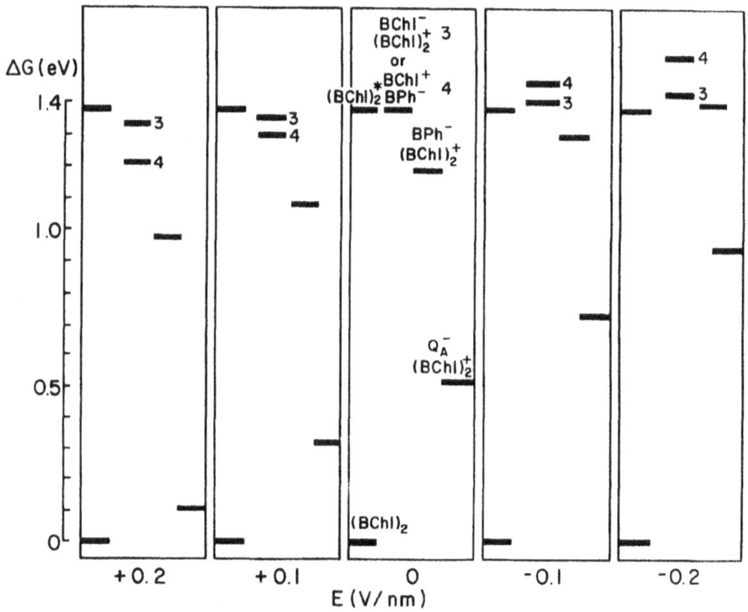

Figure 4. The Effect of Applied Electric Field on the Energy Levels in Reaction Centers.

It is clear that within the range of fields that can be applied experimentally to monolayer films of reaction centers (11), substantial alterations in the energy levels can be achieved; this predicts significantly different behavior for the free energy changes of the initial step(s) if model 1, 3, or 4 are considered. A cursory examination of this system, in the absense of further experimentally derived constraints, reveals that the number of possibilities for changing the rate constants involved with changing E becomes enormous and is beyond the scope of this brief report. However, the following points are clear with regard to the use of applied fields to obtain experimental information that hopefully will serve to diminish the number of possibilities:

a) The calculations show that small changes in the ratio of rate constants accompanying changes in $-\Delta G°$ can produce large variation in the yield of the intermediates. Figure 3 right and left show the dramatic effect of changing the k_4/k_2 ratio on the BChl intermediate concentration from 6 at $E = 0$ to 60 (Fig 3 right) or to 0.6 (Fig 3 left). A rule of thumb from our studies (see 13,14) on the effects of ΔG on rates would suggest, in the arrangement of figure 4, that k_2 would have a considerably steeper dependence on field induced changes in ΔG than would k_4. These variations should be observed in the range of fields used in our experiments.

b) With opposing fields the likelyhood of detecting an intermediate state diminishes because of thermodynamic limitations in addition to any changes in kinetics. The dependence on E of the amplitude of the transiently formed intermediate will be governed by the model and what is the energy level of the intermediate at E =0. However, it is predicted that as E becomes more negative and the intermediate state assumes a position on the energy scale higher than the $(BChl)_2^*$, the kinetics of $(BChl)_2^+$ BPh^- formation will become temperature dependent due to the introduction of a thermally accessed intermediate state. Moreover, the dependence of the activation energy (E_a) on E will provide information on this state. Thus, in figure 4 it is clear that the onset of temperature dependency and the slope of the relationship between E_a and E is governed by the $-\Delta G°$ values at E = 0 and the existance and nature of the intermediate states involved.

c) We have found that the Φ/E relationship (11) can be resonably modeled by several sets of rate constants and ΔG values. There is at present no clear view of which set of values is appropriate. Experiments are planned with picosecond time resolution of the early flash activated events in the reaction centers under the influence of electric fields. These should help to resolve many of these questions.

d) Aided by the influence of fields it may be possible to trap an intermediate state in a stable form for examination by conventional spectroscopy. Optical spectroscopic analysis of monolayer films has been accomplished (28,29) and analysis of single monolayers is now routine. More over, it can be expected that with multilayers EPR analysis should be feasible. The first successful trapping experiment of this kind in solution was the Q_A^- in _C vinosum_ analalysed by EPR (24). This involved the irreversible millisecond transfer of an electron from cytochrome _c_ 553 to $(BChl)_2^+$ which at the low temperatures stranded Q_A in the reduced state. Later, using the same approach, but starting with Q_A already reduced, "I" was trapped reduced at low temperatures; optical and EPR spectroscopy proved this to represent predominantly BPh^-; any accompanying spectral shifts associated with BChl have been interpreted as originating in electrochromic effects (25-27) rather than any redox changes in the BChl. It should be possible to trap the postulated BChl intermediate in monolayer films of reaction centers-cytochrome c complex from

C vinosum, *Rp vividis* or *Rp gelatinosa*, or, if supplied with cytochrome *c*, reaction centers from *Rp sphaeroides*.

e) It has been shown (30) that when the reaction centers are trapped with BPh^- reduced prior to flash activation, no $(BChl)_2^+$ is detected even within the 10ps time period of a flash. This was interpreted as demonstrating that no intermediate exists between $(BChl)_2^*$ and $(BChl)_2^+$ BPh^-. However, reduction of BPh in this way will eliminate any mechanism of the kind described in model 4. In the case of model 3, it is also possible to rationalize these results; if the intermediate energy level (E = 0) is equal or close to that of the $(BChl)_2^*$, then the presence of BPh^- could conceivably push the intermediate level up above $(BChl)_2^*$ by local coulombic forces, thereby preventing the reaction. This latter possiblitity can be explored in reaction center films that have the BPh^- trapped reduced but which have a positive field applied to counter the effects of BPh^-. Under these conditions a transient formation of the intermediate state (model 3) may be observed.

Thus, in summary, this report is intended to indicate the usefulness of techniques based on Langmuir-Blodgett films. Recent advances have been made in the construction of monolayer films in which the reaction centers are highly oriented, stable and active when corporated into an electric cell. The density of reaction centers in the films is sufficient to permit spectroscopic analysis of the reaction centers simultaneously with electric field manipulation. The analysis by electrical or optical methods of light-activated reactions in oriented single monolayer films under the influence of applied electric fields of magnitude equal or greater than that encountered in nature now appears to be an approach with considerable application for not only the uncertainties and problems that face studies with reaction center at present, but also for the future, more detailed studies. More generally, the approach displays considerable promise in providing information about the nature of charge transfer states in redox proteins as well as the means to evaluate the effective dielectric properties localized between protein redox cofactors.

Dr. Watson: This is indeed a mystery. What do you imagine that it means?

Mr. Holmes: I have no data yet. It is a capital mistake to theorize before one has data. Insensibly one begins to twist the facts to suit theories, instead of theories to suit facts.

Scandal in Bohemia
Conan Doyle, 1892

ACKNOWLEDGEMENTS

This work was supported by a grant from the Department of Energy DOE FG02-86-13476

REFERENCES

1. Kirmaier C, Holten, D and Parson W.W. FEBS Letters 185; 76-82 (1985)
2. Woodbury N.W., Becker M, Middendorf D and Parson W.W. Biochemistry 24; 7516-7521 (1985)
3. Martin J-L, Breton, J, Hoff A.J., Migus, A., and Antonetti, A. Proc. Natl Acad Sci; U.S.A. 83 957-961 (1986)
4. Allen, J.P., Feher, G., Yeates, T.O., Komiya, H and Rees, D.C., Proc Natl Acad Sci; U.S.A. 84 5730-5734 1987
5. Chang, C.H., Tiede, D.M., Tang, J., Smith, U., Norris, J., and Schiffer, M.; FEBS Letts. 205 82-86 (1986)
6. Shuvalov V.A., and Klenanik V.A., FEBS Letters 160 51-55 (1983)
7. Marcus R.A., Chem. Phys. Letters 133 471-477 (1987)
8. Haberkorn, R., Michel-Beyerle, M.E. and Marcus, R.A., Proc. Natl Acad Sci U.S.A. 76 4185-4189 (1979)
9. Kirmaier, C., Holten D., Parson, W.W. Biochim Biophys Acta 810 33-42 (1985)
10. Holten D., Hoganson C., Windsor M.W., Schenck C.C., Parson W.W., Migus A., Fork R.L., and Shank C.V., Biochim Biophys Acta. 592 461-473 (1980)
11. Popovic Z.D., Kovacs, G.J., Vincett, P.S., Alegria,G., and Dutton, P.L. Biochim, Biophys Acta 851 38-48 (1986)
12. Gunner, M.R., and Dutton, P.L. Accompaning manuscript, this volume.
13. Gunner, M.R., and Dutton, P.L. Submitted to Biophys J
14. Dutton, P.L., Alegria, A., and Gunner, M.R. Biophys J. Abstracts for Biophysical Meeting. February. (1988)
15. Wraight, C.A. and Clayton, R.K., Biochim Biophys Acta.
16. Cho, H.M., Mancino, L.J., and Blankenship, R.E., Biophys J 45 455-461 (1984).
17. Loach, P.A., and Sekura, D.L., Biochemistry 7, 2642-2649 (1968)
18. Popovic, Z.D., Kovacs, G.J., Vincett, P.S. and Dutton, P.L. Dutton, Chem Phys Letts 116 405-410 (1985)
19. Popovic, Z.D., Kovacs, G.J., Vincett, P.S., Alegria, G., Dutton, P.L., Chem. Phys 110 227-237 (1986)
20. Gopher, A., Blatt, Y., Schoenfeld, M., Okamura, M.Y., Feher, G., and Montal, M., Biophys J 48 311-320 (1985)
21. Packham, N.K., Mueller, P., and Dutton, P.L., Biochim Biophys Acta In press.
22. Feher, G., Arno, T.R., and Okamura, M.Y., This volume
23. Campillo, A.J., Hyer, R.C., Monger, T.G., Parson, W.W., and Shapiro, S.L., Proc Natl Acad Sci U.S.A. 74 1997-2001 (1977)
24. Leigh, J.S., and Dutton, P.L., Biochem Biophys Res. Comm 46 414-418 (1972)
25. Tiede, D.M., Prince, R.C., and Dutton, P.L., Biochim Biophys Acta 449 447-467 (1976)
26. Prince, R.C., Tiede, D.M., Thornber, J.P., and Dutton, P.L., Biochim Biophys Acat 462 731-747 (1977)
27. Prince, R.C., Dutton, P.L., Clayton, R.K., Biochim Biophys Acta 502 354-358 (1978)
28. Tiede, D.M., Mueller, P., and Dutton, P.L., Biochim Biophys Acta 681 191-201 (1982)
29. Alegria, G., and Dutton, P.L., In " Cytochrome Systems: Molecular Biology and Bioenergetics" (S. Papa, B. Chance,

L. Ernster and J. Jaz eds). Plenum Press, London. In press, (1987)
30. Netzel, T.L., Rentzepis, P.M., Tiede, D.M., Prince, R.C., and Dutton, P.L., Biochim Biophys Acta 460 467-479 (1977)

THE PRIMARY ELECTRON TRANSFER IN PHOTOSYNTHETIC PURPLE BACTERIA : LONG RANGE ELECTRON TRANSFER IN THE FEMTOSECOND DOMAIN AT LOW TEMPERATURE

J. L. Martin* , J. Breton# , J. C. Lambry*, and G. Fleming +

*Laboratoire d'Optique Appliquée, INSERM U275
Ecole Polytechnique ENSTA 91128 Palaiseau Cedex, France
#Service de Biophysique, CEN/Saclay 91191 Gif-sur-Yvette
+Department of Chemistry, The University of Chicago, USA

INTRODUCTION

The conversion of light energy into chemical free energy in the reaction center (RC) of photosynthetic purple bacteria is a highly efficient process which involves very fast initial reactions able to efficiently compete with radiative lifetimes. The primary charge separation occurs between a bacteriochlorophyll dimer (P) and a bacteriopheophytin molecule (H_L) located on the side of the L polypeptide subunit. The structure of the RC, as solved by X-ray crystallography, shows that a monomeric bacteriochlorophyll (B_L) is located in between P and H_L. The role of this molecule in the initial charge separation process is not yet understood and is the object of much current experimental and theoretical scrutinity.

Recent femtosecond absorption data revealed that direct excitation of the primary donor (P) leads to the formation of the radical pair $P^+H_L^-$ within 2.8 ps (at 295K) in RCs from both Rb. sphaeroides (1) and Rps. viridis (2). No evidence for a transient depletion of B_L , as the result of formation of a B_L^- or B_L^+ species, is found. While this observation must be reconciled with the X-ray structure data, there is a general agreement to give to the B_L molecule an essential role in the primary charge separation process which involves transfer of an electron at a rate of $3.6 \times 10^{11} s^{-1}$ over a distance of 17 Å (center - to - center).

Several mechanisms have been recently suggested which give different roles to B_L and are compatible with the absence of spectroscopic evidence for a transient bleaching of its ground state absorption. One such possibility is the direct electron transfer from P^* to H_L ($P^*B_LH_L$ ---> $P^+B_LH_L^-$) via the virtual state $P^+B_L^-H_L$ (3-5). This is the so-called " superexchange mechanism ". The existence of an actual chemical intermediate (B_L^- or B_L^+) has also been postulated (6). To be consistent with the femtosecond spectroscopy data, this model requires a rate of depletion of the intermediate (k_2) subtancially higher than its rate of formation (k_1). Using $k_2 \geq 5\ k_1$, Marcus has recently proposed that such a two-step process is more realistic than the superexchange coupling mechanism (6). His argument was based on the estimate of the electronic superexchange matrix element which value (0.7 cm^{-1}) was too low to be consistent with the fast electron transfer rate observed experimentally (1,2,7). However, Jortner, Michel-Beyerle and coworkers disputed that such a low value does not exclude the superexchange mechanism, their main argument in favor of the latter process being that the primary electron step should be an activated process with a strong temperature dependence if the two-step mechanism was involved (4). An alternative approach to the above models has been proposed by Friesner at al. (8,9). Using a set of strongly coupled vibrationnal modes, their theory is capable of modelling different experimental results such as hole-burning (8) and temperature dependence data (9).

In this report, we briefly present the results of the temperature dependence of the primary electron transfer rate following excitation of RCs from Rps. viridis and Rb. sphaeroides with fs pulses. We describe the effect of deuterium substitution as well as of borohydride-treatment on the RCs of Rb. sphaeroides. We also make a detailed analysis of the kinetic data obtained at wavelengths close to the maximum of the 834-nm band in Rps. viridis RCs at 10K which is currently assigned to the B molecules. These data confirm the absence of an observable transient depletion of the ground state of B_L in the 100fs to ps time domain and give new limits to the value of k_2/k_1 when the two-step mechanism is postulated.

RESULTS AND DISCUSSION

The Primary Electron Transfer Rate at 10K

An accurate determination of the decay of P^* and of the electron transfer rate is achieved by monitoring the stimulated emission on the long wavelength side of the near IR band of P (1,2,7). For RCs from Rps. viridis at 10K such a measurement at 1045nm yields a time constant for electron transfer of 700 ± 100 fs (Fig. 1a). The same time constant is obtained for the band shift induced absorption at 821nm (Fig. 1b). At 795nm the contribution of P^* to the induced absorption is significant as revealed by the partial relaxation with a 700-fs time constant (Fig. 1c).

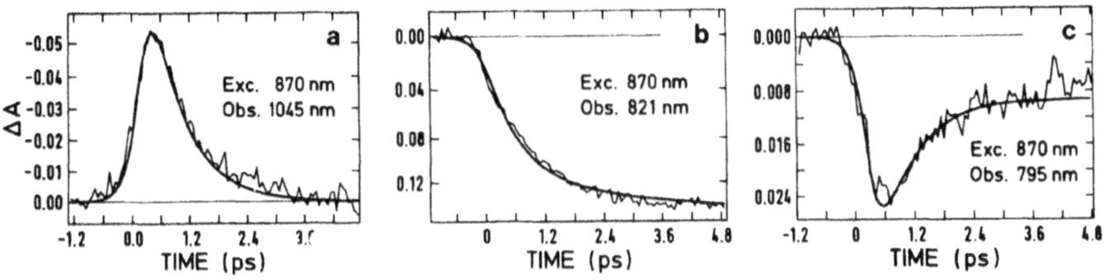

Figure 1. Primary electron transfer in Rps. viridis RCs at 10K.

Similarly upon direct excitation of P at 870nm in RCs from Rb. sphaeroides, a value of 1.2 ± 0.1ps for the electron transfer time constant is obtained at 10K by monitoring either the stimulated emission from P^* at 925nm or the band shift of the monomeric B molecules around 800nm (10).

The Effect of Chemical Modifications of the Reaction Center

In the structural models of the RCs, the cofactors are organized in two "branches" starting from the center of P which, together with the L and M polypeptides, are related by an approximate C2 symmetry axis (11-13). The electron transfer proceeds along the L branch. Borohydride treatment of RCs from Rb. sphaeroides allows the removal of the B_M monomeric bacteriochlorophyll molecule located on the M branch (14). When such modified RCs are investigated at 10K they still exhibit the same 1.2 ± 0.1ps time constant for electron transfer found for unmodified RCs. Furthermore the kinetic traces in the region of the band shift and notably close to the isosbestic point near 800nm are not perturbed by the chemical modification (10). These results thus confirm previous observations made at room temperature (15).

The Effect of Deuterium Isotopic Substitution

Fully deuterated RCs isolated from Rb. sphaeroides, generously provided by Drs. J. Norris and S. Kolaczkowski (Argonne National Laboratory and the University of Chicago), still exhibit within the 10% precision of our measurement the same primary electron transfer rate at 10K as for control RCs (16).

The Temperature Dependence of the Primary Electron Transfer Rate

The result of an investigation of the electron transfer rate over the temperature range 10K-300K for RCs from Rps. viridis and monitored by the decay of the stimulated emission from P^* is shown Figure 2.

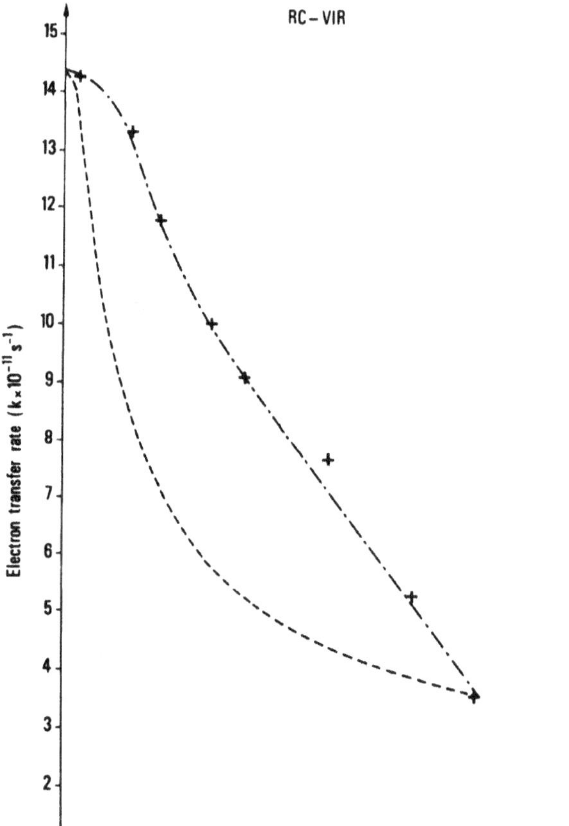

Figure 2. Temperature dependence of the primary electron transfer rate in Rps. viridis. Theoretical fit using Eq.2 and a value for $\hbar\omega$ =25 cm^{-1} (---).

The primary charge separation process is characterized by an apparent negative activation energy with an **increase of the rate by a factor of four** when the temperature is **decreased** from 300K to 10K.

According to the model of Jortner and coworkers (4), this negative activation energy behaviour is what we can expect for an activationless electron transfer process.

If we assume that the electron transfer process is non adiabatic then the conventional expression for the rate can be used

$$k = (2\pi/\hbar)\,|V|^2\,F \qquad (1)$$

where V is the electronic interaction matrix element and F is the thermally averaged Franck-Condon factor.

Assuming a coupling of the primary electron transfer to low frequency modes of the protein and in the single mode approximation k can be expressed as

$$\frac{k}{k(0)} = \left[\frac{\exp(\hbar\omega / kT) - 1}{\exp(\hbar\omega / kT) + 1}\right]^{1/2} \qquad (2)$$

where

$$k(0) = \frac{2\pi\,|V|^2}{\hbar\,\omega\,(2\pi p)^{1/2}} \qquad (3)$$

and ω is the average frequency of the coupled mode , $p = \Delta E/\hbar\omega$ and ΔE is the energy gap for the electron transfer.

Equation (2) provides a good fit to the data obtained for RCs from Rb. sphaeroides using $\hbar\omega = 80\ cm^{-1}$ (16). However the same equation gives a very poor fit to the data for the Rps. viridis RCs (Fig. 2) and a value for $\hbar\omega = 25\ cm^{-1}$ is needed to try to fit a ratio of 4 for the rates of electron transfer between 10K and 300K. This difference in the temperature dependence observed between the RCs of the two bacterial species could possibly indicate that V increases more strongly with decreasing temperature in Rps. viridis than in Rb. sphaeroides. However it also raises the question of the validity of the hypothesis that the primary electron transfer is a non-adiabatic process.

The limit on the putative depletion of B_L

If B_L is involved as a real intermediate in the electron transfer process, as it is proposed by Shuvalov et al. (17), it should be possible to detect a bleaching upon probing at the maximum of the ground state absorption of this molecule which at 10K is located at 800nm and 834nm for the RCs from Rb. sphaeroides and Rps. viridis, respectively (18, and references therein). The results of such measurements are depicted in Fig.3 for the 834-nm band of RCs from Rps. viridis at 10K. When the probe wavelength is located about 1nm (Fig. 3a; 833 nm) or 2 nm (Fig. 3c; 832-nm) on the high energy side of the 834-nm band, an initial induced absorption is observed at early time. This is followed

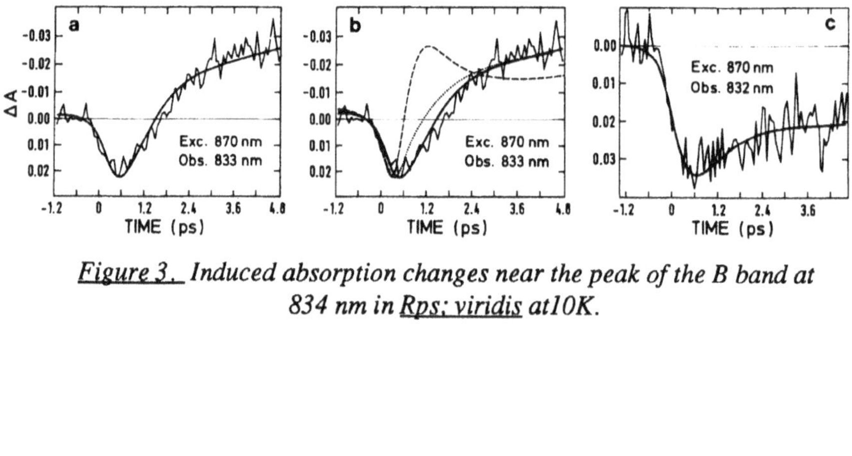

Figure 3. Induced absorption changes near the peak of the B band at 834 nm in Rps; viridis at10K.

by a delayed bleaching (Fig. 3a) or a partial relaxation (Fig. 3c). It should be pointed out that taking into account (i) the large absorption cross section of B_L in this region and (ii) the fraction of excited RCs (about 30%), an initial bleaching should be observed at these wavelengths if a significant depletion of B_L was taking place. The results of a more quantitative analysis of these kinetics is represented in Fig. 3 by the numerical fits which include the effects of a band shift and the instrumental function.

Simulation of the induced absorption changes in the 834-nm band

For small ΔA, the observed signal at wavelength λ is given by the expression

$$\Delta A(\tau) = \eta N \int_{-\infty}^{+\infty} dt' \, T(\lambda, \tau - t') \int_{-\infty}^{+\infty} I_{ex}(t) \, I_t(t + t') \, dt$$

where I_{ex} (t) and I_t (t) are the energy profiles of the excitation pulse and probe pulse respectively, τ is the delay between the excitation and the probe pulses, η is the fraction of excited molecules and N is the total population.

In the hypothesis of a two-step mechanism, the impulse response of the medium at wavelength λ can be written as

$$T(\lambda,t) = -\sigma_a^{P^*} e^{-t/\tau_1} + \sigma_a^{B} \frac{k_1}{k_2 - k_1} e^{-t/\tau_1} - \sigma_a^{B} \frac{k_1}{k_2 - k_1} e^{-t/\tau_2} + \Psi_a^{B} (\lambda_{(t,\tau_1)})$$

with $k_1 = 1/\tau_1$ and $k_2 = 1/\tau_2$

and where $\sigma_a^{P^*}$ and σ_a^{B} are the absorption cross-sections of P^* and B species respectively.

The function Ψ_a^{B} represents a "rigid" shift of the absorption band at 834 nm.

The time dependence of the band shift in the approximation of a fast second step, is given by

$$\lambda(t, \tau_1) \approx \lambda_0 + (\lambda_\infty - \lambda_0)(1 - e^{-t/\tau_1})$$

where ($\lambda_\infty - \lambda_0$) is the amplitude of the shift which is measured in a separate experiment.

The energy profile near the peak of the band is known (18) and can be well fitted by

$$\Psi_a^{B}(\lambda) = \Psi_a^{B}(\lambda_M) \exp - [(\lambda - \lambda_M)^2 / \delta]$$

The simulations have been made assuming a smooth profile for the absorption spectrum of P^* in this spectral region with a value of $\sigma_a^{P^*}$ not significantly different from the value determined at 795 nm . The only adjustable parameter is then the ratio k_2 / k_1.

The results of such a simulation are given in Fig. 3b where a two-step process is assumed. The broken line (---) represents the simulation for $k_2 / k_1 = 5$ which is clearly not fitting correctly the experimental data. The dotted line (·······) corresponds to a value of $k_2 / k_1 = 25$ which is also still significantly different from the experimental trace . Finally using $k_2 / k_1 = 70$ we get the result represented by the solid line (——) which is a satisfactory fit of the data. Such a value for the ratio of rates means a time constant of around 10 fs for the second step in Rps. viridis at 10K.

Using the **same** set of parameters but assuming we are probing at a wavelength lower by only 1 nm the simulation gives a good fit of the experimental results at 832 nm (Fig.3c).

However we must point out that the fact we obtained a satisfactory fit of the data with the hypothesis of a two-step process does not imply that we need such a model to fit the data. This is clearly shown in Fig.3a where the simulation has been made assuming a single step process.

In the hypothesis of a two-step model, a timescale of ~ 10 fs for the second step is clearly outside the regime of conventional electron transfer theory. A scheme in which non-adiabatic electron transfer occurs between P and B_L and an adiabatic process is involved between B_L and H_L is discussed by Marcus in this volume (19).

Our data are compatible with a superexchange mechanism as discussed in the present volume by Bixon et al. (4) . In this hypothesis they conclude that a single-step nonadiabatic process is likely the valid mechanism.

A way to distinguish among these models will be to study the rate of the primary electron transfer in the presence of an electrical field. Different behaviours have been predicted (4) with respect to the nature of the postulated models.

REFERENCES

1. J.-L. Martin, J. Breton, A.J. Hoff, A. Migus, and A. Antonetti, *Proc. Natl. Acad. Sci. USA* **83**, 957 (1986).
2. J. Breton, J.-L. Martin, A. Migus, A. Antonetti and A. Orszag, *Proc. Natl. Acad. Sci. USA* **83**, 5121 (1986).

3. M. Bixon, J. Jortner, M.E. Michel-Beyerle, A. Ogronik and W. Lersch, *Chem. Phys. Lett.* **140**, 626 (1987).
4. M. Bixon, J. Jortner, M. Plato, M.E. Michel-Beyerle, *This volume.*
5. A. Warshel, S. Creighton and W.W. Parson, *J. Phys. Chem.* In press.
6. R. Marcus, *Chem. Phys. Lett.* **133**, 471 (1987).
7. N. W. Woodbury, M. Becker, D. Middendorf and W.W. Parson, *Biochem.* **24**, 7516 (1985).
8. Y. Won and R.A. Friesner, *Proc. Natl. Acad. Sci. USA* **84**, 5511 (1987)
9. Y. Won and R.A. Friesner. *This volume.*
10. J. Breton, J.L. Martin, G. Fleming and J.C. Lambry. *Biochem.* (submitted).
11. J. Deisenhofer and H. Michel. *This volume.*
12. J.P. Allen, G. Feher, T.O. Yeates, H. Komiya and D.C. Rees. *This volume.*
13. D.M. Tiede, D.E. Budil, J. Tang, O. El- Kabbani, J.R. Norris, C.H. Chang and M. Schiffer. *This volume.*
14. H. Sheer, D. Beese, R. Steiner and A. Angerhofer. *This volume.*
15. J. Breton, J.L. Martin, J. Petrich, A. Migus and A. Antonetti, FEBS **209**, 37 (1986).
16. G. Fleming, J.L. Martin and J. Breton. *Nature* (Submitted)
17. V.A. Shuvalov, A.O. Ganago, A.V. Klevanik and A. Ya. Shkuropatov. *This volume.*
18. J. Breton. *This volume.*
19. R. Marcus. *This volume.*

THE PROBLEM OF PRIMARY ENERGY CONVERSION IN REACTION CENTERS OF PHOTOSYNTHETIC BACTERIA

V.A.Shuvalov, A.O.Ganago, A.V.Klevanik, and A.Ya.Shkuropatov

Institute of Soil Science and Photosynthesis
USSR Academy of Sciences
142292 Pushchino, Moscow Region, USSR

INTRODUCTION

Picosecond laser spectroscopy revealed the P^+H^- † state in charge transfer process in RCs of photosynthetic bacteria upon excitation in the visible region and selective excitation of P at 880 nm.[1-3] This state converts into $P^+Q_A^-$ during ~200 ps. There are contradictory results concerning existence of an earlier transient state between the purely excited P^* and the charge-transfer state P^+H^-.

An earlier state preceding P^+H^- was identified by bleaching of the P and B bands at 870 and 805 nm, in the absence of bleaching of the H bands at 760 and 545 nm.[3] The electron transfer to H is accompanied by the additional bleaching at 805 nm (see below). The recovery of the band at 800 nm at this time reported in[3,4] has been attributed to two-photon processes.[5]

Femtosecond laser spectroscopy has proved that formation of the P^+H^- state is delayed by several (2.8-7.0) ps with respect to the excitation of P.[6-9] It was found that the stimulated emission from P^* can be observed at 930 nm in RCs of *Rb.sphaeroides* which decays with the time constant characteristic of the P^+H^- formation.[8-10] However, in contrast with the earlier findings,[3] the femtosecond kinetics measured at several wavelengths provided no evidence in favour of the presence of an intermediary state between P^* and P^+H^-,[8,9,11,12] hence it has been suggested that the primary charge separation in RC's occurs between P^* and H with the formation of P^+H^- within 2.8 ps.

---

†Abbreviations: ΔA, absorbance changes; B, bacteriochlorophyll monomer in RC (subscripts L and M refer to polypeptide subunits); Cyt, cytochrome bound to RC; fs, femtosecond; H, bacteriopheophytin molecule in RC (subscripts L and M refer to polypeptide subunits); I, intermediate electron acceptor including B and H in RC; L and M, polypeptide subunits of RC; P, primary electron donor in RC, "special pair" of bacteriochlorophyll molecules; P^* and P^T, singlet and triplet excited states of P, respectively; Q_A, primary quinone electron acceptor in RC; *R.*, *Rhodopseudomonas*; *Rb.*, *Rhodobacter*; RC, reaction center; T_1, T_2, and T_2^*, relaxation times; CR, charge resonance; CT, charge transfer.

Here we discuss new findings concerning the process of conversion of P^* into P^+H^- state to emphasize features pertinent to a transient charge-transfer state which is formed prior to P^+H^-.

To resolve the problem of primary charge separation in bacterial RCs, we have recently employed various new techniques including hole-burning experiments at low temperatures, studies of femtosecond kinetics and spectra, and selective trapping of charge-transfer states. The same approaches as well as Stark effect measurements have been used in several laboratories.

MATERIALS AND METHODS

The RCs were isolated from *R.viridis*,[4] *Chloroflexus aurantiacus*,[13] *Rb.sphaeroides*[13] and modified to remove B_M.[13] Picosecond[10] and femtosecond[14] measurements were made at room temperature. Hole-burning experiments were made at T < 2 K using selective continuous excitation by xenon lines[15] or selective excitation by picosecond laser pulses from a parametric oscillator described earlier.[4]

RESULTS AND DISCUSSION

Absorption and Fluorescence Spectra of RCs at ~ 2 K

Fig. 1 shows absorption and fluorescence spectra of *R.viridis* RCs measured at 2 K in two different states, PIQ_A (Fig. 1,A) and $PI^-Q_A^-$ (Fig. 1,B).[16] The absorption spectrum in Fig. 1,A includes the main band with a maximum at 1000 nm and a smaller band seen as the shoulder at 1015 nm. Similar features were observed in *Rb.sphaeroides* RCs.[4] The fluorescence spectrum in Fig. 1,A is symmetrical to the smaller absorbance band rather than to the main band. Reduction of Q_A does not affect the absorption and fluorescence spectra (data not shown). On the contrary, when I is reduced, the long-wavelength shoulder disappears in the absorption spectrum, the main band shifts to 995 nm, and the fluorescence spectrum is symmetrical to the main absorbance band (see Fig. 1,B). It is noteworthy that at 77 K no shoulder is discerned in the absorption spectra of the both states, and the fluorescence spectrum of either state is symmetrical to its main absorbance band.[16]

Several alternative suggestions can be made to explain these spectral features:

1. The shoulder at 1015 nm might represent contamination of samples. In this case, fluorescence of RCs themselves should be symmetrical to the

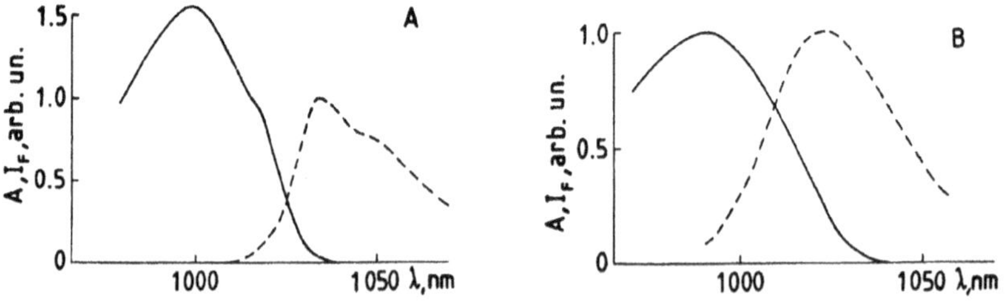

Fig. 1. Absorption (solid curves) and fluorescence (dashed curves) spectra of *R.viridis* RCs at 1.9 K measured in redox states PIQ_A (A) and $PI^-Q_A^-$ (B).[16]

main absorbance band as it is observed at 77 K.[16] This disagrees with symmetry between the fluorescence spectrum and the shoulder rather than the main band of the absorption spectrum (Fig. 1,A), and sensitivity of both the shoulder and the fluorescence spectrum to redox states of RC components (Fig. 1,B). Similarly, heterogeneity of RCs seems implausible.

2. The shoulder at 1015 nm may represent the zero-phonon line within the band of P. Again, its sensitivity to the redox state of H is not quite understandable. The hole-burning experiments would verify this assumption (see below).

3. The shoulder at 1015 nm might represent the second allowed electronic Q_Y transition that belongs to P. To treat electronic transitions of P, both the excitonic and the charge-resonance (CR) states should be considered. It can be shown[17] that transitions of P can be represented by the following:

$$|P_L P_M> \rightarrow |P_L^- P_M^+> \quad (1)$$

$$\frac{1}{\sqrt{2}} (|P_L^* P_M> + |P_L P_M^*>) \equiv |+> \text{ (state involved in (4),(5))} \quad (2)$$

$$|P_L P_M> \rightarrow \frac{1}{\sqrt{2}} (|P_L^* P_M> - |P_L P_M^*>) \equiv |-> \quad (3)$$

$$|P_L P_M> \rightarrow \cos\varphi |+> + \sin\varphi |P_L^+ P_M^-> \equiv |+ CR> \quad (4)$$

$$|P_L P_M> \rightarrow \cos\varphi |P_L^+ P_M^-> - \sin\varphi |+> \equiv |- CR> \quad (5)$$

$$\tan 2\varphi = \frac{2L}{\hbar(\omega_0 - \omega_R) + V} \quad (6)$$

$$\hbar\omega_{4,5} = \frac{\hbar}{2}(\omega_0 + \omega_R) + \frac{V}{2} \pm \frac{1}{2}\sqrt{[\hbar(\omega_0 - \omega_R) + V]^2 + 4L^2}$$

where P_L and P_M are bacteriochlorophylls in P liganded by histidines of L and M subunits, respectively; ω_0 is the frequency of a monomer transition; ω_R is the frequency of the transition $|P_L P_M> \rightarrow |P_L^+ P_M^->$; V is the negative energy of excitonic interaction in P; L is the energy of charge-transfer interaction in P.

Transitions (1) and (3) are probably forbidden, and only two allowed transitions (4) and (5) with a large energy splitting which exceeds $|V|$ ($-V \sim 500 - 1000$ cm^{-1}) can be observed. However, the small energy splitting observed in the long-wavelength band of RCs (~ 5 meV, see below) seems to show that only one of transitions (4) and (5) is allowed. Therefore, one should search for another explanation for the shoulder at 1015 nm.

4. The shoulder in Fig. 1,A may be due to an interaction of P with a neighbouring pigment molecule, e.g. B_L.[16] In agreement with Plato's calculations (see the contribution to this volume), the lowest charge-transfer (CT) state in P is $P_L^+ P_M^-$ which we have considered above (eqns. 4, 5) as being mixed with the $|+>$ excitonic component of P. According to the discussion above, only one of transitions (4) and (5) is allowed in RCs; we may assume that the former is allowed. We take into account that the energy of the lowest unoccupied molecular orbital of B is close to that of P,[10] and that the edge-to-edge distance between P and B is about 3.5 Å [19-20] which approaches van der Waals distance, and therefore consider a mixture of states $|+CR>$ and $|P^+B^->$. The expected allowed transitions are:

$$|P_L P_M B_L> \rightarrow \cos\psi |+CR> + \sin\psi |P^+B^-> \equiv |+ CT> \quad (7)$$

$$|P_L P_M B_L> \rightarrow \cos\psi |P^+B^-> - \sin\psi |+ CR> \equiv |- CT> \quad (8)$$

$$\tan 2\psi = \frac{2T}{\hbar(\omega_4 - \omega_T)}$$

$$\hbar\omega_{7,8} = \frac{\hbar}{2}(\omega_4 + \omega_T) \pm \frac{1}{2}\sqrt{[\hbar(\omega_4 - \omega_T)]^2 + 4T^2}$$

where T is the energy of charge-transfer interaction between P and B; ω_T is the frequency of the transition $|P_L P_M B\rangle \rightarrow |P^+B^-\rangle$.

The energy splitting between transitions (7) and (8) is proportional to $\sqrt{[\hbar(\omega_4 - \omega_T)]^2 + 4T^2}$ and can be small enough to explain the splitting between the two components of the long-wavelength absorption band of RCs.

The transition dipole moments are:

$$\vec{\mu}_+ = \cos\psi \cdot \vec{\mu}_{CR} + \sin\psi \cdot \vec{\mu}_{PB} \qquad (9)$$

$$\vec{\mu}_- = \cos\psi \cdot \vec{\mu}_{PB} - \sin\psi \cdot \vec{\mu}_{CR} \qquad (10)$$

These two transitions may be not discerned by LD and CD if $\vec{\mu}_{PB}$ is small compared to $\vec{\mu}_{CR}$ since the both $\vec{\mu}_+$ and $\vec{\mu}_-$ would be almost parallel to $\vec{\mu}_{CR}$, which agrees with experiment.[18]

Thus eqns. (7,8) may descibe the two electronic transitions in *R.viridis* RCs revealed by absorption bands at 1000 and 1015 nm at 1.9 K (Fig. 1,A). From comparison of absorption and fluorescence spectra measured at 1.9 and 77 K [16] one can estimate energy positions of the 0-0 transitions for the two bands and their energy difference; the latter equals ~5 meV. Thus the two electronic states are distinct at 1.9 K and thermally mixed at 77 K. Sensitivity of the band at 1015 nm to reduction of H supports the proposal that the band reflects a transition into a state mixed with P^+B^- since H is located very close to B,[19,20] and H^- prevents formation of B^-.

Analysis of eqns. (4,5,7,8) shows that optical transitions in the long-wavelength absorption band of RCs relate to delocalization of electron density between three molecules, P_M, P_L, and B_L, in an excited state. This should result in a strong electron-phonon coupling which can be monitored by hole-burning experiments. Mixing of excitonic states of P with the charge-resonance and charge-transfer states can be detected by Stark effect spectroscopy. Participation of the P^+B^- state in electron transfer processes can be observed in femtosecond kinetics and spectra since selective excitation in the long-wavelength absorption band of RCs should lead to bleaching of that band accompanied by simultaneous partial bleaching of the B band.

Hole-burning Experiments on RCs

The first hole-burning measurements were made on *R.viridis* RCs[15] using the process of photo-reduction of H under continuous light in the presence of reduced Cyt and Q_A^- at low temperature.[21] Fig. 2 shows the difference absorption spectrum for the reduction of H at 4.2 K upon illumination with white light. The same spectrum was obtained at 1.9 K upon illumination by narrow lines of a xenon lamp at 995 nm or at 1012 nm; narrow holes due to zero-phonon lines with widths determined by the 3-ps lifetime T_1 of P^* were not observed. The absence of narrow holes may indicate a strong electron-phonon coupling[15] due to formation of CT states.

Similar results were obtained by the two groups[22-25] in hole-burning experiments at ~2 K on RCs of *Rb.sphaeroides* and *R.viridis* with selective excitation in the long-wavelength absorption band to produce the $P^+ Q_A^-$ state.

Fig. 3 shows the results of hole-burning experiments on RCs of *Chloroflexus aurantiacus* selectively excited with the picosecond parametric oscillator described earlier[4] to produce the $P^+ Q_A^-$ state at < 2 K.

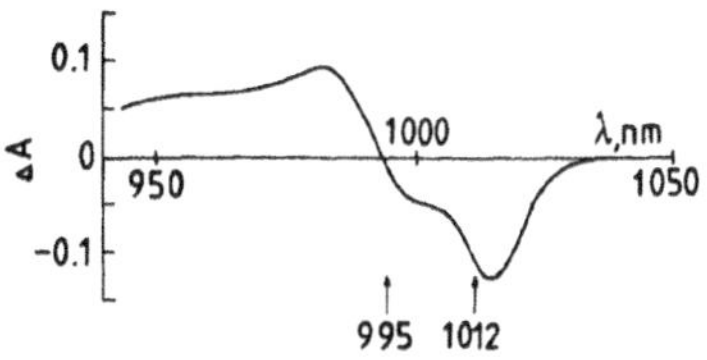

Fig. 2. Difference absorption spectrum of I^- formation induced by white light illumination of *R.viridis* RCs at 4.2 K. The same spectrum was obtained after continuous illumination of RCs with xenon lines at 995 nm and at 1012 nm at 1.9 K.[15]

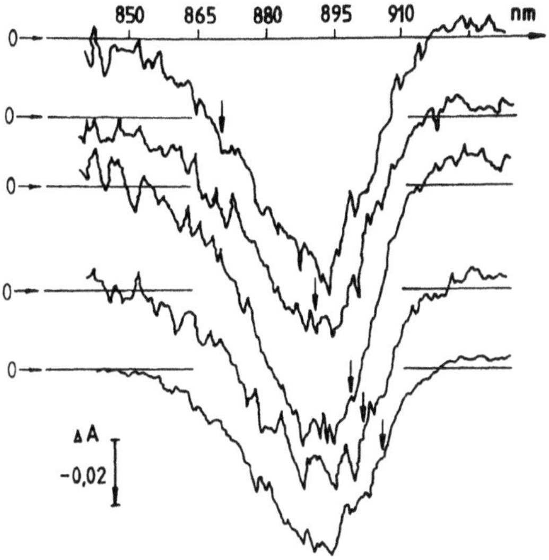

Fig. 3. Difference absorption spectrum of $P^+ Q_A^-$ formation in RCs of *Chloroflexus aurantiacus* at 1.7 K induced by laser pulses at wavelengths indicated with arrows above the spectral curves. Excitation pulses were generated by a picosecond parametric oscillator described earlier.[4] $A_{890} = 0.8$.

In accordance with the results mentioned above, a broad (~350 cm^{-1}) absorbance band is bleached independent of the excitation wavelength.

Femtosecond and Picosecond Measurements on RCs

Fig. 4 shows the kinetics of ΔA at several wavelengths induced by ~300-fs pulses at 620 nm in modified RCs of *Rb.sphaeroides* R-26[14] from which the B_M molecule had been removed[13] to avoid its contribution to ΔA measurements. Bleaching of the 870-nm band occurs simultaneously with the excitation. At a 1-ps delay when all the processes related to excitation of P, B, H terminated,[26] partial bleachings are observed at 805 nm (~35% of the maximal value) and at 815 nm (~60% of its maximal value).

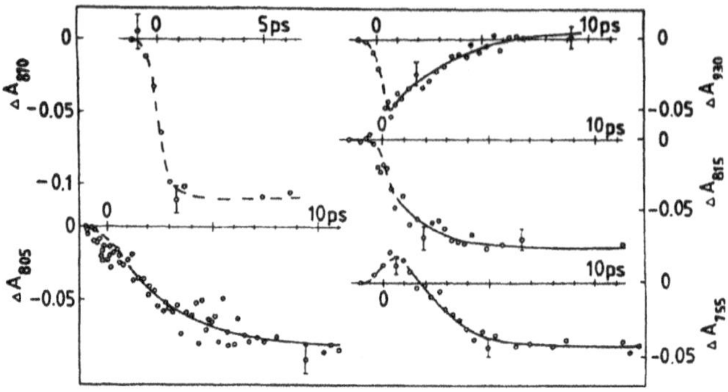

Fig. 4. Kinetics of ΔA measured at 293 K at various wavelengths, induced with ~300-fs laser pulses at 620 nm in modified RCs of *Rb.sphaeroides* R-26 from which B_M was removed.[14]

The kinetics of ΔA measured at 805 and 815 nm apparently comprise two components, fast (≲300 fs) and slow (~3 ps). We refer the state reached via the fast component as the "early state". The fast component of the bleachings at 805 and 815 nm is not observed in the kinetics of bleaching of the H band at 755 nm and in the decay of stimulated emission of RCs at 930 nm. The latter two kinetics have a time constant of ~3 ps which is also characteristic of the slow component of the bleachings at 805 and 815 nm.

The kinetics discussed above are consistent with the ΔA spectra measured with a different set-up[10] in the case when modified RCs of *Rb.sphaeroides* R-26 were illuminated for ~1.5 ps using selective excitation at 880 nm. To obtain this time resolution, the negative delay of -38 ps was used for 33-ps measuring and exciting pulses.[10] The ΔA spectrum at this delay (Fig. 5,C) includes bleaching of the 870-nm band and a partial bleaching (~60% of the maximal one measured at later delay) of the band at 810 nm. No bleaching at 760 nm could be discerned at this delay in experiments with different angles between electric vectors of the excitation and measuring beams (0° and 90°) used to exclude possible contribution of photoselection into measured spectra. A very small trough is observed at 545 nm. At the same delay, a considerable stimulated emission from P^* was measured at 930 nm.[10] The ΔA spectra measured at later delay (-20 ps) include bleaching of the bands at 870, 805, 760, 600 and 545 nm which are characteristic of the formation of P^+H^- (Fig. 5,D).

The femtosecond measurements made within an intense absorbance band with optical density greater than 2 at 800 nm reported without a light-saturation curve for ΔA[9,11,12] may yield difficulties for interpretation of the results.

The most important fact concerning the conversion of an excited state in an RC of *Rb.sphaeroides* into the P^+H^- state is the partial bleaching observed at 805-815 nm at early stages of the process when the presence of P^+H^- is not yet detected. The ΔA spectrum of this early state differs from the spectrum of P^T formation (Fig. 5,B) where the bleaching at 805 nm is much smaller. Therefore the spectrum of the early state implies participation of P^+B^- in a mixed state discussed above (eqns. 7,8) rather than bleaching of the P band at 805 nm.

Similar results were obtained for *R.viridis* RCs using selective excitation by 33-ps pulses at 960 nm.[31] At the -38 ps delay the ΔA spectrum is completely different from that of the P^+H^- state and includes bleaching

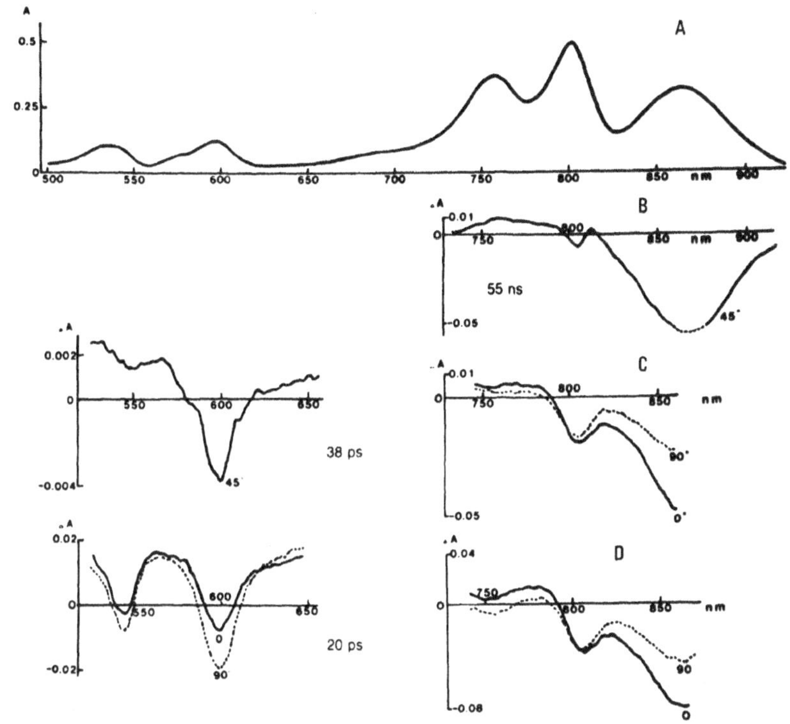

Fig. 5. Absorption spectrum (A), and difference absorption spectra (B-D) measured at various delays in modified RCs of *Rb.sphaeroides* R-26 induced by 33-ps laser pulses at 880 nm at 293 K. The ΔA spectra correspond to: P^T state (B, 55-ns delay), the "early state" (C, -38 ps), P^+I^- (D, -20 ps). Data taken from[10].

of the bands at 960 nm and at 850 nm (Fig. 6). The latter band does not belong to P since only a blue shift of that band is observed in the ΔA spectrum of P^T formation,[27] therefore the spectrum of the early state indicates participation of P^+B^- in a mixed state described in eqns. (7,8). Simultaneous bleachings of the bands at 1000 and 850 nm in *R.viridis* RCs were observed at low temperature by Martin et al. (see this volume).

The assumption concerning participation of P^+B^- in a mixed state expressed by eqns. (7,8) agrees with hole-burning experiments and photon-echo measurements on *R.viridis* RCs.[15,22-25] The latter method have shown that the upper limit for optical dephasing time T_2 is ~100 fs, and the action spectrum of echo signals includes the bands of P as well as of B.[24] This indicates a possibility for P^+B^- formation during excitation of P.

Stark Effect Measurements[28-30]

The charge-transfer features of the excited state of RCs can also be inferred from Stark effect measurements made on various RCs.[28-30] A large electric dipole moment (~8 debye) was found for the excited state of RC. Since the spatial arrangement of P and its ligands is almost symmetrical,[19,20] it seems unlikely that a charge separation within the special pair would yield a large electric dipole moment. It is more plausible to assume a contribution of P^+B^- to the formation of this dipole moment, in accordance with the data mentioned above.

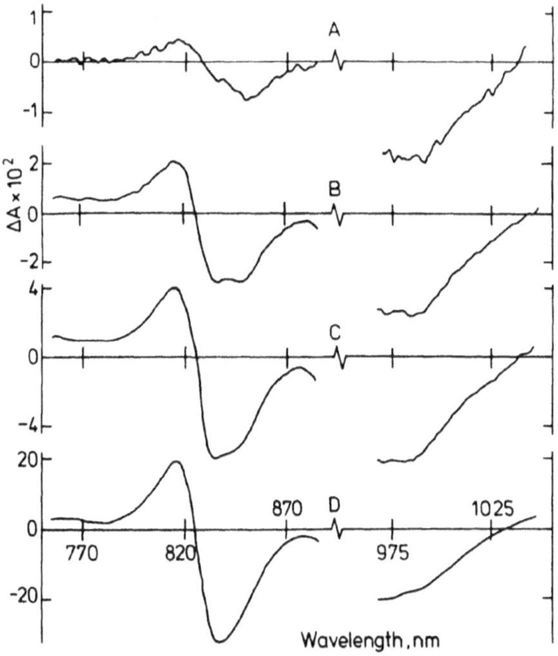

Fig. 6. Difference absorption spectra of *R.viridis* RCs at 293 K upon selective excitation with 33-ps laser pulses at 960 nm measured at delays of -38 ps (A), -30 ps (B), -10 ps (C), and 70 ps (D). Quinone acceptors were reduced in the dark. Data taken from[18].

Selective Trapping of Charge-Transfer States

It is appropriate to analyze spectral features observed at different temperatures upon formation of two different states, (i) $P^+B\,H^-\,Q_A^{(-)}$, and (ii) $P\,B\,H^-\,Q_A^-$. The ΔA spectrum pertinent to the formation of state (i) includes absorbance changes related to both P^+ and H^-, while that for state (ii) reveals only the features of H^-.

The first indication that the spectrum of H^- depends on temperature has been found in nanosecond measurements of the formation of the $P^+BH^-Q_A^-$ state.[32] Namely, the spectrum measured at 77 K includes the bleachings of the H bands at 545 and 760 nm, while the spectrum measured at 293 K shows a relative decrease of the 760 nm bleaching and a new bleaching at 800 nm (that mainly belongs to B). However, it has remained obscure if the difference is related to a thermal broadening of the H and B bands, or to another reason.

More clear data can be obtained in experiments where RCs are illuminated to trap the $P\,B\,H^-\,Q_A^-$ state (ii) at various temperatures, and all the spectra are subsequently measured at the same (low) temperature. Fig. 7 shows the absorption spectra measured at 77 K for two preparations of *R. viridis* RCs at ~-400 mV: (1) frozen in the dark; (2) frozen to 77 K in the dark and then illuminated at 77 K; (3) frozen in the dark to 77 K, illuminated at 77 K, warmed in the dark up to 230 K, frozen in the dark back to 77 K.[33] One can see that illumination at 77 K induced bleachings of the H_L bands at 545 and 810 nm, and a blue shift of the B bands at 830 and 610 nm, along with formation of the radical-anion band of H at 680 nm. These are the features of state (ii). After the warming-freezing cycle,

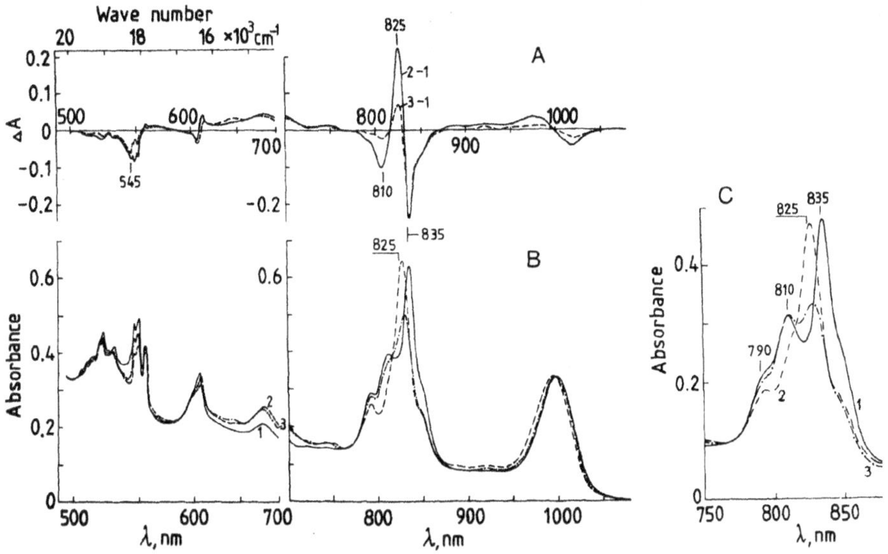

Fig. 7. Absorption spectra (B and C) measured at 77 K on two different preparations of *R.viridis* RCs at -400 mV: RCs frozen in the dark (1); illuminated for 1 h at 77 K (2); after illumination at 77 K, RCs were warmed up in the dark to 220 K for several min (3). (A) shows difference between the spectra in (B): 2-minus-1 (solid line) and 3-minus-1 (dashed line). Data taken from[33].

state (ii) is relaxed which is accompanied by a considerable recovery of the 810-nm band and a bleaching of the 830-nm band. A new radical-anion band is observed at 645 nm. No recovery of the 545-nm band is discerned.

If the RCs at -400 mV are illuminated at 293 K for 2 min, and then frozen under illumination to 77 K, a spectrum is observed (Fig. 8) very similar to that of relaxed state (ii). Subsequent illumination at 77 K induces a further bleaching of the 825 nm band, a development of the bacteriochlorophyll radical-anion band at 1060 nm (see[34]), and a bleaching of the H_M band at 790 nm. The development of the 1060-nm band clearly indicates the formation of B^-.

The band observed at 810 nm in the spectrum of relaxed state (ii) has many features (wavelength position and half-width, and amplitude) similar to those of the original H_L band. The simplest explanation to this fact is that the H_L transition is formed anew after relaxation of state (ii). This is accompanied by a decrease of the dipole strength of the B transition related to the 825-nm band.

The development of the 810-nm band and the decrease of the 825-nm band can be explained by electron hopping between $H_L^{\cdot}$ and B_L in the relaxed state (ii) facilitated by a reorganization of the medium which results in convergence of energy levels of $B_L^{\cdot}$ and $H_L^{\cdot}$. Such convergence may agree with the formation of a new radical anion band at 645 nm in Figs. 7 and 8. The fast electron hopping between $H_L^{\cdot}$ and B_L would not affect the interaction between $H_L^{\cdot}$ and $(Fe\,Q_A)^{\cdot}$ revealed by low-temperature EPR measurements in nanosecond time domain (Fig. 9) for state (ii) before and after relaxation (see[35] for discussion).

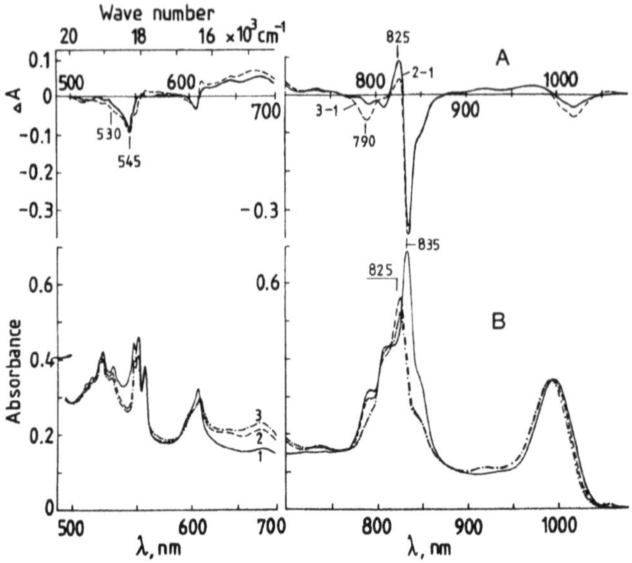

Fig. 8. Absorption spectra (B) of *R.viridis* RCs at -400 mV measured at 77 K: RCs frozen in the dark (1); RCs frozen after 2 min illumination at 293 K (2); after illumination at 293 K, RCs were additionally illuminated at 77 K for 1 h (3). (A) shows difference between the spectra of (B): 2-minus-1 (solid line) and 3-minus-1 (dashed line).Data taken from[33].

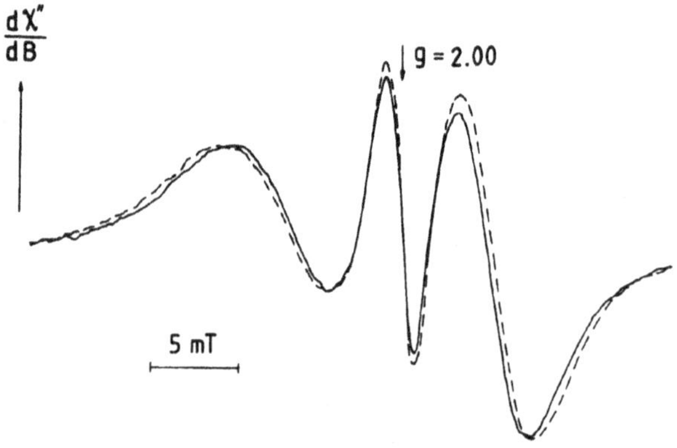

Fig. 9. EPR spectra of state P I$^-$ $(Q_AFe)^-$ in *R.viridis* RCs before (solid) and after (dashed line) relaxation, which correspond to absorption spectra in Fig. 7, curves 2 and 3, respectively. Spectra measured at 8 K; microwave power, 1 mW; modulation amplitude, 15 G.

The assumption about fast electron hopping between $H_L^{\bar{\cdot}}$ and B_L is supported by femtosecond measurements of state (i) formation in *Rb.sphaeroides* RCs at 293 K.[14] The ΔA spectrum at the 6-ps delay (Fig. 10) includes bleachings of the bands at 755 and 805 nm ascribed mainly to H and B, respectively. Absorbance increase near 780 nm which could imply formation of solely P^+H^- was not observed. Similar spectra were obtained for *R.viri-*

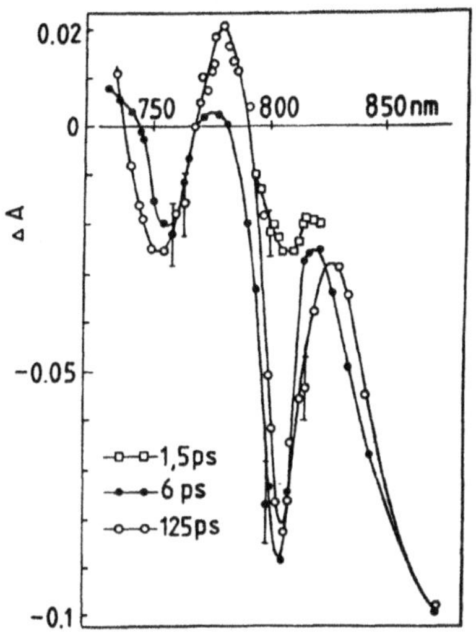

Fig. 10. Difference absorption spectra of modified RCs of *Rb.sphaeroides* R-26 at 293 K. ΔA induced by ~300-fs laser pulses at 620 nm and measured at delays of 1.5 ps, 6 ps, and 125 ps. Data taken from [14].

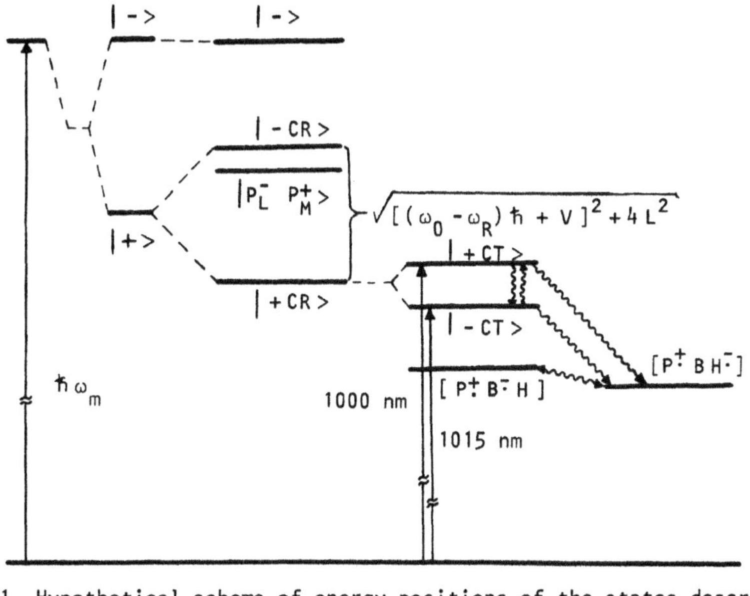

Fig. 11. Hypothetical scheme of energy positions of the states described in the text.

dis RCs at 293 K (Fig. 6). At the delay sufficient for electron transfer to H, the absorbance bleaching is observed only near 830-850 nm which is characteristic of B. These spectral features imply development of a relaxed state (i) in which electron hopping occurs between $H_L^{\bar{\cdot}}$ and B_L.

Likeness between the kinetics at 755 nm and the slow component at 805 nm in *Rb. sphaeroides* RCs measured at 293 K (Fig. 4) indicates that (a) the electron hopping between $H_L^{\bar{\cdot}}$ and B_L has a time constant less than 300 fs,[14] and (b) the relaxation of state (i) occurs within 300 fs.

CONCLUSION

Fig. 11 shows a hypothetical scheme of energy positions of the states described by eqns. (4,5,7,8) and of $P^{\dot{+}}H^{\bar{\cdot}}$. Relaxation of RCs after electron transfer to H leads to development of a state $P^{\dot{+}}B^{\bar{\cdot}}$ almost resonant to $P^{\dot{+}}H^{\bar{\cdot}}$ which results in fast electron hopping between the two states.

ACKNOWLEDGEMENTS

We thank I. I. Proskuryakov for the help in measuring EPR signals.

REFERENCES

1. M. G. Rockley, M. W. Windsor, R. J. Cogdell, and W. W. Parson, Picosecond detection of an intermediate in the photochemical reaction of bacterial photosynthesis, Proc. Natl. Acad. Sci. USA 72:2251 (1975).
2. K. J. Kaufmann, P. L. Dutton, T. L. Netzel, J. S. Leigh, and P. M. Rentzepis, Picosecond kinetics of events leading to reaction center bacteriochlorophyll oxidation, Science 188:1301 (1975).
3. V. A. Shuvalov, A. V. Klevanik, A. V. Sharkov, Yu. A. Matveetz, and P. G. Krukov, Picosecond detection of bacteriochlorophyll-800 as an intermediate electron carrier between selectively excited P870 and bacteriopheophytin in *Rhodospirillum rubrum* reaction centers, FEBS Lett. 91:135 (1978).
4. V. A. Shuvalov and A. V. Klevanik, The study of the state $P870^{+}B800^{-}$ in bacterial reaction centers by selective picosecond and low-temperature spectroscopies, FEBS Lett. 160:51 (1983).
5. S. A. Akhmanov, A. Y. Borisov, R. V. Danielius, R. A. Gadonas, V. S. Kozlowsky, A. S. Piscarskas, A. P. Razjivin, and V. A. Shuvalov, One- and two-photon picosecond processes of electron transfer among the porphyrin molecules in bacterial reaction centers, FEBS Lett. 114:144 (1980).
6. D. Holten, C. Hoganson, M. W. Windsor, C. C. Schenck, W. W. Parson, A. Migus, P. L. Fork, and C. V. Shank, Subpicosecond and picosecond studies of electron transfer intermediates in *Rhodopseudomonas sphaeroides* reaction centers, Biochim. Biophys. Acta 592:461 (1980).
7. A. V. Klevanik, P. G. Krukov, Yu. A. Matveetz, V. A. Semchishin, and V. A. Shuvalov, Subpicosecond measurements of energy and electron transfer in physical stages of photosynthesis, Pizma v GETF 32:107 (1980).
8. N. W. Woodbury, M. Becker, D. Middendorf, and W. W. Parson, Picosecond kinetics of the initial photochemical electron-transfer reaction in bacterial photosynthetic reaction centers, Biochem. 24:7516 (1985).
9. J.-L. Martin, J. Breton, A. J. Hoff, A. Migus, and A. Antonetti, Femtosecond spectroscopy of electron transfer in the reaction center of the photosynthetic bacterium *Rhodopseudomonas sphaeroides* R-26: Direct electron transfer from the dimeric bacteriochlorophyll primary donor to the bacteriopheophytin acceptor with a time constant of 2.8±0.2 psec, Proc. Natl. Acad. Sci. USA 83:957 (1986).

10. V. A. Shuvalov and L. N. M. Duysens, Primary electron transfer reactions in modified reaction centers from *Rhodopseudomonas sphaeroides*, Proc. Natl. Acad. Sci. USA 83:1690 (1986).
11. J. Breton, J.-L. Martin, A. Migus, A. Antonetti, and A. Orszag, Femtosecond spectroscopy of excitation energy transfer and initial charge separation in the reaction center of the photosynthetic bacterium *Rhodopseudomonas viridis*, Proc. Natl. Acad. Sci. USA 83:5121 (1986).
12. J. Breton, J.-L. Martin, J. Petrich, A. Migus, and A. Antonetti, The absence of a spectroscopically resolved intermediate state P^+B^- in bacterial photosynthesis, FEBS Lett. 209:37 (1986).
13. V. A. Shuvalov, A. Ya. Shkuropatov, S. M. Kulakova, M. A. Ismailov, and V. A. Shkuropatova, Photoreactions of bacteriopheophytins and bacteriochlorophylls in reaction centers of *Rhodopseudomonas sphaeroides* and *Chloroflexus aurantiacus*, Biochim. Biophys. Acta 849:337 (1986).
14. S. V. Chekalin, Ya. A. Matveetz, A. Ya. Shkuropatov, V. A. Shuvalov, and A. P. Yartzev, Femtosecond spectroscopy of primary charge separation in modified reaction centers of *Rhodobacter sphaeroides* (R-26), FEBS Lett. 216:245 (1987).
15. V. G. Maslov, A. V. Klevanik, and V. A. Shuvalov, On the rate of energy transfer between pigments in bacterial reaction centers, Biofizika (USSR) 29:156 (1984).
16. V. G. Maslov, A. V. Klevanik, M. A. Ismailov, and V. A. Shuvalov, On the nature of the longwavelength absorption band of reaction centers from *Rhodopseudomonas viridis* in relation with the primary charge separation, Dokl. Akad. Nauk SSSR 269:1217 (1983).
17. A. V. Klevanik, The study of the primary processes in reaction centers of photosynthesis, Candidate thesis, Pushchino, 1983.
18. V. A. Shuvalov and A. A. Asadov, Arrangement and interaction of pigment molecules in reaction centers of *Rhodopseudomonas viridis*, Biochim. Biophys. Acta 545:296 (1979).
19. J. Deisenhofer, O. Epp, K. Miki, R. Huber, and H. Michel, X-ray structure analysis of a membrane protein complex: Electron density map at 3 Å resolution and a model of the chromophores of the photosynthetic reaction center from *Rhodopseudomonas viridis*, J. Mol. Biol. 180:385 (1984).
20. J. Deisenhofer, O. Epp, K. Miki, R. Huber, and H. Michel, Structure of the protein subunits in the photosynthetic reaction centre of *Rhodopseudomonas viridis* at 3 Å resolution, Nature 318:618 (1985).
21. V. A. Shuvalov, I. N. Krakhmaleva, and V. V. Klimov, Photooxidation of P-960 and photoreduction of P-800 (bacteriopheophytin b-800) in reaction centers from *Rhodopseudomonas viridis*, Biochim. Biophys. Acta 449:597 (1976).
22. S. R. Meech, A. J. Hoff, and D. A. Wiersma, Evidence for a very early intermediate in bacterial photosynthesis. A photon-echo and hole-burning study of the primary donor band in *Rhodopseudomonas sphaeroides*, Chem. Phys. Lett. 121:287 (1985).
23. S. G. Boxer, D. J. Lockhart, and T. R. Middendorf, Photochemical hole-burning in photosynthetic reaction centers, Chem. Phys. Lett. 123: 476 (1986).
24. S. R. Meech, A. J. Hoff, and D. A. Wiersma, The role of charge-transfer states in bacterial photosynthesis, Proc. Natl. Acad. Sci. USA 83: 9464 (1986).
25. S. G. Boxer, T. R. Middendorf, and D. J. Lockhart, Reversible photochemical hole-burning in *Rhodopseudomonas viridis* reaction centers, FEBS Lett. 200:237 (1986).
26. Yu. A. Matveetz, S. V. Chekalin, and A. P. Yartzev, Femtosecond spectroscopy of the primary photoprocesses in reaction centers of *Rhodopseudomonas sphaeroides*, Dokl. Akad. Nauk. SSSR 292:724 (1986).
27. V. A. Shuvalov and W. W. Parson, Triplet states of monomeric bacteriochlorophyll in vitro and of bactriochlorophyll dimers in antenna and reaction center complexes, Biochim. Biophys. Acta 638:50 (1981).

28. D. de Leeuv, M. Malley, G. Butterman, M. Y. Okamura, and G. Feher, The Stark effect in reaction centers from *R.sphaeroides*, Biophys. J. 37:111a (1982).
29. D. J. Lockhart and S. G. Boxer, Magnitude and direction of the charge in dipole moment associated with excitation of the primary electron donor in *R.sphaeroides* reaction centers, Biochem. 26:664 (1987).
30. H. P. Braun, M. E. Michel-Beyerle, J. Breton, S. Buchanan, and H. Michel, Electric field effect on absorption spectra of reaction centers of *Rb.sphaeroides* and *Rps.viridis*, FEBS Lett. 221:221 (1987).
31. V. A. Shuvalov, J. Amesz, and L. N. M. Duysens, Picosecond charge separation upon selective excitation of the primary electron donor in reaction centers of *Rhodopseudomonas viridis*, Biochim. Biophys. Acta 851:327 (1986).
32. V. A. Shuvalov and W. W. Parson, Energies and kinetics of radical pairs involving bacteriochlorophyll and bacteriopheophytin in bacterial reaction centers, Proc. Natl. Acad. Sci. USA 78:957 (1981).
33. V. A. Shuvalov, A. Ya. Shkuropatov, and M. A. Ismailov, Selective reduction and modification of bacteriochlorophylls and bacteriopheophytins in reaction centers from *Rhodopseudomonas viridis*, in: Progress in Photosynthesis Research, v.1, p.1.2.161, J. Biggins, ed., Martinus Nijhoff Publishers, Dordrecht (1987).
34. J. Fajer, M. S. Davis, D. C. Brune, L. D. Spaulding, D. C. Borg, and A. Forman, Chlorophyll radicals and primary events, Brookhaven Symp. Biol. 28:74 (1977).
35. D. M. Tiede, E. Kellogg, and J. Breton, Conformational changes following reduction of the bacteriopheophytin electron acceptor in reaction centers of *Rhodopseudomonas viridis*, Biochim. Biophys. Acta 892:294 (1987).

TEMPERATURE EFFECTS ON THE GROUND STATE ABSORPTION SPECTRA AND ELECTRON TRANSFER KINETICS OF BACTERIAL REACTION CENTERS

Christine Kirmaier and Dewey Holten

Department of Chemistry
Washington University
St. Louis, MO 63130 USA

INTRODUCTION

It has been a general finding that the rates of many of the electron transfer processes in bacterial reaction centers are rather insensitive to temperature.[1,2] In fact, often the rates actually increase slightly as the temperature is reduced. This is observed, for example, in *Rb. sphaeroides* reaction centers for electron transfer from P^* to I (BPh_L),[3] from I^- (BPh_L^-) to the primary quinone Q_A,[4] and for the charge recombination reaction $P^+Q_A^- \longrightarrow PQ_A$.[5-7] (P is the dimer of BChl molecules.) These findings have been taken to suggest that these reactions are activationless, or nearly so.

Temperature effects on the ground state absorption spectrum of reaction centers are also well known. The most dramatic effect of lowering the temperature is on the dimer's long-wavelength absorption band, which narrows and shifts to lower energy. Recently, we reported the interesting observation that, in *Rb. sphaeroides* reaction centers, the form of the temperature dependence of the absorption band of P, and the form of the temperature dependence of the BPh_L^- to Q_A electron transfer rate, mirror each other.[4,8] More specifically, the rate of electron transfer from BPh_L^- to Q_A increases slightly as the temperature is lowered from room temperature to about 80 K, and then remains constant down to 5 K.[4] Remarkably, <u>both</u> the position and bandwidth of the long-wavelength absorption band of P have this same general form of temperature dependence; most of the change occurs between room temperature and about 50 K, with little further change occurring between 50 K and 5 K.[8] That the position and width of the long-wavelength band of P are highly temperature dependent has been known for many years,[5] but that the two spectral parameters have a very similar temperature dependence has not received much attention. We have speculated that the identical temperature dependence of the peak position and width of the long-wavelength band of P might be related to a change in vibronic coupling in the dimer as a function of temperature.[8]

To further examine the temperature dependence of the rate of electron transfer from BPh_L^- to Q_A, the temperature dependence of the absorption band of P, and any possible connection between these two observables, we have carried out measurements on *Rps. viridis* reaction centers analogous to those we have made on *Rb. sphaeroides*. We find that although the rate of electron transfer from BPh_L^- to Q_A appears not to change with temperature in *Rps. viridis* reaction centers, the temperature dependence of the spectral data for the dimer is virtually identical to that observed in *Rb. sphaeroides*. These data will be summarized and briefly discussed in this article.

RESULTS

Figure 1 shows the temperature dependence of the rate of the electron transfer reaction $P^+BPh_L^- \longrightarrow P^+Q_A^-$ in *Rb. sphaeroides* reaction centers (closed symbols) imbedded in a polyvinyl alcohol (PVA) film.[4] The rate increases (the time constant decreases) by about a factor of two between 295 K and about 80 K, and then remains constant down to 5 K. The solid curve is a theoretical fit of the data, and is discussed below.

In contrast, the rate of electron transfer from BPh_L^- to Q_A in *Rps. viridis* reaction centers appears to be independent of temperature. This can also be seen in Figure 1 (open symbols), where it is shown that a 170 ± 15 ps time constant for this electron transfer step is observed at 295, 76, and 5 K. (We have not been able to carry out a more complete study of the temperature dependence of the BPh_L^- to Q_A electron transfer reaction in *Rps. viridis*, owing to complications due to the presence of the bound cytochrome. Details of this, and other low temperature measurements on the primary photochemistry in *Rps. viridis* reaction centers, will be published elsewhere. (See also ref. 9.)

The temperature dependence of the long-wavelength band of P in the *Rb. sphaeroides* reaction center ground state absorption spectrum is shown in

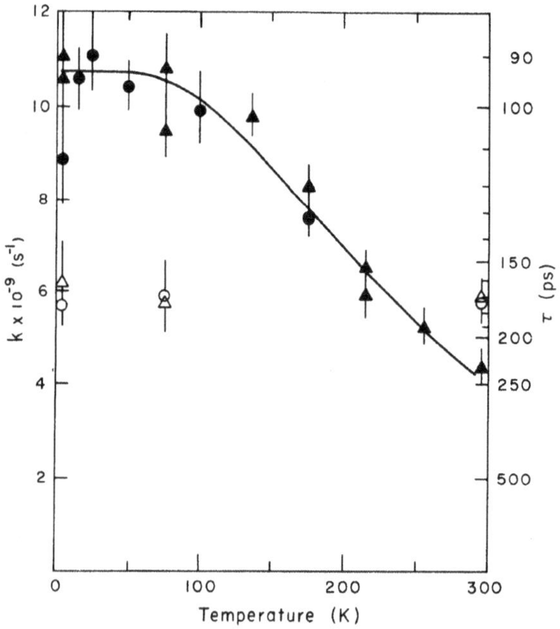

Figure 1. Temperature dependence of the rate of electron transfer from BPh_L^- to Q_A in *Rb. sphaeroides* (closed symbols) and *Rps. viridis* (open symbols). Kinetics were observed in the BPh Q_x band (closed and open circles) and in the BPh anion band (closed and open triangles).

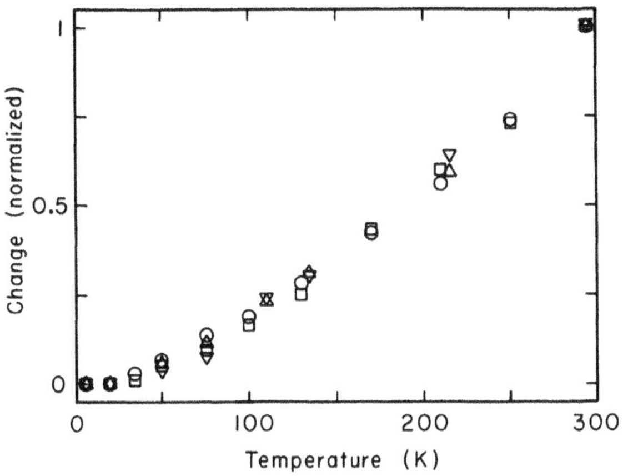

Figure 2. Temperature dependence of the peak position and width of the absorption band of P in *Rb. sphaeroides* and *Rps. viridis* reaction centers. Symbols: Δ, ∇ are position and half-width, respectively, for *Rb. sphaeroides*; ○, □ are position and full-width, respectively, for *Rps. viridis*. See also Tables 1 and 2.

Figure 2. (These data are replotted from ref. 8.) PVA was again the medium for these measurements. The upright triangles denote the peak position, and the inverted triangles correspond to the half-width at half-height on the long-wavelength side (i.e., the red side) of the band. (This half-width is plotted because the absorption band of the dimer has significant overlap with

Table 1. Position and Width as a Function of Temperature of the Long-Wavelength Absorption Band of the Dimer (P) in *Rb. sphaeroides* Reaction Centers in a PVA Film.[a]

T (K)	λ_{max} (cm^{-1})	Normalized Change[b]	Half-Width (cm-1)[c]	Normalized Change[b]
5	11,364	0	265	0
20	11,364	0	265	0
50	11,383	0.06	272	0.03
76	11,403	0.12	280	0.06
110	11,442	0.23	319	0.23
135	11,468	0.31	332	0.29
215	11,561	0.59	413	0.63
295	11,696	1	498	1

[a]Spectra were measured in wavelength units (± 1 nm), and the spectral parameters converted to wavenumbers.
[b]Normalized to the total change observed between 5 and 295 K.
[c]Half-width at half-height on the long-wavelength (red) side of the absorption band.

Table 2. Position and Width as a Function of Temperature of the Long-Wavelength Absorption Band of the Dimer (P) in *Rps. viridis* Reaction Centers in a Gelatin Film.[a]

T (K)	λ_{max} (cm^{-1})	Normalized Change[b]	Full-Width (cm-1)[c]	Normalized Change[b]
5	10,101	0	494	0
20	10,101	0	494	0
35	10,111	0.03	498	0.01
50	10,121	0.07	517	0.05
76	10,142	0.14	538	0.09
100	10,158	0.19	571	0.16
130	10,183	0.28	615	0.25
170	10,225	0.42	705	0.43
210	10,267	0.56	786	0.60
250	10,320	0.74	847	0.73
295	10,395	1	978	1

[a]Spectra were measured in wavelength units (± 1 nm), and the spectral parameters converted to wavenumbers.
[b]Normalized to the total change observed between 5 and 295 K.
[c]Full-width at half-height.

the 800-nm band, especially at higher temperatures, making it difficult to determine the full-width.) The data are normalized to the total change in each parameter between 5 and 295 K. It can be seen that both the peak position and the width of the band show essentially the same temperature dependence when plotted in this way.

We have analyzed the ground state absorption spectrum of *Rps. viridis* reaction centers as a function of temperature in the same manner, except that in this case we were able to determine the full-width (at half-height) of the dimer band. These data are also plotted in Figure 2. Again, notice that the peak position (circles) and full-width (squares) of the long-wavelength band change with temperature in a virtually identical manner. Furthermore, the data for *Rps. viridis* and *Rb. sphaeroides* are indistinguishable within the error of the measurements.

The spectral data used to construct Figure 2 are collected in Tables 1 and 2. It is important to remember that one can expect to obtain somewhat different absolute values of the parameters if another medium for the reaction centers is chosen. For example, the values we have obtained for *Rb. sphaeroides* reaction centers at 76 K in 65% glycerol, are 11,217 cm^{-1} and 210 cm^{-1} for the position and half-width of the dimer band, respectively. Similarly, it is well known that the spectral parameters can be different in various media at room temperature, or when reaction centers are dehydrated.[5] However, the medium does not appear to alter the form of the temperature dependence. For example, in ethylene glycol *Rb. sphaeroides* and *Rps. viridis* reaction centers show virtually the same temperature dependence of the position of the dimer band[10] as we have observed for the two species in films. We also believe that a change in medium will not alter the more general and more revealing conclusion that follows from Figure 2, namely that there is a correlation in the temperature dependence of both the peak position and width of the long-wavelength absorption band in both species.

DISCUSSION

Essentially all of the data presented above were acquired in the same or similar media, so that it may be particularly meaningful to compare the various results in order to learn something about the electron transfer and/or spectral properties of reaction centers. Another advantage of the data we have presented is that in film media one is not concerned with the effects of a bulk phase transition from liquid to solid state as the temperature is reduced. Here we wish to briefly discuss some possible interpretations of the data. Ultimately, such comparisons will be the most useful when the data, treated as a whole, are the subject of detailed theoretical analyses of the spectra and kinetics.

Temperature Dependence of the Rate of Electron Transfer

Let us first consider the temperature dependence of the rate of electron transfer from BPh_L^- to Q_A in *Rb. sphaeroides*. As described previously,[4] there are basically two ways to explain the results, depending on whether the temperature dependence is ascribed to the electronic factor or to the vibrational (Franck-Condon) factor.

The first extreme point of view is that the increase in rate with decreasing temperature solely reflects a contraction of the protein. In other words, BPh_L and Q_A move closer together or into a more favorable relative orientation as the temperature is reduced. Using a standard exponential dependence of the rate on distance,[2,11] we have estimated that BPh_L and Q_A would have to more closer together by about 1 Å as the temperature is changed from 295 K to 80 K in order to reproduce the observed factor of two increase in rate.[4] Since the edge-to-edge distance between these two molecules is about 10 Å,[12-14] a 1 Å change corresponds to contraction by about 10%.

Recently it was reported, on the basis of crystallographic data, that the interatomic distances in myoglobin are smaller, on the average, by about 3% at 80 K compared to 295 K.[15] Since this is an average value, it is likely that some distances are changed more than others, and that some distances are unchanged, with temperature. In the reaction center, whether there is sufficient freedom for a 10% change in distance between BPh_L and Q_A is not clear, and the ultimate answer to the question at hand awaits a low-temperature crystal structure determination. In this regard, one must also be concerned with the effect that temperature may have on the positions of protein residues or groups appended to the pigments that potentially may be important in electron transfer, e.g., the tryptophan near BPh_L and Q_A, the hydrocarbon chain of Q_A, the glutamic acid hydrogen-bonded to BPh_L, etc.[12-14] Several of these moieties could conceivably enhance the rate to some degree by providing a more favorable medium for electron transfer. If so, then their contribution to the electron transfer process could change if their positions are altered when the temperature is varied.

The second extreme point of view is to take the relevant distance (i.e. the electronic factor) to be unchanged with temperature. Then, the temperature dependence of the rate is contained in the thermal-weighting parameters of the Franck-Condon factor. Using this approach, the general form of the data can be understood within the context of many theories,[2,11,16,17] if the potential surface for the final state ($P^+ BPh_L Q_A^-$) crosses at or near the minimum of the potential surface for the initial state ($P^+ BPh_L^- Q_A$). In other words, the reaction is activationless. The solid curve in Figure 1 is a fit of the kinetic data for *Rb. sphaeroides* using the theory of Kakitani and Kakitani,[16] assuming the participation of a high-energy (1550 cm^{-1}) mode that undergoes an origin shift and a low-energy (250 cm^{-1}) mode that experiences a frequency change as a result of electron transfer.[4]

The best explanation for the data may be represented by a combination of both of these extreme points of view. In other words, it is likely both that the *Rb. sphaeroides* reaction center contracts somewhat as the temperature is reduced, and that electron transfer from BPh_L^- to Q_A is an essentially activationless process. It has been proposed that contraction of the protein is also responsible, at least in part, for the increase in the rate of the $P^+Q_A^-$ charge recombination reaction observed in *Rb. sphaeroides* when the temperature is lowered.[6,7]

Our observation that the rate of electron transfer from BPh_L^- to Q_A in *Rps. viridis* is the same within experimental error at 295, 76, and 5 K indicates that there might be a subtle difference in a least one parameter (activation energy, extent of thermal contraction, etc.) for *Rps. viridis* as compared to *Rb. sphaeroides*. Electron transfer from BPh_L^- to Q_A in *Rps. viridis* could, for example, be a slightly activated process, with thermal contraction providing a compensating effect which renders the observed rate unchanged with temperature. In this regard, a recent re-examination of the $P^+Q_A^-$ charge recombination kinetics in *Rps. viridis* has revealed an Arrhenius temperature dependence for this process, with an apparent activation energy of about 0.18 eV.[18] This too is in contrast to the non-Arrhenius temperature dependence normally found for this process in *Rb. sphaeroides*.

Water molecules may be an important parameter in understanding the thermal contraction process.[5-7] One indication of this is the finding that the rate of $P^+Q_A^-$ charge recombination in *Rb. sphaeroides* becomes essentially independent of temperature if the reaction centers are thoroughly dehydrated.[5,7] Water molecules appear to be present at various locations in the *Rps. viridis* crystal structure (J. Deisenhofer, personal communication). The difference between the observed temperature dependences of BPh_L^- to Q_A electron transfer in the two species might be associated with the freezing (and expanding) of water molecules present in different locations or quantities in the protein. Because of such possibilities, and others described above, one cannot readily infer from the temperature dependence data whether there may be some <u>small</u> differences between the two species in the activation energies for electron transfer.

<u>Temperature Dependence of the Long-Wavelength Absorption Band of the Dimer</u>

It has been known for many years that vibronic interactions within an exciton coupled dimer can affect both the position and the bandwidth of the absorption bands of a dimer.[19-21] Following along this line, we suggested previously that vibronic coupling involving a low-frequency deformation mode of P may be important in determining the position and bandwidth of the long-wavelength band in *Rb. sphaeroides*.[8] This mode would likely involve motions of the two rings of the dimer with respect to one another. Such a mode has been proposed to be important in electron transfer involving P.[22] Similarly, a low-frequency mode associated with BPh_L might be important for electron transfer from BPh_L^- to Q_A.[4] These out-of-plane motions of the atoms of P or BPh also could be coupled to protein motions via residues hydrogen bonded to the peripheral groups of the macrocycles.

Before discussing this idea further, let us address why the temperature dependence of electron transfer and the temperature dependence of absorption bands might be similar. It is really rather simple and natural to suggest that similar vibrations and electron-vibration coupling parameters might be associated both with excitation and with electron transfer for tetrapyrroles. This idea is based on the fact that the same π and π^* orbitals of the chromophores are probably involved in the two processes. This would tend to result in similar (and small) changes in bond lengths and angles in the chromophore for the excitation or redox processes. Obviously, one also must consider the internal reorganizations of the electron acceptor (another tetrapyrrole or a quinone), as well as the medium reorganizations. On electrostatic grounds, one might expect the medium reorganizations to be different for excitation or electron transfer, depending on the immediate environments of the pigments. Unfortunately, in the reaction center the changes in the protein or in the pigment-protein interactions that accompany either process are unknown. Thus, it is useful to explore similarities between the temperature dependence of excitation (i.e., of the absorption spectrum) and of the rate of electron transfer in order to obtain a more clear picture of which vibrational modes of the pigments or the protein are the most strongly coupled to the two electronic processes.

This line of reasoning allows that the temperature dependence of the absorption band(s) of P, and of electron transfer reactions involving P, might be related or even be the same. The temperature dependence of the width of the long-wavelength band of P, and of the rate of the $P^+Q_A^- \longrightarrow PQ_A$ reaction in *Rds. rubrum* and *Ectothiorhodospira sp.* have been analyzed together in this way.[23] Although the analysis of the rate and bandwidth gave reasonable agreement in the fitting parameters, the vibrational coupling strengths were rather large ($S = 20\text{-}40$). One could hypothesize that the large coupling parameters, and the vibrational energies obtained (about 300 cm^{-1}), could be associated with the surrounding protein, although whether $S = 20\text{-}40$ is appropriate for the protein is also unclear. (These values imply that there is a very large total reorganization energy of about 1 eV.)

More reasonable vibronic coupling parameters and reorganization energies have been obtained for electron transfer in the reaction center by including the participation of more than one vibrational mode, or the possibility that the modes change frequency as a result of electron transfer.[4,16,17] Similarly, one might also need to use one of these two approaches in order to obtain more physically reasonable vibronic parameters from analysis of the temperature dependence of the bandwidth data. In the ultimate analysis, one also needs to explain the fact that both the width of the long-wavelength band of P and its position have the same temperature dependence (Figure 2).

Thus we return to the idea of a temperature-dependent change in vibronic coupling within the dimer as a possible means of explaining the thermal effects on both the position and width of its main absorption band. One can imagine several ways for molecular motions to affect the spectral properties of an exciton coupled dimer such as P. The simplest approach is to consider how vibrational modes of the monomeric units affect the Franck-Condon progression of each exciton state of the dimer (and possibly also the partitioning of oscillator strength between the exciton states).[19,20] The vibrations known to be important in monomeric tetrapyrroles are usually high-frequency ($\approx$ 1500 cm^{-1}) modes, such as in-plane asymmetric stretching modes involving the atoms of the carbon skeleton. However, a change in the thermal population of these high-frequency modes with temperature would not be expected to alter the width of the long-wavelength band of P; this band has a width at half-height of about 1000 cm^{-1} at room temperature and about half this value at 5 K (Tables 1 and 2). A lower-frequency (< 250 cm^{-1}) mode must be important in determining the bandwidth. What type of mode might this be? One possibility is a new low-frequency mode of the dimer, which is not present in the monomer. The most obvious mode of this type would involve the motions of the atoms along the axis perpendicular to both macrocycles of the dimer (i.e. along the intermolecular axis). Such intermolecular modes could be decomposed into combinations of several of the relatively low-frequency (300-600 cm^{-1}) out-of-plane deformation modes of the monomers. In the reaction center, there are undoubtedly also low-frequency modes of the protein, and some of these could be sensitive to excitation of the dimer. In fact, the low-frequency "out-of-plane" motions of the rings of the dimer might be coupled to motions of the protein via residues hydrogen bonded or coordinated to the macrocycles of P.

One way to analyze the temperature dependence of the dimer band in terms of the important low-frequency mode is to consider how its thermal population in the ground electronic state dictates the distribution of intensity in the Franck-Condon progression, and thus the position and overall width of the band. As with analysis of electron transfer, it is reasonable to suggest that the relevant mode might experience a frequency change as well as an origin shift when the dimer is excited. (This means that the excited state potential surface will be both shifted from the ground state potential surface and have a different shape.) The absorption and fluorescence spectra of an anthracene dimer[21] and of P in the *Rps. viridis* reaction center[24] have been analyzed in this way. However, in the previous analysis of the dimer band in *Rps. viridis*, the similarity of the temperature dependence of the position and width was not recognized because of the limited number of data points available.[24]

We have analyzed the spectral data of Figure 2 and Tables 1 and 2 using the expressions derived by Zgierski.[21] We obtained good fits to the temperature dependence of both the peak position and the bandwidth using a mode having a vibronic coupling parameter near 1.5, and a frequency (divided by the speed of light) near 150 cm^{-1} in the ground state which increases to about 250 cm^{-1} in the excited state. This is a physically reasonable value for the vibronic coupling strength, implying relatively small changes in the equilibrium positions of the vibrating atoms upon excitation. It is not so clear whether such a large frequency change is reasonable. Including more than one mode probably would give smaller frequency changes.

An alternative way to interpret the temperature dependence of the dimer band in terms of the low-frequency mode(s) is to consider in a more explicit way the structural effects of the molecular motions on the exciton coupling. The peak position of the long-wavelength band will be very sensitive to the distance and orientation of the two macrocycles of P, since distance and orientation define the magnitude of the exciton interaction. The low-frequency intermolecular deformation modes that we have been considering will result in fluctuations in the average distance (and orientations) of the two macrocycles of the dimer, i.e., in a fluctuating exciton coupling. This will strongly influence the bandwidth.

The temperature dependence of the long-wavelength band of the dimer can then be understood in terms of how a temperature-dependent structure determines both the equilibrium exciton coupling (i.e., the peak position), and the fluctuating exciton coupling (i.e., the bandwidth). If a decrease in temperature causes the average distance of the two rings of the dimer to move closer together, this would increase the exciton coupling, and cause a red shift, as is observed (Tables 1 and 2). Similarly, if the rings move closer together, or if the protein surrounding the dimer contracts, then the extent of the structural fluctuations in the dimer will be reduced on steric grounds. The resulting smaller time-dependent exciton coupling would lead to a reduced bandwidth at low temperature, which is also observed. One can also explain the medium dependence (glycerol, films, etc.) of the position and width of the long-wavelength band of P by invoking changes in structure.

As for electron transfer, the best explanation for the temperature dependence of the spectral data probably derives from both points of view. In other words, there is undoubtedly both a change in the thermal population of the important low-frequency modes with temperature, as well as a change in structure; the latter effect would result in a change in the equilibrium and fluctuating exciton coupling. This discussion suggests the importance of considering both dynamic as well as static pigment-pigment and pigment-protein interactions in analyzing the electron transfer kinetics and the absorption spectrum of the reaction center.

REFERENCES

1. C. Kirmaier and D. Holten, Photochemistry of reaction centers from the photosynthetic purple bacteria, Photosynth. Res. 13:225 (1987).
2. R. A. Marcus and N. Sutin, Electron transfer in chemistry and biology, Biochim. Biophys. Acta 811:265 (1985).
3. N. W. Woodbury, M. Becker, D. Middendorf and W. W. Parson, Picosecond kinetics of the initial photochemical electron-transfer reaction in bacterial photosynthetic reaction centers, Biochem. 24:7516 (1985).
4. C. Kirmaier, D. Holten and W. W. Parson, Temperature and detection-wavelength dependence of the picosecond electron transfer kinetics measured in *Rhodopseudomonas sphaeroides* reaction centers. Resolution of new spectral and kinetic components in the primary charge separation process, Biochim. Biophys. Acta 810:33 (1985).
5. R. K. Clayton, Effects of dehydration on reaction centers from *Rhodopseudomonas sphaeroides*, Biochim. Biophys. Acta 504:255 (1978).
6. M. R. Gunner, D. E. Robertson and P. L. Dutton, Kinetic studies on the reaction center protein from *Rhodopseudomonas sphaeroides*: The temperature and free energy dependence of electron transfer between various quinones in the Q_A site and the oxidized bacteriochlorophyll dimer, J. Phys. Chem. 90:3783 (1986).
7. G. Feher, M. Okamura and D. Kleinfeld, Electron transfer reactions in bacterial photosynthesis: Charge recombination kinetics as a structure probe, in: "Protein Structure: Molecular and Electronic Reactivity", R. Austin, E. Buhks, B. Chance, D. DeVault, P. L. Dutton, H. Frauenfelder and V. I. Gol'danskii, eds., Springer-Verlag, New York (1987).
8. C. Kirmaier, D. Holten and W. W. Parson, Picosecond photodichroism studies of the transient states in *Rhodopseudomonas sphaeroides* reaction centers at 5 K. Effects of electron transfer on the six bacteriochlorin pigments, Biochim. Biophys. Acta 810:49 (1985).
9. C. Kirmaier, D. Holten and W. W. Parson, Picosecond study of the $P^+I^-Q \longrightarrow P^+IQ^-$ electron transfer reaction in *Rps. viridis* reaction centers, Biophys. J. 49:586a (1986).
10. J. M. Hayes, J. K. Gillie, D. Tang and G. J. Small, Theory for spectral hole burning of the primary electron donor state of photosynthetic reaction centers, Biochim. Biophys. Acta (in press).
11. J. Jortner, Dynamics of the primary events in bacterial photosynthesis, J. Am. Chem. Soc. 102:6676 (1980).
12. J. Deisenhofer, O. Epp, K. Miki, R. Huber and H. Michel, Structure of the protein subunits in the photosynthetic reaction centre of *Rhodopseudomonas viridis* at 3 Å resolution, Nature 318:618 (1985).
13. C.-H. Chang, D. Tiede, J. Tang, U. Smith, J. Norris and M. Schiffer, Structure of *Rhodopseudomonas sphaeroides* R-26 reaction center, FEBS Lett. 205:82 (1986).
14. J. P. Allen, G. Feher, T. O. Yeates, H. Komiya and D. C. Rees, Structure of the reaction center from *Rhodobacter sphaeroides* R-26: The cofactors, Proc. Natl. Acad. Sci. USA 84:5730 (1987).
15. H. Frauenfelder, H. Hartman, M. Karplus, I. D. Kuntz Jr., J. Kuriyan, F. Parak, G. A. Petsko, D. Ringe, R. F. Tilton Jr., M. L. Connolly and N. Max, Thermal expansion of a protein, Biochem. 26:254 (1987).
16. T. Kakitani and H. Kakitani, A possible new mechanism of temperature dependence of electron transfer in photosynthetic systems, Biochim. Biophys. Acta 635:498 (1981).
17. A. Sarai, Possible role of protein in photosynthetic electron transfer, Biochim. Biophys. Acta 589:71 (1980).
18. R. J. Shopes and C. A. Wraight, Charge recombination from the $P^+Q_A^-$ state in reaction centers from *Rhodopseudomonas viridis*, Biochim. Biophys. Acta 893:409 (1987).

19. R. L. Fulton and M. Gouterman, Vibronic coupling. II. Spectra of dimers, J. Chem. Phys. 41:2280 (1964).
20. M. Gouterman, D. Holten and E. Lieberman, Porphyrins XXXV. Exciton coupling in μ-oxo scandium dimers, Chem. Phys. 25:139 (1977).
21. M. Z. Zgierski, Fluorescence-absorption energy gap in the stable anthracene dimer spectra, J. Chem. Phys. 59:1052 (1973).
22. A. Warshel, Role of the chlorophyll dimer in bacterial photosynthesis, Proc. Natl. Acad. Sci. USA 77:3105 (1980).
23. T. Mar, C. Vadeboncoeur and G. Gingras, Different temperature dependencies of the charge recombination reaction in photoreaction centers isolated from different bacterial species, Biochim. Biophys. Acta 724:317 (1983).
24. P. O. J. Scherer, S. F. Fischer, J. K. Horber and M. E. Michel-Beyerle, On the temperature-dependence of the long-wavelength fluorescence and absorption of *Rhodopseudomonas viridis* reaction centers, in: Antennas and Reaction Centers of Photosynthetic Bacteria," M. E. Michel-Beyerle, ed., Springer-Verlag, Berlin (1985).

ENDOR OF EXCHANGEABLE PROTONS OF THE REDUCED INTERMEDIATE ACCEPTOR IN REACTION CENTERS FROM *RHODOBACTER SPHAEROIDES* R-26

G. Feher, R. A. Isaacson, and M. Y. Okamura
University of California, San Diego, La Jolla, California 92093

W. Lubitz
Freie Universität, Berlin, Germany

I. INTRODUCTION

To understand quantitatively the electron transfer kinetics in reaction centers (RCs) one needs to know both the *spatial*, three-dimensional, structure as well as the *electronic structure* of the reactants. The advances made in the determination of the three-dimensional structure of RCs in *Rp. viridis* and *Rb. sphaeroides* were presented earlier at this Conference by H. Deisenhofer, D. Tiede and our group. In this communication we would like to report on the results of investigations of the electronic structure of the intermediate acceptor $I^{\cdot}$. The acceptor, I, is believed to be a bacteriopheophytin a (Bphe a), that receives an electron from the singlet excited primary donor in ~ 4 picoseconds and passes it on to a quinone acceptor with a characteristic time of ~ 200 ps (for a review see ref. 1). In general, the charge transfer in photosynthesis is a one electron process that results in the formation of donor cation and acceptor anion radicals.

The technique of choice to investigate electronic structures of radicals is Electron Paramagnetic Resonance (EPR) and Electron Nuclear Double Resonance (ENDOR). In an ENDOR experiment one induces NMR transitions and monitors them via the EPR signal. Since the magnetic moment of the electron is three orders of magnitude larger than that of the nucleus, one gains many orders of magnitude in sensitivity over regular NMR experiments and many orders of magnitude in resolution over regular EPR experiments. ENDOR spectroscopy has been extensively used to investigate the structure of the primary donor as well as the primary and secondary quinone acceptors. A summary of these investigations has been reported, for example, at the 1985 Feldafing meeting (2,3).

In an ENDOR experiment one determines the NMR frequency of the nuclei that interact with the electron spin. The shift of the NMR frequency from the unperturbed nuclear Larmor frequency in the applied magnetic field depends on the strength of the electron-nuclear interaction (called the hyperfine, hf, interaction); it is related to the electron spin density at the nucleus, which depends on the electronic wave function of the unpaired electron. In some instances, to be discussed later, one can determine from the hf couplings distances between nuclei. This is particularly important for protons since they are not observable by x-ray diffraction. Thus, ENDOR provides a technique that is complementary to x-ray diffraction in determining structure.

The first ENDOR spectra from I· were reported over ten years ago (4,5). At that time only one set of two broad lines was observed. Progress in instrumentation resulted in improved sensitivity and resolution. As a consequence a much richer spectrum was obtained (6). In this presentation we shall not discuss the entire ENDOR spectrum of $I^{\cdot}$ but focus only on part of it, namely the exchangeable protons (7).

II. EXPERIMENTAL PROCEDURES

RCs from *Rb. sphaeroides* R-26 were purified by standard procedures (8). I^- was trapped at low temperatures by illuminating an RC sample at ambient temperature in the presence of cytochrome c_2^+ and dithionite followed by quick freeze in liquid nitrogen (9). Protons were exchanged by incubating the sample in D_2O at 23°C for various lengths of time prior to illumination. In the model studies the anion radicals of Bchl and Bphe were generated electrolytically in 2-methyl tetrahydrofurane (MTHF) using tetra n-butyl ammonium perchlorate as a supporting electrolyte. In protic solvents (e.g. pyridine/H_2O) the radicals were prepared as described by Fajer et al. (10). ENDOR spectra were taken on frozen solution (~ 100-150K) at 9 GHz with a spectrometer of local design utilizing a loop gap resonator (11).

III. EXPERIMENTAL RESULTS AND DATA ANALYSIS

A. The ENDOR Spectrum

The central (4 MHz) region of the ENDOR spectrum from $I^{\overline{\cdot}}$ is shown in Fig. 1. The solid line was obtained from RCs in H_2O; the dashed line is the ENDOR spectrum of RCs incubated in D_2O for 70 hours. The difference between the two spectra is due to protons that exchanged with deuterons and is shown at twice the gain in the lower part of Fig. 1. Three main splittings labelled A_1, A_2, A_3 were observed. A set of smaller lines was tentatively labelled A_4. At shorter incubation times (10 minutes) the three main lines appeared with approximately the same amplitudes, whereas the smaller lines seemed to have reduced amplitudes, indicating a longer exchange time. In addition the hfcs A_2 and A_3 seemed to increase by ~ 5% after 70 hours incubation suggesting the presence of another, unresolved, line originating from slower exchanging protons. The values of the hyperfine splittings were determined to be:

$$A_1 = 0.38\ \text{MHz} \quad ; \quad A_2 = 0.93\ \text{MHz} \qquad (1)$$
$$A_3 = 1.92\ \text{MHz} \quad ; \quad A_4 = 2.36\ \text{MHz}$$

where the values quoted for A_2 and A_3 were determined after the 10 minute incubation time (data not shown). The rest of the ENDOR spectrum extending over a frequency interval of ~ 12 MHz (6) did not exhibit the presence of exchangeable protons.

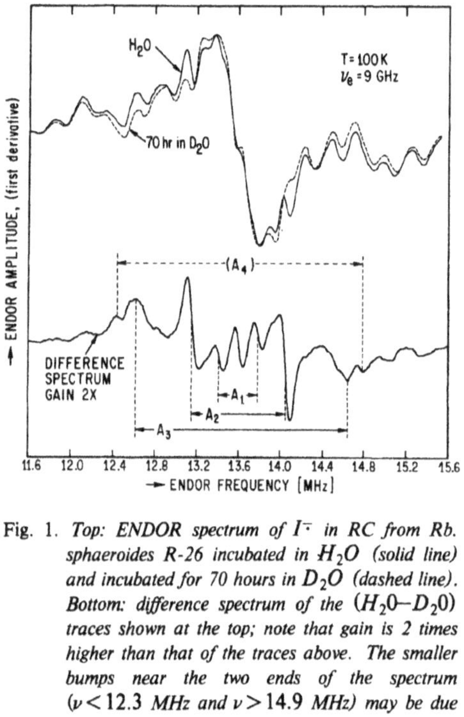

Fig. 1. *Top: ENDOR spectrum of* $I^{\overline{\cdot}}$ *in RC from Rb. sphaeroides R-26 incubated in* H_2O *(solid line) and incubated for 70 hours in* D_2O *(dashed line). Bottom: difference spectrum of the* $(H_2O{-}D_2O)$ *traces shown at the top; note that gain is 2 times higher than that of the traces above. The smaller bumps near the two ends of the spectrum* ($\nu < 12.3$ *MHz and* $\nu > 14.9$ *MHz) may be due to artifacts associated with the subtraction.*

B. Identification of the ENDOR lines

To obtain meaningful information from an ENDOR spectrum each line has to be identified with the corresponding nucleus. This is in general a difficult problem that is usually solved by isotopic substitutions (e.g. deuteration) or by other, more indirect methods. There are three main classes of exchangeable protons on the bacteriopheophytin anion radical, $I^{\overline{\cdot}}$ (see shaded areas in Fig. 2). These are: the protons bound to the pyrrole nitrogens of rings I and III, the proton on carbon C_{10} of ring V and the proton from the glutamic acid residue of the L subunit (Glu L104) bonded to 0_1 of ring V. In principle, hydrogen bonds could also be formed to the other carbonyl oxygens of Bphe (see Fig. 2). Of particular relevance to ENDOR spectroscopy may be 0_6 (acetyl group) which has a non-negligible spin density (22,23). However, there is no evidence so far from x-ray structure analysis or from other measurements that hydrogen bonds are formed to these oxygens (12-15). The proton from Glu L104 is of particular interest since it is not present in the second, symmetry related, Bphe. It may, therefore, contribute to the asymmetry in the electron transfer properties of the two branches. We shall start with an identification of the lines associated with this proton.

There are three pieces of independent evidence that implicate the splittings A_2 and A_3 as arising from the interaction of the unpaired electron with the proton of Glu L104. These are:

1) *The dipolar form of the hf interaction*: From theoretical considerations one expects the interaction of the electron with the hydrogen bonded proton to be dipolar in nature. This is similar to the situation encountered for the protons hydrogen bonded to the carbonyl oxygens of the quinones (3,16). The hfc of such a proton will be given to a first approximation by the simple point dipole formula:

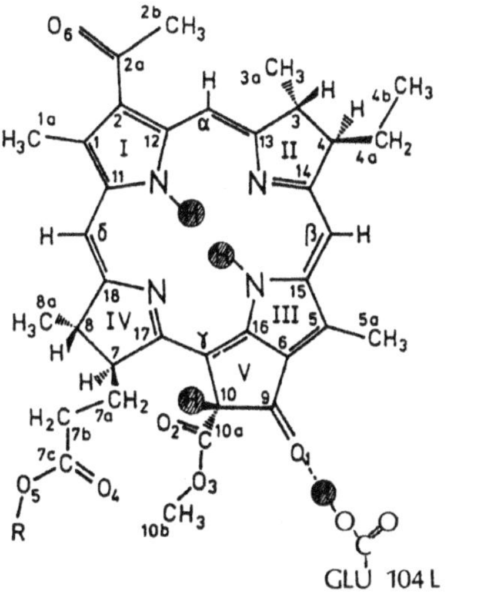

Fig. 2. *Structure of bacteriopheophytin a (Bphe a). Shaded circles represent exchangeable protons. Of particular interest is the proton from Glu L104 hydrogen bonded to the carbonyl O_1 on ring V.*

$$A_H = \frac{(g_e\mu_e)(g_H\mu_H)}{r^3}\rho(3\cos^2\Theta - 1) = \frac{79}{r^3}\rho(3\cos^2\Theta - 1)\ \text{MHz} \qquad (2)$$

where $g_e\mu_e$ and $g_H\mu_H$ are the electron and proton g-values and magnetic moments, respectively; Θ is the angle between the applied magnetic field and the line joining O and H; ρ is the spin density on the oxygen and r is the distance in Å between O and H. For randomly oriented stationary molecules one obtains first derivative ENDOR signals at the inflection points of the absorption where $\Theta = 0$ and $\Theta = 90°$. They have a characteristic line shape (see Fig. 3 of ref. 3) and are related to each other by a factor of 2 (i.e. $|A_\parallel|/|A_\perp| = 2.0$). The lines giving rise to the splittings A_2 and A_3 satisfy these two requirements and are therefore identified with $A_\perp$ and $A_\parallel$, i.e.

$$|A_\perp| = A_2 = 0.93\ \text{MHz} \quad ; \quad |A_\parallel| = A_3 = 1.92\ \text{MHz} \qquad (3)$$

The signs of A_2 and A_3 were not determined directly. However, from the shapes of the lines (3) we conclude that A_2 and A_3 have opposite signs. Thus, the isotropic hfc, given by (1/3) (2 $A_\perp + A_\parallel$), $= \pm 0.02$ MHz; i.e. it is within experimental accuracy zero as expected from a dipolar interaction.

2. *The temperature dependence of the hf interaction*: The hfcs of the protons associated with the ring structure of $I^{\bar{\cdot}}$ are not expected to be as temperature dependent as those of the protons hydrogen bonded to the amino acid residues of the RC since the latter is sensitive to the thermal expansion of the protein. For a fractional change in spacing between two points of a solid with an anharmonic interatomic potential one has the expression (17,18):

$$\frac{\Delta r}{r} = \beta\,\frac{T_o}{2}\left(\coth\frac{T_o}{2T} - 1\right) \qquad (4)$$

where β is a linear expansion coefficient and T_o is the characteristic frequency of vibration. For small changes in the bond length, Δr, the change in the hyperfine coupling ΔA (see Eq. 2) is:

$$\frac{\Delta A}{A} \simeq -3\,\frac{\Delta r}{r} \qquad (5)$$

From Eqs. 4 and 5 one obtains the predicted temperature dependence of the hyperfine interaction, A(T):

$$A(T) = A(0)\left[1 - \frac{3}{2}\beta\, T_o\left[\coth\frac{T_o}{2T} - 1\right]\right] \qquad (6)$$

where A(0) is the hyperfine interaction at T = 0.

The experimental dependence of the hfc on temperature is shown in Fig. 3 (circles). The solid line represents the theoretical fit (Eq. 6) with $T_o = 200K$ and an expansion coefficient $\beta = 2.2 \times 10^{-4}K^{-1}$. These values are approximately the same as found for the temperature dependence of the hfcs of the protons hydrogen bonded to the carbonyl oxygens of the quinone anion radical in RCs from *Rb. sphaeroides* (18).

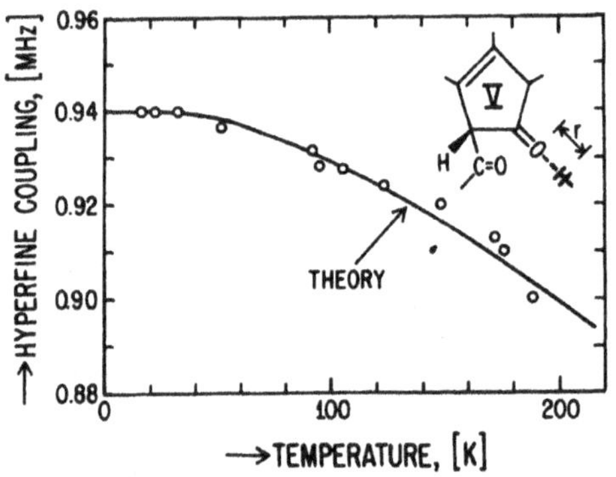

Fig. 3. *Temperature dependence of the hyperfine coupling of the hydrogen bonded proton from Glu L104 to the carbonyl oxygen on ring V of Bphe$^{\bar{\cdot}}$ in RCs from Rb. sphaeroides. Solid line represents theoretical fit with $T_o = 200K$ and expansion coefficient $\beta = 2.2 \times 10^{-4}K^{-1}$.*

3. *The absence of the A_2 splitting in Bphe$^{\bar{\cdot}}$ in several solvents*: The ENDOR spectrum of Bphe$^{\bar{\cdot}}$ in pyridine/10% H_2O is shown by the solid line in Fig. 4 and in pyridine/10% D_2O by the dashed line. The bottom trace shows the difference spectrum (H_2O–D_2O), which is due to exchangeable protons. The arrows indicate the positions of the two prominent lines associated with the hfc A_2 in I$^{\bar{\cdot}}$ of RCs. These lines are conspicuously absent in Bphe$^{\bar{\cdot}}$. Similar results were observed for Bphe$^{\bar{\cdot}}$ in isopropanol and MTHF (data not shown). The lack of a pronounced line near the position of the postulated proton from Glu-L104 in RCs is interpreted as an absence of a well ordered specific hydrogen bond to the carbonyl oxygen of ring V. However, we cannot exclude the possibility that in a protic medium a proton is hydrogen bonded to the carbonyl oxygen with a spread of configurations that broadens the ENDOR line which makes detection difficult. Possible assignments of the other lines are discussed in the next section.

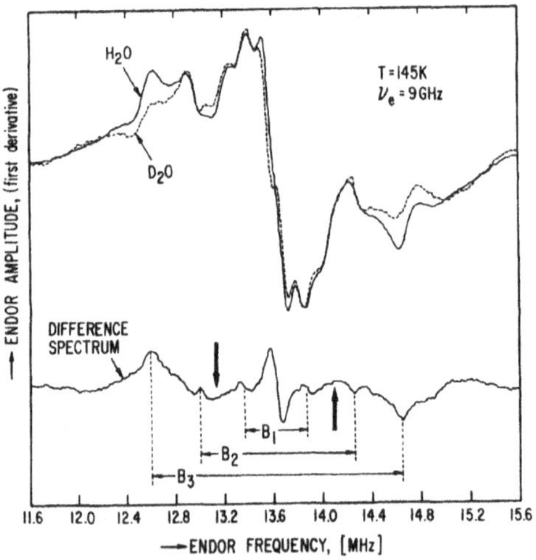

Fig. 4. *Top: ENDOR spectrum of Bphe$^{\bar{\cdot}}$ in pyridine/10% H_2O (solid line) and in pyridine/10% D_2O (dashed line). Bottom: difference spectrum of the (H_2O–D_2O) traces shown at the top. Arrows indicate positions of the prominent lines (A_2) in Fig. 1.*

It should be noted that neither of the three pieces of evidence discussed above provide an iron-clad proof of the assignment. However, taken together, we believe that they form a reasonably good basis for the assignment. The assignment could be more definitely confirmed by ENDOR spectra of I$^{\bar{\cdot}}$ from either a mutant in which Glu L104 has been replaced by another amino acid residue (19) or from RCs in which the Bphe in the B-branch (which does not have an acidic residue in the homologous position) has been reduced (20). Work along these lines is in progress.

4. *Assignment of the other exchangeable ENDOR lines*: Having tentatively assigned the splittings A_2 and A_3 to the protons from Glu L104, we need to assign ENDOR lines to the N-H protons and the proton on C_{10} (see Fig. 2). To aid us in these assignments we refer to the model compound studies shown in Fig. 4. The three splittings, due to exchangeable protons, observed in $Bphe^-$ in pyridine/10% H_2O are:

$$B_1 = 0.52 \text{ MHz} \; ; \quad B_2 = 1.26 \text{ MHz} \; ; \quad B_3 = 2.05 \text{ MHz} \qquad (7)$$

A splitting close to B_3 was also observed in a sample of $Bphe^{\overline{\cdot}}$ in the non protic solvent, MTHF (no hydrogen bonds present) and in $Bchl^{\overline{\cdot}}$ (no N-H groups). A theoretical estimate (22) of the hfc tensor components of the C_{10} proton in $Bphe^{\overline{\cdot}}$ yields one large value (~ 2 MHz) and two smaller ones (< 1 MHz). Consequently, we assign B_3 to the largest component of the hf tensor and, similarly, identify A_4 with this component. The relatively small difference between A_4 and B_3 can be rationalized by postulating different environments for $I^{\overline{\cdot}}$ and $Bphe^{\overline{\cdot}}$. The splitting B_1 and A_1 could arise from the second tensor component of the C_{10} proton, the third of this non-axial tensor may be buried under the central (matrix) line (Fig. 1 and 4). The splitting B_2 has so far not been assigned.

In the above assignment we have not accounted for the N-H protons in either the $I^{\overline{\cdot}}$ or the model system $Bphe^{\overline{\cdot}}$. From model studies at higher temperatures (21) and theoretical calculations (22), it has been concluded that the N-H protons have a small (< 1 MHz) hfc. Their magnitude can be further changed due to a deviation from planarity of the tetrapyrrole structure of the RC (15). At present we cannot exclude, therefore, that A_1 belongs to the NH protons.

IV. DETERMINATION OF THE O--H DISTANCE

Having assigned the splittings A_2 and A_3 to $A_\perp$ and $A_\parallel$ of the hfcs of the hydrogen bonded proton we can estimate from Eq. 2 the distance, r, of the proton of Glu L104 to the carbonyl oxygen O_1 on ring V. With the theoretically evaluated spin density on O_1, which ranged from 0.05 (22) to 0.10 (23) and the experimental value of A_2 (or A_3) we obtain a hydrogen bond distance of 1.6 Å - 2.0 Å. These values are consistent with the accepted lengths of hydrogen bonds. It should be noted that despite the uncertainty in ρ by a factor of two, the distance, r, has an error of only ~ ± 10%. If we were to assume an enol form of ring V, the calculated spin density on O_1 is 0.040 (23) which results in a distance $r \cong 1.5$ Å. This value is not consistent with the accepted distance for a *covalent* O-H bond (~ 1 Å). However, a small admixture of an enol form as suggested by Bocian et al (24) cannot be excluded.

V. SUMMARY AND DISCUSSION

We have identified the hfc of the proton from Glu L104 with the carbonyl oxygen O_1 on ring V of the intermediate electron acceptor, $I^{\overline{\cdot}}$ and have determined the O--H distance to be 1.8 ± 0.2 Å. This hydrogen bond has also been postulated recently by Nabedryk et al. from FTIR measurements (25). From the temperature dependence of the hfc we have estimated the characteristic temperature T_0 of the vibrational mode to be 200K. The assignment of the splittings of the other exchangeable protons have been highly speculative.

Exchangeable protons are of particular interest in trying to understand electron transfer mechanisms. Electron transfer theory postulates a vibrationally coupled tunneling mechanism (for a review see 26,27). If the relevant vibrations involve an exchangeable proton then its replacement by deuterons should in principle affect the electron transfer. We have observed such an isotope effect when the protons on the primary quinone were replaced by deuterons (28). The kinetics of electron transfer from I^- to Q_A or from the excited dimer state to I may similarly show an isotope effect. C. Kirmaier and D. Holten looked for this effect but, within their accuracy of ~ 10%, did not observe it (29) (note that for quinones the effect was only 6% (28)).

VI. ACKNOWLEDGEMENT

We thank L. K. Hanson, J. Fajer, and M. Plato for communicating their theoretical results, J. P. Allen for discussions of the RC structure, and E. Abresch for the preparation of the RCs. The work was supported by grants from the NSF (DMB 85-18922), the NIH (GM13191) the DFG (SFB 312) and NATO (RG86/0029).

REFERENCES

1. C. Kirmaier, and D. Holten, Primary photochemistry of reaction centers from the Photosynthetic Purple Bacteria, *Photosynth. Res.* **13**:225-260 (1987).
2. W. Lubitz, F. Lendzian, M. Plato, K. Möbius, and E. Tränkle, ENDOR studies of the primary donor in bacterial reaction centers, *in*: "Antennas and Reaction Centers of Photosynthetic Bacteria - Structure, Interactions and Dynamics," Michel-Beyerle, ed., Springer-Verlag, Berlin, pp. 164-173 (1985).
3. G. Feher, R. A. Isaacson, M. Y. Okamura, and W. Lubitz, ENDOR of semiquinones in RCs from *Rhodopseudomonas sphaeroides*, *in*: "Antennas and Reaction Centers of Photosynthetic Bacteria - Structure, Interactions and Dynamics," Michel-Beyerle, ed., Springer-Verlag, Berlin, pp. 174-189 (1985).
4. J. Fajer, M. S. Davis, and A. Forman, ENDOR and ESR characteristics of bacteriopheophytin and bacteriochlorophyll anion radicals, *Biophys. J. Abstr.* **17**:150 (1977a).
5. G. Feher, R. A. Isaacson, and M. Y. Okamura, Comparison of EPR and ENDOR spectra of the transient acceptor in reaction centers of *Rhodopseudomonas sphaeroides* with those of bacteriochlorophyll and bacteriopheophytin radicals, *Biophys. J.* (Abstr.) **17**:149 (1977).
6. G. Feher, R. A. Isaacson, M. Y. Okamura, and W. Lubitz, ENDOR of the reduced intermediate electron acceptor in RCs of *R. sphaeroides*, *Biophys. J.* (Abstr.) **51**:377a (1987).
7. G. Feher, R. A. Isaacson, M. Y. Okamura, W. Lubitz, ENDOR of exchangeable protons of the reduced intermediate acceptor in RCs from *Rb. sphaeroides* R-26, *Biophys. J. Abstr.*, in press (1988).
8. G. Feher and M. Y. Okamura, Chemical compositon and properties of reaction centers, *in*: "The Photosynthetic Bacteria," R. K. Clayton and W. R. Sistrom, eds., Plenum Press, New York, pp. 349-386 (1978).
9. M. Y. Okamura, R. A. Isaacson, G. Feher, Spectroscopic and kinetic properties of the transient intermediate acceptor in reaction centers of *Rhodopseudomonas sphaeroides*, *Biochim. Biophys. Acta* **546**:394-417 (1979).
10. J. Fajer, A. Forman, M. S. Davis, L. D. Spaulding, D. C. Brune and R. M. Felton, Anion radicals of bacteriochloropyll a and bacteriopheophytin a. Electron spin resonance and electron nuclear double resonance studies, *J. Am. Chem. Soc.* **99**:4134-4140 (1977).
11. W. Froncisz and J. Hyde, The loop-gap resonator: A new microwave lumped circuit ESR sample structure, *J. Mag. Res.* **47**:515-521 (1982).
12. H. Michel, O. Epp, and J. Deisenhofer, Pigment-protein interactions in the photosynthetic reaction center from *Rhodopseudomonas viridis*, *EMBO J.* **5**:2445-2451 (1986).
13. J. P. Allen, G. Feher, T. O. Yeates and D. C. Rees, Structure analysis of the reaction center from *Rhodopseudomonas sphaeroiodes*: Electron density map at 3.5Å resolution, *in*: "Progress in Photosynthesis Research," Vol. I, pp. I.4.375-I.4.378, J. Biggins, ed., Martinus Nijhoff, Boston (1987).
14. J. P. Allen, G. Feher, T. O. Yeates, H. Komiya, and D. C. Rees, Structure of the reaction center from *Rhodobacter sphaeroides* R-26. II. The protein subunits, *Proc. Natl. Acad. Sci.* **84**:6162-6166 (1987).
15. J. P. Allen, G. Feher, T. O. Yeates, H. Komiya and D. C. Rees, Structure of the reaction center from *Rhodobacter sphaeroides* R-26 and 2.4.1. These proceedings.
16. P. I. O'Malley, T. K. Chandrashekar, and G. T. Babcock, ENDOR characterization of hydrogen-bonding to immobilized quinone anion radicals, *in*: "Antennas and Reaction Centers of Photosynthetic Bacteria - Structure, Interactions and Dynamics," Michel-Beyerle, ed., Springer-Verlag, Berlin, pp. 339-344 (1985).

17. R. P. Feynman, *Statistical Mechanics: A Set of Lectures*, pp. 53-55, W. A. Benjamin, Reading, PA. (1972).

18. G. Feher, M. Y. Okamura, and D. Kleinfeld, Electron transfer reactions in bacterial photosynthesis: charge recombination kinetics as a structure probe, *in*: "Protein Structure: Molecular and Electronic Reactivity," Robert Austin, Ephraim Buhks, Britton Chance, Don DeVault, P. Leslie Dutton, Hans Frauenfelder and Vittallii I. Goldanskii, eds., Springer Verlag, New York, pp. 399-421 (1987).

19. E. J. Bylina and D. C. Youvan, Site specific mutagenesis of the bacterial reaction center from Rhodobacter capsulata. These Proceedings.

20. B. Robert, M. Lutz and D. M. Tiede, Selective photochemical reduction of either of the two bacteriopheophytins in reaction centers from *Rhodopseudomonas sphaeroides* R-26, *FEBS Lett.* **183**:326-330 (1985).

21. W. Lubitz, F. Lendzian, and K. Möbius, The bacteriopheophytin a anion radical. A solution ENDOR and TRIPLE resonance study, *Phys. Chem. Letters* **84**:33-38 (1981).

22. M. Plato, personal communication.

23. L. K. Hanson and J. Fajer, personal communication.

24. D. F. Bocian, N. J. Boldt, B. W. Chadwick, and H. A. Frank, Near-infrared-excitation resonance Raman spectra of bacterial photosynthetic reaction centers, *FEBS Lett.* **214**:92-96 (1987).

25. E. Nabedryk, W. Mäntele and J. Breton, FTIR spectroscopic investigations of the intermediate electron acceptor photoreduction in purple photosynthetic bacteria and green plants. These Proceedings.

26. D. D. DeVault, "Quantum-Mechanical Tunnelling in Biological Systems," Cambridge University Press (1984).

27. R. A. Marcus and N. Sutin, Electron transfers in chemistry and biology, *Biochim. Biophys. Acta* **811**:265-302 (1985).

28. M. Y. Okamura and G. Feher, Isotope effect on electron transfer in reaction centers from *R. sphaeroides*, *Proc. Natl. Acad. Sci. USA* **83**:8152-8156 (1986).

29. C. Kirmaier and D. Holten (personal communication).

FTIR SPECTROSCOPIC INVESTIGATIONS OF THE INTERMEDIARY ELECTRON ACCEPTOR PHOTOREDUCTION IN PURPLE PHOTOSYNTHETIC BACTERIA AND GREEN PLANTS

E. Nabedryk, S. Andrianambinintsoa, W. Mäntele* and J. Breton

Service de Biophysique, Département de Biologie, CEN Saclay
91191 Gif-sur-Yvette cedex, France

*Institut für Biophysik und Strahlenbiologie, Universität Freiburg, Albertstr. 23, D-7800 Freiburg, FRG

INTRODUCTION

We have recently demonstrated that light-induced Fourier transform infrared (FTIR) spectroscopy can be applied to bacterial reaction centers -RCs- (1-3) as well as to photosystems from green plants (4) in order to detect changes in the molecular interactions between the pigments involved in the primary charge separation and their anchoring sites in the protein. Using FTIR difference spectroscopy the sensitivity is high enough to detect perturbations in the vibrational modes of chlorophylls and protein groups in a large complex such as a RC. For both primary electron donor photooxidation and intermediary electron acceptor photoreduction we have reported very specific absorbance changes (1-4). However, due to the non selectivity of IR spectroscopy, the FTIR signals might arise from the pigments, the protein (peptide and side chains groups), the lipids and even bound water molecules. Thus, a precise assignment of the FTIR absorbance changes to chemical bonds requires further investigations using model compounds radicals of the isolated chlorophylls as well as isotope-substituted material.

In this report, molecular changes associated with the light-induced reduction of the intermediary electron acceptor in purple photosynthetic bacteria (bacteriopheophytin - BPheo a or b) and in green plants (pheophytin - Pheo a) were compared by means of FTIR difference spectroscopy. More specifically, in order to determine the possible contribution of exchangeable protons to the FTIR signals, we have investigated the effect of ^{1}H-^{2}H substitution in RCs and chromatophores of _Rps. viridis_ on the IR signals due to the BPheo b photoreduction. These data, together with those inferred from the comparison between the light-induced IR spectra obtained _in vivo_ and the spectra of the BPheo b anion radical generated electrochemically (5) lead to a tentative description of the molecular interactions of the neutral and radical intermediary electron acceptor with the protein environment.

EXPERIMENTAL

^{1}H-^{2}H exchange of _Rps. viridis_ RCs and chromatophores was performed by incubating the samples in $^{2}H_2O$ (99.8% deuterium enrichment) for a variable time. In another set of experiments, chromatophores were isolated from _Rps. viridis_ cells grown in 60% $^{2}H_2O$ medium.

In order to trap photochemically the intermediary electron acceptor in its reduced state, films of _Rps. viridis_ RCs and chromatophores and of _C. vinosum_ RC-B875 complexes (prepared according to ref. 6) as well as photosystem II (PSII)-enriched particles (4) were prereduced with sodium dithionite in a borate buffer (2) either prepared with $^{1}H_2O$ or $^{2}H_2O$. Illumination of these films leads to the photoaccumulation of the (B)$Pheo^{-}$ state (2-4). FTIR difference spectra were recorded before and during continuous illumination with actinic light _in situ_ on a Nicolet 60SX FTIR spectrophotometer. Further experimental details can be found in (2,4).

Electrolysis of the isolated BPheo b (prepared according to ref.7) in fully deuterated tetrahydrofuran (THF D8) was performed in the electrochemical cell described in (8). FTIR spectra were recorded before, during and after anion formation.

RESULTS

Using well-dried THF D8 as a solvent, the BPheo b anion radical was generated at U=-0.8V in an electrochemical cell transparent in the visible and IR (8). The optical absorbance spectrum of the BPheo b anion (data not shown) is in close agreement with published data (9).

An IR absorbance spectrum of BPheo b in THF D8 in the electrochemical cell before electrolysis is shown in Fig. 1 together with the IR spectrum of the BPheo b anion after electrolysis. The three highest frequency bands (at 1741, 1703 and 1671cm^{-1} in the neutral form) can be assigned to the ester, keto and acetyl C=O vibrations respectively, of the BPheo b molecule. The frequency maxima of these groups are characteristic for non-interacting groups as it is expected for a non hydrogen-bonding solvent (10). The BPheo b anion-minus-BPheo FTIR

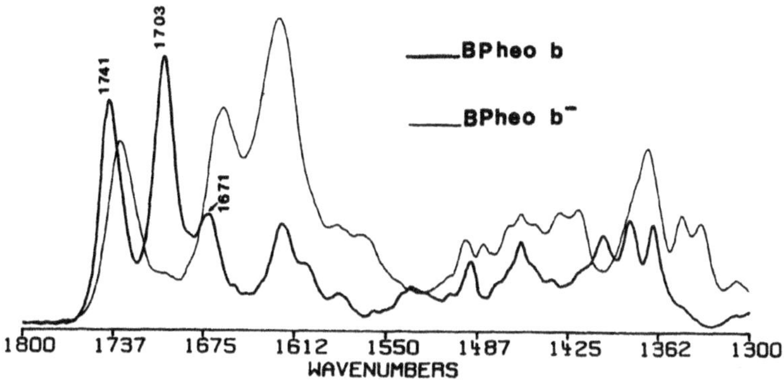

Figure 1 : FTIR absorbance spectra of the neutral BPheo b (▬) in THF D8 in the spectroelectrochemical cell before electrolysis and of the fully evolved anion (—) after electrolysis at U=-8V. T=295K. 64 interferograms added. Resolution : 4cm^{-1}.

difference spectrum (termed BPheo$^-$ spectrum) displayed on Fig. 2a exhibits strong difference bands, especially at 1743cm^{-1} and 1703cm^{-1} which can be assigned to the ester and 9 keto C=0 groups of the neutral form. Both the ester and keto carbonyls of BPheo b appear shifted to lower energy upon anion formation (see also Fig. 1). Such spectra of the BPheo b anion radical generated electrochemically can be used as model compound spectra for the interpretation of the light-induced difference spectra obtained *in vivo*.

The FTIR difference spectrum of the photoreduction of the BPheo b intermediary electron acceptor (also termed H^- spectrum) in *Rps. viridis* RCs is shown in Fig. 2b. It corresponds to the difference between a sum of spectra taken under continuous illumination (leading to the accumulation of a stable H^- state) and a sum of spectra taken before illumination, i.e. positive bands are due to the appearing H^- state while negative bands correspond to the disappearing H neutral state. This H^- spectrum exhibits quite different spectral features in the C=0 region, compared to the BPheo b^- spectrum (Fig. 2a). Instead of a single negative band in the ester C=0 region observed at $1743cm^{-1}$ in the model compound, two negative bands are found at $1747cm^{-1}$ and $1732cm^{-1}$ in the RC after photoreduction of H (2,3). These two bands might be assigned to the 7c and 10a ester C=0 bonds of the intermediary electron acceptor provided that no contribution of amino acid side chains appears in this spectral region. They would thus be interpreted by a difference in the *in vivo* bonding of both propionic and carbomethoxy ester carbonyls. Unequivalent ester C=0 vibrations have also been reported in the case of BChl a and chlorophyll a hydrated in micellar form (10). Comparing with the negative $1743cm^{-1}$ band in the BPheo b^- spectrum, the 1747 and $1732cm^{-1}$ negative signals in *Rps. viridis* RC would then correspond to a free 7c and bonded 10a C=0 groups, respectively. In addition, recent resonance Raman spectra of *Rb. sphaeroides* RC using near Q_Y excitation show a band at $1726cm^{-1}$ which was attributed to the 10a C=0 group of the intermediary electron acceptor (11).

One should also consider possible contributions from amino acid side chains, i.e. from carboxylic C=0 bonds in the $1760\text{-}1700cm^{-1}$ spectral region of the FTIR H^- spectrum. Indeed, bands due to the C=0 stretches of ASP or GLU protonated side chains are observed in this region of the IR spectrum. A band due to a protein carboxylic C=0 group can be distinguished from one of the pigment C=0 by a downshift of approx. $10\text{-}15cm^{-1}$ after $^1H\text{-}^2H$ exchange (12). In order to test the possibility that a C=0 from an amino acid side chain contributes to the 1747 and/or $1732cm^{-1}$ signals observed in Fig. 2b, the photoreduction of BPheo b was performed in *Rps. viridis* RCs which have been deuterated by incubation in ^2H_2O for 3 days at room temperature (Fig. 2c) leading to about 70% exchanged peptide hydrogens as deduced from the decrease of the amide II/amide I ratio (Fig. 3). In these conditions, a very reproducible ~ 30% decrease of the intensity of the $1747cm^{-1}$ band with

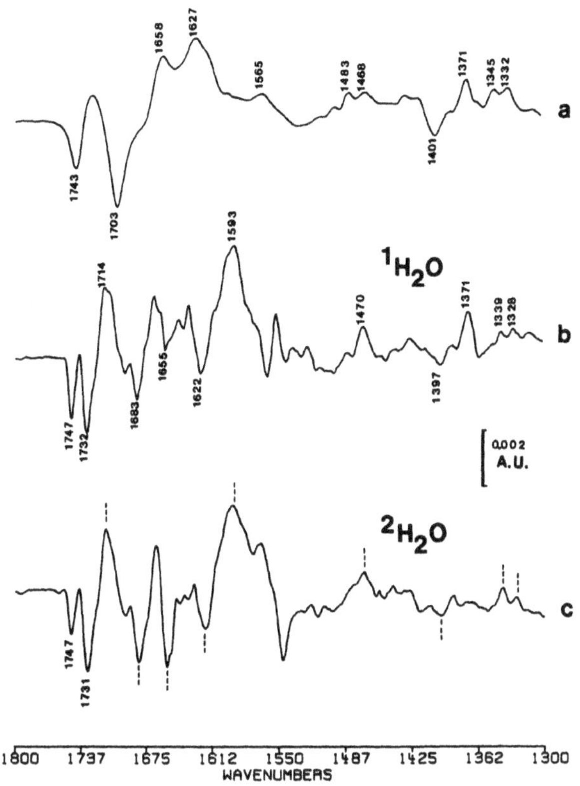

Figure 2 : a) FTIR difference spectrum ($BPheo^-$ - BPheo) of the anion formation in THF D8. Measuring conditions as in Fig. 1. Light-induced FTIR difference spectra of films of Rps. viridis RCs prepared b) from ^1H_2O suspension and covered with dithionite redox ^1H_2O buffer c) from ^2H_2O suspension after three days incubation in ^2H_2O, and covered with ^2H_2O dithionite redox buffer. T=240K, 1024 interferograms added. Resolution, $4cm^{-1}$. 715nm $<\lambda$ excitation $<$ 1100nm.

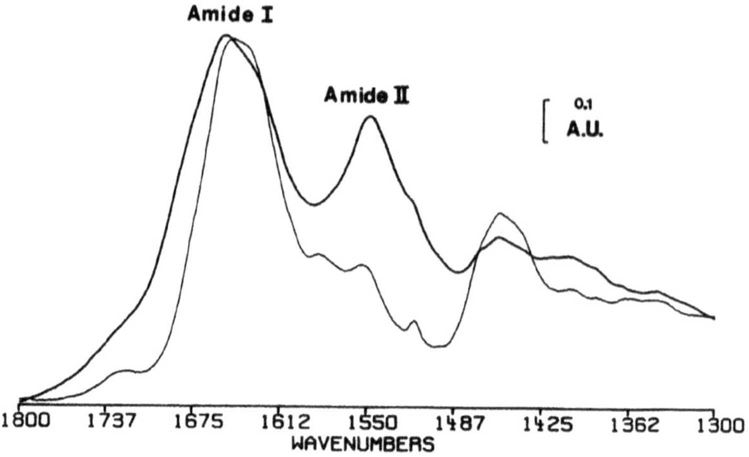

Figure 3 : FTIR absorbance spectra of films of Rps. viridis RCs prepared either from ^1H_2O suspension (▬) or after incubation in ^2H_2O for three days at room temperature (——). T=290K - 256 interferograms added. Resolution: 4cm^{-1}.

respect to the 1732cm^{-1} band was observed (Fig. 2c). This might be explained by assuming a partial exchange of a proton at a carboxylic group absorbing at 1747cm^{-1}. The corresponding COO^2H band displaced to lower frequency would be masked by the 1732cm^{-1} band but add to its intensity and contribute to its small but very reproducible shift to 1731-1730cm^{-1}. In another experiment, films of RCs prepared in ^1H_2O were prereduced with a deuterated redox buffer and left to equilibrate for 1.5 h. A spectrum identical to that shown in Fig. 2c was again obtained suggesting that the exchangeable proton is not too deeply buried in the protein cage of the RC. In addition, the original amplitude of the 1747cm^{-1} signal can be recovered after incubating the deuterated RCs in ^1H_2O, demonstrating that this proton can be reversibly back-exchanged. One can also notice that in the C=0 region of the H^- spectrum, other signals are not affected by the deuteration of RCs. In particular, a strong negative band at 1683cm^{-1} is observed on H^- spectra obtained either in ^1H_2O or ^2H_2O (Fig. 2b,c). This signal is assigned to the 9 keto C=0 of the intermediary electron acceptor. Its downshift with respect to the 1703cm^{-1} keto C=0 band of the BPheo b^- model compound spectrum in THF (Fig. 2a) indicates that it is hydrogen-bonded in the neutral state of the intermediary electron acceptor. It is interesting to note that similar deuteration conditions do not lead to any modifications of the 1750-1670cm^{-1} C=O signals in the light-induced FTIR spectra of the primary donor photooxidation in both Rps. viridis and Rb. sphaeroides RCs (Fig. 4), indicating that no accessible protein group contributes to the IR signals in this region.

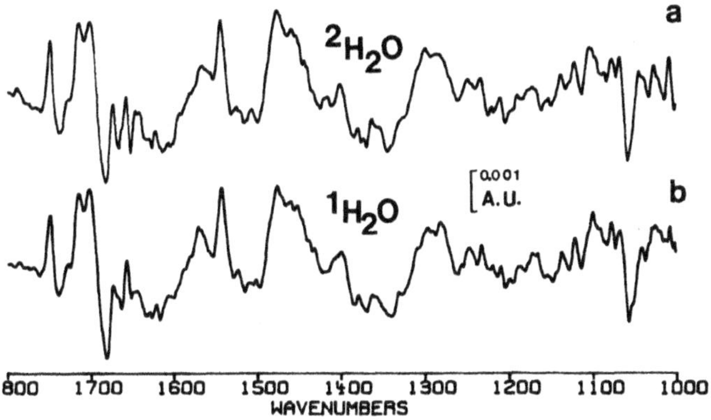

Figure 4 : Light-induced FTIR difference spectrum of a film of Rb. sphaeroides RCs prepared from a) ^2H_2O solution after three days incubation in ^2H_2O at room temperature b) ^1H_2O solution. 715nm $< \lambda$ excitation $<$1100nm. T=255K. 512 interferograms added. Resolution, 4cm^{-1}.

In agreement with previously reported spectra (2), the H^- spectrum of Rps. viridis chromatophores in ^1H_2O (Fig. 5a) compared to that of RCs already shows a smaller 1747/1732 cm^{-1} band ratio than in native RCs. Furthermore, as for RCs, the H^- spectra obtained from Rps. viridis chromatophores which have been deuterated by incubation in ^2H_2O (Fig. 5b) or which have been isolated from Rps. viridis cells grown in 60% ^2H_2O (data not shown) also show a decrease of the 1747cm^{-1} band, as

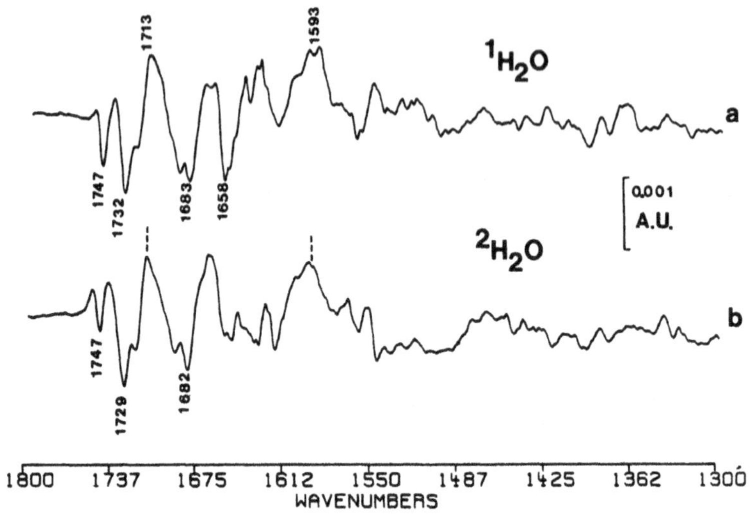

Figure 5 : Light induced FTIR difference spectra of films of Rps. viridis chromatophores covered with dithionite redox buffer. a) film from ^1H_2O suspension, ^1H_2O buffer. b) film from ^2H_2O suspension after three days ^2H_2O incubation, ^2H_2O buffer. T=240K. 1024 interferograms added. Resolution : $4cm^{-1}$.

well as a marked shift of the $1732cm^{-1}$ band to $1729cm^{-1}$. In this case, the $1729cm^{-1}$ band can be assigned to the contribution of the native $1747cm^{-1}$ and $1732cm^{-1}$ modes both being downshifted as an effect of the extensive isotope substitution.

It thus appears that in both RCs and chromatophores, the decrease of the $1747cm^{-1}$ band upon ^1H-^2H exchange favours the assignment of at least part of this band to a protein side chain C=O group protonated in the neutral state of the intermediary electron acceptor.

DISCUSSION

Comparison of FTIR and X-ray structural data

The X-ray structural analysis of the Rps. viridis RC has provided structural details of the pigment-binding sites to the protein (13). In particular, several C=0 groups of the BChls and BPheos show specific interactions with polar amino acid and side chains of the L and M protein subunits. Moreover, a striking environment asymmetry in the distribution of the protein polar goups with respect to the L and M

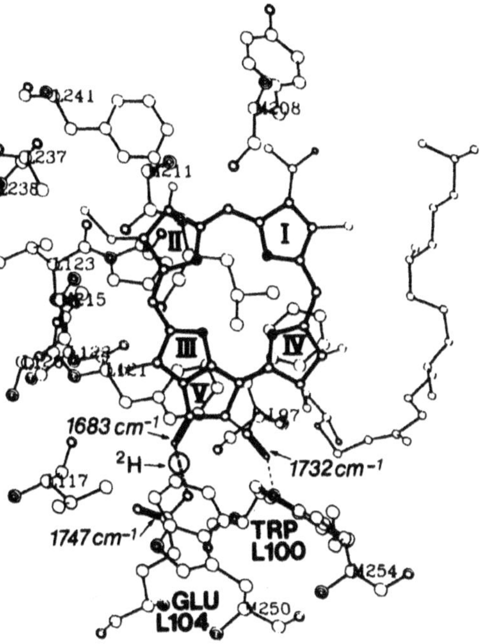

Figure 6 : The protein residues surrounding BPheo L. Possible hydrogen bonds are indicated with dashed lines. From Michel et al., ref. 13. Tentative IR bands assignments (Fig. 2c) to specific C=0 groups are also indicated.

branches of the pigments is clearly exhibited at the level of both the dimer and the BPheos (13). The environment of the BPheo on the L branch (BPheo L) is depicted on Fig. 6. BPheo L (BPheo M) has a single TRP group L100 (M 127) which appears hydrogen-bonded to the 10a ester C=0 group of ring V. According to IR and X-ray data, we thus assign the negative band at 1732cm^{-1} in the H^- spectra to the 10a ester C=0 of BPheo L whose frequency is downshifted by hydrogen-bonding to TRP L100 side chain.

The most important difference in the binding sites of BPheo L and BPheo M is the presence of one GLU residue (L104) close to the ring V keto of BPheo L with the correct distance to form an hydrogen bond (Fig. 6). It is replaced in the M branch by a non polar residue, VAL (M131). Michel et al. (13) have inferred that this GLU L104 is protonated because a negatively charged group would prevent the light-driven reduction of the BPheo L. Indeed, our FTIR H^- spectra indicate an interacting keto C=0 group (at 1683cm^{-1}) for the intermediary electron acceptor, in good agreement with the high resolution X-ray structure. Furthermore, in view of the isotope effect on the 1747cm^{-1} band which strongly suggests that this mode is, at least in part, due to a protonated proteic group, we propose that the exchangeable proton is the one from the carboxylic group of GLU L104. An identical proposal has been put forward at this Conference by Feher et al. (14) from ENDOR studies of the reduced intermediary electron acceptor in Rb. sphaeroides RCs. Our FTIR data on Rps. viridis RCs and chromatophores imply that this residue is protonated in the neutral state of the intermediary electron acceptor. We can reasonably exclude that the exchangeable proton detected here is the one from TRP L100 NH side chain as this proton is bonded to the 10a C=0 of BPheo L which thus cannot absorb at 1747cm^{-1}, a frequency too high for a bound ester carbonyl. In this model, upon photoreduction of BPheo b, the bleaching of the 1747cm^{-1} band is assigned to a change in the strength of the hydrogen bond between the proton of GLU L104 and the oxygen of the 9 keto C=0 of BPheo L. Although our FTIR data alone cannot exclude that the residue GLU L104 becomes fully deprotonated upon photoreduction of the intermediary acceptor, this appears unlikely in view of the conclusion from the ENDOR study (14) which indicates that the exchangeable proton is located 1.6 Å - 2 Å away from the 9 keto carbonyl. We thus conclude that the formation of H^- alters the hydrogen bond between GLU L104 and the 9 keto carbonyl of BPheo L.

Furthermore, amino acid sequences analysis indicates that the GLU residue is conserved between the subunits of RCs from various BChl a-

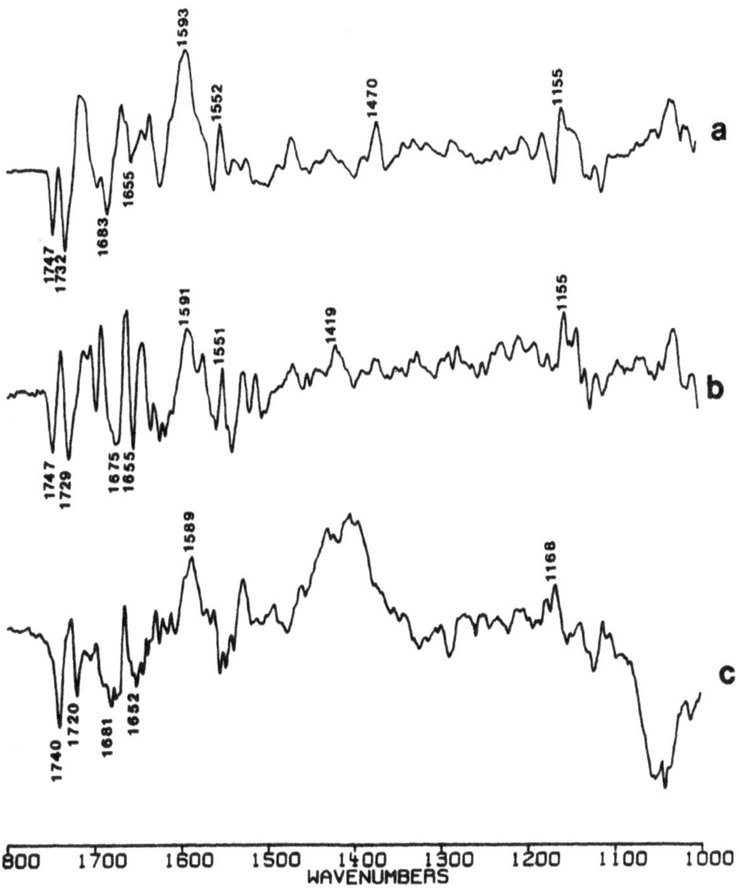

Figure 7 : Comparison of the light-induced difference spectra obtained on films of (a) Rps. viridis RCs. (b), C. vinosum RC-B875 complexes and (c) photosystem II particles. The films were covered with dithionite $^{1}H_2O$ redox buffer. 1024 interferograms added. Resolution : $4cm^{-1}$. T=240K, $\lambda >$ 715nm (a,b) ; T=260K, $\lambda >$ 665nm (c).

and BChl b-containing bacteria and the D_1 polypeptide of PSII. Indeed, a comparison of H^- spectra obtained from Rps. viridis RCs, RC-B875 complexes from C. vinosum and PSII particles containing respectively BPheo b, BPheo a and Pheo a as intermediary electron acceptors, shows for the three types of membranes (Fig. 7) similar spectral features with a decrease in the absorbance strength of two C=0 groups in the high frequency region, at 1740 and 1720cm^{-1} in PSII, at 1747 and 1729cm^{-1} in C. vinosum, at 1747 and 1732cm^{-1} in Rps. viridis. These two bands are apparently specific for the photoreduction of the intermediary electron acceptor in bacteria as well as in green plants. It can thus be assumed that the conserved GLU amino acid would give rise to the equivalent signals at 1740cm^{-1} in PSII and at 1747cm^{-1} in C. vinosum. In addition, we suggest that the 10a carbomethoxy ester C=0 of the BPheo a (at 1729cm^{-1} in C. vinosum) and Pheo a (at 1720cm^{-1} in PSII) is hydrogen bonded in the neutral state of these intermediary electron acceptors.

The remarkable similarity between the spectra obtained on two different types of photosynthetic bacteria and on PSII suggests that the specific changes in the bonding interactions between the intermediary electron acceptor and some defined residue(s) of the proteic cage which accompany the photoreduction is a general process. As our FTIR data also demonstrate that no large conformational change of the RC polypeptide backbone takes place during the photoreduction of the intermediary electron acceptor, it thus appears that the first photoinduced conformational change in the protein which can be detected is the one involving GLU L104 in Rps. viridis RC.

ACKNOWLEDGEMENTS

We would like to thank Dr. G. Berger and J. Kléo for the preparation of the purified bacteriopheophytin as well as A. Wollenweber for her help with the electrochemistry. Part of this work was funded by the European Economic Community (contract ST2J-0118-2-D).

REFERENCES

1. W. Mäntele, E. Nabedryk, B. A. Tavitian, W. Kreutz and J. Breton (1985) Light-induced Fourier transform infrared (FTIR) spectroscopic investigations of the primary donor oxidation in bacterial photosynthesis. FEBS Lett. 187, 227-232.

2. E. Nabedryk, W. Mäntele, B. A. Tavitian and J. Breton (1986) Light-induced Fourier transform infrared spectroscopic investigations of the intermediary electron acceptor reduction in bacterial photosynthesis. Photochem. Photobiol. 43, 461-465.

3. E. Nabedryk, B. A. Tavitian, M. Mäntele, W. Kreutz and J. Breton (1987) Fourier transform infrared (FTIR) spectroscopic investigations of the primary reactions in purple photosynthetic bacteria. In :"Progress in Photosynthesis Research Vol. I, pp. 177-180 (Biggins, J. ed.) Martinus Nijhoff Publishers, Dordrecht.

4. B. A. Tavitian, E. Nabedryk, W. Mäntele and J. Breton (1986) Light-induced Fourier transform infrared (FTIR) spectroscopic investigations of primary reactions in photosystem I and photosystem II. FEBS Lett. 201, 151-157.

5. W. Mäntele, A. Wollenweber, E. Nabedryk, J. Breton, F. Rashwan, J. Heinze and W. Kreutz (1987) Fourier-transform infrared (FTIR) spectroelectrochemistry of bacteriochlorophylls. In :"Progress in Photosynthesis Research, Vol. I, pp 329-332 (Biggins, J. ed) Martinus Nijhoff Publishers, Dordrecht.

6. S. Andriañambinintsoa, A.-M. Bardin, G. Berger, A. Bourdet, J. Breton, G. Hervo and E. Nabedryk. Colloque de Photosynthèse, "Caractérisation du quantasome de Rps. viridis, Rsp. rubrum et Rps. sphaeroides", Saclay 24-25 April 1986.

7. A. Wollenweber (1986) Diplomarbeit Universität Freiburg, FRG, "Spektroelektrochemische Untersuchungen an den Bakteriochlorophyllen a und b, sowie den Bakteriophäophytinen a und b".

8. W. Mäntele, A. Wollenweber, F. Rashwan, J. Heinze, E. Nabedryk, G. Berger and J. Breton (1987) Fourier transform infrared spectroelectrochemistry of the bacteriochlorophyll a anion radical. Photochem. Photobiol., in press.

9. M. S. Davis, A. Forman, L. K. Hanson, J. P. Thornber and J. Fajer (1979) Anion and cation radicals of bacteriochlorophyll and bacteriopheophytin b. Their role in the primary charge separation of Rhodopseudomonas viridis. J. Phys. Chem. 83, 3325-3332.

10. K. Ballschmiter and J. J. Katz (1969) An infrared study of chlorophyll-chlorophyll and chlorophyll-water interactions. J. Am. Chem. Soc. 91, 2661-2677.

11. D. F. Bocian, N. J. Boldt, B. W. Chadwick and H. A. Frank (1987) Near-infrared-excitation resonance Raman spectra of bacterial photosynthetic reaction centers. Implications for path-specific electron transfer. FEBS Lett. 214, 92-96.

12. F. Siebert, W. Mäntele and W. Kreutz (1982) Evidence for the protonation of two internal carboxylic groups during the photocycle of bacteriorhodopsin. FEBS Lett. 141, 82-87.

13. H. Michel, O. Epp and J. Deisenhofer (1986) Pigment-protein interactions in the photosynthetic reaction centre from Rhodopseudomonas viridis. The Embo Journal 5, 2445-2451.

14. G. Feher, R.A. Isaacson, M.Y. Okamura and W. Lubitz, ENDOR of exchangeable protons of the reduced intermediate acceptor in reaction centers from Rhodobacter sphaeroides R-26, these proceedings.

CHARGE RECOMBINATION AT LOW TEMPERATURE IN PHOTOSYNTHETIC BACTERIA REACTION CENTERS

Pierre Parot, Jean Thiery* and André Vermeglio

Association pour la Recherche en Bioénergie Solaire et
*Service de Radioagronomie, C.E.N. de Cadarache
B.P. n° 1, 13115 Saint-Paul-lez-Durance, France

INTRODUCTION

The first photochemical event in purple bacterial photosynthesis occurs in an integral membrane protein-pigment complex, the so-called reaction center. The reaction center is composed of three polypeptide subunits and a number of cofactors : four bacteriochlorophylls, two bacteriopheophytins, one (or two) ubiquinones, and one non-heme iron atom (Fe^{2+}) (see ref. 1 for a review). The three-dimensional structure of the reaction center from the photosynthetic bacterium, Rhodopseudomonas viridis, has been recently determined by X-ray diffraction at a resolution of 2.9 Å (2,3). Upon light absorption, a bacteriochlorophyll dimer (P) is raised to an excited state (P^*). In a few picoseconds, an electron leaves the primary donor P^* and a transient radical pair state (P^+Bpheo^-) is formed (4, 5, 6). The electron then migrates from the reduced Bpheo molecule to one of the quinone molecule, Q_A, in about 200 ps. (7, 8, 9). The possibility of structural changes linked to this light-induced charges separation has been discussed by several authors. NOKS et al (10) found that electron transfer kinetics are affected when chromatophores are incubated with the cross-linker glutaraldehyde, but only if the incubation is performed under illumination. ARATA and PARSON (11, 12) have presented evidences for a decrease in the volume of the reaction center-solvant system from calorimetric studies. WOODBURY and PARSON (13) suggest that there is a discrete distribution of states involved on the time scale of the early electron-transfer steps. VERMEGLIO and PAILLOTIN (14) have proposed from photodichroism studies at low temperature, that the voyeur bacteriochlorophyll molecule framework moves in the reaction center complex after charge separation. Similar results have been obtained by VASMEL et al (15) in their linear dichroism study on oriented Chloroflexus aurantiacus reaction centers in squeezed polyacrylamide gel. KIRMAIER et al (16, 17) have observed different kinetics components as a function of probe wavelength for the formation of state $P^+Q_A^-$ in both Rhodobacter sphaeroides and Chloroflexus aurantiacus reaction centers. They postulated that this behaviour is due to readjustments of the pigments and/or the protein following the charges separation process. KLEINFELD et al (18) have modeled the non-exponential behaviour for the charges recombination kinetics observed at low temperature in terms of a distribution of structural configurations. These authors also observed that, when reaction centers are cooled under continuous illumination, the recombination time of a subsequent light-induced charge separation at low temperature was lengthened by a factor of 5. They proposed (18) that the light-pretreatment induces structural changes which lead to an increase of the donor-acceptor electron-transfer distance.

In the present paper, we have investigated the low temperature charge recombination of state $P^+Q_A^-$ for isolated reaction centers and intact chromatophores membranes of the species Rhodobacter sphaeroides, Rhodospirillum rubrum and Rhodopseudomonas viridis.

RESULTS AND DISCUSSION

Figures 1 and 2 show the light-induced difference spectra linked to state $P^+Q_A^-$, observed in the near infra-red region at 10°K, for a suspension of G_9 chromatophores (Fig. 1) or a suspension of isolated reaction centers from Rhodopseudomonas viridis (Fig. 2).

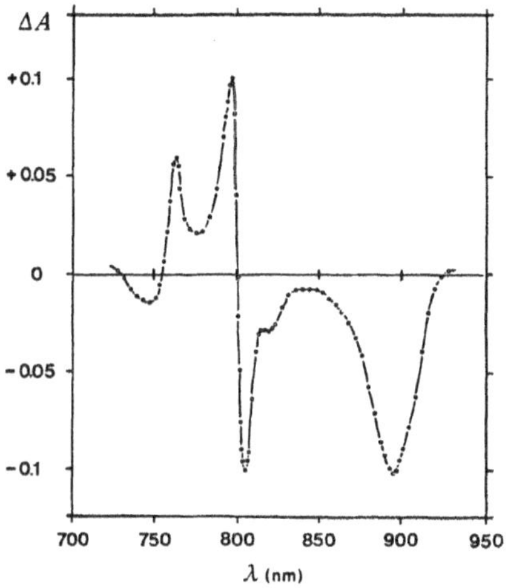

Fig.1 - Light-induced difference spectrum measured at 10°K for a suspension of chromatophores prepared from R.rubrum strain G_9. The absorbance changes were induced by a laser actinic flash.

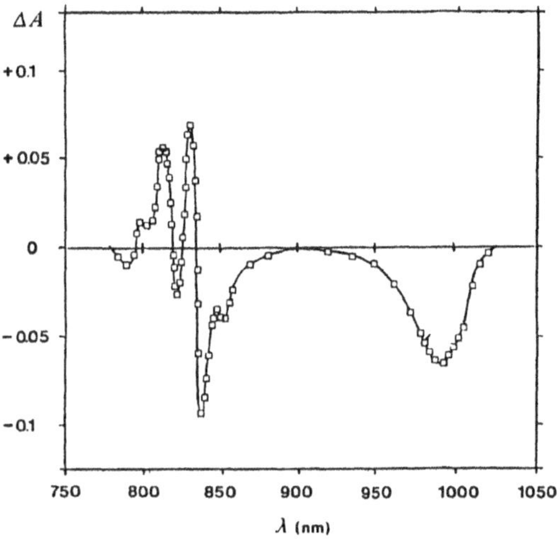

Fig.2 - Same as Fig.1 for a suspension of Rhodopseudomonas viridis reaction centers.

The decay of the photo-induced charges separated state can be monitored in the long-wavelength band of the primary donor, i-e. 890 nm for G_9 chromatophores or 1000 nm for Rhodopseudomonas viridis reaction centers. Such kinetics are depicted in Fig.3 A and B. To prevent electron transfer from the primary (Q_A) to the secondary electron acceptor (Q_B), orthophenanthroline was added. Addition of this chemical suppressed a slow decaying component of small amplitude as shown in Fig. 3 A. All subsequent kinetics were therefore performed in presence of 5 mM orthophenanthroline.

The apparent half-time ($t_{1/2}$) is equal to 15 ms for G_9 chromatophore (Fig. 3 A). The half-time for the back reaction, for Rhodopseudomonas viridis reaction centers is found to be equal to 5 ms (Fig. 3 B).

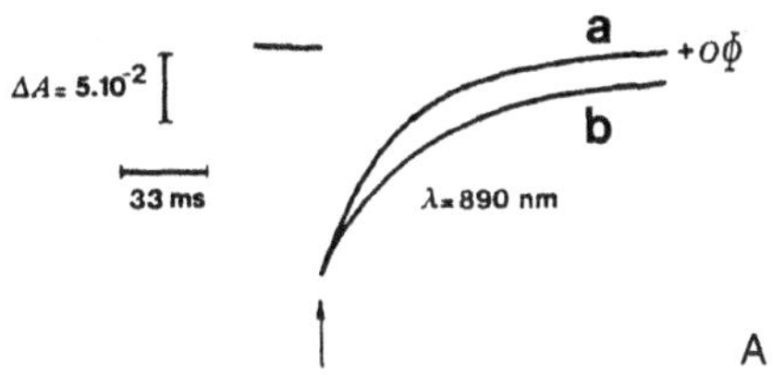

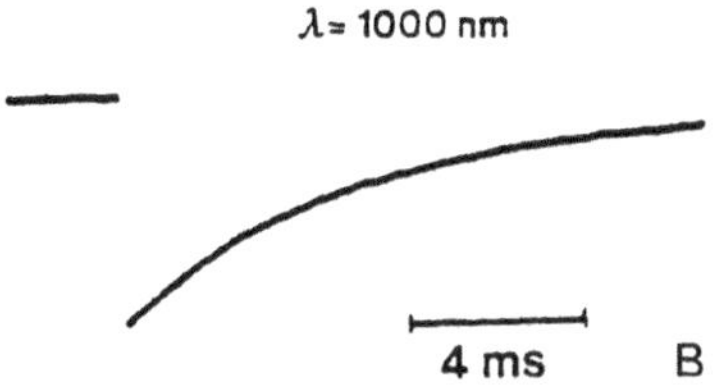

Fig. 3 A - Kinetics of the laser-induced absorbance changes occuring in the long-wavelength band of the primary electron donor at 890 nm. G_9 chromatophores in presence (a) or absence (b) of 5 mM orthophenanthroline. Temperature 10°K.

Fig. 3 B - Same as part A but for <u>Rhodopseudomonas viridis</u> reaction centers. wavelength detection 1000 nm. 5mM orthophenanthroline was present in the suspension.

Careful examination of the absorption changes occuring around the crossing points of the light-induced difference spectrum (approx 801 and 757 nm in the case of G_9 chromatophores reveals important differences in their kinetics) (Fig. 4). A striking indication of the complexity of the kinetics of the charges recombination is given at 801,2 nm (Fig. 4) where the absorbance change is negative in the first milliseconds and subsequently positive. A similar complex behaviour is also observed for reaction centers isolated from <u>Rhodobacter sphaeroides</u> (Fig. 5) or <u>Rhodopseudomonas viridis</u> (Fig. 6).

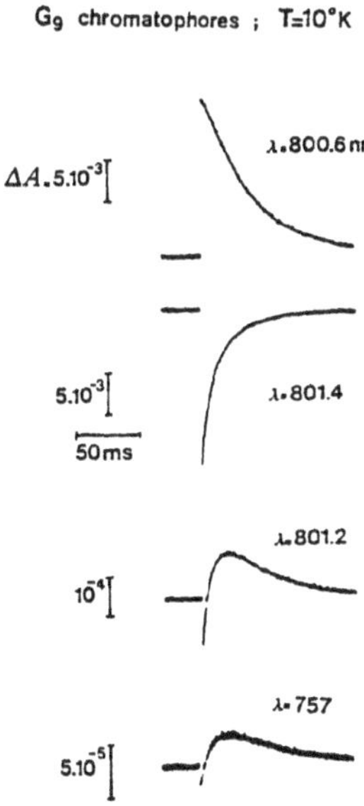

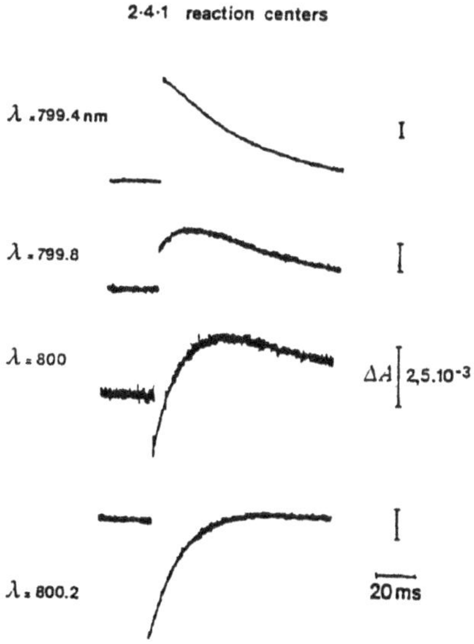

Fig. 4 - Comparison of the kinetics occuring around the crossing points of the light-induced difference spectrum (∼ 801 nm and ∼ 757 nm) for a suspension of G_9 chromatophore.

Fig. 5 - Same as Fig. 4 but for a suspension of isolated reaction centers from <u>Rhodobacter sphaeroides</u> strain 2-4-1.

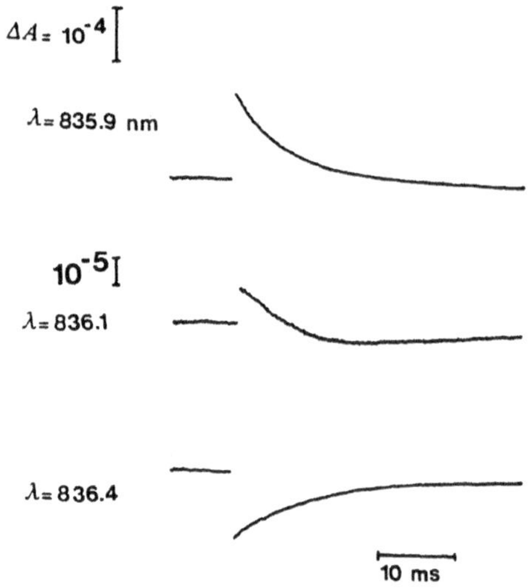

Fig. 6 - Same as Fig. 4 but for a suspension of isolated reaction centers from Rhodopseudomonas viridis.

Table I summarized the results obtained with different preparation and species. No important differences could be detected when comparing the kinetics of the main absorbance changes at 765, 795, 810 and 890 nm, for both chromatophores or reaction centers. However significative differences are observed for the half-times measured at wavelength between 795 and 810 nm (Table II).

Table I - Summary of the half-times and wavelength occurence of the fast and slow phases of the back reaction at 10°K.

Preparation	Fast phase		Slow phase		$t_{1/2}$ at 890 nm (ms)
	(nm)	$t_{1/2}$ (ms)	(nm)	$t_{1/2}$ (ms)	
G_9 chromatophores	801,4	10	800,6	30	15
2-4-1 Reaction centers	800,2	9	799,4	32,5	17,5
R_{26} Reaction centers	801,5	11,5	800,4	36,5	18
Viridis Reaction centers	835,9	4,5	836,4	7,5	$t_{1/2}$ at 1000 nm (ms) 5

Table II - Summary of decay kinetics of state $P^+Q_A^-$ in *Rb sphaeroides* 2-4-1 RC's at 10°K.

(nm) observation	765	794	796	798	799,4	800,2	801	802	806	810	890
$t_{½}$ (ms)	17,5	17,5	19	20	32,5	9	15,5	16,5	17,5	17,5	17,5

The rather restricted wavelength range (Fig. 4, 5 , 6) where the different kinetic components of the decay of the state $P^+Q_A^-$ are distinguishable explains why this phenomenon has not been reported previously. KIRMAIER et al (16, 17) have reported variation in the kinetic time constant, as a function of probe wavelength, but for the formation of state $P^+Q_A^-$. In the experiments we present here the wavelength dependence is observed for the decay of state $P^+Q_A^-$.

One explanation for the complex wavelength dependence of the decay of state $P^+Q_A^-$ is that this reflects the presence of two main populations of reaction centers. Each of these populations possesses its own rate of charge recombination and slightly different light-induced spectra. If at one particular wavelength the contribution of one population is equal to zero, only the kinetic component of the other population will be observed. This is approximatively the case at 800,6 and 801,4 nm for G_9 chromatophores (Fig. 4) and at 799,4 and 800,2 nm for 2-4-1 reaction centers.

Before discussing the different possible origins of the two main populations of reaction centers, we want first to emphasize their general occurence in different species and preparations including intact chromatophore membranes. This excludes the artefactural formation of the two types of reaction centers during the purification process for example.

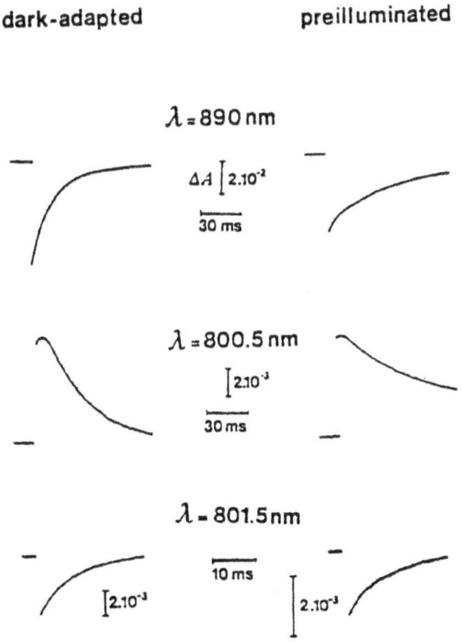

Fig. 7 - Comparison of the flash-induced absorbance changes observed around 801 nm and in the long-wavelength region (890 nm) for R26 purified reaction centers. Left traces: dark-adapted reaction centers. The half-times are equal to 18 ms at 890 nm, 30 ms at 800.5 nm and 11 ms at 801.5 nm. Right traces: reaction centers preilluminated at room temperature. The half-times are equal to 51 ms at 890 nm, 105 ms at 800.5 nm and 11 ms at 801.5 nm. The temperature of the experiment was 10°K.

We have performed several experiments with reaction centers subjected to different modifications or treatments to obtain more information on the factors which may govern these two states. We found no influence of pH, in the range 6 to 9, on the complex kinetic behaviour around the crossing points of the light-induced difference spectra. Removal of the H polypeptide slows down by a factor of 2.5 the half-time of the back reaction (19) but the complex wavelength dependence of the kinetics was still observed. Removal, by borohydride treatment, of the accessory bacteriochlorophyll molecule (20) affects neither the rate of the back reaction nor the complex behaviour in the 801 nm region. The only treatment which affects the kinetics around the crossing point (801 nm) of the light-induced difference spectrum in a strong illumination of the reaction centers during the cooling process (Fig. 7).

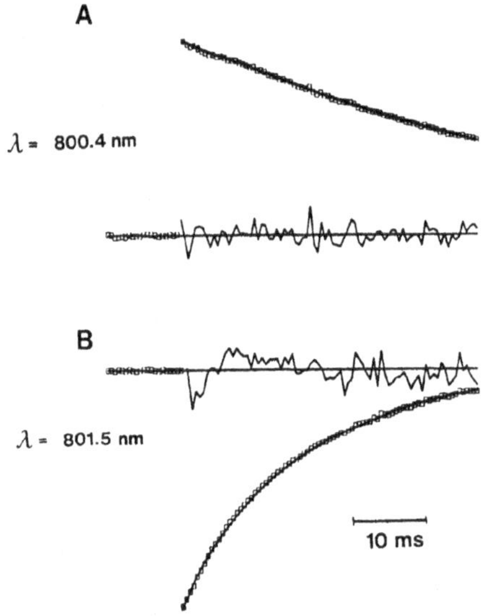

Fig. 8 - Fitting of computed exponential (full line) and experimental decay curves (square) measured respectively at 800.4 (A) and 801.5 nm (B) for a suspension of R26 reaction centers. The half-times of the two exponential components are equal to 36.5 ms (800.4 nm) and 11.5 ms (801.5 nm). The difference between the computed and experimental curves, enlarged by a factor of 10, represents a standard deviation of 4/1000 (curve A) and 5/1000 (curve B).

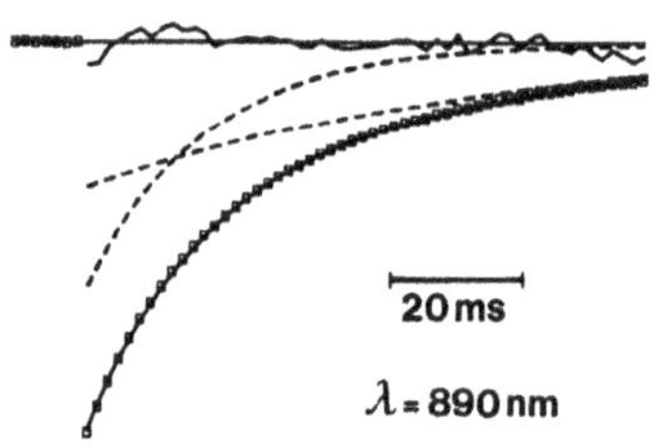

Fig. 9 - Decomposition of the absorbance changes occurring in the long-wavelength band (890 nm) of a suspension of R26 reaction centers. The experimental curve (squares) is compared to the computed sum (full line) of the two exponentials (broken line) fitting the absorbance changes observed at 800.4 and 801.5 nm (see Fif. 8). The relative weights are 62 and 38 for the fast and slow components, respectively. The standard deviation is equal to 2/1000. The difference between computed and experimental curves by a factor 10.

This light-pretreatment has been shown by KLEINFELD et al (18) to slow down by a factor of 4-5 the recombination kinetics measured at cryogenic temperature. We confirmed their results (Fig. 7).
In addition we have found that the light-pretreatment lengthens by a factor of 3.5 the already slow component while the kinetics of the fast component was unaffected (Fig. 7).

The careful spectral analysis we have performed on the light-induced absorbance changes linked to the charge recombination process at low temperature demonstrates the complexity of its kinetics. We interpret this complexity as reflecting the presence of two main states for the reaction centers at low temperature.

These two states could pre-exist before the charge separation or be the consequence of nuclear relaxations induced by the photochemistry. The first hypothesis is similar to the one proposed by KLEINFELD et al (18), but in contrast with these authors who supposed a distribution of different structural states, we interpret our results as evidence of only <u>two</u> discrete states for the reaction centers. The second hypothesis is derived from the proposal of KIRMAIER et al (16, 17) that re-adjustments of pigments and/or the protein can follow the charge separation mechanism. One may also consider a third hypothesis where the electron can occupy two distinct sites after the charge separation. Preliminary experiments, with reaction centers of <u>Rhodospirillum</u> <u>rubrum</u> G_9 isolated in presence of Triton X-100, show that only the fast kinetic is observed for the back reaction. As the iron is easily removed from this type of reaction centers (21), the involvment of this atom in the two decaying states could be envisaged. Experiments are in progress to check that hypothesis.

Whatever is the origin of the two different rates observed for the charge recombination process, the occurence of these two decays may give a straight-forward explanation of the non-exponential time dependence observed in the main absorption band of the primary donor (18). Fig. 8 shows that the charge recombination kinetics of each state, monitored at 800.4 and 801.5 nm in the case of R26 reaction centers can be well fitted with a single exponential. A good fit in the decomposition of the absorbance changes occurring in the long-wavelength band (Fig. 9), by imposing the number of exponential components, 2, and the values of the half-times as determined from the analysis of Fig. 8, could be obtained. We therefore, conclude that the non exponential character of the relaxing process of state $P^+Q_A^-$ measured in the longwavelength band of the primary electron donor is due to the presence of mainly two states decaying with different rates.

ACKNOWLEDGMENTS

We thank Daniel GARCIA for his participation in the deconvolution of kinetics in single exponential components.

REFERENCES

(1) OKAMURA, M.Y., FEHER, G. and NELSON, N. (1982) in Photosynthesis.I. Energy Conversion by Plants and Bacteria (Govindjee, ed), pp 195229, Academic Press, New York.
(2) DEISENHOFER, J., EPP, O., MIKI, K., HUBER, R. and MICHEL, H. (1985) Nature (London) 318, 618-624.
(3) MICHEL, H., EPP, O. and DEISENHOFER, J. (1986) EMBO J, 5, 2445-2451.
(4) PARSON, W.W., SCHENCK, C.C., BLANKENSHIP, R.E., HOLTEN, D., WINDSOR, M.W. and SHANK, C.V. (1978) in Frontiers of Biological Energetics, Vol 1, pp 37-44, Academic Press, New York.

(5) HOLTEN, D., HOGANSON, C., WINDSOR, M.W., SCHENCK, C.C., PARSON, W.W., MIGUS, A., FORK, R.L. and SHANK, C.V. (1980) Biochim. Biophys. Acta 592, 461-477.
(6) MARTIN, J.L., BRETON, J., HOFF, A., MIGUS, A. and ANTONETTI, A. (1986) Proc. Natl. Acad. Sci. USA 83, 957-961.
(7) ROCKLEY, M., WINDSOR, M.W., COGDELL, R.J. and PARSON, W.W. (1975) Proc. Natl. Acad. Sci. USA 72, 2251-2255.
(8) KAUFMANN, K.J., DUTTON, P.L., NETZEL, T.L., LEIGH, J.S. and RENTZEPIS, P.M. (1975) Science 188, 1301-1304.
(9) SHUVALOV, V.A., KLEVANIK, A.V., SHARKOV, A.F., MATWEETZ, J.A. and KRUBOV, P.G. (1978) FEBS Lett. 91, 135-139.
(10) NOKS, P.P., LUKASHEV, E.P., KONONENKO, A.A., VENEDIKTOV, P.S. and RUBIN, A.B. (1977) Mol. Biol. (Moscou) 11, 1090-1099.
(11) ARATA, H. and PARSON, W.W. (1981) Biochim. Biophys. Acta 636, 7081.
(12) ARATA, H. and PARSON, W.W. (1981) Biochim. Biophys. Acta 638, 201209.
(13) WOODBURY, N.W.T. and PARSON, W.W. (1984) Biochim. Biophys. Acta 767, 345-361.
(14) VERMEGLIO, A. and PAILLOTIN, G. (1982) Biochim. Biophys. Acta 681, 32-40.
(15) VASMEL, H., MEIBURG, R.F., KRAMER, H.J.M., DE VOS, L.J. and AMESZ, J. (1983) Biochim. Biophys. Acta 724, 333-339.
(16) KIRMAIER, C., HOLTEN, D. and PARSON, W.W. (1985) Biochim. Biophys. Acta 810, 33-48.
(17) KIRMAIER, C., HOLTEN, D. and PARSON, W.W. (1986) Biochim. Biophys. Acta 850, 275-285.
(18) KLEINFELD, D., OKAMURA, M.Y. and FEHER, G. (1984) Biochem. 23, 5780-5786.
(19) DEBUS, R.J., FEHER, G. and OKAMURA, M.Y. (1985) Biochem. 24, 24882500.
(20) DITSON, S.L., DAVIS, R.C. and PEARSTEIN, R.M. (1984) Biochim. Biophys. Acta 766, 623-629.
(21) LOACH, P.A. and HALL R.L. (1972) Proc. Natl. Acad. Sci. USA 69, 786-790.

TEMPERATURE AND $-\Delta G°$ DEPENDENCE OF THE ELECTRON TRANSFER TO AND FROM Q_A IN REACTION CENTER PROTEIN FROM RHODOBACTER SPHAEROIDES

M.R. Gunner and P. Leslie Dutton

Department of Biochemistry
University of Pennsylvania
Philadelphia, Pa. 19104

INTRODUCTION

Photosynthetic reaction center proteins (RC) have provided an important system for the study of biological electron transfer mechanisms. Initially, the investigation focused simply on the oxidation-reduction of the cofactors that function as donor and accepter. As details of both the reaction pathways and protein structure emerge, attention can now turn toward the role of the protein matrix in electron transfer. This not only provides the scaffolding for the redox sites, but also a medium that should be energetically coupled to the electron transfer event. On a broader front, the reactions occurring in the RC, with its defined structure and characterized reactions, offer a unique opportunity to test current, general theoretical models for electron transfer.

DeVault and Chance (1) first directly addressed the question of the mechanism of electron transfer in biological systems. They recognized that the photooxidation of a cytochrome c bound to the reaction center protein of Chromatium vinosium involved quantum mechanical tunneling. This was established by their observation that, although the reaction shows Arrhenius behavior at room temperature, below 120K the rate becomes independent of temperature. Since then a similar pattern has been shown for the photooxidation of cytochromes associated with several other bacterial species (2; see 3 and 4 for reviews).

This remarkable temperature dependence of physiological cytochrome oxidation has excited a considerable number of theoretical treatments (1,5,7-9; and for more recent analyses see 10-13). Each is characterized by differences in the assumptions made, the ways of handling the calculations, the number of reactions considered to be involved, the treatment of the vibrations considered to be coupled to the electron transfer, and the reorganization energy of the reaction (λ).

However, despite data covering a substantial temperature range, all these theoretical approaches fit the data adequately. Thus, it has not been possible to discriminate between them.

Soon afterwards, a different, even more remarkable pattern of temperature dependence was found for several of the intra-RC electron transfers, which are completely activationless. This includes the electron transfer from the light activated special pair of bacteriochlorophylls ($(BChl)_2$) to the bacteriopheophytin (BPh) (14,15), from BPh^- to the primary ubiquinone-10 (Q_A) (13), and from Q_A^- to $(BChl)_2^+$ (17-20). With knowledge of the behavior only at the $-\Delta G°$ in the native protein, it has been a common practice to explain the non-classical temperature dependence of all of these reactions as arising from the $-\Delta G°$ of these biological electron transfers being equal to their reorganization energy, λ.

Many of the issues required for understanding these electron transfer reactions were discussed at a meeting held in Philadelphia in 1978 (5). Also, at this meeting new experimental approaches to the problem were introduced. In particular, Miller described a different style of experiment where electron transfer was studied at a single temperature, but over a wide range of reaction $-\Delta G°$. The rate of tunneling of trapped electrons to a series of accepters, immobilized in a glass was measured at 77 K. As the reaction $-\Delta G°$ was increased, the rate rose and then fell. These general features were predicted by the work of Marcus (21,22) and others (11,23-25). However, it was clear from the fairly slow decrease in rate at large exothermicity, that at least some of the vibrations coupled to the electron transfer were of sufficiently high energy that it was necessary to treat them quantum mechanically. Thus, this experiment showed that simple treatments of electron transfer, where the nuclei are treated classically, are not adequate to model this simple electron transfer reaction. A second meeting held in Philadelphia in 1985 (6) reported on progress in the area and continued the discussions.

It is apparent that studies exploring electron transfer mechanisms have generally been characterized by two kinds of experiments. There are those that measure the temperature dependence of electron transfer reactions at a fixed $-\Delta G°$, and there are those that measure the $-\Delta G°$ dependence at a single temperature. It is suprising that prior to studies of electron transfer in Q_A-replaced RCs (26) electron transfer rates have not been examined as a function of both $-\Delta G°$ and temperature. Studied alone it is clear that there is much less stricture placed on the analysis of the experimental findings.

In the work reported here, we have measured the rate of electron transfer from BPh^- to Q_A and from Q_A^- to $(BChl)_2^+$. The $-\Delta G°$ of these reactions has been altered by removing the native Q_A, ubiquinone-10, and reconstituting function with a variety of other quinones (26-30). The temperature dependence of the electron transfer rates of the two reactions that involve Q_A in each of the Q_A-replaced RCs has been determined from 300 to 10 K. We have modeled the results by current electron theories to obtain information about the vibrations that are coupled to the electron transfer reaction.

METHODS

Replacement of Q_A with other quinones. The bound quinones, Q_A and Q_B, were removed from purified RC by the method of Okamura et. al. (27). Q_A function was reconstituted with a variety of quinones as previously described (26,28).

Measurement of the rate of electron transfer from Q_A^- to $(BChl)_2^+$. $(BChl)_2^+Q_A^-$ was formed by excitation of the RC with a 30 μs Xenon flash. The rate of return of an electron from Q_A^- to $(BChl)_2^+$ was monitored by the decay of the EPR signal of $(BChl)_2^+$ at g=2.0026 as described in ref 26. This rate constant is designated k_3 in figure 1.

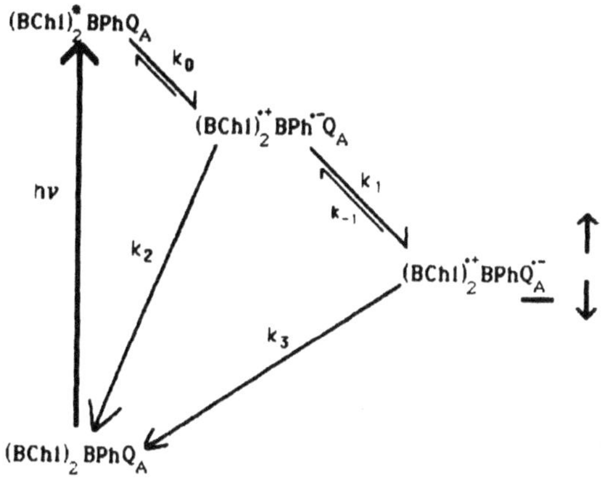

Figure 1 Simplified reaction scheme for electron transfer in isolated RCs.

Measurement of the rate of electron transfer from BPh^- to Q_A. Two methods are available to measure the rate of electron transfer from BPh^- to Q_A, designated k_1 in figure 1.

(1) The decay of $(BChl)_2^+BPh^-$ can be monitored directly by picosecond spectroscopy (31,32). It occurs at $k_1 + k_2$ (see figure 1). The rate k_2 represents the sum of the rates for decay of $(BChl)_2^+BPh^-$ by all processes other than electron transfer to Q_A. This method provides the most accurate measurements when $k_1>k_2$. The rate k_2 has been independently established in RC where reduction of Q_A is blocked by the removal of the quinone or by prior chemical reduction (33,34).

(2) The rate k_1 can also be determined from the quantum yield of $(BChl)_2^+Q_A^-$ (Φ) (35,36). Where:

$$\Phi = \frac{k_1}{k_1 + k_2} \quad (1)$$

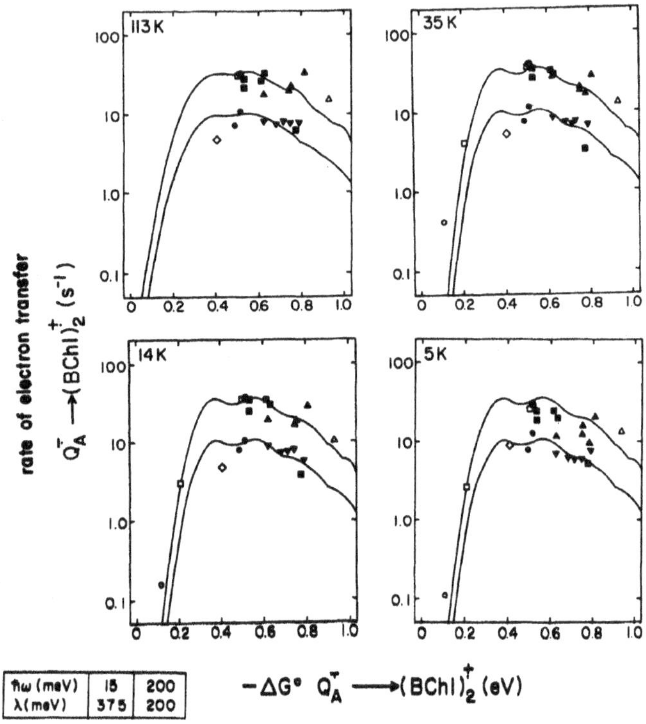

Figure 2 Dependence of the rate of $(BChl)_2^+$ reduction on the $-\Delta G°$ between $(BChl)_2^+ \; Q_A^-$ and the ground state. The reaction $-\Delta G°$ was measured _in situ_ (filled symbols) or estimated from _in vitro_ $E_{1/2}$ values of the quinones . Data is from reference 26. The various symbols are (o) BQs; (□) 1,4-NQs; (◇) 1,2-NQ; (▽) AQs substituted in the 1-position; (△) AQs substituted in the 2-position. Theoretical line calculated using equation 2 with $h\omega_M$=15 meV, λ_M=375 meV, $h\omega_L$=200 meV and λ_L=200 meV. Values of V(r) are 4.8×10^{-8} eV (upper curve) and 1.4×10^{-8} eV (lower curve).

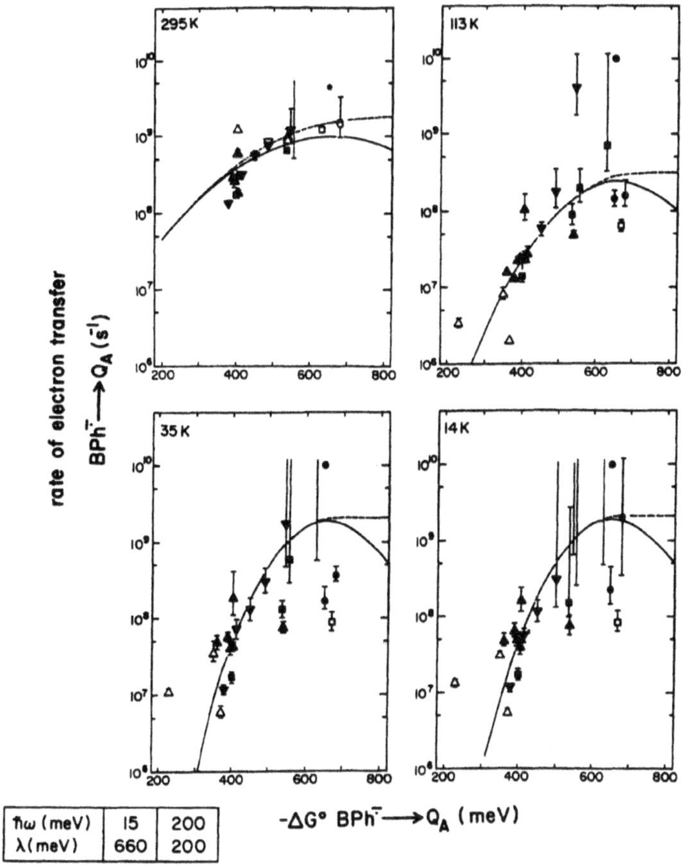

Figure 3 Dependence of k_1 on the $-\Delta G°$. The rates are calculated from Φ_Q using equation 1. At 295K k_2 was assumed to be 7.7×10^7 s^{-1}, at lower temperature a value for k_2 of 3.3×10^3 s^{-1} was used. Data is from reference 30. The quinones functioning as Q_A were (o) BQs; (□) NQs; (▽) AQ and 1-substituted AQs; (△) 2-substituted AQs. The error bars represent the variation in k_1 given the standard deviation of Φ_Q. (a) Data at 295K. Open symbols represent values derived from picosecond measurements of the decay of $(BChl)_2{}^{+}BPh^{-}$. Filled symbols are for k_1 calculated from Φ_Q. All $-\Delta G°$s were calculated with Q_A $E_{1/2}$s determined in situ. (b) Data at 113K; (c) 35K; (d) 14K. For (b-d) filled symbols imply the $-\Delta G°$ was calculated using Q_A $E_{1/2}$ values determined in situ while for open symbols in vitro values were used. The theoretical lines were calculated with equation 2 with $h\omega_M$=15 meV, λ_M=660 meV, $h\omega_L$=200 meV, λ_L=200 meV, $V(r)=4\times10^{-1}$ meV at 295K, 35K and 14K and 1×10^{-1} at 113K.

This provides the most accurate values for k_1 when it is similar to or less than k_2. Most data reported here was determined by this method.

Measurement of the in situ midpoint for Q_A^-/Q_A in the Q_A-replaced RCs. The in situ $E_{1/2}$ for quinones, relative to the value of the native Q_A, UQ-10, were obtained as described in references 26 and 29. These values were used to calculate the reaction $-\Delta G°$ for each Q_A-involved reaction.

RESULTS

Figures 2 and 3 show the variation of rate with $-\Delta G°$ for both Q_A-involved reactions at four different temperatures. Currently, the data for the electron transfer from Q_A^- to $(BChl)_2^+$, k_3, is most complete for the regin of increasing $-\Delta G°$ beyond the 520 meV found in the native protein. The data-set for the reduction of Q_A by BPh^-, k_1, highlights the region of $-\Delta G°$ smaller than 660 meV, the value found in the native RC. There is currently some controversy as to the exact value of the $-\Delta G°$ for the latter reaction in the native RC (see reference 29 and 30 for a discussion). The value chosen was derived from fluorescence measurements, which should be on the same time scale as the electron transfer from BPh^- to Q_A. This uncertainty does not effect the conclusions drawn here.

The theoretical lines drawn through the data in figures 2 and 3 utilize the general model of electron transfer reactions as non-adiabatic multi-phonon decay processes. These were calculated using:

$$k = \frac{2\pi}{\hbar^2 \omega_m} |V(r)|^2 \left(\sum_{q=0}^{\infty} \frac{e^{-S'} S'^q}{q!} \right) e^{-S(2\bar{n}+1)} \left(\frac{\bar{n}+1}{\bar{n}} \right)^{P/2} I_P \left[2S\sqrt{\bar{n}(\bar{n}+1)} \right] \quad (2)$$

where S is $\lambda_m/h\omega_m$, S' is $\lambda_L/h\omega_L$, P is $(-\Delta G°-qh\omega_L)/h\omega_L$, $\bar{n}$ is $[\exp(h\omega_m/k_bT) - 1]^{-1}$, and $I_P(Z)$ is the modified Bessel function. See DeVault (37) for a more complete description of this equation. This expression treats the electron transfer as being coupled to two vibrations. For one, the quantum vibrational energy ($h\omega_m$) is similar to the thermal energy (k_bT). The energy of the second, higher frequency mode ($h\omega_L$) is assumed to be much greater than k_bT, so that it is frozen into its lowest vibrational level at all temperatures. Since the theory is complex, a range of parameters can be used to model the data. However, the results do put several constraints on the model that provides new information about the nature of electron transfer in the RC. In particular, it allows a general characterization of the changes in the motions of the redox sites and the surrounding protein that accompany electron transfer. These are considered to be the source of the temperature and $-\Delta G°$ dependence in this analysis.

Trends as the $-\Delta G°$ is decreased relative to the reaction found in the native RC. It was found for the electron transfer from BPh^- to Q_A, that the reaction slows as the $-\Delta G°$ is diminished relative to that found in the native protein. The reaction remains essentially temperature independent over this free energy range. A similar trend is seen for the electron transfer from Q_A^- to $(BChl)_2^+$, with a smaller number of Q_A-replaced RCs. These observations permit the following conclusions to be drawn:

(1) The fall-off in rate as the reaction $-\Delta G°$ is diminished supports the established view that the reorganization energy of the reaction is similar to the $-\Delta G°$ of the native reactions. However, the maxima for the two reactions studied are quite broad so λ cannot be precisely determined.

(2) There is significant coupling of the reaction to modes where 10 meV<$h\omega$<30 meV. The upper limit is established by the sensitivity to decreasing $-\Delta G°$. The lower limit is set by the lack of strong temperature dependence in this free energy region.

(3) Significant coupling to very small energy vibrations ($h\omega_S \ll k_bT$) should be seen by the reaction showing classical activation energies for $-\Delta G°$ smaller than λ for these modes. Therefore, since the reaction is temperature independent at $-\Delta G°$ of as little as 200 meV, λ must be less than 200 meV for vibrations where $h\omega$<10 meV.

Trends as the $-\Delta G°$ is increased relative to that found in the native RC. It was found that the rate of electron transfer from Q_A^- to $(BChl)_2^+$ does not slow significantly as the $-\Delta G°$ is increased beyond that found in the native RC. (There is little data for the electron transfer from BPh^- to Q_A in this free energy region.) This shows that:

(1) Vibrations with 150 meV>$h\omega$ are coupled to the reaction.

DISCUSSION

The conclusions derived from the temperature and $-\Delta G°$ dependence of both k_1 and k_3 will be discussed as providing one general picture of electron transfer in RC. Further work, extending the $-\Delta G°$ range covered for each reaction is in progress.

The picture emerging from this work is that the reorganization energy for both Q_A reactions is approximately 500 to 800 meV. The data allows this to be apportioned between vibrations of different energies relative to k_bT.

λ_L is the reorganization energy of vibrations where $h\omega_L \gg k_bT$. Skeletal motions of the redox sites themselves should fall into this class (10,38,39). Their coupling to the electron transfer is seen by the relative $-\Delta G°$ independence of the reaction as the reaction free energy is increased. However, a limit can be placed on λ_L as being $\leq 2h\omega_L$ since the rate does not increase significantly with increasing exothermicity in this $-\Delta G°$ region.

Because the reaction is coupled to vibrations of this energy, "exothermic rate restriction" cannot be invoked as a a mechanism to control the rates of intra-RC electron transfers. This has been advanced as a reason for the useful charge separating electron transfer from BPh^- to Q_A being faster that the more exothermic competing electron transfer from BPh^- to $(BChl)_2^+$ to reform the ground state (see figure 1). Such a kinetic arrangement is required for the quantum yield for formation of $(BChl)_2^+Q_A^-$ to be high. However, coupling the reaction to high frequency vibrations moves the fall-off in rate at $-\Delta G°>\lambda$ to regions of free energy not relevant on the scale of biological reactions. Thus, other mechanisms will be needed to understand the relative slowness of k_2 if the finding that λ is between 500 and 800 meV, and if the possibility that the reaction is weakly coupled to high frequency vibrations holds for all intra-RC reactions (see discussion in ref 26).

The reaction appears to be significantly coupled to vibrations in the range where $h\omega_M \approx k_bT$. These have been calculated at the higher energy end of the protein normal modes (40) as well as the lower energy vibrations of the RC redox sites (10). As they begin to become activated over the temperature region of the experiments, these modes should impart some activation energy to the reaction. The amount of temperature dependence should increase as the $-\Delta G°$ becomes much smaller than λ_m. Over the free energy region measured, only small effects are predicted. More precise measurements of the temperature dependence, focusing on the smaller $-\Delta G°$ region, may provide additional indications of the involvement of these vibrations.

So far there is no evidence that electron transfer is coupled to modes where $h\omega_s << k_bT$. For the temperature range studied this should represent vibrations with $h\omega_s < 1$ meV. These have been calculated as part of the spectrum of normal modes of proteins (40). Semi-classical Marcus theory suggests electron transfer is coupled only to these vibrations (i.e. λ_M and λ_L are both zero). In addition, analysis of the $-\Delta G°$ dependence of electron transfer in chemical systems has utilized models where the reaction is strongly coupled to these vibrations (41,42). However, none of these experiments determined the temperature dependence of the rate. The data for each of the reactions presented here, if analysed at only a single temperature, could be modeled by the reaction being coupled to only $h\omega_S$ and $h\omega_L$. It is only the knowledge that there is no classical, Arrhenius activation energy that rules out strong coupling to $h\omega_S$.

Because the electron transfers are not coupled strongly to vibrations where $h\omega_S << k_bT$, the reactions can be essentially temperature independent over a substantial $-\Delta G°$ range. Thus, λ cannot be assigned simply by measurement of a single free energy at which it shows no temperature dependence. Rather, since λ_{Total} can be operationally defined as the $-\Delta G°$ for which the rate is maximal, it must be obtained from the free energy dependence of the rate. The observed temperature independence of the rate comes simply from the reaction being coupled to vibrations that are frozen out or only weakly activated over the temperature range measured.

Future experiments may allow these vibrations to be better characterized. To finally establish that any lower energy modes are coupled to the electron transfer, temperature dependence would need to be seen at smaller $-\Delta G°$s. A more detailed study of the dependence of the rates on temperature in the intermediate $-\Delta G°$ range ($\lambda_S < -\Delta G° < \lambda_M$) could provide additional information about the nature of $h\omega_M$. In addition, I.R. or Raman spectroscopy might allow the high frequency vibrations to be assigned.

Thus, it is seen that by measurements of both the temperature and $-\Delta G°$ dependence of the rate of an electron transfer reaction, considerable information can be obtained about the requirements for long range electron transfer reactions. These measurements on Q_A-replaced RCs have established that several vibrations with different energies are coupled to these reactions. The measurements have demonstrated several simple assumptions that are common in treatments of biological electron transfer to be without foundation. Two in particular deserve mention. It is now clear that a temperature independent reaction does not necessarily imply that $-\Delta G° = \lambda$; the corrollary of this is that if a reaction is demonstrated to be dependent on $-\Delta G°$, it does not follow that the rate will be sensitive to temperature. Secondly, if electron transfer is generally coupled to high frequency vibrations in proteins, the rate is unlikely to be sufficiently sensitive to $-\Delta G°$ above λ such that exothermic rate restriction could be invoked as a source of control (see also ref 26). These experiments provide a much more elaborate picture of electron transfer in protein; but for all the complexity, it is probable that the picture is more realistic then was evident previously; it also offers considerably potential for future investigation.

ACKNOWLEDGEMENTS

This work was supported by a grant from the National Science Foundation (DMB 85-18433) and Department of Energy grant (DOE FG02-86-13476).

Cheshire Cat: Bye-the-bye, what became of the baby? I'd nearly forgotten to ask.

Alice: It turned into pig.

Cheshire Cat: I thought it would.
(The cat slowly vanishes and then after a while reappears)

Cheshire Cat: Did you say "pig", or "fig"?

Alice's Adventures in Wonderland

Caroll

REFERENCES

1. DeVault, D. and Chance, B. Biophys. J. 6; 825-847 (1966).
2. Kihara, T. and McCray, J. A. Biochim. Biophys. Acta. 292 297-309 (1973).
3. Dutton, P. L. and Prince R. C. "The Photosynthetic Bacteria"; Clayton, R. K. and Sistrom W. R. (eds); Plenum, N.Y., London; 525-570 (1978).
4. Dutton, P. L. "Encyclopedia of Plant Physiology" New Series Vol. 19; Staechelein, L. A. and Arntzen, C. J. (eds); Springer Verlag, N.Y., Berlin; 197-237 (1986).
5. "Tunneling in Biological Systems"; Chance, B., DeVault, D., Frauenfelder, H., Marcus, R. A., Schrieffer, J. R., and Sutin, N. (eds); Acad. Press; N.Y., London (1979).
6. "Protein Structure, Molecular and Electronic Reactivity"; Austin, R., Buhks, E., Chance, B., DeVault, D., Dutton, P. L., Frauenfelder, H. and Goldanskii, V. I. (eds); Springer Verlag, N.Y., Berlin (1987).
7. Jortner, J. Biochim. Biophys. Acta. 594; 193-230 (1980).
8. Marcus, R. A. and Sutin, N. Biochim. Biophys. Acta. 811; 265-322 (1985).
9. Hopfield, J. J. Proc. Natl. Acad. Sci. U.S. 71; 3640-3644 (1974).
10. Warshel, A. Proc. Natl. Acad. Sci. U.S. 77; 3105-3109 (1980).
11. Sarai, A. Chem. Phys. Lett. 63; 360-366 (1979).
12. Bixon, M. and Jortner, J. FEBS Letts. 200; 303-308 (1986).
13. Knapp, E. W. and Fischer, S. F. J. Chem. Phys. 87; 3880-3887 (1987).
14. Woodbury, N. W. T., Becker, M., Middendorf, D. and Parson, W. W. Biochemistry 24; 7516-7521 (1985).
15. Martin J.-L., Breton, J., Hoff, A. J., Migus, A. and Antonetti, A. Proc. Natl. Acad. Sci. U.S. 83; 957-961 (1986).
16. Kirmaier, C., Holten, D. and Parson, W. W. Biochim. Biophys. Acta. 810; 33-48 (1985).
17. Clayton, R. K. and Yau, H. F. Biophys. J. 12; 867-881 (1972).
18. Clayton, R. K. Biochim. Biophys. Acta. 504; 255-264 (1978).
19. Hsi, E. S. P. and Bolton, J. R. Biochim. Biophys. Acta. 347; 126-133 (1974).
20. McElroy, J. D., Mauzerall, D. C. and Feher, G. Biochim. Biophys. Acta. 333; 261-277 (1974).
21. Marcus, R. A. J. Chem. Phys. 24; 966-978 (1956).
22. Marcus, R. A. Ann. Rev. Phys. Chem. 15; 155-196 (1964).
23. Ulstrup, J. and Jortner, J. J. Chem. Phys. 63; 4358-4368 (1975).
24. Fischer, S. F. and van Duyne, R. P. Chem. Phys. 26; 9-15 (1977).
25. Jortner, J. J. Chem. Phys. 64; 4860-4867 (1976).
26. Gunner, M. R., Robertson, D. E. and Dutton, P. L. J. Chem. Phys. 90; 3783-3795 (1985).
27. Okamura, M. Y., Isaacson, R. A. and Feher, G. Proc. Natl. Acad. Sci. U.S. 72; 3492-3496 (1975).

28. Gunner, M. R., Tiede, D. M., Prince, R. C. and Dutton, P. L. in "Functions of Quinones in Energy Conserving Systems"; Trumpower, B. L. (ed); Acad. Press; N.Y., London; 271-276 (1982).
29. Woodbury, N. W. T., Parson, W. W., Gunner, M. R., Prince, R. C. and Dutton, P. L. Biochim. Biophys. Acta. 851; 6-22 (1986).
30. Gunner, M. R. and Dutton, P. L. Biochim. Biophys. Acta. Submitted (1988).
31. Liang, Y., Nagus, D. K., Hochstrasser, R. M., Gunner, M. R. and Dutton, P. L. Chem. Phys. Letts. 84; 236-240 (1981).
32. Gunner, M. R., Liang, Y., Nagus, D. K., Hochstrasser, R. M. and Dutton P. L. Biophys. J. 37; 226a (1982).
33. Schenck, C. C., Blankenship, R. E. and Parson, W. W. Biochim. Biophys. Acta. 680; 44-59 (1982).
34. Chidsey, C. E. D., Kirmaier, C., Holten, D. and Boxer, S. G. Biochim. Biophys. Acta.; 424-437 (1984).
35. Cho, H. M., Mancino, L. J. and Blankenship, R. E. Biophys. J. 45; 445-461 (1984).
36. Mauzerall, D. in "Biological Events Probes by Ultrafast Lasers Spectroscopy"; Alfano, R. R. (ed); Acad. Press; N.Y., London; 215-235 (1978).
37. DeVault,D. Q. Revs. Biophys. 13; 387-564 (1980).
38. Hadzi, D., Sheppard, N. J. Am. Chem. Soc. 13; 5460-5465 (1951).
39. Tripathi, G. N. R. J. Chem. Phys. 74; 6044-6049 (1981).
40. Go, N., Noguti and Nishikawa, T. Proc. Natl. Acad. Sci. U.S. 80; 3696-3700 (1983).
41. Miller, J. R., Beitz, J. V. and Huddleston, R. K. J. Amer. Chem. Soc. 106; 5057-5068 (1984).
42. Closs, G. L., Calcaterra, L. T., Green N. J., Penfield, K. W. and Miller, J. R. J. Phys. Chem. 90; 3673-3683 (1986).

THE EFFECT OF AN ELECTRIC FIELD ON THE CHARGE RECOMBINATION RATE OF $D^+Q_A^- \rightarrow DQ_A$ IN REACTION CENTERS FROM *RHODOBACTER SPHAEROIDES* R-26

G. Feher, T. R. Arno, and M. Y. Okamura

University of California, San Diego
La Jolla, California 92093

I. INTRODUCTION

The name of the game in photosynthesis is effective charge separation across the plasma membrane of photosynthetic organisms. The main focus of this conference has been to understand this process quantitatively. In a sense one can view the different contributions from spectroscopy, x-ray structure determinations, dynamics, the effect of mutagenesis etc. as providing the experimental and theoretical frameworks upon which electron transfer (ET) theories are built. To critically test the validity of ET theories, it is important to devise experiments that focus on particular predictions of the theory. This has been done in the present work whose aim was to test the free energy dependence of electron transfer.

The electron transfer rate that we studied in reaction centers (RCs) from *Rb. sphaeroides* is the charge recombination rate k_{AD} of the reaction

$$D^+Q_A^- \xrightarrow{k_{AD}} DQ_A$$

where D^+ is the primary donor (a bacteriochlorophyll dimer) and Q_A is the primary electron acceptor (ubiquinone-10). Although this is admittedly not the most exciting electron transfer step in photosynthesis,[‡] it has the advantage of being easily monitored and that it has been investigated in great detail in the past (1-12). It should therefore be considered in the spirit of a model system against which different ET theories can be tested.

One of the important parameters in all ET theories is the free energy gap, $-\Delta G^o$, between the reactant and product state, (i.e. in our case between $D^+Q_A^-$ and DQ_A). All ET theories, from the original work of Marcus (13) to the more current developments (14-20, for reviews see 21,22) predict a characteristic, parabola-like shaped, curve relating the transfer rate to $-\Delta G^o$. The curve peaks (highest rate) when $-\Delta G^o = \lambda$, the reorganization energy, which is associated with the configurational energy changes of the reactant and product states. The general feature of this dependence has recently been confirmed by several groups (10,11,23-25). The basic approach was to measure electron transfer rates between donor-acceptor pairs having different values of $-\Delta G^o$. The *discrete* points obtained in such a procedure makes it difficult to observe possible "fine structure" effects in the recombination kinetics.

Our approach was to vary $-\Delta G^o$ *continuously* by applying a variable voltage across a planar lipid bilayer into which RCs had been incorporated (26,27). The change in $-\Delta G^o$ is caused by

[‡]There is, however, the intriguing result that the charge recombination rate, k_{BD}, from the secondary quinone, Q_B, is several orders of magnitude smaller than k_{AD} (9). In view of the apparent, approximate, symmetry of Q_A and Q_B with respect to D this finding is difficult to reconcile with a mere change in $-\Delta G^o$. It points to the importance of the intervening amino acid residues and cofactors in the paths between D & Q_A and D & Q_B in determining the electron transfer rate.

the electric field that appears across the charge separated species. Experiments of this nature had been performed by Gopher et al. (26) who, however, reported no changes in k_{AD} with applied voltage. We have repeated these experiments with higher precision and higher voltages and have found a significant change in k_{AD}.. We have observed variations in the results obtained on different membranes that could explain the failure to observe the effect in the previous work (26). Possible origins of the variability will be discussed. In an alternate approach we have investigated the changes in k_{AD} by varying the *internal* electric field. This was accomplished by changing the pH, thereby creating charges on titratable amino acid residues (28). The result of these two approaches will be compared.

Two other sets of experiments are closely related to the present work. Gunner et al (8,10) performed an extensive and systematic study of the effect of replacing the native ubiquinone in RCs from *Rb. sphaeroides* with a variety of quinone having different midpoint potentials. This is an important extension of previous work in which only a few quinones had been used (11,29). In a second set of experiments Popovic et al. (12) incorporated RCs into Langmuir-Blodgett films that were closely packed, stacked, and placed between electrodes to which an external voltage was applied. The advantage of this system over planar lipid bilayers is that larger voltages can be applied and that the kinetics can also be monitored optically. The disadvantage is the lack of accessibility of the RCs to exogenous reactant and the non-exponential decay that requires a deconvolution procedure to arrive at a set of rates. The results of these experiments will be compared with ours.

We shall start with a brief review of ET theories with special emphasis on their applicability to our system (section II) followed by a description of the experimental procedures (section III) and results with their analysis (section IV). In section V we examine the data more critically and compare them with other results as well as with prevailing ET theories.

II. THEORETICAL CONSIDERATIONS

A. The Recombination Kinetics in the Absence of an Applied Field; Direct and Indirect Pathways

The electron transfer reactions that we are concerned with are illustrated in the energy level diagram of Fig. 1. The charge recombination from $D^+IQ_A^-$ to DIQ_A can proceed either directly (with rate constant k_{AD}) or, by thermal activation, via the intermediate state $D^+I^-Q_A$ (8,26,30). Which of these two pathways dominates at a given temperature will depend on the free energy difference, ΔG_1^o, between the states $D^+I^-Q_A$ and $D^+IQ_A^-$. For RCs at room temperature, with $k_{ID} = 8 \times 10^7$ s^{-1} (31) and $k_{AD} \cong 10$ s^{-1}, the rates of the two pathways become comparable for $\Delta G_1^o \cong 0.40$eV (26). When $e^{\Delta G_1^o/(k_BT)} >> 1$, the observed recombination rate becomes (26):

$$k_{obs} = k_{ID}e^{-\Delta G_1^o/(k_BT)} + k_{AD} \qquad (1)$$

where the first term represents the indirect and the second the direct pathway. For RCs containing ubiquinone, $\Delta G_1^o \cong 0.55$ eV (26) and the indirect pathway at 276K contributes only ~ 0.1% to the observed recombination rate. For reaction centers containing antraquinone (AQ) the situation is radically different; $\Delta G_1^o = 0.34$eV (26) and equation 1 predicts that ~ 90% of the recombination proceeds via the indirect pathway.

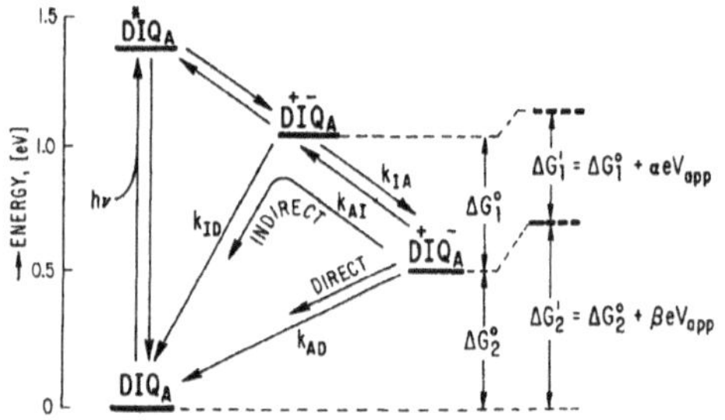

Fig. 1. *Simplified energy level scheme showing electron transfers (arrows) discussed in this work. The state* $D^+IQ_A^-$ *can decay either via the direct pathway (with rate* k_{AD}*) or via the intermediate state* $D^+I^-Q_A$ *depending on the value of the energy difference,* ΔG_1^o*. In RCs containing UQ the direct pathway predominates; with AQ the indirect pathway predominates. The changes in the energy levels in an applied electric field is shown for a positive voltage as defined in the inset of Fig. 5.*

B. The Effect of an Applied Electric Field on the Free Energy Difference ΔG_1^o

When a voltage is applied across the lipid bilayer into which RCs were incorporated, the energy levels of the charge separated excited states are changed due to the interaction of their respective dipole moment with the electric field. This is illustrated on the right side of Fig. 1. Since the state $D^+IQ_A^-$ has a larger dipole moment (i.e. a larger distance between the charge separated species) than $D^+I^-Q_A$, its energy will be affected more than that of D^+I^-Q. The change in energy due to the applied voltage can be written as:

$$\delta G_1^o = eV_{app} \frac{d_{IQ}}{L} = \alpha \, eV_{app} \tag{2}$$

$$\delta G_2^o = eV_{app} \frac{d_{DQ}}{L} = \beta \, eV_{app} \tag{3}$$

where e is the electromic charge, d_{IQ} and d_{DQ} are the projections of the distances between the respective cofactors perpendicular to the plane of the membrane (i.e. along the applied electric field), L is the width of the membrane. This expression assumes that the RC is embedded in the membrane and that its average dielectric constant is uniform. From x-ray analysis d_{IQ} and d_{DQ} (measured between centers of the cofactors) is 11 and 25 Å respectively (32,33). This results in a ratio of $\alpha/\beta = 0.44$. If the dielectric constant is not uniform one needs to consider the dielectrically weighted distances. These have been determined by H.-W. Trissl from a measurement of the electrogenicitis of the various steps in RCs oriented at the heptane/water interface (34) and RCs embedded in the natural membrane (35). His recent value of $\alpha/\beta = .59$ (35).*

C. The pH Dependence of ΔG^o and the Concommittant Change in the Charge Recombination Kinetics

In the previous section we discussed the effect of an *applied* electric field on ΔG^o. We shall now discuss the effect of *internal* fields on k_{AD}. These are created by charged groups of the RC. Since we do not calculate accurate, absolute, values of k_{AD} we need to vary the internal field and correlate it with the resulting change in k_{AD}. A convenient way to vary the internal field is to change the pH thereby varying the charge on the titratable amino acid residues (e.g. Glu, Asp, Agr, Lys, His) of the RC. In the charge separated state, $D^+Q_A^-$, residues close to D^+ will tend to lose protons (i.e. decrease their pK), whereas those near Q_A^- will take up protons (increase their pK). Since a net uptake of protons is observed when reaction centers are illuminated (36,37), the protons interact on the average more strongly with Q_A^- than with D^+. This net proton uptake results in a lowering of the free energy ΔG_2^o (see Fig. 1), i.e. a stabilization of $D^+Q_A^-$. The change in free energy can be obtained directly from the observed proton uptake (36,37) and it is given by (28):

$$-\Delta G^o(pH_1) + \Delta G^o(pH_2) = 2.3k_BT \int_{pH_1}^{pH_2} (H^+/e^-)\, d(pH) \tag{4}$$

where H^+/e^- is the proton uptake per electron (i.e. per photon absorbed).

D. The Effect of ΔG^o on the Charge Recombination Kinetics

1) *The indirect pathway*: If the indirect pathway predominates and $e^{\Delta G_1^o(k_BT)} >> 1$ one obtains for the observed decay rate, k_{obs} the expression:

$$k_{obs} = k_{ID} e^{-\Delta G_1^o/(k_BT)} e^{-\delta G_1^o/(k_BT)} = k_{obs}^o \, e^{-\alpha eV_{app}/(k_BT)} \tag{5}$$

where k_{obs}^o is the recombination rate in the absence of an applied electric field. This relationship has been experimentally verified for RCs containing AQ (26). The value α is an experimentally determined parameter; it represents the fraction of the applied voltage that occurs across the projection, d_{IQ}, along the normal of the membrane (see Eq. 2).

2) *The direct pathway*: As discussed in Section II.A, this is the predominant pathway for the charge recombination $D^+Q_A^- \rightarrow DQ_A$ in RCs containing ubiquinone (see Fig. 1). In this section

*Trissl's original value of 0.33 (34) is believed to be due to a strong tilt of the D—Q_A axis with respect to the normal of the interface in the artificial system (35).

we briefly review the main features of electron transfer (ET) as applied to the charge recombination kinetics, k_{AD}. We start out with a qualitative, pictorial representation of the electron transfer process to be followed with quantitative results derived by Marcus, Hopfield, Jortner, and others (14,16,17, for reviews see 21,22).

The reactant state, R, ($D^+Q_A^-$) and product state, P, (DQ_A) can be represented by a pair of harmonic oscillators whose potential energy surfaces (parabolas) are displaced along a nuclear coordinate as shown in Fig. 2. It should, of course, be kept in mind that we are presenting only a one dimensional profile, although we know that in reality the large number of coordinates produce multi-dimensional potential energy surfaces. For simplicity we assume a single vibronic mode with energy $\hbar\omega = k_BT_o$ for both the reactant (R) and product (P) state. The vibronic quantum numbers of the two states are denoted by n_R and n_P respectively; the difference in energy between the bottoms of the two parabolas, ΔG^o, is the energy difference between the redox couples D^+/D and Q_A/Q_A^- (as shown, the sign of ΔG^o is negative).

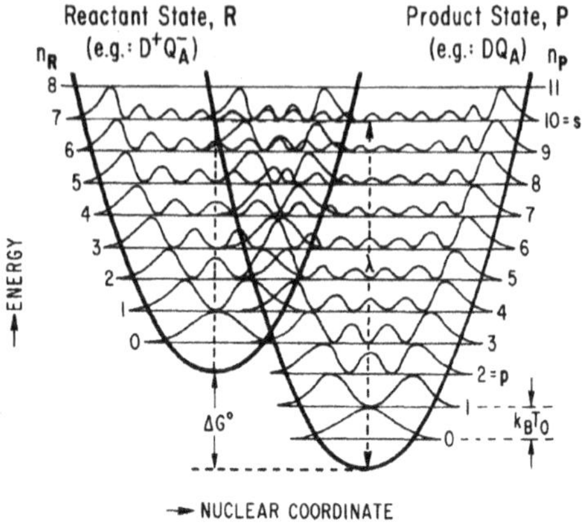

Fig. 2. *Potential energy vs. nuclear coordinate of the reactant states, R and product state, P. A single mode with the same characteristic energy* ($k_B T_o$) *for both parabolas has been assumed. The vibronic (nuclear) wave functions are sketched; the vibrational quantum numbers are indicated adjacent to the parabolas.* ΔG^o *is the energy difference between the redox couples* D^+/D *and* Q_A/Q_A^-; λ *is the reorganization energy. The electron transfer rate depends on the overlap of the vibronic wave function and is given by Eqs. 6 and 7. Note that the values of* ΔG^o, λ *and* T_o *are not drawn to scale for the charge recombination* $D^+Q^- \rightarrow DQ$ *in RCs. For this case the values, to be discussed later in the text, are:* $-\Delta G^o = 0.52 eV$; $\lambda = 0.64 eV$ *and* $T_o = 200°K$ *(i.e. p=30; s=37).*

How do transitions from R to P occur? Two conditions have to be satisfied simultaneously during the transfer: i) the total energy of the system must be the same immediately before and after the transfer and ii) the nuclear coordinates do not change since nuclear rearrangements are slow compared to electron transfer. This is the Frank-Condon principle. These conditions are satisfied for electron transfer between isoenergetic states* at positions where the vibronic wave functions (sketched in Fig. 2) overlap. Note that most transitions shown in Fig. 2 are classically forbidden. This process has, therefore been called vibronically coupled tunneling. The strength of the electron-nuclear coupling is characterized by the vibronic coupling parameter, λ, which is also called the reorganization energy (see dashed vertical line in Fig. 2).

From the above qualitative discussion one can predict the main features of electron transfer:

i) As shown in Fig. 2 large overlaps of the nuclear wave functions occur near the crossing of the two parabolas. In classical ET theory the crossing point is called the transition state and is located $(\Delta G + \lambda)^2/(4\lambda)$ above the bottom of the R-parabola. Thus, one can think of the transfer process as being thermally activated with an activation energy $(\Delta G + \lambda)^2/(4\lambda)$.

ii) At low temperatures ($T < T_o$) only the ground state R ($n_R = 0$) will be populated and the transfer rate to P will be temperature independent.

iii) When the intersection occurs at the bottom of the R parabola, i.e. for $|\Delta G| = \lambda$ the activation energy goes to zero and one expects approximately a temperature independent rate.

*Note that we have shown the energy levels to coincide for the R and P states (i.e. $nk_B T_o = \Delta G^o$). If the energy levels have a width, some mismatch between the energy levels can be tolerated. Since we have many nuclear coordinates, there are bound to be some vibrational excitation (involving many modes) that satisfy energy conservation.

Actually, since the main contribution to the transfer rate comes from the overlap with the ground state of R ($n_R = 0$) *an increase* in temperature depletes the ground state resulting in a *decrease* in the rate.

iv) The transfer rate is maximum for $|\Delta G| = \lambda$ (maximum overlap as discussed in iii)) The rates decrease for $|\Delta G| < \lambda$ and $|\Delta G| > \lambda$.

We now proceed to describe more formally the results of ET theory. The electron transfer can be described as a product of an electronic and vibronic part as follows:

$$k = \frac{2\pi}{\hbar}\left(|T_{RP}|^2\right)\left(\sum_{n_R n_P} (FC)_n B(n_R)\right) \tag{6}$$

where the first parenthesis represents the electronic matrix element $|T_{RP}|^2$ and the second parenthesis is the sum of Frank-Condon factors. T_{RP} is a measure of the overlap of the tails of the electronic wave functions of the R and P states. The overlap is assumed to depend exponentially on the distance between the reactants, i.e, $T_{RP}(r) = T_{RP}(0)e^{-r/r_o}$, where the characteristic lengths, $r_o \cong 1$ Å. The second parenthesis is the sum of overlap integrals of the *vibronic* wave functions of the R state (designated by quantum numbers n_R) with those of the P state (with quantum numbers n_P). They are weighted by the appropriate Boltzman factors, B, i.e. the fraction of oscillators in state n_R at temperature T. The electronic part, T_{RP}, is not expected to be affected to first order by T, ΔG, or λ. The dependence of k on these parameters is, therefore, attributed to changes in the FC factors.

Making the assumption that the donor and acceptor are coupled to the same single vibrational mode, Jortner has derived for any temperature, T, the following expression for the charge transfer rate (16,17):

$$k_{AD}(T) = \frac{(2\pi)}{\hbar}|T_{RP}|^2 \frac{1}{k_B T_o}\left[\frac{\nu+1}{\nu}\right]^{p/2} e^{-s(2\nu+1)} I_p\left(2s\sqrt{\nu(\nu+1)}\right) \tag{7a}$$

$$\text{where}: \quad s = \frac{\lambda}{k_B T_o}\ ; \quad p = \frac{-\Delta G^o}{k_B T_o}\ ; \quad \nu = \frac{1}{e^{T_o/T} - 1} \tag{7b}$$

and I_P (x) refers to the modified Bessel function of order p, k_B is Boltzman's constant and $\hbar$ Planck's constant. Note that energy conservation requires p to be an integer corresponding to a specific quantum numbers, n_P, as indicated in Fig. 2 (see footnote on previous page).

Equation 7 simplifies in the limit of either high or low temperatures. At high temperatures the rate is given by (14,16,17)

$$\text{For } T >> T_o; \qquad k_{AD} = \frac{2\pi}{\hbar}|T_{RP}|^2 \frac{1}{\sqrt{4\pi\lambda k_B T}}\, e^{\frac{-(\Delta G^0 + \lambda)^2}{4\lambda k_B T}} \tag{8}$$

This approximation is closely related to the transition state theory of transfer reactions, which has been known to chemists for several decades (13,38). It represents an activated process that has been described in a previous section (i).

In the low temperature limit the reactant state is in the lowest vibrational mode ($n_R = 0$) and thus the recombination rate is independent of temperature. In this limit the rate is given by a Poisson distribution of the rearrangement energy, i.e. (16,17):

$$\text{For } T << T_o; \qquad k_{AD} = \frac{2\pi}{\hbar}|T_{RP}|^2 \frac{1}{k_B T_o}\, s^p \frac{e^{-s}}{p!} \tag{9}$$

This expression has a maximum for $s = p$ (i.e. $-\Delta G^o = \lambda$) and decreases for $p > s$ and $p < s$, as discussed in section (iv).

For the special, activationless case, for which $-\Delta G^o = \lambda$ and $T < T_o$, k_{AD} is temperature independent; for $T > T_o$, the exponent in Eq. 8 vanishes and k_{AD} is proportional to $T^{-1/2}$. From Eq. 7 one obtains for the temperature dependence of k_{AD}:

$$-\Delta G^o = \lambda\ ;\ T >> T_o\ ;\ p! >> 1: \quad k_{AD} = k_{AD}(0)\left(\frac{T_o}{2T}\right)^{1/2} \tag{10}$$

where $k_{AD}(0)$ is the low temperature limit of the recombination rate.

III. EXPERIMENTAL PROCEDURES

Planar bilayers were formed from monolayers as previously described (26,39-41). The basic steps in the preparation of monolayers are: i) preparation of liposomes into which RCs are incorporated, ii) the transformation of liposomes into monolayer at the air-water interface, and iii) formation of bilayer by apposing two monolayer across an aperture in a teflon septum.

RCs containing 0.7 UQ/RC were prepared as described (29). The RCs were concentrated in an Amicon filtration cell to $A_{865}^{1cm} = 200$ and diluted in a buffer solution of 10 mM KCl, 10 mM TRIS pH8 and 20 mg/ml partially purified soybean azolectin (42) to a final concentration of $A_{865}^{1cm} = 1 - 2$. Liposomes were formed by sonicating the solution for ~ 10 minutes (until optically clear). The liposomes were diluted 1:33 into 10 mM KCl, 10 mM TRIS pH8, 10 mM $CaCl_2$ and added to the two compartments separated by a 12.5 micron teflon septum containing a 0.16mm diameter hole. Prior to membrane formation, approx 0.2 μl off 1% tetradecane in hexane was applied to the rim of the hole with a 1 μl syringe. After evaporation of the hexane, bilayers are formed by sequentially raising the level of solution above the hole in each compartment (see Fig. 3).

After the membrane was formed, reduced cytochrome (cyt c) was added to one compartment to a concentration of ~ 15 mM to functionally orient the RCs (41). Terbutryne (~ 20 μM) was routinely added to retard the leakage of electrons from the fraction of the RCs that were inactivated by the cytochrome.

For bilayers with RCs containing both UQ and AQ as electron acceptors the following procedure was adopted: RCs containing 0.2 UQ/RCs were prepared and bilayers formed as described above. After taking a set of kinetics data, exogenous AQ was added to the chamber to a final quinone content of 0.5 AQ/RC and 0.2 UQ/RC as judged from the value of the integrated current.

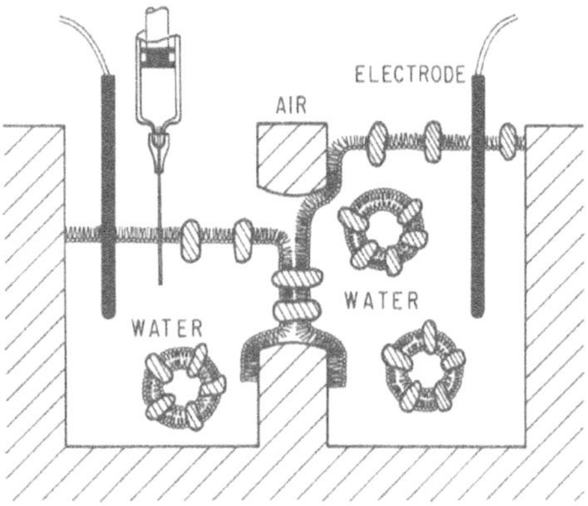

Fig. 3. *Idealized, schematic representation of the formation of a planar lipid bilayer by the method of Montal and Müller (39). A teflon partition with a small aperture (~ 0.16 mm) separates two aqueous compartments. Liposomes containing RCs are introduced into the solution; they break at the surface forming a lipid monolayer containing RCs (40). The water level is raised by injecting buffer with a syringe; electric fields were applied across the bilayer by connecting the electrodes to a voltage source: Note that drawing is not to scale (e.g. thickness of teflon septum ≅ 12μm, diameter of aperture in septum ≅ 0.16 mm, width of the bilayer ≅ 40 Å, typical liposome diameter ≅ 500 Å).*

The light source was a 50 milliwatt helium neon laser (λ = 6328 Å) (Model 125 Spectra Physics Inc.). The beam was focused to the size of the aperture in the teflon to avoid generation of photocurrents from RCs adsorbed to the teflon septum. The beam passed through a shutter with a 1 ms opening time; the intensity was reduced with neutral density filters to a value just sufficient to bleach the RCs.

Two calomel electrodes (Ingold Corp.) shielded from light, were introduced into the two compartments and connected to a variable voltage source. The current was measured with an operational amplifier (Burr-Brown OP 128) with a feedback resistance of 10^{10} ohms. The signal was amplified to 1 volt/picoamp and passed through an RC filter with a variable time constant. Data obtained from UQ and AQ containing RCs were filtered with time constant of 10 ms and 1 ms respectively. Signal amplitudes were typically 3 picoamps. To improve the signal-to-noise ratio several traces were averaged with a Nicolet model 1180 data system. The temperature of the cell containing the lipid bilayer was held at 3 ± 1°C with a circulating water bath.

The pH dependence of k_{AD} was measured on a sample with 0.8 UQ/RC (3 μM) in the presence of 0.1% LDAO to which the following buffers in 10 mM concentration each were added: MES (2-N-morpholino-ethane sulfonic acid; pK=6.2); MOPS (3-N-morpholino-propane sulfonic acid; pK=7.2); TRIS (tris-hydroxy-methyl-amino methane; pK=8.3); CHES (3-cyclohexyl amino-propane sulfonic acid; pK=9.5); CAPS (cyclohexyl amino-ethane sulfonic acid; pK=10.4). The experiments were started at the lowest pH (6) value; successive measurements at higher pH were made by adding aliquots of KOH. To test reversibility the pH was varied in the opposite direction by adding KCl. The recombination kinetics were measured optically as described (2).

The temperature dependence of k_{AD} was measured on RCs embedded in polyvinyl alcohol (PVA). RCs were mixed with a 10% (w/v) solution of PVA, poured into a teflon mold and dried under a flow of nitrogen. The optical absorbance of the film was $A_{800} \cong 1$. The thoroughly dehydrated sample was exposed to a stream of dry nitrogen for 40 days.

IV. EXPERIMENTAL RESULTS AND DATA ANALYSIS

A. The Basic Photo Response in Functionally Oriented RCs

When RCs were incorporated into a planar lipid bilayer no photocurrents were observed in the absence of exogenous secondary reactants. Apparently RCs with opposite orientation are inserted into the membrane with equal probability resulting in a zero net photocurrent. To observe a photocurrent, the symmetry of the system has to be broken by introducing secondary reactants into one compartment. This modifies preferentially RCs pointing in one direction thereby obtaining a net functional orientation. We have used cyt c^{2+} to reduce D^{+} in those RCs whose donors are exposed to the reductant (26,41). Packham et al have used Fe $(CN)_6^{3-}$ to selectively oxidize D (43).

Figure 4 shows the photocurrent generated by RCs in the presence of cyt c^{2+}. The light pulse sequence is indicated at the bottom and the charge states of the 2 RC populations is schematically represented at the top of the figure. Following the charge separation, D^+ is reduced by cyt c^{2+} in one of the RC populations; the observed photocurrent shows that the reduction of D^+ by cyt c^2 is electrogenic as has been previously observed (26,43,44). The RCs that were reduced by cyt c^{2+} are in the state DQ_A^- and are effectively inactivated. This situation persists as long as the electron is on Q_A^-, i.e., ~ 10 minutes in the absence, and ~ 20 minutes in the presence of terbutryne. Thus, during this time interval we have a functionally oriented sample with only those RCs that did not interact with cyt c^{2+} being active.

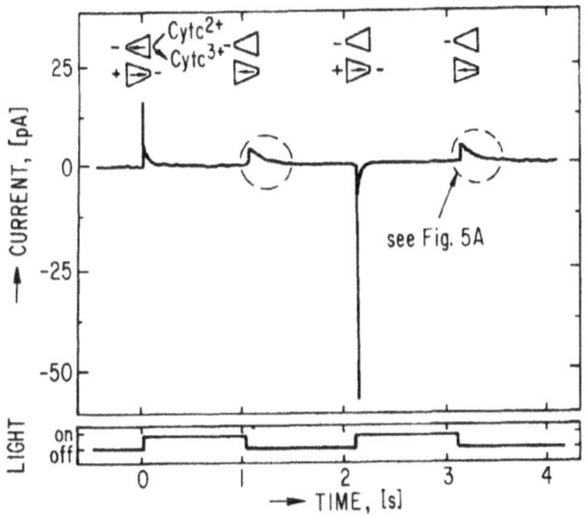

Fig. 4. *Photocurrents obtained from RCs (0.7 UQ/RC) incorporated in a planar lipid bilayer. Cyt c^{2+} was added to one side of the bilayer. The charge separation events following the periodic light pulses are shown schematically at the top of the trace. The recombination kinetics were obtained from the decay rate of the current after the light was turned off (see circled region).*

The currents produced at the onset of the second and all subsequent flashes are due to the charge separation process $DQ_A \rightarrow D^+Q_A^-$ occurring in those RCs whose donors were not reduced by cytochrome c^{2+}. The current observed when the light is turned off is due to the charge recombination process $D^+Q_A^- \xrightarrow{k_{AD}} DQ_A$ (see circled region in Fig. 4).

B. The Effect of an Electric Field on the Charge Recombination Kinetics

The charge recombination kinetics in the presence of an applied external voltage is shown in Fig. 5; the inset is a schematic representation of the RC-lipid bilayer system showing only the functionally oriented RC. The compartment containing cyt c^{2+} (left) is defined as being at zero potential. Consequently, a positive voltage raises the energy of the charge separated state (see Fig. 1). The effect of the voltage on k_{AD} is clearly discernable; the time integral of the current is proportional to the total charge transferred and is, therefore, the same for positive and negative potentials. In Fig. 5B the data are replotted semilogarithmically. The good straight lines that are obtained are indicative of a single exponential recombination kinetics. From the slopes of the line values of k_{AD} were obtained.

The dependence of the recombination rate on applied voltage is shown for three different bilayers in Fig. 6. All of them show qualitatively the same behavior, i.e. an increase in recombination rate with increasing voltage (i.e. with increasing energy gap between excited and ground state) and a significant curvature as predicted by theory. However, there are serious quantitative differences between the three curves. We interpret this to mean that the environment of the reaction centers varies from membrane to membrane; consequently the relation between applied voltage and internal field (i.e. α and β in Eqs. 2 and 3) seen by the charge separated species varies between membranes. Occasionally we obtained membranes in which no change in k_{AD} with voltage was observed. This probably explains the failure to observe the effect of an electric field on k_{AD} in previous work (26). It should be noted, however, that in Fig. 6 one can adjust (i.e. expand or contract) the scale of the abscissa for each membrane to have all three curves coincide.

We next compare the experimentally observed field dependence of k_{AD} with the theoretical prediction of section II.D. The experimental results on two membranes are shown by circles and crosses in Fig. 7. The bottom scale of the abscissa represents the voltage applied across the bilayer. The solid line

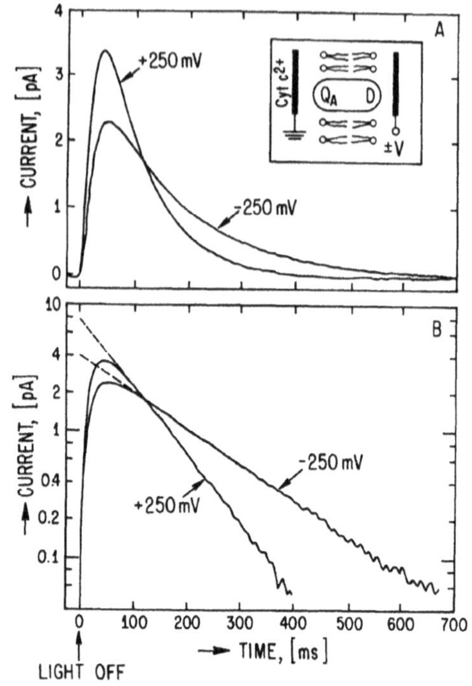

Fig. 5. *Effect of an applied voltage on the charge recombination kinetics in RCs with 0.7 UQ/RC. Inset in (A) shows the polarity of the voltage with respect to a functionally oriented RC. In (B) the data are plotted semilogarithmically; from the slope of the lines, values of k_{AD} were obtained.*

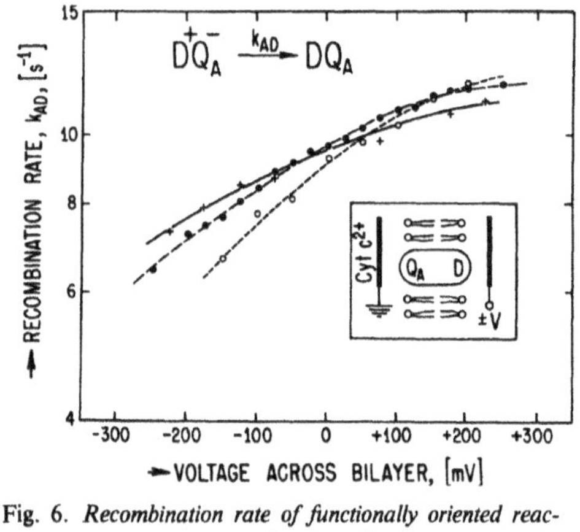

Fig. 6. *Recombination rate of functionally oriented reaction centers as a function of voltage across the bilayer for three different membranes. We interpret the differences in the results obtained on the three membranes as arising from differences in the mode of incorporation or the environment of the RCs in the bilayer. By appropriately scaling the abscissas for each membrane one can make the three curves coincide. Inset shows functionally oriented RC.*

represents the theoretical fit to the data for the high temperature case (Eq. 8) with ΔG^o (the upper scale of the abscissa) and λ as adjustable parameters. The value of ΔG^o in the absence of an electric field was taken as 0.52 eV, the redox potential difference between Q_A/Q_A^- and D/D^+ (for a detailed discussion see ref. 10). The peak of the curve occurs at $-\Delta G^o = \lambda = 0.64$ eV.

What is the significance of the difference between the upper and lower scale of the abscissa? If $D^+ Q_A^-$ were to span the entire membrane one would expect the two scales to be the same. Thus, the ratio of the upper to the lower scale represents the fraction of the applied voltage that occurs across $D^+Q_A^-$, i.e. it is the value of β in Eq. 3. For our case $\beta = 0.3$, an unexpectedly low value. From the x-ray structure the projection of the distance between D^+ and Q_A^- on the perpendicular to the plane is 25 Å (32,33). Assuming a bilayer thickness of 35 Å one would expect, under the simplest assumptions, $\beta = 25/35 = 0.7$. Possible explanations for the discrepancy of this value and the observed one are given in section V. In summary, the following parameters in Eq. 8 were derived from the fit:

$$\begin{aligned} \lambda &= 0.64\text{eV} \\ \Delta G^o(V_{app} = 0) &= -0.52\text{eV} \\ \beta &= 0.30 \\ T_{RP} &= 2.4 \times 10^{-8}\text{ev} \end{aligned} \tag{12}$$

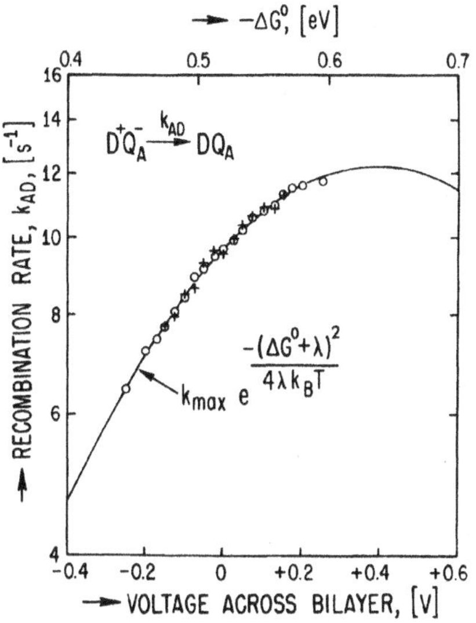

Fig. 7. *Recombination rate, k_{AD}, as a function of voltage across the bilayer. The experimental results on two membranes are shown by circles and crosses. The solid line represents a theoretical fit to Eq. 8 with ΔG^o and λ as adjustable parameters. The difference in the scales of the upper and lower abscissa is due to β (Eq. 3), the fraction of the applied voltage that occurs across D^+Q^-.*

To test the validity of the high temperature approximation we fitted the experimental results with the expression valid for any temperature (Eq. 7). We found a fit as good as the one shown in Fig. 7 (data not shown) with $T_o = 200$K. The value of T_o was arrived at from independent consideration (11,45). The parameters differ only slightly from those obtained in the high temperature approximation (i.e. $\lambda = 0.65$ eV, $\beta = 0.31$). The values for p and s obtained from Eqs. 7b and 12 were 30 and 37, respectively. Similarly, the data could also be fitted well with the low temperature approximation (Eq. 9) by choosing the following parameters: $T_o = 1000$K, $\lambda = 0.7$ eV, $\beta = 0.48$. Although the high value of T_o does not seem to be realistic, even in this limiting case the value of λ changed only by ~ 10%.

C. An Attempt to Calibrate the Internal Field with Antraquinone Containing RCs

To obtain an independent value of β, we made use of the fact that RCs containing AQ decay via the indirect, thermally activated, pathway (see section II.D). From the field dependent decay rate one obtains α (see Eq. 5) which is related to β as discussed in section II.B. In view of our findings that different membranes give different results (see Fig. 6) it is important to measure k_{AD} of AQ and UQ containing RCs in the *same* membrane. Bilayers containing both populations of RCs was prepared as described in section III. The charge recombination times of D^+AQ^- and D^+UQ^- are sufficiently different to make it possible to kinetically resolve their respective k_{AD}s.

The experimental results obtained on the bilayer having a mixed population of RCs are shown in Fig. 8. The lower curve is the same as analyzed in Fig. 7. The data from the upper curve obtained from RCs containing AQ were fitted with Eq. 5 with a value of $\alpha = 0.065$. From the ratio $\alpha/\beta = 0.44$ as obtained from the x-ray structure (see section II.B), a value of $\beta = 0.15$ is derived. This differs by a factor of 2 from the value (0.3) determined from Fig. 7. Using Trissl's ratio of $\alpha/\beta = 0.6$ (35), a value of $\beta = 0.11$ is obtained. Possible origin of these disagreements will be discussed in section V.

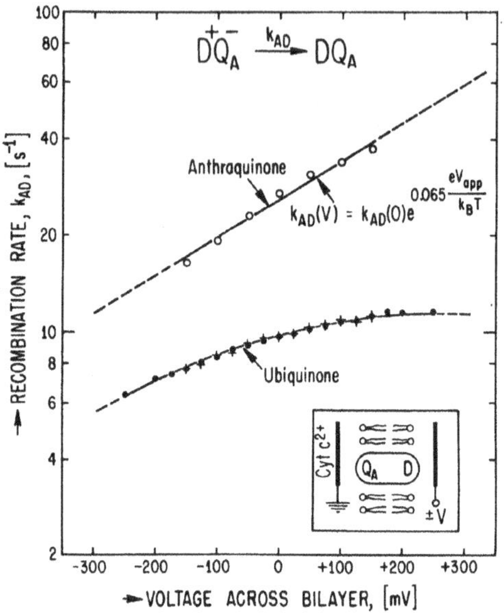

Fig. 8. *Dependence of the charge recombination rate, k_{AD}, on applied voltage for RCs containing AQ and UQ, respectively. Both types of RCs were incorporated in the same membrane; their respective k_{AD}s were kinetically resolved. The AQ data were obtained on the membrane for which the UQ data are shown by crosses. From the AQ data a value of $\alpha = 0.065$ was obtained (Eqs. 2,3,5). From the relation of α and β (Eq. 2, 3) the fraction of the applied voltage that appears across $D^+Q_A^-$ was determined. Inset shows the polarity of the voltage with respect to a functionally oriented RC.*

D. The pH-Dependence of k_{AD}

As discussed in section II.C changes in pH affect the internal electric field at the site of $D^+Q_A^-$. These fields will in turn affect k_{AD}. The kinetics measurements were performed on RCs in a buffered solution whose pH was varied by adding aliquots of KOH. To test whether the RCs were not denatured during the experiment, we cycled the pH between the lowest (pH6) and highest (pH11) values. Complete reversibility of k_{AD} was observed.

The experimental results are shown in Fig. 9A. The value of k_{AD} changed from $8s^{-1}$ to $\sim 12s^{-1}$ in the pH range from 6 to 11. To correlate the change in k_{AD} with the free energy, $-\Delta G^o$, we could, in principle, calculate the electrostatic interaction of $D^+Q_A^-$ with all the titratable residues. This is a formidable task. An alternate procedure is to measure the proton uptake (36,37) and use Eq. 11 to determine ΔG^o. This has been done with the results shown in Fig. 11b. We see that the shapes of the curves in 9A and 9B are very similar. Indeed, if k_{AD} were linearly proportioned to $-\Delta G^o$, the two curves should be approximately superimposable. However, the deviation from linearity as seen in Fig. 7 will tend to make the k_{AD} values saturate (i.e. curve more strongly) at high pH as is observed in Fig. 9A.

We next inquire about the quantitative agreement between the dependence of k_{AD} with applied voltage (Fig. 7) and the pH dependence of k_{AD} of RCs in solution. From Fig. 9 we see the change of k_{AD} from pH6 to pH11 (Fig. 9A) is caused by a free energy change of $\sim$ 50 meV (Fig. 9B). From Fig. 7, on the other hand, we find that the same change in k_{AD} is caused by a free energy change of $\sim$ 100 meV (see upper scale). What are the possible reasons for the discrepancy of a factor of two? In the analysis of the pH dependence we assumed that the sole contribution to the changes of k_{AD} is due to the internal fields of the titratable residues. However, other possible effects associated with variations in pH, e.g. conformational changes may contribute to the observed changes in k_{AD}. Another possibility is that the upper scale in Fig. 7 is not correct; this will be discussed in the next section.

V. DISCUSSION

A. Summary of Results

We have investigated the dependence of the charge recombination rate, k_{AD}, in RCs from Rb.sphaeroides R-26 in the presence of an applied electric field as well as on the internal field created by charges on titratable amino acid residues. The applied field was created by a voltage applied across a planar lipid bilayer into which RCs were incorporated. The results were fitted with the theories of vibronically coupled electron tunneling of Hopfield (14) and Jortner (16,17). The main free parameters with which the experimental data were fitted with theory were the reorganization energy, λ, and the free energy scale, ΔG^o. The fit was relatively insensitive to the characteristic temperature, T_o. The high temperature approximation which is equivalent to the original Marcus theory (13) gave as good a fit with experiments and with approximately the same parameters for λ and ΔG^o as the more elaborate quantum theory of Jortner et al (with $T_o = 200K$). The results show that recombination $D^+Q_A^- \rightarrow DQ_A$ takes place close to the peak of the Marcus curve with $-\Delta G^o$ $(E=0) = 0.52eV$ and $\lambda = 0.64eV$, i.e. there is a small activation energy. The scale of ΔG^o was correlated with applied voltage via the adjustable parameter β that reflects the fraction of the applied voltage that appears across the charge separated state $D^+Q_A^-$. It had a relatively low value of 0.3. An independent determination of α in an experiment with RCs containing AQ gave a value of $\sim$ 0.11 - 0.15, i.e. approximately a factor of two smaller than obtained from a fit of the data with theory.

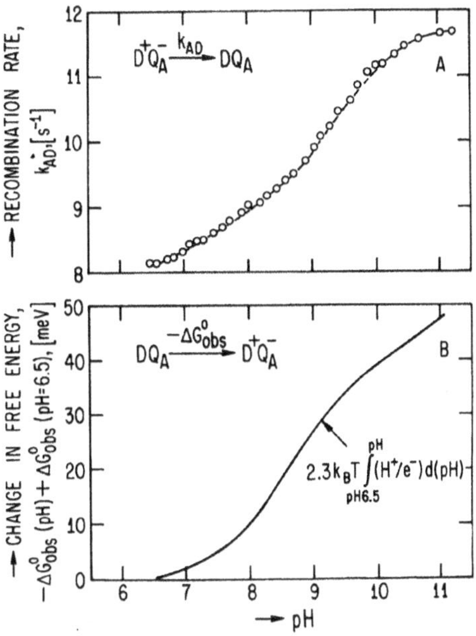

Fig. 9. *Variation of the recombination rate, k_{AD}, (A) and free energy change, ΔG^o, (B) as a function of pH for RCs in solution. Figure B was obtained from the observed proton uptake (37) and Eq. 10. If k_{AD} were linearly proportional to the electric field, (a rough approximation for the interval investigated), (see Fig. 7) the curves in A and B should be approximately superimposable.*

The pH dependence of k_{AD} which arises from the internal fields due to the titratable residues agrees qualitatively with the data obtained in the presence of an applied electric field. However, the scale of ΔG^o for the two cases differed by approximately a factor of two. The two results could be made to coincide for a value of $\beta \cong 0.15$, the same value as obtained from the AQ experiment.

B. Possible Origins of the Quantitative Discrepancies

All quantitative discrepancies between the different experiments discussed in this work can be described by variations in the value of β (see Table 1). This parameter represents the fraction of applied voltage that appears across $D^+Q_A^-$. There are several effects and approximations that may be responsible for the observed discrepancies.

1) *Local field corrections*: The local field seen by $D^+Q_A^-$ involves the effective dielectric constant in the interior of the RC as well as the dielectric constant of the lipid bilayer. To evaluate this effect accurately is a difficult, and recurring problem; it has been discussed recently in connection with the Stark effect (46). If we approximate the RC by a sphere having a dielectric constant ϵ_{RC} embedded in a lipid[‡] with a dielectric constant ϵ_L, the *macroscopic* field inside the RC is given by

$$E_{inside} = E_{app} \frac{3\epsilon_L}{2\epsilon_L + \epsilon_{RC}} \quad (13)$$

Thus, for $\epsilon_{RC} >> \epsilon_L$, a large reduction of the applied field is obtained.

2) *Non-monotonic potential distribution across the bilayer*: This effect may arise from several causes discussed by Kuznetsov and Ulstrup (48). They include: Charge separation at the membrane-electrolyte interface resulting for instance from double layer effects; charge effects arising from adsorption of ions at interface; spatial dispersion of the dielectric permitivity and image charges of ions outside the membrane. In addition, electrostriction can affect the dimensions of the bilayer-RC system.

Table 1. Values of β obtained from different experiments

Experiment	$\beta = \dfrac{V_{across} D^+Q_A^-}{V_{applied}}$
Appl. E-field & theo. fit (Fig. 7)*	0.30
pH-dependence (Figs. 7 and 9)	0.15
Indirect, temp.-activated decay, of AQ containing RCs (Fig. 8)	0.11-0.15
Naive prediction from the three-dimensional structure	0.70

*For the two membranes analyzed in Fig. 7. For the three membranes shown in Fig. 6 the values of β were: 0.20, 0.30 and 0.35, respectively.

3) *Mode of incorporation into the membrane and the local environment of the RC*: We have assumed that the RC is incorporated into the membrane with its twofold symmetry axis perpendicular to the membrane surface and with its hydrophobic part immersed in the lipid. Although this is, thermodynamically, the lowest energy state (47), this equilibrium configuration may not be reached within the time of the experiment and other configurations may exist. It is also known that lipid bilayers deviate from their idealized structure, e.g. they can incorporate solvent molecules that can form "lenses" inside the membrane (49). These may distort the local field.

4) *Changes in the energy levels of the RCs when they are incorporated in the bilayer*: We have assumed throughout this work that the energy gaps ΔG_1^o and ΔG_2^o (see Fig. 1) in the absence of an electric field are the same for RC in the membrane as there are for RCs in solution. Although this seems like a justified assumption in view of the approximate equivalence of k_{AD} in the membrane and in solution, a more critical test would be to determine ΔG_1^o directly from the temperature dependence of k_{AD}, in AQ containing RCs (26).

5) *Approximations of the ET theory*: In fitting our experimental results (see Fig. 7) we have used either the high temperature (Eq. 8) or the single mode approximation of the theory (Eq. 7). This is likely to represent an oversimplification of the real situation.

How do the above considerations affect β and which of the values listed in Table 1 should be considered most reliable? The first thing to note is that the experimental values of β differ from membrane to membrane. The three membranes shown in Fig. 6 could be fitted with values of $\beta = 0.20$, 0.30 and 0.35, respectively. The presence of these variations implicate the mode of incorporation of the RCs into the membrane or the local environment of the RC (see point 3 above). The cause of this variability may be for instance the pre-treatment of the teflon septum with tetradecane which is not a very reproducible procedure (see point 3). In the subsequent discussion, we shall focus on the membranes analyzed in Fig. 7 with values of β=0.30 which is twice the value obtained from the pH dependence and the results on AQ containing RCs.

One possibility for the cause of the discrepancy is the oversimplification of the ET theory (see point 2 above). To assess this possibility one would need to use a more sophisticated theory

‡This is clearly an oversimplification since part of the RC protrudes from the lipid bilayer (47).

that includes several modes (18-21) and check the fit with experiments performed at different temperatures; this is a difficult task with our present experimental setup. Gunner et al. were unable to fit their quinone substitution experiments with a two mode model (10).

The value of β predicted from the three-dimensional structure (see table 1) is most likely overestimated because of the neglect of points 1 and 3 discussed above. The results from the pH dependence ought to be reliable since it is based on a thermodynamic argument that does not depend on the objections raised in part 1-5. It should be noted, however, that the pH dependence does not give explicitly a value of β but has to be used in conjunction with the results of Fig. 7. The results obtained from the AQ experiments are sensitive to changes in ΔG_1^o when RCs are incorporated in the membrane (see point 4). The value of 0.15 was obtained by assuming a uniform dielectric constant inside the RC; the value of 0.11 takes into account the dielectrically weighted distances determined by Trissl (35). In summary, the value of β is at present known only to within a factor of ~ 2.

C. The Predicted Temperature Dependence of k_{AD}

With the value of λ determined in this work one can *calculate* the expected temperature dependence of k_{AD} from Eq. 7. The results for our case ($\Delta G^o = -0.52$eV, $T_o = 200$K, and $\lambda = 0.64$, i.e. p=30, s=37) is shown in Fig. 10 by the full line. It can be seen that although we are removed from the peak of the Marcus curve by 0.12 eV, the rate still decreases, (albeit mildly) with increasing temperature. The other, hypothetical cases, for $-\Delta G^o = \lambda$ is shown by a dashed line.

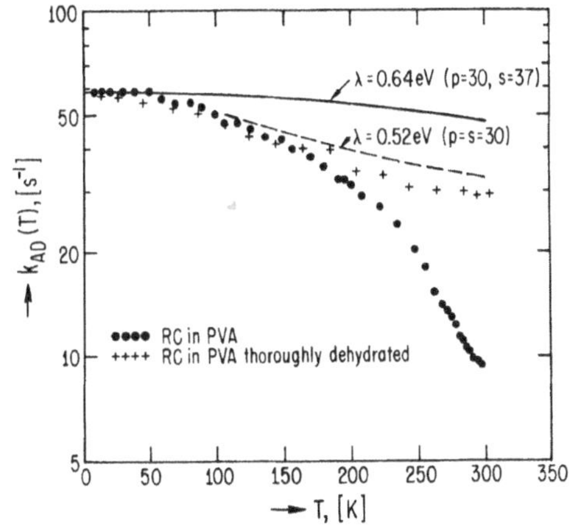

Fig. 10. *Comparison of the theoretically predicted and experimentally observed temperature dependence of the recombination kinetics k_{AD}. The theoretical curve (solid lines) were calculated from Eq. 7 for two values of λ with $-\Delta G^o = 0.52eV$ and $T_o = 200K$. The full line represents the result for $\lambda = 0.64$ as determined in this work. The dashed line represents the hypothetical, activationless case: $-\Delta G^o = \lambda = 0.64eV$. The experimental results were obtained on RCs in their native state (11) and for dehydrated RCs (50). The discrepancy between the experimental and theoretical results have been ascribed to thermal expansion of the protein (11).*

The *experimental* temperature dependences of k_{AD} are shown for two samples. One for RCs in, what is believed to be, their approximate native state (11) and the other in thoroughly dehydrated RCs (50). Both samples show a stronger temperature dependence than predicted even for the activationless (p=s) case. This discrepancy has been explained by postulating a thermal expansion of the RC (11). In addition it is clear that the water inside the RC plays an important part in the recombination rate as was first pointed out by Clayton (51). Until these additional effects are sorted out, a detailed quantitative comparison of the temperature dependence of the experimentally observed recombination rate with theory is not possible.

D. Comparison of our Experimental Results with other Measurements

As has already been pointed out in previous sections, Gunner et al. (10) have measured the temperature and free energy dependent of k_{AD} in RCs form *Rb. sphaeroides* in which the native quinone (UQ-10) was replaced by a large number of other quinones that have different midpoint potentials, thereby changing the value of the energy gap $-\Delta G_2^o$ (see Fig. 1). Gunner et al. arrived from their results at the value of $\lambda = 0.6 \pm 0.1$ eV which is an agreement with our value of 0.64 eV. However, they were unable to fit all of their results with current ET theories. The addition of a low frequency mode resulted in a good fit at a single temperature but failed to explain the observed temperature dependence. They suggested that a distribution of high frequency modes may alleviate this problem.

An interesting set of experiments that is more closely related to the spirit of the work presented here has been reported by Popovic et al. (12). These authors used a stack of close-packed Langmuir-Blodgett films into which RCs from *Rb. sphaeroides* had been incorporated. The films were placed between two electrodes to which a voltage was applied. The recombination kinetics was monitored optically as a function of applied voltage. The observed decay did not follow first order kinetics and they deconvoluted, therefore, the decay into several exponentials. The results are shown in Fig. 11. Superimposed on their experimental points is the predicted result from our work obtained under the following assumption: the electric field shown on the abscissa of Fig. 11 was multiplied by the distance between D^+ and Q_A^- (25 Å) to obtain the free energy $-\Delta G^o$. The recombination rate was calculated with Eqs. 7 and 5. The sharp increase in k_{AD} at high fields is due to the onset of the indirect recombination path (Eq. 5); if this process is omitted, the dashed curve is obtained. The disagreement of our result with theirs is most likely due to the neglect of the local field correction (see section VB1). The most disturbing part of their result is the approximate linear relation between log k_{AD} and E. The failure to observe the expected curvature may in part be due to the inherent difficulties in accurately deconvoluting the decay rate into several exponentials. It may be of interest to measure the Stark effect in these samples to get an estimate of the local field (46).

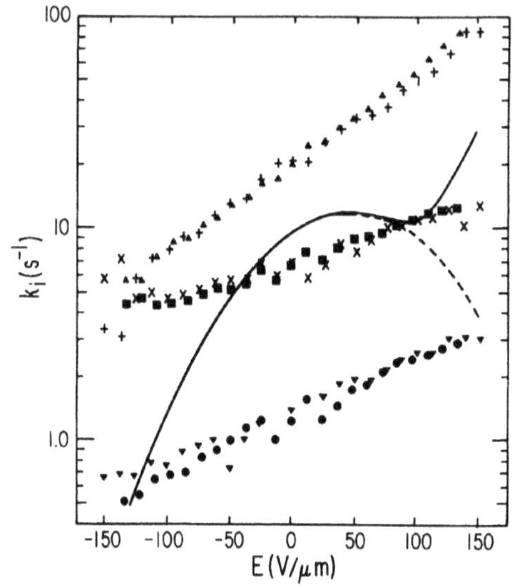

Fig. 11. *Comparison of the dependence of the recombination rate on the electric field obtained by Popovic et al. (12) in RCs incorporated into Lagmuir-Blodgett films (2 samples +, x, ▼ and ▲, ■, and ●) with the results predicted from the values obtained in this work ($\lambda = 0.64 eV$; $-\Delta G^o = 0.52 eV$), and Eqs. 8 and 5 (solid line). The sharp upturn of the rate is due to the onset of the indirect path (see Fig. 1). The dashed curve neglects the contribution from the thermally activated, indirect, path.*

E. Suggestions for Future Work

An important point that needs to be cleared up experimentally is the variability of the results obtained from the RC-bilayer system. A thorough systematic investigation of the steps taken in forming the bilayer should get to the source of this problem.

Another goal is to extend the range of the applied electric field to reach the so-called inverted, high exothermiticity, region (to the right of the peak of Fig. 7). So far, we were unable to apply more than ± 250 mV; at higher voltages either excess noise or a breakdown of the membrane occurred. A possible approach to extend the range would be to add to the applied field an internal field which would shift the experimental point of Fig. 7 to the right. This could be accomplished by buffering the system at high pH (see Fig. 9).

An exciting possibility would be to pick up oscillation in the recombination rate as the applied voltage is varied. This should, in principle, occur when the energy levels of the reactant and product states coincide (see Fig. 2). An optimistic eye may discern a faint hint of such oscillations in Fig. 7 (see the experimental points at $-\Delta G^o = \sim 0.53$ and 0.55 eV); more work needs to be done to confirm the reality of these speculative observations.

There is always the lingering question concerning the extent to which the model planar bilayer system corresponds to the in vivo situation (see sections VB3,4). It would be nice to put this question to rest by exploring more native systems. One of these are chromatophores in which the electric field could be produced by applying an ion gradient across their membrane. Another would be the patch clamp technique (52) applied to a native photosynthetic membrane. Finally there is hope of applying the methodology described in this paper to other reactions of the electron transfer chain.

VI. ACKNOWLEDGMENTS

We are indebted to A. Gopher for his contributions to the early part of this work, to P. McPherson for many discussions and his help with taking the data of Figs. 9 and 10, to D. Fredkin for stimulating discussions on the theoretical part, to R. A. Isaacson for his expert help with the electronics and to H. Schindler for several teflon septa. The work was supported by a grant from the NSF (DMB 85-18922).

REFERENCES

1. W. W. Parson, The role of P870 in bacterial photosynthesis, *Biochim. Biophys. Acta* **153**:248-259 (1967).
2. J. D. McElroy, D. C. Mauzerall, and G. Feher, Characterization of primary reactants in bacterial photosynthesis. II. Kinetic studies of the light-induced EPR signal (g=2.0026) and the optical absorbance changes at cryogenic temperatures, *Biochim. Biophys. Acta* **333**:261-277 (1974).
3. E. S. P. Hsi, and J. R. Bolton, Flash photolysis-electron spin resonance study of the effect of *o*-phenanthroline and temperature on the decay time of the ESR signal B1 in reaction-center preparations and chromatophores of mutant and wild strains of *Rhodopseudomonas sphaeroides* and *Rhodospirillum rubrum*, *Biochim. Biophys. Acta* **347**:126-133 (1974).
4. P. A. Loach, M. Kung, and B. J. Hales, Characterization of the phototrap in photosynthetic bacteria, *Ann. N. Y. Acad. Sci.* **244**:297-319 (1975).
5. B. J. Hales, Temperature dependency of the rate of electron transport as a monitor of protein motion, *Biophys. J.* **16**:471-480 (1976).
6. L. E. Morrison, and P. A. Loach, Complex charge recombination kinetics of the phototrap in *Rhodospirillum rubrum*, *Photochem. Photobiol.* **27**:751-757 (1978).
7. T. Mar, C. Vadeboncoeur, and G. Gingras, Different temperature dependencies of the charge recombination reaction in photoreaction centers isolated from different bacterial species, *Biochim. Biophys. Acta* **724**:317-322 (1983).
8. M. R. Gunner, D. M. Tiede, R. C. Prince, and P. L. Dutton, Quinones as prosthetic groups in membrane electron-transfer proteins 1: Systematic replacement of the primary ubiquinone of photochemical reaction centers with other quinones, *in*: "Function of Quinones in Energy Conserving Systems," B. L. Trumpower, ed., Academic Press, Inc., New York, pp. 265-269 (1982).
9. D. Kleinfeld, M. Y. Okamura, and G. Feher, Electron-transfer kinetics in photosynthetic reaction centers cooled to cryogenic temperatures in the charge-separated state: evidence for light-induced structural changes, *Biochemistry* **23**:5780-5786 (1984).
10. M. R. Gunner, D. E. Robertson, and P. L. Dutton, Kinetic studies on the reaction center protein from *Rhodopseudomonas sphaeroides*: The temperature and free energy dependence of electron transfer between various quinones in the Q_A site and the oxidized bacteriochlorophyll dimer, *J. Phys. Chem.* **90**:3783-3795 (1986).
11. G. Feher, M. Y. Okamura, and D. Kleinfeld, Electron transfer reactions in bacterial photosynthesis: charge recombination kinetics as a structure probe, *in*: "Protein Structure: Molecular and Electronic Reactivity," Robert Austin, Ephraim Buhks, Britton Chance, Don DeVault, P. Leslie Dutton, Hans Frauenfelder and Vitallii I. Goldanskii, eds., Springer Verlag, New York, pp. 399-421 (1987).

12. Z. D. Popovic, G. J. Kovacs, P. S. Vincent, G. Alegria, and P. L. Dutton, Electric field dependence of recombination kinetics in reaction centers of photosynthetic bacteria, *Chem. Phys.* **110**:227-237 (1986).

13. R. A. Marcus, On the theory of oxidation-reduction reactions involving electron transfer. I, *J. Chem. Phys.* **24**:966-978 (1956).

14. J. J. Hopfield, Electron transfer between biological molecules by thermally activated tunneling, *Proc. Natl. Acad. Sci. USA* **71**:3640-3644 (1974).

15. J. Ulstrup, and J. Jortner, The effect of intramolecular quantum modes on free energy relationships for electron transfer reactions, *J. Chem. Phys.* **63**:4358-4368 (1975).

16. J. Jortner, Temperature dependent activation energy for electron transfer between biological molecules, *J. Chem. Phys.* **64**:4860-4867 (1976).

17. J. Jortner, Dynamics of the primary events in bacterial photosynthesis, *J. Am. Chem. Soc.* **102**:6676-6686 (1980).

18. A. Sarai, Possible role of protein in photosynthetic electron transfer, *Biochim. Biophys. Acta* **589**:71-83 (1980).

19. T. Kakitani, and H. Kakitani, A possible new mechanism of temperature dependence of electron transfer in photosynthetic systems, *Biochim. Biophys. Acta* **635**:498-514 (1981).

20. T. Kakitani, and N. Mataga, New energy gap laws for the charge separation process in the fluorescence quenching reaction and the charge recombination process of ion pairs produced in polar solvents, *J. Phys. Chem.* **89**:8-10 (1985).

21. D. D. DeVault, "Quantum-Mechanical Tunnelling in Biological Systems," Cambridge University Press (1984).

22. R. A. Marcus, and N. Sutin, Electron transfers in chemistry and biology, *Biochim. Biophys. Acta* **811**:265-302 (1985).

23. J. R. Miller, L. T. Calcaterra, and G. L. Class, Intramolecular long-distance electron transfer in radical anions. The effects of free energy and solvent on the reaction rates, *J. Am. Chem. Soc.* **106**:3047-3049 (1984).

24. J. R. Miller, J. V. Beitz, and R. K. Huddleston, Effect of free energy on rates of electron transfer between molecules, *J. Am. Chem. Soc.* **106**:5057-5068 (1984).

25. A. D. Joran, B. A. Leland, P. M. Felker, A. H. Zewail, J. J. Hopfield, and P. B. Dervan, Effect of exothermicity on electron transfer rates in photosynthetic molecular models, *Nature* **327**:508-511 (1987).

26. A. Gopher, Y. Blatt, M. Schönfeld, M. Y. Okamura, G. Feher, and M. Montal, The effect of an applied electric field on the charge recombination kinetics in reaction centers reconstituted in planar lipid bilayers, *Biophys. J.* **48**:311-320 (1985).

27. T. Arno, A. Gopher, M. Y. Okamura, and G. Feher, Dependence of the recombination rate $D^+Q_A^- \rightarrow DQ_A$ on the electric field applied to reaction centers from *Rb. sphaeroides* R-26 incorporated into a planar lipid bilayer, *Biophys. J.* (Abstract), February 1988, in press.

28. P. H. McPherson, T. Arno, M. Y. Okamura, and G. Feher, pH-Dependence of the charge recombination rate $D^+Q_A^- \rightarrow DQ_A$ in reaction centers from *Rb. sphaeroides* R-26, *Biophys. J.* (Abstract), February 1988, in press.

29. M. Y. Okamura, R. A. Isaacson, and G. Feher, Primary acceptor in bacterial photosynthesis: obligatory role of ubiquinone in photoactive reaction centers of *Rhodopseudomonas sphaeroides*, *Proc. Natl. Acad. Sci. USA* **72**:3491-3495 (1975).

30. H. Arata, and W. W. Parson, Delayed fluorescence from *Rhodopseudomonas sphaeroides* reaction centers enthalpy and free energy changes accompanying electron transfer from P-870 to quinones, *Biochim. Biophys. Acta* **638**:201-209 (1981).

31. W. W. Parson, and B. Ke, Primary photochemical reactions, *in*: "Photosynthesis," Vol. I, Govindjee, ed., Academic Press, Inc., New York, pp. 331-385 (1982).

32. J. P. Allen, G. Feher, T. O. Yeates, H. Komiya, and D. C. Rees, Structure of the reaction center from *Rhodobacter sphaeroides* R-26: The cofactors, *Proc. Natl. Acad. Sci. USA* **84**:5730-5734 (1987).

33. J. P. Allen, G. Feher, T. O. Yeates, H. Komiya, and D. C. Rees, Structure of the reaction center from *Rhodobacter sphaeroides* R-26: The protein subunits, *Proc. Natl. Acad. Sci. USA* **84**:6162-6166 (1987).

34. H. W. Trissl, Spatial correlation between primary redox components in reaction centers of *Rhodopseudomonas sphaeroides* measured by two electrical methods in the nanosecond range, *Proc. Natl. Acad. Sci. USA* **80**:7173-7177 (1983).

35. H. W. Trissl, (personal communication).

36. P. Maroti, C. A. Wraight, Light induced proton binding-unbinding dynamics in reaction centers from *Rhodobacter sphaeroides*, *in*: "Progress in Photosynthesis Research," Vol. 2, p. II.6.401, J. Biggins, ed., Martinus Nijhoff, Boston (1986).

37. P. H. McPherson, M. Y. Okamura, G. Feher, and M. Schönfeld, Light induced proton uptake by RCs from *R. sphaeroides* R-26.1, *Biophys. J.* (Abstr.), p. 125a (1987).

38. H. Eyring, J. Walter, and G. E. Kimball, "Quantum Chemistry," J. Wiley and Sons, Inc., New York (1944).

39. M. Montal, Formation of bimolecular membranes from lipid monolayers, *Methods Enzymol.* **32**:545-554 (1974).

40. H. Schindler, Formation of planar bilayer from artificial or native membrane vesicles, *FEBS (Fed. Eur. Biochem. Soc.) Lett.* **122**:77-79 (1980).

41. M. Schönfeld, M. Montal, and G. Feher, Functional reconstitution of photosynthetic reaction centers in planar lipid bilayers, *Proc. Natl. Acad. Sci. USA* **76**:6351-6355 (1979)

42. Y. Kagawa, and E. Racker, Partial resolution of the enzymes catalyzing oxidative phosphorylation: XXV. Reconstitution of vesicles catalyzing $^{32}P_1$-ATP exchange, *J. Biol. Chem.* **246**:5477-5487 (1971).

43. N. K. C. Packham, P. Packham, P. Mueller, D. M. Tiede, and P. L. Dutton, Reconstitution of photochemically-active centers in planar phospholipid membranes: light induced electric currents under voltage clamped conditions, *FEBS (Fed. Eur. Biochem. Soc.) Lett.* **110**:101-106 (1980).

44. G. Feher and M. Y. Okamura, Structure and function of the reaction centers from *Rhodopseudomonas sphaeroides*, *in* "Advances in Photosynthesis Research," Vol. II, pp. 155-164, C. Sybesma, ed., M. Nijhoff/W. Junk, The Netherlands (1984).

45. M. Bixon, and J. Jortner, Coupling of protein modes to electron transfer in bacterial photosynthesis, *J. Phys. Chem.* **90**:3795-3800 (1986).

46. M. Lösche, G. Feher, and M. Y. Okamura, The Stark effect in reaction centers from *Rhodobacter sphaeroides* R-26 and *Rhodopseudomonas viridis*, *Proc. Natl. Acad. Sci. USA*, **84**:7537-7541 (1987).

47. T. O. Yeates, H. Komiya, D. C. Rees, J. P. Allen, and G. Feher, Structure of the reaction center from *Rhodobacter sphaeroides* R-26: Membrane-protein interactions, *Proc. Natl. Acad. Sci. USA* **84**:6438-6442 (1987).

48. A. M. Kuznetsov, and J. Ulstrup, The effect of temperature and transmembrane potentials on the rates of electron transfer between membrane-bound biological redox components, *Biochim. Biophys. Acta* **636**:50-57 (1981).

49. S. H. White, Temperature-dependent structural changes in planar bilayer membranes: solvent "freeze-out", *Biochim. Biophys. Acta* **356**:8-16 (1974).

50. T. Arno, P. H. McPherson, and G. Feher (unpublished results).

51. R. K. Clayton, Effects of dehydration on reaction centers from *Rhodopseudomonas sphaeroides*, *Biochim. Biophys. Acta* **504**:255-264 (1978).

52. B. Sakman and E. Neher (eds.) "Single Channel Recording," Plenum Press, New York (1983).

PRESSURE EFFECTS ON ELECTRON TRANSFER

IN BACTERIAL REACTION CENTERS

Maurice W. Windsor

Department of Chemistry
Washington State University
Pullman, Washington 99164-4630 USA

INTRODUCTION

Studies of the effects of pressure on the various electron transfer reactions, both forward and backward, that mediate charge separation and recombination in bacterial reaction centers, can provide useful information on changes in conformation and distances involving the various chromophores and their protein scaffolding that may accompany these movements of charge. We have recently reported the effects of pressure on the rates of one forward reaction and three back reactions for two species of photosynthetic bacteria, Rb. sphaeroides and Rps. viridis (1). In this brief report I review these results and compare them with earlier studies of the temperature dependence of these processes (2,3). A simple model is proposed that appears to account for both pressure and temperature data and which can be tested by additional experiments.

SUMMARY OF PRESSURE RESULTS

The pressure data are summarized in Figures 1 and 2. Experimental details on the laser spectrometer, the pressure cell and on sample preparation have been reported previously (1).

In Rb. sphaeroides R-26 the back reaction $P^+Q_A^- \rightarrow PQ_A$ is about twice as rapid at 3000 atm as at 1 atm (Fig. 1A,B). In addition, less Q_B^- is formed at high pressure (Fig. 1C). A plausible interpretation of these data is that application of high pressure causes Q_A to move away from Q_B and closer to P. Evidence already exists that, on receiving an electron from H, Q_A moves away from P and closer to Q_B as part of the normal functioning of the reaction center. Kleinfeld, et al. (4) observed that for RC's cooled to 77K in the dark, charge recombination $P^+Q_A^- \rightarrow PQ_A$ is faster and, in addition, further forward transfer of the electron from Q_A^- to Q_B essentially does not occur (τ_{AB}light/τ_{AB}dark $> 10^8$). On the other hand, RC's cooled to 77K while under illumination exhibited a slower back reaction and a faster forward transfer to Q_B.

The temperature dependence of the kinetics of charge separation is also relevant to understanding the pressure results. The time constant for the $H \rightarrow Q_A$ forward step in Rb. sphaeroides drops from 250 ps at room temperature to about 100 ps at temperatures of 100K or less (3).

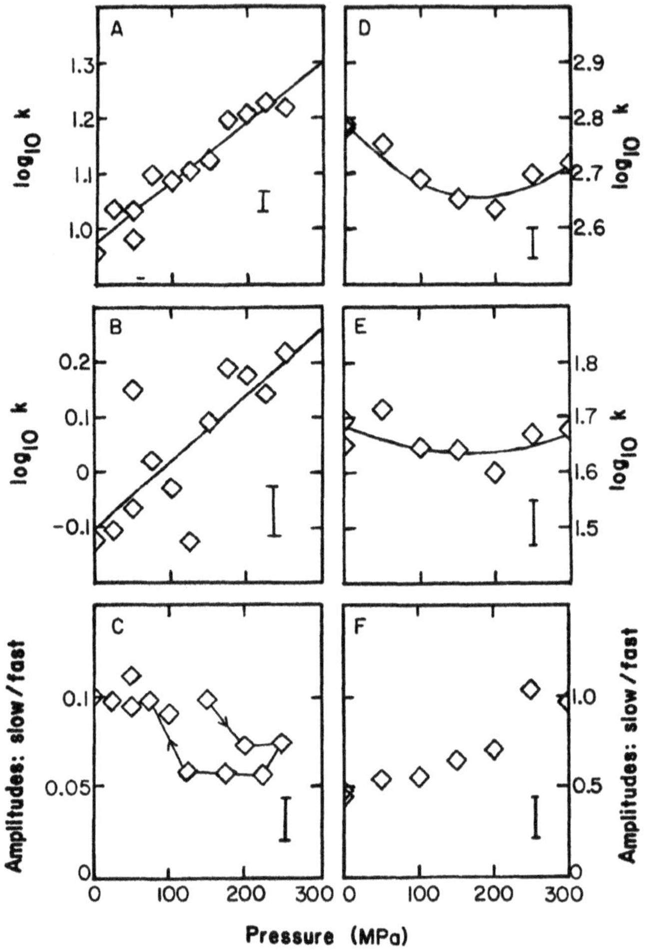

Fig. 1. Effect of pressure on the decay of P^+. A, B, and C, Rb. sphaeroides R-26, detected at 600 nm; D, E, and F, Rps. viridis, detected at 450 nm, with 75 uM $K_3Fe(CN)_6$ and 20 mM, 1,10 o-phenanthroline present. Excitation was at 857 nm. A and D, decay of $P^+Q_A^-$ (fast component); B and E, decay of $P^+Q_B^-$ or other slow component; C and F, ratio of initial amplitudes of the slow component and fast component. Least-squares lines (A and B) or parabolas (D and E) are also shown. C also indicates the hysteresis in Q_B participation. The uncertainty in the data is shown as an error bar, I, on each panel.

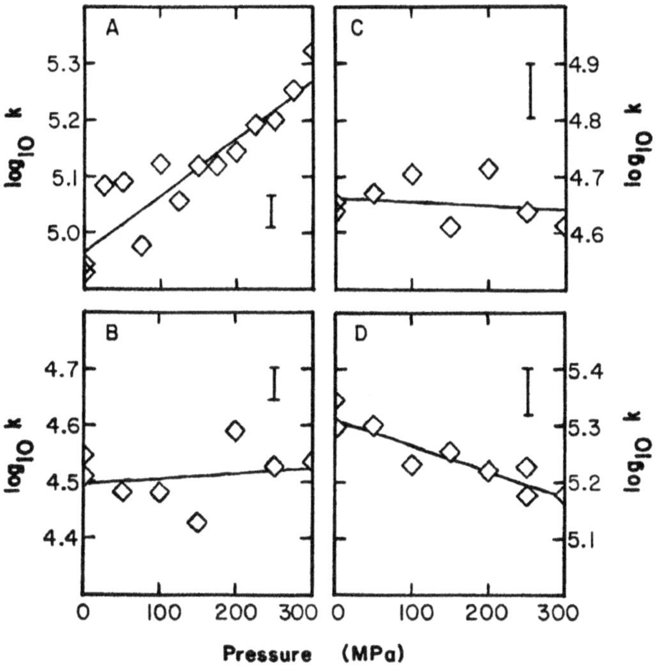

Fig. 2. Effect of pressure on triplet decay rates. A, Rb. sphaeroides R-26 reaction centers reduced with $Na_2S_2O_4$; B, Rb. sphaeroides R-26 reaction centers depleted of quinone; C, Rps. viridis reaction centers reduced with $Na_2S_2O_4$; D, bacteriochlorophyll a in pyridine. Excitation was at 857 nm, A, or at 694 nm, B, C, and D. Detection was at 680 nm, A; 420 nm, B; 570 nm, C; or 530 nm, D. The uncertainty in the data is shown as an error bar, I, on each panel.

This result suggests that lowering the temperature causes the reaction center to shrink, thus statically reducing the HQ_A separation. We have not yet made a detailed study of the effects of pressure on the 200 ps $H \rightarrow Q_A$ forward step. However, preliminary results for the back reaction $P^+H^-Q_A^- \rightarrow PHQ_A^-$ in RC's reduced with sodium dithionite prior to excitation so as to block further forward transfer from H to Q_A, show that high pressure (3000 atm) does not materially change the rate, but does appear to inhibit the forward transfer $P \rightarrow H$, as indicated by reduced amplitudes for the bleaching at 870 nm and the excited state absorption at 680 nm (5). From these observations we may infer that the 3 ps forward transfer step $P \rightarrow H$ in reduced RC's is slower and therefore less effective at higher pressure. Pressure-induced movement of Q_A^- towards H again would account for this observation, since the negative charge carried by Q_A, being brought closer to H by pressure, would be expected to inhibit the forward transfer of an electron from P* to H.

All of the above results, the light versus dark cooling experiments of Kleinfeld, et al., the studies of the temperature dependence of the 200 ps step, and the pressure studies, are consistent with the following picture. In reaction centers of Rb. sphaeroides, the first quinone acceptor, Q_A, is somewhat loosely bound in the protein pocket in which it resides. In normal operation, Q_A, on receipt of an electron from H, moves away from H and P and towards Q_B. This motion facilitates the continued forward electron transfer from Q_A to Q_B and inhibits back reactions that result in charge recombination. Low temperatures or high pressures confine the system to its initial configuration and prevent such motion from occurring, thus enhancing the back reaction and inhibiting further forward transfer.

Our data on the pressure dependence of the rate of decay of the triplet state P^R in Rb. sphaeroides R-26 provide additional support for the above model (Fig. 2). The decay rate of $P^R = {}^3[P^+H^-Q_A^-]$, in RC's with Q_A reduced with dithionite, increases by about a factor of 2 when the pressure is increased from 1 atm. to 3000 atm. (Fig. 2A). But, in RC's with Q_A removed by treatment with o-phenanthroline, the rate is unchanged by pressure (Fig. 2B). Again, the enhanced rate in the former case can be attributed to pressure-induced movement of Q_A^- closer to H^-, thus increasing the magnitude of the hyperfine coupling. When Q_A is absent, as in the latter case, no such effect can occur.

Turning now to the effects of pressure on the corresponding processes in RC's of Rps. viridis, we note (Fig. 1D) that the decay rate of $P^+Q_A^-$ is about 25% slower at 3000 atm., while the decay of the triplet P^R in RC's with Q_A reduced with dithionite is essentially unchanged (Fig. 2C). In the context of the model, we can understand these results by recognizing that Q_A (menaquinone) in viridis is more strongly bound than Q_A (ubiquinone) in sphaeroides (6). The reasons for these differences are not yet known, but it would surely be of interest to study both pressure and temperature dependences of the corresponding decay rates in Rb. sphaeroides RC's in which ubiquinone has been removed and replaced with menaquinone or other quinones (7,8). Studies with modified RC's in which the non-heme iron has been either removed or replaced by other metal ions (9,10) would also be of interest. Extension of the Kleinfeld, et al. light-dark cooling experiments to viridis would also be helpful. The effect of pressure on the formation and decay of the "doubly reduced" trapped state, $cyt^+PBH^-Q_A^-$, produced by continuous illumination of $cytPHQ_A^-$ could also be investigated by applying pressure either before or after illumination and then following up with kinetic laser excitation studies at higher pressures.

ACKNOWLEDGEMENT

This work was supported by National Science Foundation Grant PCM-8310167. I am grateful to M. E. Michel-Beyerle for helpful discussions.

REFERENCES

1. C. W. Hoganson, M. W. Windsor, D. I. Farkas, and W. W. Parson, Biochim. Biophys. Acta 892:275 (1987)
2. C. C. Schenck, W. W. Parson, D. Holten, M. W. Windsor, A. Sarai, Biophys. J. 36:479 (1981).
3. C. Kirmaier, D. Holten, W. W. Parson, Biochim. Biophys. Acta, 810:33 (1985).
4. D. Kleinfeld, M. Y. Okamura, and G. Feher, Biochemistry, 23:5780 (1984).
5. R. Menzel and M. W. Windsor, unpublished results.
6. A. Ogrodnik, W. Lersch, M. E. Michel-Beyerle, J. Deisenhofer, and H. Michel, in Antennas and Reaction Centers of Photosynthetic Bacteria, ed. M. E. Michel-Beyerle, Springer Series in Chemical Physics 42, Springer-Verlag, 1985, p. 198.
7. M. Y. Okamura, R. A. Isaacson, and G. Feher, Proc. Natl. Acad. Sci. USA 72:3492 (1975).
8. N. W. Woodbury, W. W. Parson, M. R. Gunner, R. C. Prince, and P. L. Dutton, Biochim. Biophys. Acta, 851:6 (1986).
9. C. Kirmaier, D. Holten, R. Debus, M. Y. Okamura, and G. Feher, Proc. Natl. Acad. Sci. USA, 83:6407 (1986).
10. M. W. Windsor, J. Chem. Soc., Faraday Trans. II, 82:2237 (1986).

THE SPECTRAL PROPERTIES OF CHLOROPHYLL AND BACTERIOCHLOROPHYLL DIMERS; A COMPARATIVE STUDY

A. Scherz* and V. Rosenbach-Belkin

Department of Biochemistry
The Weizmann Institute Of Science, Rehovot, 76100, Israel
*Ricaneti Career Development Chair

INTRODUCTION

Reaction centers (RCs) of photosynthetic bacteria generally contain four bacteriochlorophylls (Bchls) and two bacteriopheophytins (Bphs) that have three main absorption bands in the near infrared (NIR) (1). The transitions occur at 860-870, 803-814 and 760 nm in species that contain Bchla, and at 960-1000, 820-850 and 790-810 nm in species that contain Bchlb. They are all optically active with a net positive rotation for the lowest energy transition (2). The total oscillator strength in the NIR region is larger by 10-20% than the oscillator strength of the chromophore extract in that region (3). In the visible region, the reaction centers absorb at ≈ 600 nm with a double Cotton effect (2) and an oscillator strength which is smaller than that of the *in vitro* extract by $\approx 10\%$ (3). There is another band at ≈ 540 nm, which splits at low temperature. A strong absorption with a complicated circular dichroism pattern is seen in the UV; Its oscillator strength is weaker and it is red shifted relative to the maximum absorption of the pigment extract in that region.

At the atomic coordinates given by X-ray data to the chromophores of the bacterial RCs (4,5), their ground and photo-excited states should perturb each other. Therefore, the absorption and emission bands should reflect transitions among linear combinations of local states. The coefficients of the local states can be solved by diagonalization of the RCs' Hamiltonian but for that end one needs the excitation energies of the locally excited states (diagonal terms) and the perturbation terms (off-diagonal terms) *in vivo*.

In one mode of calculation, primarily suggested by Scherz and Parson for aggregates of Bphs in vitro (3), the off-diagonal terms result from the interactions among transition dipoles of degenerate and non-degenerate states. Parson, Scherz and Warshel applied this principle to the four Bchls in RCs of *R.viridis* and figured out the increase in oscillator strength relative to *in vitro* monomers, the non conservative CD and a Qy shift of 800-900 cm^{-1}. For the additional shift that was required to adjust the calculated spectra to the experimental one, Parson, Scherz and

Warshel considered interactions of the locally excited states with intramolecular charge transfer (CT) states (6).

The groups of Fischer (7) and Hoff (8) also used the transition dipoles (or extended monopoles) to calculate the off-diagonal terms, but in the zero exciton theory level where mixing of the Soret and Qx transitions with the NIR region is not considered. The transition dipoles and the energies of the locally excited states were adjusted to fit the spectra lineshapes. For the *ad hoc* adjustment Knapp *et al* (7) and Hoff *et al* (8) assumed chromophore-protein interactions of a non-defined nature.

From these calculations it was deduced that the major contributions to the off-diagonal terms in the RC's Hamiltonian come from interactions among the transition dipoles and that the lowest energy transition in the NIR region is a linear combination (lower energy exciton band) of the local Qy transitions of the special pair Bchls (with some contribution from By and Qx according to Scherz and Parson (3)). According to Fischer (7), the higher energy exciton band should be at around 810 nm, but its dipolar strength is distributed in the 830-850 nm region. The rest of the absorption at 830-850 nm should be due to linear combinations of the local Qy transitions of the accessory Bchls with small contributions from the Bchls of the special pair and the Bphs. Most of the absorption in the 800-830 nm is attributed to the Qy transitions of the Bphs.
In the present meeting, Scherer and Fisher presented PPP/CI calculation in which they obtained similar off-diagonal terms from transition monopoles interactions. In that calculation they also obtained contributions from CT states to the shift of the local states *in vivo,* although smaller than the contribution of the dipolar interaction.

Parson and Warshel used a different approach. They calculated the transition monopoles by the QCFF/CI method (9) and scaled them to the observed dipoles. The calculated interactions among the transition monopoles were much smaller than those calculated in the point dipole approximation. Parson and Warshel concluded that the locally excited states are mostly perturbed by intermolecular CT states as previously suggested by Warshel (10) for dimers of Chls. The calculated energy for the local Qy transitions (15,000 cm^{-1}), was significantly higher than the experimental energy for the Qy transition of Bchl monomer *in vitro* and lowered parametrically to about 11800 cm^{-1}. The energy of the lowest CT state was adjusted until the lowest energy exciton of the Bchls in the special pair was moved to 960 nm. The off-diagonal terms for the interactions among other transitions were calculated in a similar manner.
An extensive, all-valence-electrons method had been recently elaborated by Plato *et al* for the lowest transition of the special pair (11).

Interestingly, the simulations of the RCs' spectra (optical absorption, circular dichroism, linear dichroism and Stark effect) by four methods that propose fundamentally different *ad hoc* adjustments appear to result in similar reasonable agreement with the experimental results. However, these adjustments should mainly affect the calculated spectral properties of the "special pair" Bchls. The mutual separation of other chromophores is large enough to give equal interaction matrix elements in the point dipole or the point monopole approximations. Since the Bchls of the special pair are only two out of six chromophores with a similar range of absorption, their spectral properties are "diluted" and a reasonable fit of the overall spectra may be quite misleading. For example, the experimental dipolar strength in the NIR region for *R.viridis* and *R.pfinegii* (330 debye^2) is higher than the *in vitro* total dipolar strength of 4 Bchls and 2 Bphs monomers (b type) in the NIR region (280 debye^2). However the *in vitro* strength for the NIR region was conserved in the calculations of Fischer group and Parson and Warshel (7,9). The discrepancy in dipolar strength between these calculated spectra and the experimental ones is reflected in a slight (although apparent) deviation over the full wavelength range and may therefore not be appreciated. The enhanced strength *in vivo* indicates that there must be an intensity borrowing from upper bands to the NIR region that results from strong dipolar coupling of non-degenerate states (e.g. the Qy and By of the special pair chromophores as predicted by Scherz and Parson (3)) or from a change in the configuration interaction (CI) as suggested by Pearlstein in the present meeting. In either case the theoretical predictions have to account for the intensity redistribution. Other significant misfits in the simulation of the RCs' spectra include the calculated energy for the *in vivo* Soret band (ca. 395 vs. ex. 420 (12)) and the CD in the 550-780 nm band (9).

Because of these limitations we have been looking for well-defined systems of Bchls and Chls that enable a systematic study of the variations in the spectral properties of Bchls when monomers form dimers and dimers organize into larger aggregates. Possible systems are:

1) LHCs of the purple bacteria where the Bchls form dimers with the same bathochromically shifted Qy transition and enhanced oscillator strength as observed for the Bchls in the RCs (13,14);

2) *In vitro* dimers of Bchls and Chls that have the spectral patterns of the *in vivo* Bchls.

The possibility of having families of Chls and Bchls with selected chemical modifications seems to be especially attractive as it should enable a systematic study of the variations in spectra of dimers due to variations in the electronic structure of the participating monomers and thereby a better quantification of the orbitales that participate in the electron transfer process.

MATERIALS AND METHODS

Bchla and Bchlb were extracted from liophilized cells of *R. rubrum* and *R. viridis* with methanol, purified by chromatography over DEAE-sepharose and checked by reversed phase HPLC.

Chla and Chlb were extracted from spinach and liophilized cells of blue-green algae, purified over HP Sephasorb (Pharmcia) or sugar and checked for purity by reversed-phase HPLC.

PyroBchla was prepared as described (15) and 10-OH Bchla was separated by HPLC from Bchla that was exposed to methanol in DEAE-Sephrose under prolonged illumination.

The corresponding Phs or Bphs were prepared by adding acetic acid to a dry extract of purified Chls or Bchl.

LHC B850 was extracted from *R. sphaeroides R-26.1* by a modification of Austin and Sauer (16).

In vitro oligomers of different sizes were formed when 3 volumes of Formamide (FA) and then one volume of water were added to the relevant chromophore solution in 0.04 volume of pyridine while stirring. The solution was cosonicated with an equal volume of FA:Water that contained micelles of Triton X-100 (Tx-100). The ratio ([chromophore] / [micelle]) was alwayes lower than 4 (the critical micelle concentration ($3x10^{-7}$) and the number of Tx-100 molecules per micelle (1000), were calculated as described in ref. 17).

Circular dichroism measurements were done with home built and fully computerized dichrograph with very high sensitivity.

For optical absorption and fluorescent measurements we used Cary 118, fully computerized Milton-Roy spectronic 1200 and Perkin-elmer 44M spectrometers.

RESULTS

The optical absorption spectra of different Chls and Bchls in 3:1-Formamide:water (Fw) that contain Tx-100 are shown in Figs. 1,2,5. For each chromophore there are two spectral forms; one of them has Qy transition at longer wavelength. When the concentration of [Tx-100] was increased relative to the total concentration of the chromophore, the long wavelength absorption decreased and the shorter wavelength absorption incresed with three isosbestic points (not shown), indicating an equilibrium between two and only two, spectral forms. The concentrations of single chromophores in each spectral form were calculated from their extinction coeffients as previously described (17). The composition of the oligomers in a particular [Tx-100] value was dertermined by following the relative concentrations of single chromophores in the two spectral forms as the total concentration of the chromophores was increased. Typical curves are shown in Fig. 3 which illustrates the dependence of [Bchl-780] and [Bchl-853] on the total concentration of Bchla.

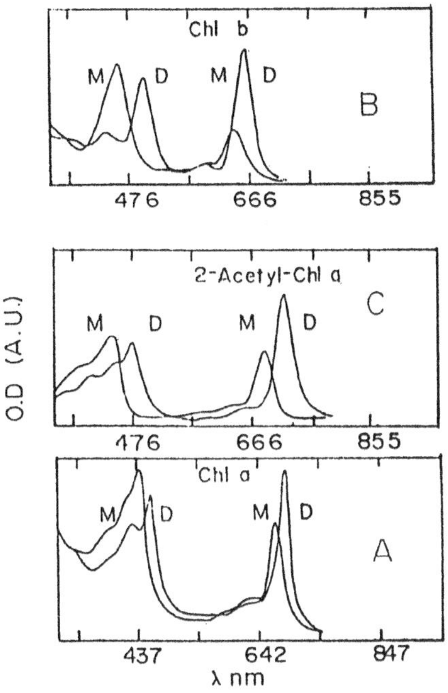

Figure 1. *The two spectal forms of different Chls in Fw solutions that contain $5.77x10^{-3}$M [Tx-100]. Upper curves - Chlb; Middel curves - 2-acetyl-Chla; lower curves - Chla.*

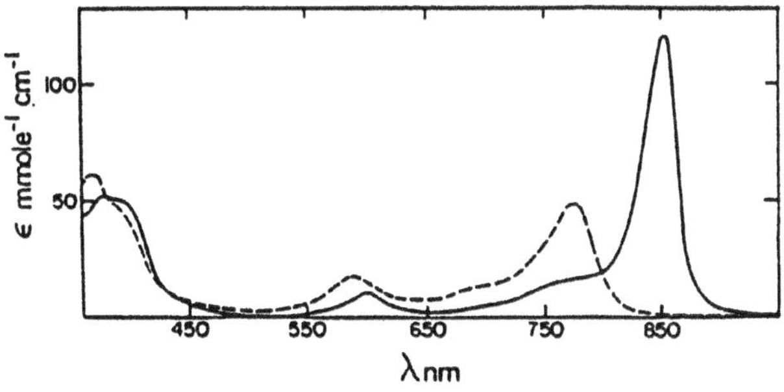

Figure 2. *The two spectral forms of Bchla in Fw that contains $5.77x10^{-3}$ M [Tx-100].*

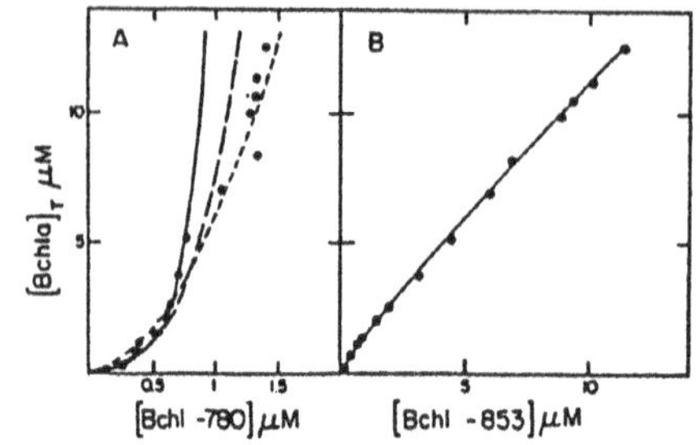

FIGURE 3. *(A) - The* $[Bchla]_T$ *as a function of [Bchl-780] in Fw that contains* $5.77x10^{-3}M$ *[Tx-100]. (•) - experimental values; (- - -) - Values calculated for* $[Bchl\text{-}853] = K_2[Bchl\text{-}780]^2$*; (– --) - Values calculated for* $[Bchl\text{-}853] = K_3[Bchl\text{-}780]^3$*; (-----) - Values calculated for* $[Bchl\text{-}853] = K_4[Bchl\text{-}780]^4$*. (B) - The* $[Bchla]_T$ *as a function of the [Bchl-853]; (•) - experimental values; (—) - values calculated for* $[Bchla]_T = 4.9x[Bchl\text{-}853]^2$

The equilibrium between Bchl-780 and Bchl-853 is best described by the quadratic equation:

$$\text{(I)} \qquad [\text{Bchl-853}] = 4.9\text{x}10^6\text{x}[\text{Bchl-780}]^2$$

Since the Bchl-780 have the spectral characteristics of Bchla monomers (17), we suggest that there is an equilibrium between Bchla monomers absorbing at 780 nm and Bchla dimers absorbing at 853 nm (17). A similar procedure was applied to Bphea, Chla, Chlb, Pha and Phb, PyroBchla and 10-OH-Bchla. All these molecules maintained an equilibrium between monomers and dimers at low enough concentrations relative to the micelles. All dimers had bathochromically shifted Qy and By transitions. The Qy transitions had enhenced oscillatore strengths relative to the *in vitro* monomers and positive optical activities that were compensated by a decreased oscillator strength and negative optical activities in the Soret band region. Fig. 4 illustrates the dependence of the Qy shift (in cm^{-1}) on the dipolar strength of the monomeric transition. We also included some other transitions (e.g. By). Substitution of the C-10 carbomethoxy group by a hydrogen atom (in PyroBchla) had a negligible effect on the spectral properties of the monomers but increased the shift of the Qy transition in the aggregated form to 922 nm (fig. 5). The substituted aggregate also had a new transition at 625 nm and the Soret band maximum was red shifted relative to the Soret band of the non-substituted dimer to 405 nm.

Substitution of the C-10 hydrogen by OH (in the 10-OH-Bchla) decreased the shift of the Qy transition in the aggregated form to 838 nm. The spectra of 2-acetyl-Chla and Chla dimers are compared in fig. 5. The Qy transition of the substituted Chla is shifted by ≈ 100 cm^{-1} more than the non-substituted compound.

The optical absorption of the LHC B850 that was extracted from R-26.1 in 1% SDS polyacrylamide gels are compared with the optical absorption and CD of the *in vitro* dimer Bchl-853 in fig. 6

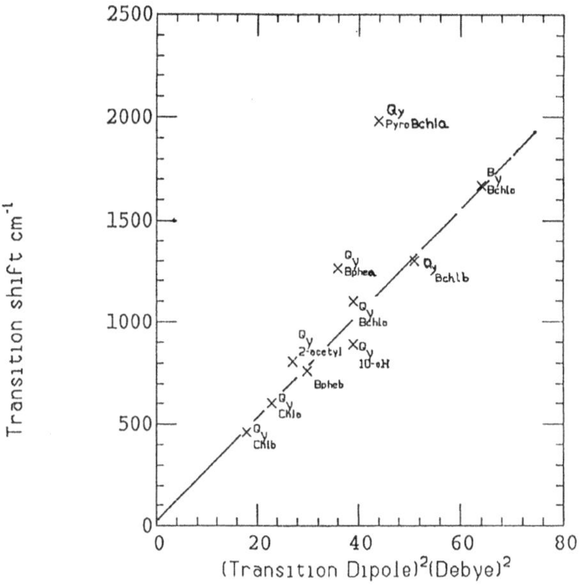

Figure 4. *The shift of the Qy transitions in several Chls and Bchls as a function of its dipolar strength.*

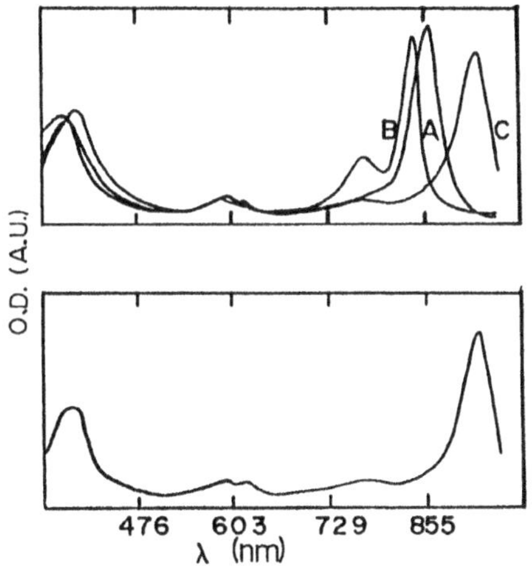

Figure 5. *Upper curves: A - The optical absorption of Bchla dimer (Bchl-853); B - 10-OH-Bchla dimer; C - PyroBchla dimer; Lower curve- The optical absorption of PyroBchla dimer.*

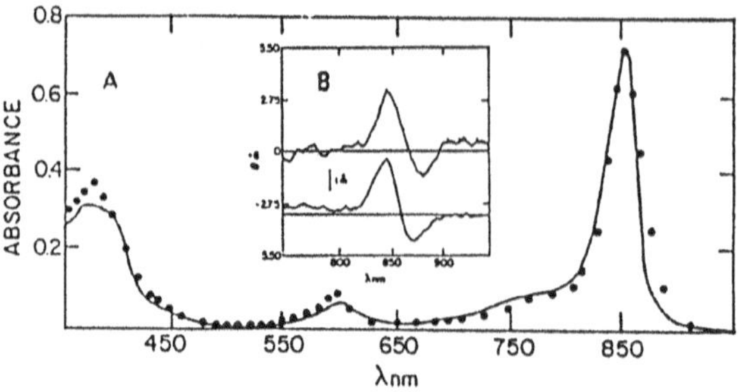

Figure 6. *The optical absorption of Bchl-853 (solid line) and B850 in 1% SDS PAGE (dots). Inset - The circular dichroism of Bchl-853 (lower curve) and B850 (upper curve).*

DISCUSSION

The thermodynamic parameters for the aggregation of Bchla, Bpha and Chla were studied in some detail by following the monomer-dimer equilibria at different temperatures (preliminary data were presented in ref.18). From these studies we deduced that ΔH for the pairing of two Bchls is about -8 Kcal/mole and that ΔG^{o} is about -3.8 Kcal/mole. The enthalpy change is too small to involve covalent bonding of the interacting monomers and the overwhelming concentration of protons (the Fw hydrogens) probably prevents intermolecular hydrogen bonding of the peripheral carbonyls. Thus, the exothermic self-organization probably involves electrostatic interactions between the participating monomers.

Since the monomers of Chls and Bchls have negligible permanent dipoles (9,19) we may either assume dispersive interactions among individual chromophores (Jortner, private communication) or interactions among partially positive atoms of one molecule (e.g. peripheral hydrogens, central Mg^{++}) and partially negative groups of the other molecule (e.g. acetyl groups, ring V carbonyls). Interactions of the second type were proposed for Phenylalanine pairs that prepherred a geometry similar to the "special pair" in different polypepetides with known structure (e.g. table 1. in ref. 20 and refs.21-23)

For preliminary calculations of the dispersive interactions among two Bchla molecules we expressed the dispersive interaction (24) in terms of the interactions V_d among the transition dipoles of the paired molecules

$$\text{(II)} \qquad E'' = -\Sigma\Sigma\ V^2_{damobno}/(E_{amo}-E_{bno})$$

where E_{amo} is the energy for transition from |o> to |m> in molecule a and E" is the second order perturbation to the ground state of the dimer. The interaction among the transition dipoles was calculated in the point monopole approximation:

(III) $V_{damobno} = \Sigma\Sigma\ [\Phi_{ajmo} \times (1/R_{ij}) \times \Phi_{bino}]$

where $V_{damobno}$ is the energy contributed by interactions among transition monopoles Φ formed on atoms j and i of molecules a and b respectively.

Point monopoles for calculating the interactions among the Qy transition dipoles were taken from Weiss (19) and splitted into two halvs: 1 Å above and below the Bchls macrocycle as suggested by Philipson and Sauer (25).

The angular and spatial dependence of the interactions among the monompoles of the Bchla Qy transitions are shown in Fig.7.

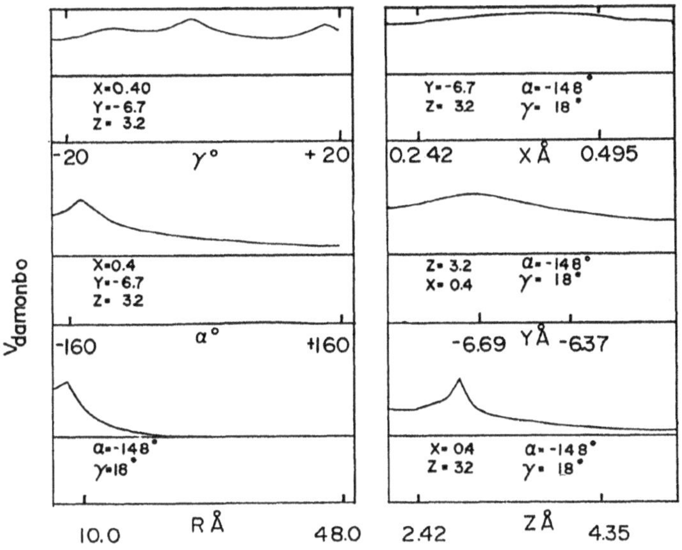

Figure 7. *$V_{damobmo}$ (cm^{-1}) as a function of Bchls geometry in a dimer. in the upper two curves X,Y,Z were held constant. In the lower curve α, β, γ, X and Y/Z were kept constant.*

The maximum interaction is found for two monomers that are separated by 0.4 Å , -6.7 Å and 3.2 Å to the Y, X and Z directions respectivly (Fig. 8) with $\alpha = 165^0$ $\beta = 0^0$, and $\gamma = 20^0$ (see Fig. 8 for nomenclature). The dispersive energy for this configuration is ≈ -1000 cm^{-1} (- 4 Kcal/mole), close to the experimental enthalpy change for the formation of Bchl-853. Interestingly, this geometry is very close to the one found for the special pair of *Rb. sphaeroides* by X-rays studies (5) and may therefore indicate that the geometry of the Bchls in P-860 is partly determined by their self-organization.

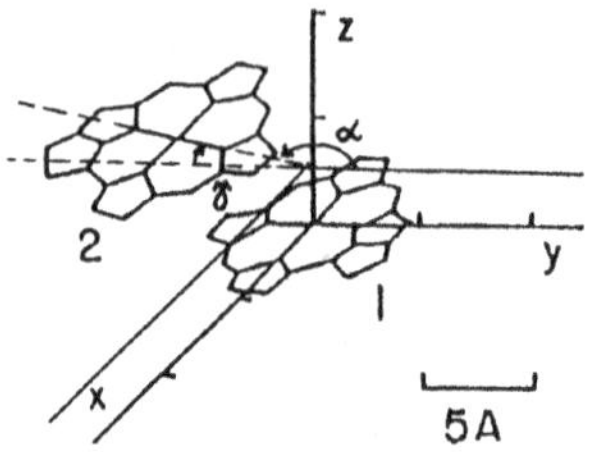

Figure 8. *The coordinate system used to describe the configuration of the Bchla dimer. The separation between the carbon centers was calculated by using the X-ray coordinates (5)*

The tendency of the Bchls to self-orgenize into dimers with geometry that maximize the interaction between their transition dipoles probably accounts for the similar spectra of the LHC B850, the *in vitro* dimer Bchl-853 and the "special pair" P-860 ; it also implies that other Chls or Bchls with similar polarizabilities will self orgenize into the same geometry if allowed by their environment.

The calculated interaction between the Qy monopoles are much larger than the values given by Parson and Warshel for the Qy monopoles' interactions in P-960 although the atomic coefficients of the participating orbitals are almost identical (compare refs. 9 and 19). Its magnitude relaxes to the value of the interactions among point dipoles at large separation with no further scaling. it can account for the Qy shift in P-860 , B850 (13), Bchl-853 (17), the Bchlb dimer *in vitro* and the *in vitro* dimers of Chla, Chlb, Phea, Pheb and Bpheb (fig. 3)
A clue for the additional shift in the P-960 may be found in the spectrum of the PyroBchla dimer (Fig. 5). where modification at the C-10 position increased the Qy shift dramatically. A possible correlation with the similar shift in P-960 is that the substitution of the C-10 Carbomethoxy group by hydrogen atom effects the spatial distribution of the Π electrons and thereby the dipolar interactions. The selective hydrogen bonding of the C-10 carbomethoxy to the protein in P-960 (and not in P-860) may have a similar effect. In further support to this assumption the PyroBchla dimer have a new transition at 625 nm as found for P-960 (fig. 5).

The major contribution of the dipolar interaction to the Qy shift is further demonstrated by the lineshapes of the different dimers. For example, the Qy_{0-1}

band shift in Chla dimer is smaller than the Qy_{0-0} as expected for intermediate exciton coupling (26) (fig. 1) whereas in Bchla dimer the complete Qy band shifts (strong coupling limit).

Acknowledgment

The study was financially supported by the U.S.-Israel Binational Foundation (Grant No. 84-00144)

REFERENCES

1. Okamura, M.Y., Feher, G. and Nelson, N. in: Photosynthesis; Energy conversion in Plants and Bacteria. ed. Govindjee, Academic Press, N.Y. (1982), 195. and photosynthetic bacteria, ed. Govindjee, Academic Press, N.Y. (1982), 331.

2. Reed, D.W. and Ke, B. (1973) J. Biol. Chem. *248*, 3041.

3. Scherz, A. and Parson, W.W. (1984) Biochim. Biophys. Acta *766*, 666.

4. Deisenhofer, J., Epp, O., Miki, K., Huber, R. and Michel, H. (1985) Nature *318*, 618.

5. a. Allen, J.P., Feher, G., Yeates, T.O., Komiya, H. and Rees, D.C. (1987) Proc. Natl. Acad. Sci. USA *84*, 5730.
   b. Allen, J.P., Feher, G., Yeates, T.O., Komiya, H. and Rees, D.C., *ibid* 6162

6. Parson, W.W., Warshel, A. and Scherz, A. in: "Antennas and Reaction Centers of Photosynthetic Bacteria Structure, Interactions and Dynamics" ed. Michele-Beyerle, M.E., Springer-Verlag, pp. 122.

7. Knapp, E.W., Scherer, P.O.J., Fischer, S.F.,(1986) Biochim. Biophys. Acta *852*, 295

8. Hoff, A., Lous, E.J., Moehl, K.W. and Dijkman, J.A., (1985) Chem. Phys. Lett. *114*, 39

9. Parson, W.W. and Warshel, A. (1987) J. Am. Chem. Soc. *109*, 6154

10. Warshel, A., (1979) J. AM.Chem. Soc. 1*01*,744

11. Plato, M., Trankle, E., Lubitz, W., Lendzian, F. and Mobius, K. (1986) Chem. Phys. *107*,185

12. Prince, R.C. Tiede, D.M., Thornber, J.P. and Dutton, P.L. (1977) Biochim. Biophys. Acta, *462*,467

13. Rosenbach-belkin, V., Braun, P. Kovatch, P., Scherz, A. (1988) in "Photosynthetic Light-HarvestingSystems; Structure and Function." eds. Scheer, H. and Schneider.000.

14. Bolt, J. (1980) Thesis, Univ. California, Berkeley.

15. Wsielewsky, M.R and Svec, W.A. (1980) J. Org. Chem. *45*, 1969

16. Sauer, K. and Austin, L.A., Biochem. (1978) *17*, 2012.

17. Scherz, A., Rosenbach-Belkin, V. (1987) submitted.

18. Scherz, A., Rosenbach, V. and Malkin, S. in: Antennas and reaction centers of photosynthetic bacteria, ed. M.E. Michel-Beyerle, Springer Verlag (1985), 314.

19. Weiss, R. (1972), J. Mol. Spect.

20. Singh, J. and Thornton, J.M. (1985) FEBS Lett. *191*, 1-6.

21. Burley, S.K. and Petsko, G.A. (1985) Science *229*, 23-28.

22. Burley, S.K. and Petsko, G.A. (1986) J. Am. Chem. Soc. *86*, 7995-8001.

23. Thomas, K.A., Smith, G.M., Thomas, T.B. and Feldman, R.J. (1982) Proc. NAtl. Acad. Sci. USA *79*, 4843-4847.

24. W. Kauzmann, Quantum Chemistry, Ch. 13 (1957), Academic Press.

25. Philipson, K.D. and Sauer, K. (1971) J. chem. Phys. *75*,1440

26. Simpson, T.W. and Paterson, D.L. (1957), J. Chem. Phys. *26*, 588

As the Cadarache workshop went on, God finally learned how He should have built the perfect reaction center...

SPECTROSCOPIC PROPERTIES AND ELECTRON TRANSFER DYNAMICS OF REACTION CENTERS

W. Parson[1], A. Warshel[2], S. Creighton[2] and J. Norris[3,4]

[1]Dept. of Biochemistry, University of Washington, Seattle WA 98195
[2]Dept. of Chemistry, University of Southern California, Los Angeles CA 90007
[3]Chemistry Division, Argonne National Laboratories, Argonne, IL 60439; and
[4]Dept. of Chemistry, University of Chicago, Chicago IL 60637

INTRODUCTION

With the elucidation of the crystal structures of reaction centers from several species of bacteria [1-3], we no longer need to depend on spectroscopic properties for hints on the positions and orientations of the pigments. Nor do we have to be content with fitting the electron transfer kinetics to phenomenological expressions with freely adjustable parameters. Instead, we can reverse these challenges. Is it possible now to start with a crystal structure and to calculate the reaction center's absorption, CD and linear dichroism spectra and the spectroscopic effects of external fields? And can we predict how rapidly an electron will move from one component of the reaction center to another? If so, we can look forward to a clearer understanding of the mechanism of the primary electron transfer reaction, and to having a sharpened set of tools for exploring other complexes that still resist crystalization.

One difficulty in carrying out such calculations is that the pigments in the reaction center are very close together. This means that a point-dipole treatment of the exciton interactions among the pigments is inherently unreliable. In addition, charge-transfer (CT) transitions in which an electron moves from one molecule to another can contribute strongly to the spectroscopic properties by mixing with the local excitations of the individual molecules. Hopefully, the effects of CT transitions on the spectroscopic properties will tell us something useful about the photochemical electron transfer reaction. However, any analysis that aims to dispense with point-dipole approximations and to include CT transitions must be quantum mechanical in nature. The problem here is that the reaction center is far too large and complicated to be amenable to an *ab-initio* quantum mechanical treatment.

Our approach to these problems has been to start by writing molecular orbitals for BChl and BPh, and to adjust these wavefunctions so that they account as well as possible for the spectroscopic properties of the *monomeric pigments in solution.* We then seek to describe the excited states of the reaction center as linear combinations of the local excited states of the six pigments, along with the intermolecular CT transitions [4,5]. The interaction matrix elements that mix these different transitions can be evaluated by using semiempirical Coulomb and resonance integrals that have been parametrized in previous studies of other π-electron systems. Our aim is not simply to adjust the molecular orbitals of the monomers or the semiempirical integrals to *fit* the properties of the reaction center. Instead, we hope to use the agreement (or disagreement) between the calculated and observed properties as a test of the approach. If the calculated matrix elements lead to a gcod account of the measured spectroscopic properties, we can have some confidence that they also will serve in calculations of quantities that cannot be measured directly, such as the terms that govern the rate of electron transfer.

CALCULATIONS OF SPECTROSCOPIC PROPERTIES

In the formalism outlined above, one can describe an excited state of the reaction center as

$$\Psi_\alpha = \Sigma_i B^\alpha_i \psi_i \qquad (1)$$

The coefficients B^α_i are obtained by diagonalizing the interaction Hamiltonian of the special pair of Bchls (P) and the local π-π* excited states of the other BChls and BPhs, and the ψ_i are given by

$$\psi_i = \Sigma_N C_{i,N} \phi_N = \Sigma_N C_{i,N} \phi_{n1\rightarrow n2} \qquad (2)$$

Here $\phi_{n1\rightarrow n2}$ is a wavefunction for singlet excitation from molecular orbital ϕ_{n1} to ϕ_{n2}. For local transitions of the chromophores, the coefficients $C_{i,N}$ are obtained by diagonalizing configuration-interaction matrices for the isolated monomers. If one considers the top two filled molecular orbitals and the first two empty orbitals, each molecule has four such local transitions, Q_y, Q_x, B_x and B_y. The pair of BChls that make up P also has 8 CT transitions, for which ϕ_{n1} and ϕ_{n2} are on separate molecules.

At the present stage of the spectroscopic calculations, we still need to introduce some adjustable parameters at the level of the reaction center. These are the energies of the local and CT transitions, which make up the diagonal terms of the Hamiltonian matrix. Although one can put limits on the energies of the CT transitions [5], these energies are particularly sensitive to interactions of the pigments with the protein and solvent. We therefore have varied the CT transition energies and examined how this affects the calculated spectroscopic properties. Starting with the *Rhodopseudomonas viridis* crystal structure [1], good agreement between the calculated and observed spectra was obtained by putting the energy of the lowest CT transition of the special pair of BChls (P_L and P_M) near 14,500 cm^{-1} [5]. This is well above the local Q_y transitions of the BChls, which are at about 12,000 cm^{-1} for BChl-b, but the energies are close enough so that the two types of transitions can mix extensively in the reaction center's excited states. The calculated matrix elements that mix the CT and Q_y transitions are actually larger than the Q_y exciton interactions, and can account for most of the red shift of the long-wavelength absorption band to 960 nm.

It also appears necessary to assume that the CT transitions in which an electron moves from P_L to P_M lie somewhat above the corresponding transitions in the opposite direction [5]. Interchanging the energies of the CT transitions in the two directions gives a linear dichroism spectrum that is at odds with the experimentally measured spectrum [6] in the region between 600 and 700 nm. An asymmetry of the CT transitions also helps to account for the sensitivity of the long-wavelength band to external fields [7,8]. If the CT transiitions in opposite directions were degenerate, the lowest excited state would contain similar contributions from the two transitions and would involve only a small change in permanent dipole moment.

One conclusion that emerges from the calculations is that the *Rps. viridis* reaction center's absorption bands in the region from 810 to 850 nm contain extensively mixed contributions from all six pigments. For example, the contributions of the accessory BChls to the 850-nm band appear to be comparable to those of P_L and P_M. It thus is an oversimplification to identify this band solely with a symmetric Q_y exciton transition of the special pair, as is often done.

The same treatment accounts reasonably well for the absorption and CD spectra of *Rhodobacter sphaeroides* reaction centers (Fig. 1). These calculations were done just as those for *Rps. viridis* [4,5] except that the *Rb. sphaeroides* crystal structure [2,9] was used, and molecular orbitals and transition energies appropriate for BChl-a and BPh-a replaced those for BChl-b and BPh-b. (We assumed that all four of the individual BChls in the *Rb. sphaeroides* reaction center would absorb at 800 nm if they were bound to the protein but did not interact with each other.) To make the calculated spectra agree with the experimental spectra, the lowest CT transition had to be put in the range of 13,000 cm^{-1}. This is somewhat lower than the value used for *Rps. viridis*, but is still above the Q_y transition energy. We emphasize that the estimates of the CT energies for both species should be viewed as tentative, because the crystal structures are still under refinement. As the refinement progresses, we hope that the microscopic simulation approach described below will allow us to obtain the energies of the CT states directly.

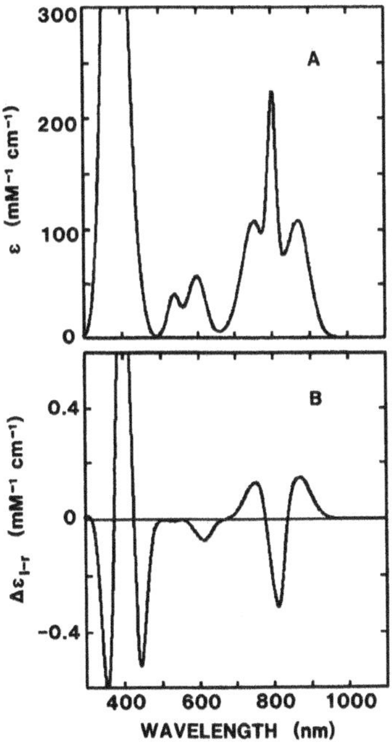

Fig. 1. Absorption (A) and circular dichroism (B) spectra of *Rb. sphaeroides* reaction centers, calculated as described in the text and in [4,5].

The resonance integrals that mix local and CT transitions also underlie the electronic interaction terms $H_{\alpha\beta}$ that control the rate of electron transfer from P to the neighboring molecules. To discuss these terms, one first needs to define the initial and final states for the electron transfer reaction, α and β. We shall take these to be "diabatic" states, which means that they do not diagonalize the complete Hamiltonian of the system, but we want them to conform to physical intuition. The initial excited state for the primary electron transfer reaction, P*, can be defined simply as the lowest-energy Ψ_α of eqns. 1 and 2. Similarly, the CT state P^+B^-, which could be an intermediate in the transfer of an electron from P to the BPh on the L side of the reaction center, is made up of the lowest-energy combination of transitions in which an electron moves from P_L or P_M to the accessory BChl on the L side (B). A CT state of this nature can be described as

$$\Psi_\beta = \Sigma_\omega B^\beta{}_\omega \Omega_\omega = \Sigma_\omega B^\beta{}_\omega \phi_{\omega 1 \to \omega 2} \qquad (3)$$

where the coefficients $B^\beta{}_\omega$ are obtained by solving the interaction submatrix [4] for the relevant CT transitions. The dominant components of P^+B^- involve electron transfer from the highest filled molecular orbital of P_L or P_M to the first empty orbital of B. P^+H^- can be defined similarly as a combination of the 8 CT transitions in which an electron moves from P_L or P_M to the BPh (H). The distinction between the diabatic *states* (Ψ_α and Ψ_β) and the individual orbitals and transitions that underlie these states is an important one that is discussed in more detail elsewhere [4,5,10,11].

The Hamiltonian matrix elements that mix Ψ_α and Ψ_β are given by [4,10]

$$H_{\alpha\beta} = \Sigma_{i,\omega} B^\alpha{}_i B^\beta{}_\omega U_{i,\omega} = \Sigma_{i,N,\omega} B^\alpha{}_i B^\beta{}_\omega C_{i,N} A_{i,\omega} \qquad (4)$$

where $U_{i,\omega} = \langle\psi_i|H|\Omega_\omega\rangle$. The $A_{i,\omega}$ are matrix elements that connect the individual local π-π^* transitions, or the individual CT transitions of P that contribute to Ψ_α, with the individual CT transitions that make up Ψ_β:

$$A_{i,\omega} = \Sigma_{s,t}(\delta_{n1,\omega 1} v_{n2,t} v_{\omega 2,s} - \delta_{n2,\omega 2} v_{n1,t} v_{\omega 1,s})\beta_{st} \qquad (5)$$

where the v's are the atomic expansion coefficients for the donor and acceptor molecular orbitals and β_{st} is the resonance integral between atomic p_z orbitals on atoms s and t.

We have used these expressions to evaluate the overall matrix elements that mix P* with P^+B^- and with the alternative intermediate state B^+H^-, as well as the matrix elements that mix P^+B^- and B^+H^- with P^+H^- [10,11]. Figure 2 shows the results obtained for *Rps. viridis*. The largest matrix elements are obtained with the pathway in which P* first transfers an electron to B, generating P^+B^-, and B^- then passes an electron on to H. As we shall discuss below, the calculated matrix elements are reasonably close to the values that are required in order to account for the experimental observations [12-15] that an electron arrives on H about 3 ps after the reaction center is excited and that there is no detectable accumulation of any intermediate states. Electron transfer by a superexchange mixing of P* with P^+B^- has smaller matrix elements than the direct "hopping" pathway [10,11,16]. The alternative pathway via B^+H^- also gives somewhat smaller matrix elements (Fig. 2), but this difference may not be significant in view of the uncertainties in the structure and in the details of the resonance integrals at large distances. However, the transfer of an electron from P* directly to H would be much too slow to explain the experimental observations.

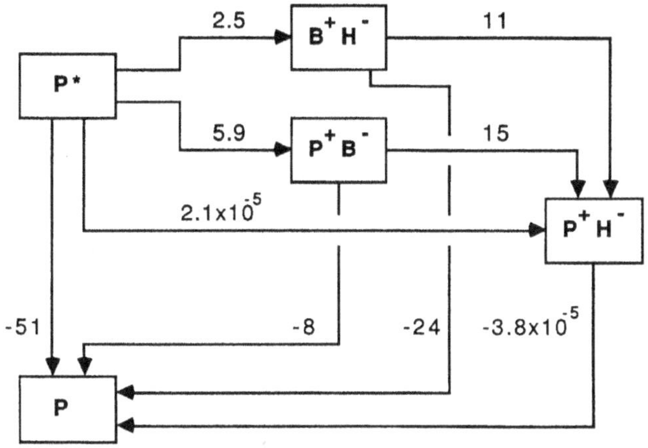

Fig. 2. Hamiltonian matrix elements (in cm^{-1}) connecting diabatic states of the *Rps. viridis* reaction center. The rate constants for electron transfer depend on the squares of the matrix elements (see eq. 6). The decay paths to the ground state (P) are slowed by their small Franck-Condon factors, which are not shown here. For further details see the text and [10,11].

Because the atomic resonance integrals fall off exponentially with distance, the matrix elements given in Fig. 2 depend critically on structural details. Movements of the molecules by ±0.5 Å, which would be within the resolution of the X-ray data, can change the calculated matrix elements by a factor of 2 or 3. With the crystallographic coordinate data set that we have used (a May, 1985 data set kindly provided by Drs. H. Deisenhofer and H. Michel), the major contributions to the matrix element that mixes P* with P^+B^- come from the interactions of the ethylidine group on ring II of B with carbons of rings III and V of P_L. It is therefore puzzling that the kinetics measured experimentally in *Rb. sphaeroides* are very similar to those measured in *Rps. viridis* [14,15], because the BChl-a that is present in *Rb. sphaeroides* has a saturated ethyl group instead of the ethylidine group. The matrix elements calculated for the formation of P^+B^- or B^+H^- using the *Rb. sphaeroides* structure are about a factor of 10 smaller than those obtained with *Rps. viridis*. This discrepancy could mean that the resonance integrals fall off more slowly at large distances than we have assumed, that orbital overlaps with amino acid side chains contribute to the pathway for electron transfer, or that nonconjugated (σ) atoms of the chromophores need to be considered. Our treatment considers only the π atoms explicity, because one expects these to make the largest contributions in many cases. Theoretical treatments that include the σ atoms and the amino acid side chains will need to be calibrated carefully in simpler systems. This is essential, because all-valence-electrons calculations of very large systems have an inherent instability that can lead to an overestimate of long-range coupling strengths unless one includes extensive configuration interactions.

ENERGETICS AND DYNAMICS OF ELECTRON TRANSFER

In addition to depending on the interaction matrix element $H_{\alpha\beta}$, the rate of an electron transfer reaction also depends on the energies of the initial and final states. In order for energy to be conserved in the reaction, the energies of the two states must be identical at the instant of electron transfer. This resonance must not depend on thermal activation of the reactant state in the reaction center, because the rate of electron transfer is even faster at cryogenic temperatures than it is at room temperature [13,17]. Since we have concluded above that the internal CT transitions of P lie above the reaction center's lowest excited state, the question arises whether either P^+B^- or B^+H^- can be sufficiently close to P^* in energy for electron transfer to occur rapidly. This question is particularly germane to the superexchange mechanism, in which the mixing of P^* and P^+B^- decreases abruptly as the two states move apart in energy. Unfortunately, one cannot determine the energies of P^+B^-, P^+H^- or B^+H^- reliably from the reaction center's spectroscopic properties, because these states do not mix strongly enough with the locally excited states.

We have evaluated the energies of the CT states of the *Rb. sphaeroides* reaction center by following an approach that has been used previously in studies of charged groups in other proteins. A central feature of the approach is to take as reference states the corresponding states of the electron carriers when the charged species are separated to an infinite distance in a polar solvent [18]. The free energy change for transferring an electron from one carrier to another in such a reference state can be obtained simply from the experimentally measured midpoint redox potentials. The free energy changes in the reaction center then can be calculated from the reference values by subtracting the difference between the solvation energies of the ions in solution and in the reaction center [16]. Solvation energies can be found by a perturbation approach in which one essentially carries out an adiabatic "charging" of the ions [19]. The time dependence of the energies can be explored by evaluating the energies as part of a molecular dynamics simulation [20]. In order to obtain meaningful results, such a calculation must include the effects of polarization of the protein by the charged carriers, and the effects of loosely bound water molecules in the reaction center. If one allows water molecules to fill the cavities in the *Rb. sphaeroides* crystal structure, there are approximately 5 molecules of water in a shell of 4. 5 Å around P and B [16].

The result of these calculations [16] is that P^+B^- appears to lie 700 ± 1700 cm^{-1} below P^* in free energy. This number is for a relaxed state, in which the nuclear coordinates of the electron carriers, the protein, and the associated water atoms have been allowed to adjust for several picoseconds to accomodate the charges on P^+ and B^-. If one considers the coordinates during a molecular dynamics trajectory on the uncharged state P^*, P^+B^- is found to lie 700 ± 1700 cm^{-1} above P^*. The "solvent" reorganization energy associated with electron transfer from P to B thus appears to be only about 1400 cm^{-1}, which is much smaller than it would be if the ions were formed in aqueous solution. (A similar calculation for the ions in water gives a reorganization energy of about 10,500 cm^{-1}.) It is important to note that these energy values were obtained directly from the crystal structure and the experimentally measured redox potentials, without *any* adjustable parameters. The microscopic simulation approach frees one from having to use an arbitrary macroscopic dielectric constant. By using the experimental redox potentials and the calculated solvation energies, one can avoid unreliable quantum mechanical calculations of the gas-phase energies of the large chromophores.

Figure 3A shows the time dependence of the energy difference between P^+B^- and P^*, as calculated during a trajectory on P^*. Whenever the energy difference goes through zero, there is an opportunity for electron transfer, with a probability that depends on the square of the Hamiltonian matrix element. The probability of a transition during the small time interval between $t_n-\delta$ and $t_n+\delta$ can be evaluated from the expression [16,20]

$$P_{\alpha\beta}(t_n) = (1/2\delta)\left|\int_{t_n-\delta}^{t_n+\delta}(-i/\hbar)H_{\alpha\beta}\exp\{-(i/\hbar)\int_0^t(\varepsilon_\beta-\varepsilon_\alpha)dt'\}dt\right|^2_\alpha \qquad (6)$$

Here ε_α and ε_β are the energies of the initial and final states, with the subscript α indicating that the energies are calculated during a trajectory in the initial state. To find the time-dependence of the populations of P*, P^+B^- and P^+H^-, one solves the coupled set of equations with the boundary condition that only P* is populated initially. While a rigorous simulation would require averaging the results over many different trajectories, the desired average can be approximated by taking the time-dependent energy gap from a single trajectory and averaging the results obtained by starting at many randomly chosen initial points. (This procedure will be rigorous if the autocorrelation function of the energy gap along the single trajectory converges.)

Figure 3B shows the calculated decay of P*, the formation of the transient intermediate state P^+B^-, and the development of the final state P^+H^-. Because the interaction matrix elements cannot yet be obtained reliably for *Rb. sphaeroides*, we left them as adjustable parameters in these calculations. For the calculations shown in Fig. 3B, we used a matrix element of 10 cm^{-1} for the mixing of P* and P^+B^-, and 25 cm^{-1} for the mixing of P^+B^- with P^+H^-; these values are within a factor of 2 of the matrix elements that were calculated above for *Rps. viridis* (Fig. 2). The results agree well with the electron transfer kinetics seen experimentally [12-15].

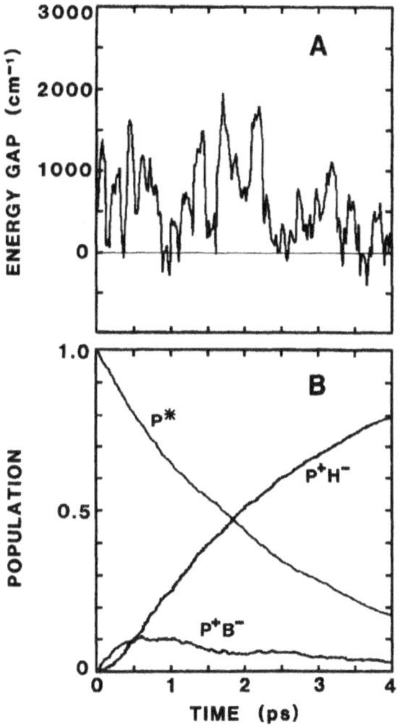

Fig. 3. (A) Calculated energy gap between P^+B^- and P* in *Rb. sphaeroides* reaction centers, for a trajectory on the potential surface of P*. The trajectory was calculated for a temperature of 100 K, following an equilibration period of about 10 ps. (B) Populations of P*, P^+B^- and P^+H^- as a function of time, as calculated from the energy gap shown in (A) with averaging over 100 randomly chosen starting points. This simulation includes superexchange in addition to the pathway in which P^+B^- forms as a discrete intermediate; superexchange is calculated to contribute about 10% of the rate. For details, see the text and [16].

The present estimates of the average energy difference betwen P* and P^+B^- ($\langle \varepsilon_\beta - \varepsilon_\alpha \rangle_\alpha$) have an uncertainty of approximately $\pm 1700\ cm^{-1}$, which comes mainly from the uncertainties in the crystallographic coordinates. In Fig. 3, the average energy difference was set in the center of the calculated range. If the energy difference is increased to the upper end of the calculated range, intersections of P* and P^+B^- are prevented, and the transfer of an electron to B becomes too slow to account for the experimental results no matter how much one increases the interaction matrix elements. In this situation, superexchange could become the dominant mechanism of electron transfer. However, the superexchange pathway also is found to be relatively slow under these conditions, unless the matrix elements that mix P*, P^+B^- and P^+H^- are increased substantially. If the matrix element between P* and P^+B^- is raised to $35\ cm^{-1}$, and that between P^+B^- and P^+H^- is raised to $80\ cm^{-1}$, the calculated rate of electron transfer by the superexchange mechanism is still less than 1/5 the rate shown in Fig. 3B [16].

The relative merits of the alternative mechanisms for the electron transfer reaction have been discussed recently from a variety of points of view [10,11,16,21-25]. In calculations like those illustrated in Fig. 3 [16], B^+H^- is estimated to be somewhat above P^+B^- in energy, which would appear to put it in a less favorable position for electron transfer from P*. As noted above, the interaction matrix elements also may be less favorable for the pathway through B^+H^- than they are for the pathway via P^+B^-, at least in *Rps. viridis* (Fig. 2). However, we shall need more refined calculations of both the matrix elements and the energies in order to choose between the two mechanisms.

A remarkable feature of the pathways through either P^+B^- or B^+H^- is that, although the calculated reorganization energies associated with the formation of the CT states are much smaller than the reorganization energies in aqueous solution, the calculated energies of the CT states are only 1-2 kcal/mol higher than the corresponding energies in solution. The reaction center appears to be able to stabilize either of these CT states very effectively, but at the same time to keep the reorganization energies small. This may suggest a hint to the longstanding puzzle of why the reaction center uses a BChl dimer. The oxidized dimer (P^+) presents a highly delocalized charge, which implies a relatively small solvation energy in addition to a moderate oxidation potential. This may allow the permanent and induced dipoles of the protein to provide nearly as much solvation to the ion as water could, but with a smaller reorganization energy.

ACKNOWLEDGEMENTS

We thank Drs. M. Schiffer and C. Chang for providing the X-ray coordinates of the *Rb. sphaeroides* reaction center, National Science Foundation grants PCM-8303385, PCM-8312371 and PCM-8316161 and the Dept. of Energy for support, and Digital Equipment Corp. for help with the purchase of a computer.

REFERENCES

1. Deisenhofer, J., Epp, O., Miki, K., Huber, R., and Michel, H., 1985, Nature, 318: 618-624.
2. Chang, C. -H., Tiede, D., Tang, J., Smith, U., Norris, J., and Schiffer, M., 1986, FEBS Lett., 186: 201-203.
3. Allen, J. P., Feher, G., Yeates, T. O., Komiya, H., and Rees, D., 1986, Proc. Natl. Acad. Sci., 83: 8589-8593.

4. Warshel, A., and Parson, W. W., 1987, J. Am. Chem. Soc., 109: 6143-6152.
5. Parson, W. W., and Warshel, A., 1987, J. Am. Chem. Soc., 109: 6152-6153.
6. Breton, J., 1985, Biochim. Biophys. Acta, 810: 235-245.
7. De Leeuv, D., Malley, M., Butterman, G., Okamura, M. Y., and Feher, G., 1982 Biophys. J. 37: 111a.
8. Lockhart, D. J., and Boxer, S. G., 1987, Biochem., 26: 664-668.
9. Tiede, D., Budil, D., Gast, P., Tang, J., Wasielewski, M., Chang, C. -H., Schiffer, M., and Norris, J. R., 1988, This Volume.
10. Warshel, A., Creighton, S., and Parson, W. W., 1988, J. Phys. Chem., submitted.
11. Parson, W. W., Creighton, S., and Warshel, A., 1987, in: "Primary Reactions in Photobiology," T. Kobayashi, ed., Springer-Verlag, Berlin, pp. 43-51.
12. Holten, D., Hoganson, D., Windsor, M. W., Schenck, C. C., Parson, W. W., Migus, A., Fork, R. L., and Shank, C. V., 1980, Biochim. Biophys. Acta 592: 461-477.
13. Woodbury, N. W., Becker, M., Middendorf, D. and Parson, W. W., 1985, Biochem. 24: 7516-7521.
14. Martin, J. -L., Breton, J., Hoff, A., Migus, A. and Antonetti, A. (1986) Proc. Natl. Acad. Sci. 83: 957-961.
15. Breton, J., Martin, J. -L., Migus, A., Antonetti, A. and Orsag, A., 1986, Proc. Natl. Acad. Sci., 83: 5121-5125.
16. Creighton, S., Hwang, J. -K., Warshel, A., Parson, W., and Norris, J., 1987, Biochem., submitted.
17. Breton, J., Fleming, G. R., and Martin, J. -L., 1988, This Volume.
18. Warshel, A., and Russell, S., 1984, Quart. Rev. Biophys., 17: 283-422.
19. Warshel, A., Sussman, F., and King, G., 1986, Biochem. 25: 8368-8372.
20. Warshel, A., and Hwang, J. -K., 1986, J. Chem. Phys., 84: 4938-4957.
21. Marcus, R., 1987, Chem. Phys. Lett., 133: 471-477.
22. Parson, W. W., Scherz, A., and Warshel, A., in: "Antennas and Reaction Centers of Photosynthetic Bacteria," M. E. Michel-Beyerle, ed., Springer-Verlag, Berlin, pp. 122-130.
23. Fischer, S. F., and Scherer, P. O. J., 1987, Chem. Phys. 115: 151-158.
24. Michel-Beyerle, M. E., Plato, M., Deisenhofer, J., Michel, H., Bixon, M., and Jortner, J., 1987, J. Chem. Phys., in press.

ANALYSIS OF A,LD,CD,ADMR AND LD-ADMR SPECTRA FOR THE REACTION CENTERS OF RPS.VIRIDIS, RB.SPHAEROIDES, C.AURANTIACUS AND MODIFIED RB.SPHAEROIDES

P.O.J. Scherer and Sighart F.Fischer

Physikdepartment der Technischen Universität München
D-8046 Garching

INTRODUCTION

Optical spectra of different kinds are very important tools for the investigation of the reaction center. Steady state spectra are sensitive to excitonic interactions between the pigments whereas the electron transfer process is reflected by time dependent spectroscopy. Since the structure has been resolved for the RC's of Rps.viridis /1/ and Rb.sphaeroides /2/ it has become possible to study models for the molecular interactions based on the structural information and to check the quality of the model by comparing calculated model spectra with experimental data. In this way approximations to the molecular states as well as to the intermolecular interactions can be obtained which are essential for the understanding of the electron transfer mechanism.

Theoretical models are approaching the problem on different levels of complexity. On the one hand a simple exciton model is very successful in modelling the experimental spectra /3-7/ but it does not explain the energy positions of the exciton states satisfactorily. Also it cannot describe the interaction within the special pair dimer adequately. On the other hand quantum chemical calculations /8/ are very complicated for such a large molecular system. So far only semiempirical methods (PPP,CNDO, INDO) seem feasible. These methods, however, have to be modified as they were developed to calculate molecular wavefunctions for smaller systems and do not incorporate the long range behaviour of the electronic wavefunctions properly which is essential for intermolecular interactions.

In the first part of this paper we present a detailed study of several reaction centers. On the basis of the Rps.viridis structure /1/ the exciton model is applied to the structurally similar reaction centers of Rb.sphaeroides, C.aurantiacus and the modified Rb.sphaeroides. For each reaction center the parameter set is optimized by a simultaneous fit of absorption, linear and circular dichroism as well as the triplet minus singlet absorption difference spectrum together with its linear dichroism.

In the second part we discuss results from PPP calculations where the reaction center of Rps.viridis is treated as a hexamer consisting of four bacteriochlorophyll and two bacteriopheophytine molecules.

THE EXCITON MODEL

A system of interacting molecules can often be described using excitations of the isolated molecules as a starting point. In the simplest version of such a model only one excitation $S_0 \rightarrow S_1$ is considered for each molecule and the wavefunction of the interacting system is approximated by a linear combination of local excitations

$$\Psi = \sum C_n \mid S_1^{(n)} \rangle \tag{1}$$

The local excitations are coupled via the Coulomb interaction V between electrons on different molecules. In the restricted basis (1) the hamiltonian of the exciton model is represented by a symmetric matrix of dimension 6 :

$$H = \sum V_{m,n} \mid m\rangle\langle n| + \sum E_m \mid m\rangle\langle m| \tag{2}$$

$$= \begin{bmatrix} E_1 & V_{1,2} & \cdots & V_{1,6} \\ V_{2,1} & E_2 & \cdots & V_{2,6} \\ \vdots & & & \vdots \\ V_{6,1} & V_{6,2} & \cdots & E_6 \end{bmatrix}$$

The energy positions E_m are parameters of the model which can be in principle calculated using more elaborate theoretical methods including higher excitations and environmental shifts (see later). The coupling matrix elements

$$V_{m,n} = \int \varphi^*_m(1) \quad \varphi_m(1) \quad r^{-1}_{mn} \quad \varphi^*_n(2) \; \varphi_n(2) \quad dr^3_1 dr^3_2 \tag{3}$$

can be approximately calculated using a multipole expansion and retaining only the lowest order contribution which is of the dipole-dipole type.

$$V_{m,n} = \text{const} \times \frac{3\vec{\mu}_m\vec{\mu}_n R^2_{mn} - (\vec{R}_{mn}\vec{\mu}_m)(\vec{R}_{mn}\vec{\mu}_n)}{R^5_{mn}} \tag{4}$$

where $\vec{\mu}_n$ is the transition dipole moment of the n-th pigment and $\vec{R}_{mn}$ the vector connecting the centers of pigments m and n. The absolute value of $\vec{\mu}_n$ is correlated with the extinction coefficient of the Q_y band of isolated molecules. It can be determined from experimental data /9,10/. There are, however, major uncertainties due to the vibronic structure of the Q_y band which overlaps with the Q_x transition and also due to the unknown local dielectric constant within the protein environment. We took intensities of $40D^2$ ($37D^2$) for the bacteriochlorophyll-b(a) and $30D^2$ ($23D^2$) for the bacteriopheophytine-b(a) molecules. The direction of $\vec{\mu}_n$ is assumed to be parallel to the N_I - N_{III} axis of the molecule as suggested by INDO calculations.

Approximation (4) is appropriate for coupled molecules at distances large compared to the spatial extension of the local excitation as measured by the transition dipole ($R_{mn} \gg \mu/e \approx 1.4$ Å for BCHL molecules).

Care has to be taken in the case of the special pair .It seems more appropriate to treat the latter as a supermolecule and to replace the local excitations of its constituents by two excitations of the dimer which will be denoted as P*(-) and P*(+) in the following. Neglecting minor asymmetries of the dimer and possible reduction of the dimer intensity by charge transfer contributions we assume that the transition dipoles of the P*(±) transition are given by $\vec{\mu}_{\pm} = 2^{-1/2}(\vec{\mu}_{MP} \pm \vec{\mu}_{LP})$.These transition dipoles are also used to calculate the coupling of the dimer states to the accessory monomers and the pheophytines. Formally this treatment of the dimer states can be incorporated into (2) taking $E_{BC_{LP}} = E_{BC_{MP}}$ and treating the coupling $V(BC_{LP},BC_{MP})$ as a parameter similar to the diagonal energies.

EXCITON MODEL FOR THE TRIPLET STATE ^3P

Microwave resonance techniques (ODMR,MIAC) /11/ provide a lot of experimental data on a variety of reaction centers. Such experiments are probing the absorption spectrum of the reaction center with the primary donor in the triplet state ^3P and therefore provide additional information which is not accessible to normal optical spectroscopy.

The exciton model can be extended to describe the state ^3P and its excitations. In a very simple model the lowest triplet ^3P is approximated by an excitonic wavefunction which also contains some charge transfer character /12/.

$$(5) \qquad {}^3P = A\ {}^3BC_{MP} + B\ {}^3BC_{LP} + C\ {}^3(BC^+_{LP}BC^-_{MP}) + D\ {}^3(BC^-_{LP}BC^+_{MP})$$

The contribution of charge separated states can be estimated from the experimental zfs tensor. Typical values of C^2+D^2 are between 0.1 and 0.3 /13/.

In the Q_Y region the excitation spectrum of ^3P contains four contributions which are very similar to the corresponding singlet excitations. These are the transitions

$$(6) \qquad {}^3P \rightarrow {}^3P\ BC^*_{MA}\ ,\ {}^3P \rightarrow {}^3P\ BC^*_{LA}\ ,\ {}^3P \rightarrow {}^3P\ BP^*_L \text{ and } {}^3P \rightarrow {}^3P\ BP^*_M\ .$$

Neglecting minor changes due to the presence of the triplet excitation on the special pair these transitions have the same energies and intensities as the corresponding singlet transitions. In addition we consider two doubly excited states of the dimer

$$(7) \qquad \begin{aligned} T^*(+) &= \alpha\ BC^*_{LP}\ {}^3BC_{MP} + \beta\ BC^*_{MP}\ {}^3BC_{LP} \\ T^*(-) &= \beta\ BC^*_{LP}\ {}^3BC_{MP} - \alpha\ BC^*_{MP}\ {}^3BC_{LP} \end{aligned}$$

The oscillator strength of the transitions $^3P \rightarrow T^*(\pm)$ and the coupling to the excitations (6) are calculated from the transition dipoles of the singlet excitations. The energy positions of $T^*(\pm)$ enter as new model parameters together with the delocalization of ^3P and $T^*(\pm)$.

RESULTS

The low temperature (~4K) experiments on Rps.viridis /14-17/ can be reproduced satisfactorily if we assume that the splitting of the two dimer excitations is ~ 2000 cm^{-1},i.e. the higher dimer excitation is close to the excitation of the accessory monomers. Table 1 shows the calculated interaction matrix elements and the diagonal energies which have been obtained from a simultaneous fit of absorption, LD, ADMR and LD-ADMR spectra as shown in figure 1. The polarization axis was taken parallel to the C_2 symmetry axis. The stick spectrum of the exciton model was dressed with asymmetric line profiles. The same profile was used for the accessory monomers and the pheophytines in all spectra. For the dimer bands the width had to be increased. Unfortunately no CD data were available at 4K. Figure 1 shows data /16/ at 77K together with two simulations. The full line was calculated with the 4K parameter set. It shows a large amplitude in the monomer region and very little at the special pair band. The discrepancy with the experimental data is mainly due to the increase of linewidth at higher temperature. The dashed line shows a second calculation, where the positions of the dimer bands and the line widths were adjusted to fit experimental absorption and LD data at 77K. The agreement with the experimental data is improved. It cannot be expected to become much better since the experimental data are nonconservative in the Q_y region whereas in the exciton model the sum of all CD intensities within the Q_y region is zero as long as higher excitations are neglected. For the simulation of the T-S spectrum we had to assume that one of the two triplet excitations is below the monomers and the other above. We need not shift the monomer excitations in the triplet spectrum. The calculated interactions in the triplet state are shown in table 5. In contrast to Hoff's analysis /17/ we get acceptable results not only if we assume localization of the triplet on the L-side of the dimer. A fit of similar quality is achieved with partial localization on the M-side with $(A^2-B^2)/(A^2+B^2)=0.25$, $\alpha^2-\beta^2=0.28$ and exchanged positions of the states $T^*(\pm)$. This second parameter set is closer related to our original simulation which did not allow for different delocalization for ^3P and the $T^*(\pm)$ states.

For Rb.sphaeroides and C.aurantiacus the splitting of the dimer states has to be reduced to $\sim$1400cm^{-1} at 4K . The interaction matrices are shown in tables 2-4. The smaller dimer splitting correlates with the smaller transition dipoles of the a-type pigments. For both reaction centers the calculated LD spectrum does not agree with the experimental data if the polarization axis is taken parallel to the C2 axis. The agreement is strongly improved if the polarization axis is taken nearly perpendicular ($\sim$80°) to the C2-axis pointing from one accessory monomer to the other (see figures 2-3) .In comparison to the results for Rps.viridis the experimental CD spectra are nearly conservative. Agreement with the calculations is quite perfect, especially for C.aurantiacus. For the T-S spectra partial delocalization of the triplet states had to be assumed. In the calculated spectra (figures 2-4) the width of the higher excitation $T^*(-)$ was increased to optimize the fit. Figure 4 shows spectra of the modified Rb.sphaeroides reaction center /18,19/. For the calculation the same line profiles were used as in figure 2. The quality of the fit can be improved, however, if the line profiles are slightly readjusted / 5/. The calculation was performed using the Rps.viridis structure also for Rb.sphaeroides and C.aurantiacus. For the modified Rb.sphaeroides we had to assume that the structure relaxes after the removal of BC_{MA}. The fit of the experimental absorption and T-S spectra ,especially the position and amplitude of the monomer band could be reproduced quite well with the assumption that the center of the special pair is translated by 2-3 Å. Table 4 shows the calculated interactions. Unfortunately at present no experimental data have been published on the LD and LD-TS spectra. The agreement with preliminary experimental results /20/ seems quite acceptable.

TABLE 1

Singlet interactions in cm^{-1} and diagonal energies relative to 12500 cm^{-1} for Rps.viridis at 4K (77K)

	BC_{MA}	BC_{LP}	BC_{MP}	BC_{LA}	BP_M	BP_L
BC_{MA}	-350	-157	-41	23	158	-11
BC_{LP}	-157	-1490(-1340)	980	-54	31	-8
BC_{MP}	-41	980	-1490(-1340)	-178	-10	36
BC_{LA}	23	-54	-178	-510	-10	162
BP_M	158	31	-10	-10	140	6
BP_L	-11	-8	36	162	6	-100

TABLE 2

Singlet interactions in cm^{-1} and diagonal energies relative to 12500 cm^{-1} for Rps. sphaeroides at 4K (77K)

	BC_{MA}	BC_{LP}	BC_{MP}	BC_{LA}	BP_M	BP_L
BC_{MA}	30	-146	-38	22	133	-9
BC_{LP}	-146	-570(-470)	680(600)	-50	26	-7
BC_{MP}	-38	680(600)	-570(-470)	-165	-8	30
BC_{LA}	22	-50	-165	30	-8	137
BP_M	133	26	-8	-8	760	5
BP_L	-9	-7	30	137	5	630

TABLE 3

Singlet interactions in cm^{-1} and diagonal energies relative to 12500cm^{-1} for C.aurantiacus

	BP_{MA}	BC_{LP}	BC_{MP}	BC_{LA}	BP_M	BP_L
BP_{MA}	760	-111	-75	13	113	-6
BC_{LP}	-111	-500	680	-50	26	-7
BC_{MP}	-75	680	-500	-165	-8	30
BC_{LA}	13	-50	-165	-90	-8	137
BP_M	113	26	-8	-8	880	5
BP_L	-6	-7	30	137	5	770

TABLE 4

Singlet interactions in cm^{-1} and diagonal energies relative to $12500 cm^{-1}$ for the modified Rb.sphaeroides The monomer BC_{MA} is removed and the dimer pigments are shifted relative to the viridis structure by 2Å.

	BC_{LP}	BC_{MP}	BC_{LA}	BP_{M}	BP_{L}
BC_{LP}	-570	680	45	36	-18
BC_{MP}	680	-570	-120	-2	28
BC_{LA}	45	-120	30	-8	137
BC_{MP}	36	-2	-8	760	5
BC_{LP}	-18	28	137	5	630

TABLE 5

Triplet parameters and calculated couplings

Parameters	Rps.viridis	Rb.spher.R26	C.aurant.	modified R26R
CT contribution to T				
$C^2 + D^2$	35%	20%	12%	20%
Localisation of T				
$(A^2-B^2)/(A^2+B^2)$	-1.0	0.45	0.62	0.45
Localisation of $T^*(\pm)$				
$\alpha^2-\beta^2$	-0.82	0.45	0.45	0.45
Energy positions in cm^{-1}				
$E_{T^*}(+)$	-290	-190	-190	-190
$E_{T^*}(-)$	-980	420	370	420
Calculated coupling in cm^{-1}				
$V(T^*(+),BX_{MA})$	-31	-103	-89	—
$V(T^*(+),BC_{LA})$	-137	-73	-71	0
$V(T^*(+),BP_{M})$	-7	15	17	23
$V(T^*(+),BP_{L})$	28	3	2	-5
$V(T^*(-),BX_{MA})$	-10	43	-41	—
$V(T^*(-),BC_{LA})$	-43	-46	-35	-66
$V(T^*(-),BP_{M})$	-2	-14	-14	-15
$V(T^*(-),BP_{L})$	9	15	14	18

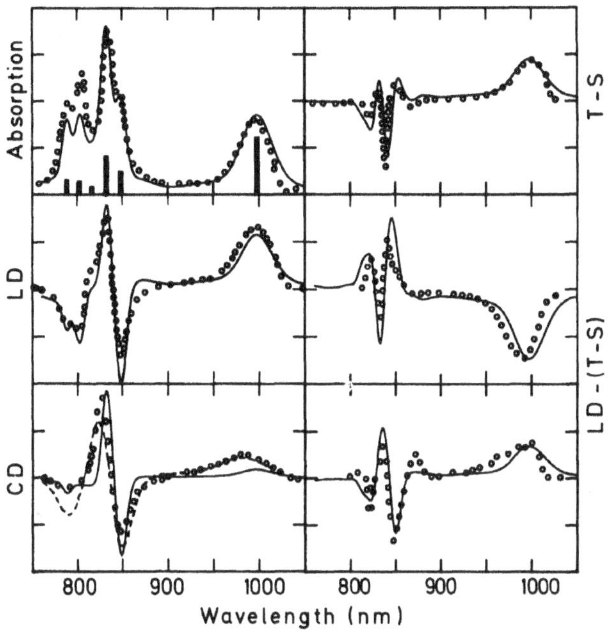

FIGURE 1. Model calculations for Rps.viridis (curves) are compared with experimental data /14-17/ (points). Exciton transitions are shown as bars.

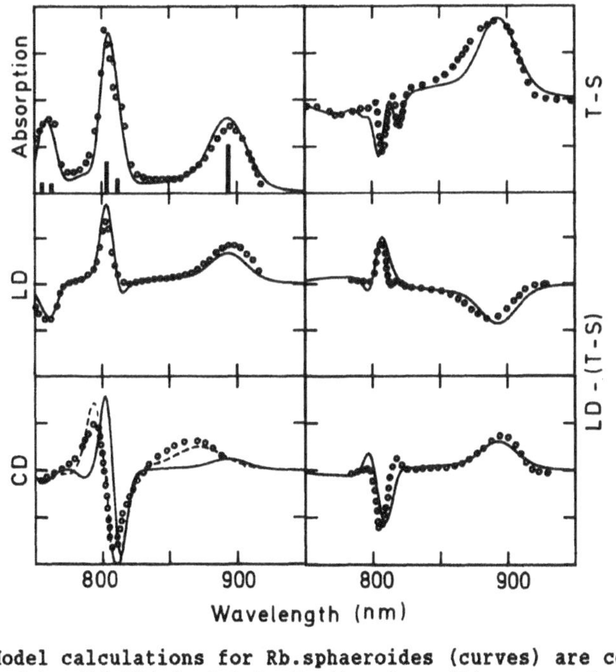

FIGURE 2. Model calculations for Rb.sphaeroides (curves) are compared with experimental data /14-17/ (points). Exciton transitions are shown as bars.

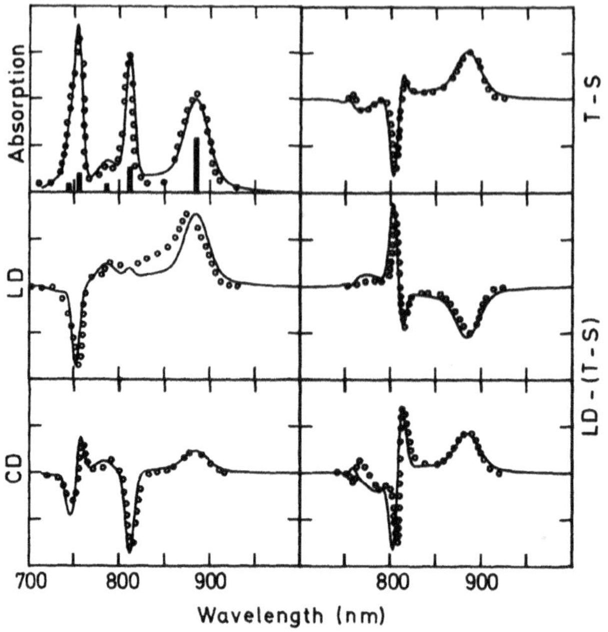

FIGURE 3. Model calculations for C.aurantiacus (curves) are compared with experimental data /6,21/ (points). Exciton transitions are shown as bars.

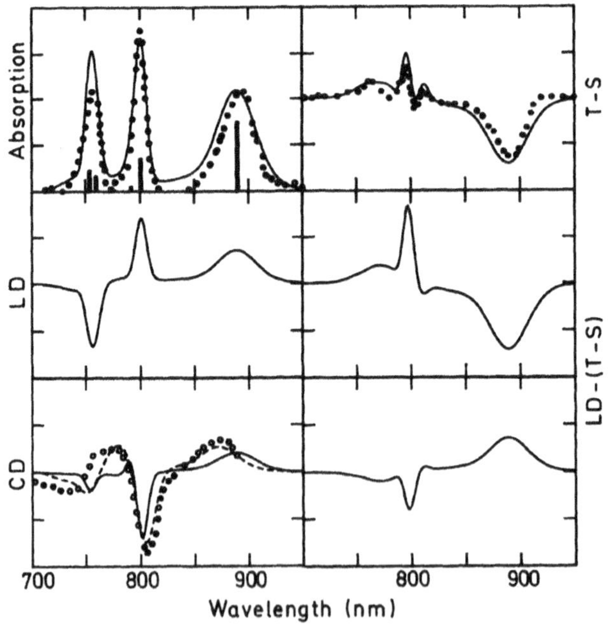

FIGURE 4. Model calculations for modified Rb.sphaeroides (curves) are compared with experimental data /18,19/ (points).

QUANTUM CHEMICAL CALCULATIONS

A more detailed understanding of the chromophore-chromophore interactions necessitates the investigation of wavefunctions. For a system as large as the reaction center only very restricted quantum chemical methods seem applicable at present.

METHODS

The excitations of single pigments (~100 atoms) and also of the special pair /22/ have been successfully treated with CNDO and INDO programs which treat all valence orbitals of the problem (~200 orbitals per pigment). With the present state of computer resources calculations on the whole hexamer (>10^3 orbitals) should be possible in principle. Eventually also part of the surrounding protein can be taken into account. However, the standard programs have been optimized for small molecules, where the long range behaviour of the atomic orbitals is less important. Additional effort has to be put into the development of reliable calculations on very large supermolecules. As an intermediate approach the pi-electron approximation is applicable to the single pigments. It reduces the number of basis orbitals by a factor of 8. Comparison with INDO calculations shows that the HOMO and LUMO orbitals are very well represented within the PPP approximation. This motivates extended calculations on the whole hexamer.

Eventhough the single pigments can be described as nearly planar pi-electron systems, the reaction center as a whole cannot be treated in terms of one single pi-electron system. Instead we have to consider six coupled electron systems, each of which is of the pi-electron type, i.e. contrary to the INDO supermolecule approach we are dealing with an aggregate of six coupled molecules /8/. Intramolecular interactions are included in the standard PPP calculation scheme. Corrections are made to take into account small deviations from planarity and differences in bond lengths and angles /23,24/. The influence of the binding histidine groups is modelled by a shift of the BCHL orbitals of 0.3eV. This value results from INDO calculations. For the coupling of the six pi-systems two interactions have to be considered. The electrostatic coupling which has been already considered in the exciton model and the resonance interaction between orbitals on different molecules. For the latter we use the parametrization given by Warshel /25/. Finally the excited states of the hexamer are calculated as a mixture of singly excited transitions (SCI method).

RESULTS

The orbitals as calculated for the hexamer are localized on the monomers or on the special pair (as a supermolecule) . This result supports the exciton model where any delocalization is neglected. The dimer orbitals are partly delocalized over the two dimer halves. Therefore the excitations of the dimer are not pure excitonic excitations of the two pigments but contain also charge transfer transitions within the dimer. The calculated excitation spectrum correlates with the experimental data and the successfull description within the exciton model only if we introduce the following corrections which seem to be necessary for PPP calculations in principle:

i) The calculated transition energies are generally too high as the interaction of the pi-system with the sigma-electrons and the environment are neglected. For the isolated BCHL-molecules the calculated transitions are between 693nm and 712nm. In the Rps.viridis spectrum the monmer excitations are at longer wavelength of ~ 840nm. Therefore we have to shift all calculated transition energies by -0.28 eV. With this correction the calculated excitations of the isolated pheophytines are at 796nm and 814nm.

ii) PPP calculations usually give much too large transition dipoles. Since the excitonic interaction between the pigments is mainly of the dipole-dipole type the calculated matrix elements also are too large. For the isolated pigments the calculated intensities are $\sim 165\ D^2$. . Therefore we introduced a reduction factor of 0.25 for the intensities and for the CI matrix elements between excitations on different molecules (not within the dimer).

iii) The dimer excitations in the Q_Y region are a mixture of nearly (anti-)symmetric excitonic transitions $2^{-1/2}$ ($BC_{MP}^* \pm BC_{LP}^*$) and charge transfer states $2^{-1/2}$ ($BC_{MP}^+BC_{LP}^- \pm BC_{MP}^-BC_{LP}^+$). Since the experimental spectrum shows only one strong dimer band with the approximate intensity of two monomers we conclude that the calculated mixture of P*(-) with the charge separated states is too strong This can be corrected for if we shift all charge transfer transitions within the dimer to higher energies.

With these corrections the resulting transitions correlate with the experimental spectrum of Rps.viridis (see figure 5). Especially the large red shift of the dimer band and the splitting of the two pheophytine transitions are in agreement. In figure 5 the internal CT states of the dimer have been shifted by 0.1 eV to higher energies.

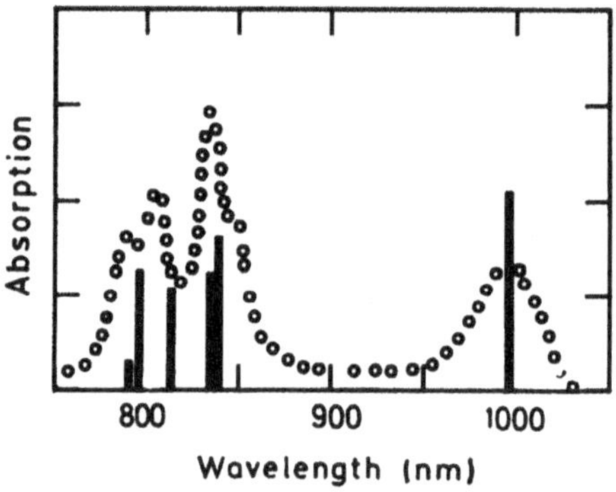

FIGURE5.
Q_Y transitions of the PPP model (bars) are compared with the experimental absorption spectrum of Rps.viridis /14/ (points).

CONCLUSION

The exciton model describes the steady state spectra of Rps.viridis and the structural similar Rb.sphaeroides and C.aurantiacus in a satisfying way. It may be very useful for the development of reliable quantum chemical methods for such a large molecular system. PPP calculations seem to be applicable but need some corrections which might be unnecessary in future INDO calculations.

ACKNOWLEDGEMENT

The authors like to thank Prof. E.W. Knapp for stimulating discussions. This work has been supported by the Deutsche Forschungsgemeinschaft SFB143.

REFERENCES

/ 1/ J. Deisenhofer, O. Epp, K. Miki, R. Huber, and H. Michel, J. Mol. Biol. 180 (1984) 385

/ 2/ J. P. Allen, G. Feher, T. O. Yeates, H. Komiya, and D. C. Rees, Proc. Natl. Acad. Sci. USA 84 (1987) 5730

/ 3/ E.W. Knapp,P.O.J. Scherer and S.F. Fischer, Biochim.Biophys.Acta 852 (1986) 295

/ 4/ P.O.J. Scherer and S.F. Fischer, Biochim.Biophys.Acta 891 (1987) 157

/ 5/ P.O.J. Scherer and S.F. Fischer,Chem.Phys.Letters 137(1987) 32

/ 6/ H.Vasmel,J.Amesz and A.J.Hoff, Biochim.Biophys.Acta 852 (1986) 159

/ 7/ E.W.Knapp, S.F.Fischer,W.Zinth, M.Sander, W.Kaiser, J.Deisenhofer and H.Michel Proc.Natl.Acad.Sci. USA 82 (1985) 8463

/ 8/ A.Warshel and W.W. Parson, J.Am.Chem.Soc. (1987) in press

/ 9/ A.Scherz and W.W.Parson, Biochim.Biophys.Acta 766 (1984) 653

/10/ Lester L. Shipman, Photochem.Photobiol. 26 (1977) 287

/11/ A.J. Hoff (1985) in Encyclopedia of Plant Physiology (New series) Vol. 19:Photosynthesis III (Stachelin A.C. and Arntzen C.J.,eds.) Springer Berlin.

/12/ A.J. Hoff in :Triplet state ODMR spectroscopy, ed.R.H. Clarke (Wiley,New York ,1981)

/13/ P.O.J Scherer and S.F.Fischer Chem.Phys.Letters 136 (1987) 431

/14/ J.Breton , Biochim.Biophys.Acta 810 (1985) 235

/15/ J. Breton (1985) in Antennas and reaction centers of photosynthetic bacteria, ed. M. E. Michel-Beyerle (Springer, Berlin, 1985) 109

/16/ V.A.Shuvalov and A.A.Asadov Biochim.Biophys.Acta 545 (1979) 296

/17/ E.J. Lous and A.J.Hoff, Proc.Natl.Acad.Sci. 84(1987) 6147

/18/ D.Beese,R.Steiner,H.Scheer and B.R.M. Lutz, Photochem.Photobiol. (1987) in press

/19/ D.Beese, A.Angerhofer,R.Steiner and H.Scheer,Photochem.Photobiol. (1987) in press

/20/ J.Breton and A.J.Hoff, private communication

/21/ H.J. den Blanken, H.Vasmel, A.P.J.M. Jongenelis,A.J.Hoff and J.Amesz FEBS 161 (1983) 185

/22/ M.PLato, private communication

/23/ G. Wagniere, Introduction to elementary molecular orbital theory and to semiempirical methods (Springer, Berlin, 1976)

/24/ I. Fischer-Hjalmas and M. Sundbom, Acta Chem. Scand. 22 (1968) 607

/25/ A.Warshel in: Modern theoretical chemistry,Vol. 7, ed. G.Segal (Plenum Press, New York 1977)

MODIFIED-CI MODEL OF PROTEIN-INDUCED RED SHIFTS OF REACTION CENTER PIGMENT SPECTRA

Robert M. Pearlstein

Physics Department
Indiana-Purdue University
1125 East 38th Street
P.O. Box 647
Indianapolis, IN 46223 U.S.A.

INTRODUCTION

The idea that its optical spectra are related to the structure of a photosynthetic pigment-protein complex first arose about two decades ago.[1] The advent of structural models based on x-ray diffraction analysis[2-4] greatly stimulated this field, and changed its nature. Initially the intent was to derive structural information from optical data. Now the major purpose is to elucidate details of pigment-pigment and pigment-protein interactions.

The x-ray structural model[3,5] of the *Rps. viridis* reaction center (RC) has inspired several independent theoretical descriptions[6-17] of RC optical spectra. Each of these has its own set of assumptions, and all give approximate agreement with experiment. They also differ in regard to several fundamental questions. Three major issues that have emerged are the involvement of charge-transfer states, the origin of the large special-pair exciton interaction energy, and the mechanism underlying the nonexcitonic portion of spectral red shifts.

The multiplicity of optical spectra theories of the RC stands in contrast to the situation with the Fenna-Matthews (F-M) complex, the only other (distinct) pigment-protein complex containing bacteriochlorophyll (BChl) for which there is an x-ray structural model.[2,18,19] In the past 12 years there has only been one theory of the F-M optical spectra, and that was based on an assumption that has proved impossible to justify.[20] Now, a new theory of the F-M spectra has been put forward that avoids the earlier assumption, and introduces a hypothesis that is also applicable to the RC.[21] The hypothesis is that configuration interaction (CI), which affects the intra-BChl electronic states, is modified by specific interactions of BChl with aromatic amino acid residues. The purpose of this paper is to show that the modified-CI hypothesis is helpful in answering some of the remaining questions.

The modified-CI hypothesis has been developed into a simple theoretical model, a detailed description of which is given in the next section. An outline of the exciton-theoretical methods used is given in the third section. The RESULTS include modified-CI model calculated parameters, exciton interaction matrix, and theoretical absorption, circular dichroism (CD), and linear dichroism (LD) spectra. The DISCUSSION compares these results with those of other optical spectra theories and with experiment. Suggestions for improvement of the theory are also given.

MODIFIED-CI MODEL

The four lowest-energy excited electronic transitions of BChl are identified, in order of decreasing energy, as B_Y, B_X, Q_X, and Q_Y.[22] These transitions result from symmetry-preserving configuration interaction (CI) of one-electron transitions. In the simplest description,[15] only two of these elementary transitions interact to form the X-polarized states, and two more to form the Y-polarized. The CI Hamiltonian for the latter may be written as

$$H_{(Y)} = \begin{bmatrix} \nu_o + \Delta & C \\ C & \nu_o \end{bmatrix}, \qquad (1)$$

where ν_o is the energy of the lower-lying Y-polarized one-electron transition, $\nu_o + \Delta$ is the energy of the higher-lying one, and C is the CI energy. The two eigenstates of this simple Hamiltonian have energies,

$$\nu_\pm = \nu_o + \frac{1}{2}\Delta \pm \frac{1}{2}\sqrt{\Delta^2 + 4C^2}. \qquad (2)$$

Here, ν_+ is to be identified as the B_Y-transition energy, ν_- as that of the Q_Y. The eigenstates themselves can be represented as

$$|\pm\rangle = a_\pm |1\rangle + b_\pm |2\rangle, \qquad (3)$$

where $|1\rangle$ is the higher-energy and $|2\rangle$ the lower-energy one-electron state,

$$a_\pm = (\nu_\pm - \nu_o) / \sqrt{(\nu_\pm - \nu_o)^2 + C^2} \qquad (4)$$

$$\text{and} \quad b_\pm = C / \sqrt{(\nu_\pm - \nu_o)^2 + C^2}. \qquad (5)$$

If μ_1 is the magnitude of the dipole strength of the transition to state $|1\rangle$, and μ_2 is that for state $|2\rangle$, then the squared dipole strengths of the eigenstates are

$$\mu_\pm^2 = (a_\pm \mu_1 + b_\pm \mu_2)^2. \qquad (6)$$

That is, μ_+^2 is the dipole strength of the B_Y transition, μ_-^2 that of the Q_Y.

In this simple picture, all of the interactions left out of the one-electron Hamiltonian are lumped into the single CI parameter, C. A principal part of these omitted interactions consists of electron-electron repulsion terms, which certainly affect the distribution of pi electron density over the porphyrin macrocycle in each state. For monomeric BChl in a solvent, the only contributions to the repulsion terms that need to be considered come from the BChl pi electrons themselves. Random interactions with solvent electrons precludes the latter contributing much to spectral shifts or intensity borrowing. However, in a protein structure significant additional pi-pi interactions can occur. This obviously happens when the ring of an aromatic amino acid residue is stacked in van der Waals contact with a BChl macrocycle. In this situation, the CI is said to be "modified"; the parameter C of the simple picture has a different value than it does in the solvent situation.

Although the altered value of C is in general difficult to calculate, it is possible to develop a model that takes account of CI modification. In this model, the quantities ν_o, Δ, μ_1, and μ_2 are first determined empirically from the spectral properties of monomeric BChl in a solvent. Equation (6) is then expressed in terms of the other parameters to give, for the strength of the lower-energy eigenstate,

$$\mu_-^2 = (\sqrt{x}\,\mu_1 + \sqrt{x+1}\,\mu_2)^2 / (2x + 1) \qquad (7)$$

$$\text{where} \quad x = (\nu_o - \nu_-)/\Delta. \qquad (8)$$

In Eqs. (7) and (8), ν_- is to be re-interpreted as the red-shifted value of the Q_Y-transition energy that results from CI modification. [Note that C has been eliminated from these equations by solving for C in terms of $\nu_- - \nu_o$ from Eq. (2).]

Thus, while in this model ν_- must still be treated as a parameter -- the BChl Q_Y diagonal energy -- Eq. (7) prescribes a quantitative correlation of dipole strength with red shift. This may be seen more readily by noting that, since usually $x \ll 1$,

$$\mu_-^2/\mu_2^2 \cong 1 + 2\sqrt{x}\,(\mu_1/\mu_2). \tag{9}$$

In the special case, $\mu_1 = \mu_2$, which holds approximately for BChl,

$$\mu_-^2/\mu_2^2 \cong 1 + 2\sqrt{x} \tag{10}$$

Equation (10) states that the fractional borrowed (integrated) intensity is twice the square-root of the red shift energy expressed as a fraction of (essentially) the Soret-red energy difference.

To lend further credibility to the model, the choice of each red-shifted energy, i.e., the value of ν_- for each BChl, should be related to the extent of interaction of that BChl with contacting aromatic amino acids. In standard notation for RC bacteriochlorins, BC_{LA} and BC_{MA} each has one such contact, whereas BC_{LP} and BC_{MP} together have 11 contacts.[24] For BChl *b* in a solvent, the value of $\mu^2_{Q_Y} + \mu^2_{B_Y}$ is taken[8] to be 92 D^2, and for simplicity it is assumed that $\mu_1^2 = \mu_2^2$; thus, $\mu_1^2 = \mu_2^2 = 46\ D^2$. Also,[25] $\nu_o = 12{,}610\ cm^{-1}$ (corresponding to 793 nm), and $\Delta = 14{,}400\ cm^{-1}$. For each bacteriopheophytin (BPh),[8] $\mu_{Q_Y}^2 = 38\ D^2$. (The BPhs are left out of the modified-CI calculation here.)

EXCITON-THEORETICAL METHODS

Exciton calculations are done by standard numerical techniques.[20,26] Input parameters include the coordinates and orientation vectors of the 6 bacteriochlorin macrocycles, the diagonal transition energies and strengths of the individual pigments as determined from the modified-CI procedure, and the distribution of transition monopoles over each macrocycle. (See RESULTS regarding the use of monopoles.) The absolute value of each transition monopole is then determined by the computer on the basis of the input diagonal transition strength for each BChl. For each pair of BChls the computer determines the exciton interaction energy by summing over the Coulomb energies of all pairs of interacting monopoles. The exciton interaction energies are the off-diagonal elements of a matrix whose diagonal elements are the input diagonal transition energies. The computer then diagonalizes this matrix, and from the resulting eigenvalues and eigenvectors determines the transition wavelengths, total dipole strengths, dipole strengths parallel and perpendicular to an input symmetry axis, and total rotational strengths of the 6 exciton transitions. Only the Q_Y transition is used in these calculations.

Theoretical spectra are produced first as tabular values generated from sums of symmetric Gaussians. Each spectrum includes 6 Gaussian components, one for each exciton transition. For (low-temperature) absorption and LD spectra, the full-width-at-half-maximum (FWHM) is taken as 350 cm^{-1} for each component except the lowest-energy one, for which the FWHM is set at 1000 cm^{-1}. For CD, the 5 higher energy components are given widths of 600 cm^{-1}, to facilitate comparison with the experimental[27] room-temperature spectrum. For the ith Gaussian component, the peak molar extinction per RC is given by[20]

$$\varepsilon_i = 1.023 \times 10^6\ \mu_i^2/(W_i E_i). \tag{11}$$

In Eq. (11), the exciton transition dipole strength μ_i^2 is in D^2, the exciton transition energy is in kcm^{-1}, and W_i is the FWHM in nm. A similar formula relates peak linear dichroic extinction to linear dichroic dipole strength. The peak difference extinction for CD, per RC, is given by[20]

$$(\Delta\varepsilon)_i = 3.796 \times 10^4 \, R_i/(W_i E_i), \qquad (12)$$

where the rotational strength R_i is in DBM. The theoretical spectra in Figs. 1-3 are drawn by computer from the generated tabular values.

RESULTS

The Q_Y dipole strengths of the 4 BChls obtained from Eq. (7) are listed in Table 1. As explained above, these values are based on a combined dipole strength of 92 D^2 for the Q_Y + B_Y transitions of BChl *b* in a solvent. Also shown are the dipole strengths of BP_M and BP_L (standard notations for RC bacteriochlorins), assumed to be the same as BPh in a solvent (i.e., unmodified in CI). The diagonal energies of BC_{MP} and BC_{LP} are taken equal for simplicity, the value having been chosen to optimize the spectral fit. Other diagonal energies are those of Eccles *et al.*[14]

Table 2 gives the Q_Y-band exciton interaction energies, calculated in the point monopole approximation, for all pairs of bacteriochlorin pigments in the RC. The transition monopoles used here are those calculated for BChl *a*,[22] renormalized to the dipole strengths given in Table 1. It can be estimated from the results of Eccles *et al.*[14] that the use of BChl *a* rather than BChl *b* monopoles yields interaction energies that are too small by ~10-20%. No correction has been made for this effect in Table 2, except in the case of the BC_{LP}-BC_{MP} interaction. The latter has proven very difficult to calculate accurately in the past, and has usually been treated as an adjustable parameter.[11,12,14] The value of this interaction energy shown in Table 2, 870 cm^{-1}, is the result of a calculation using "split" monopoles (half of the charge 1 Å above, half 1 Å below, the BChl ring plane)[28] increased by 20% to correct for the use of BChl *a* monopoles. Split monopoles were not used for the other interactions, which take place over much larger distances.

When the symmetric matrix whose diagonal elements are the energies listed in Table 1, and whose off-diagonal elements are those given in Table 2, is diagonalized, the stick spectra shown in Table 3 are obtained. Note the > 1500-cm^{-1} spacing between the two lowest-energy exciton transitions (5 and 6), which means that transition 6 is predicted to occur at the correct *low temperature* wavelength (~990 nm) with transition 5 correctly predicted to be near 850 nm (see DISCUSSION). The total absorption dipole strength for the 6 transitions is equal to the total Q_Y strengths of 2 BPhs + *5.5* BChls in a solvent. The "extra 1.5 BChls" represents the intensity borrowed from the Soret by CI modification. The LD is for 100% alignment of the RCs parallel to their C2-symmetry axis, but with the usual inverted sign convention for compressed gels (i.e., positive LD corresponds to polarization *perpendicular*, not parallel, to the alignment direction).[29] Exciton transition 6 is virtually a pure exciton-dimer state (of the special pair); but transition 5, although mainly a dimer state of the pair, has a non-negligible admixture of accessory BChl amplitude. Each of the 4 highest energy transitions includes both BChl and BPh contributions.

Theoretical absorption, LD, and CD spectra are shown in Figs. 1-3. Each is the corresponding stick spectrum of Table 3 dressed in 6 symmetric Gaussian envelopes, whose individual widths are set as described in EXCITON-THEORETICAL METHODS. Note the absolute ordinate scales, which, at least for absorption and CD, provide a more stringent comparison with experiment than is the case when arbitrary units are used (see DISCUSSION).

Table 1. Diagonal Energies and Strengths

Pigment	E(kcm^{-1})	λ(nm)	μ^2(D^2)
BP_M	12.47	802	38.0
BC_{MA}	12.15	823	62.4
BC_{MP}	11.05	905	76.3
BC_{LP}	11.05	905	76.3
BC_{LA}	12.15	823	62.4
BP_L	12.72	786	38.0

Table 2. Q_Y-Exciton Interaction Matrix (a)

BC_{MA}	BC_{MP}	BC_{LP}	BC_{LA}	BP_L	
140	-13	52	-14	8	BP_M
	-62	-263	40	-15	BC_{MA}
		870	-289	57	BC_{MP}
			-26	-19	BC_{LP}
				175	BC_{LA}

(a) Elements in cm^{-1}

Table 3. Calculated Q_Y Stick Spectra

Exciton Transition	E (km^{-1})	λ (nm)	Absorption (D^2)	CD (DBM)	LD (D^2)	Contributing Pigments (a)
1	12.774	782.8	28.1	-1.2	-16.9	BC_{LA}, BP_L
2	12.531	798.0	21.7	-0.7	-7.7	BP_M, BC_{MA}
3	12.361	809.0	16.4	-7.5	-13.6	all but BP_L
4	12.085	827.5	77.4	16.8	38.6	BP_M, BC_{MA}, BC_{LA}
5	11.694	855.1	46.5	-13.7	-46.1	all but BP_L and BP_M
6	10.149	985.3	163.2	6.2	81.4	BC_{MP}, BC_{LP}

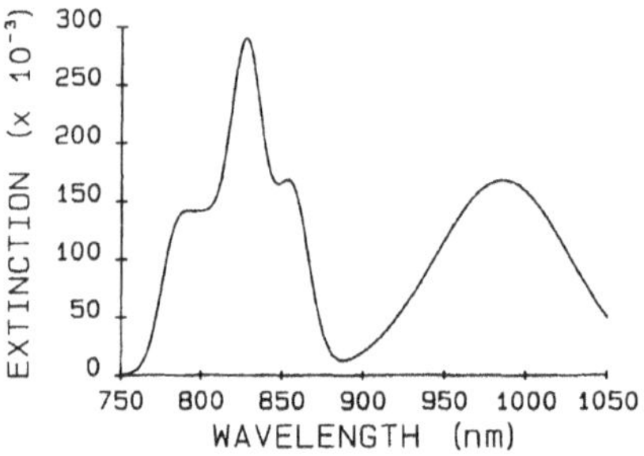

Fig. 1 (top). Theoretical low-temperature absorption spectrum in the Q_Y band of the RC. Ordinate units are molar extinction per RC.

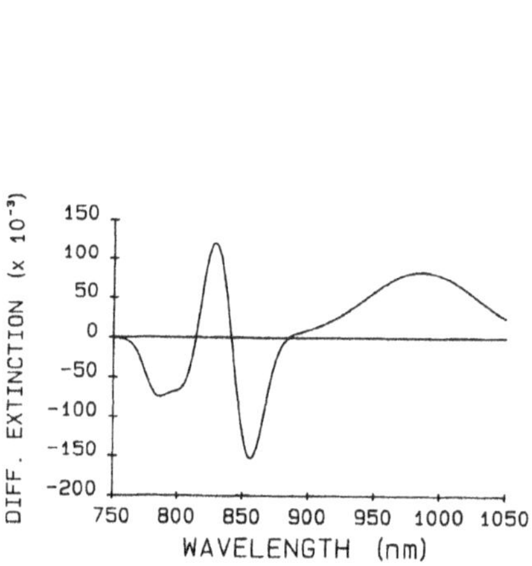

Fig. 2 (bottom): Theoretical low-temperature LD spectrum. Ordinate is molar extinction per RC for 100% alignment parallel to the C2-symmetry axis (with the usual inverted sign convention -- see text).

DISCUSSION

No theoretical model of the optical spectra of the RC bacteriochlorins yet put forward explains all experimentally observed features from first principles, i.e., with no *ad hoc* assumptions. The present model, while no exception, has some advantages. Being similar to "Model A" of Eccles *et al.* (EHS-A),[14] it reproduces the observed spectra almost as faithfully, but without parameterizing the exciton matrix element connecting BC_{MP} and BC_{LP}, as do both EHS-A[14] and the model of Knapp *et al.* (KSF).[12] A reasonable value of this matrix element is obtained by combining modified-CI and split monopole calculations. The modified-CI procedure increases the overall Q_Y dipole strengths of all 4 BChls (i.e., aside from the redistribution of intensity within the Q_Y band induced by the exciton interactions), with increases of about 50% for each of the special pair BChls. Interestingly, EHS-A also has such a large Q_Y intensity increase for the pair, but this (apparently fortuitously) results from molecular orbital approximations made in that calculation -- EHS would presumably give the same high intensity for BChl in a solvent, which would be erroneous in that case. In the modified-CI model, the intensity increase occurs only in the protein complex, as a result of specific aromatic amino acid interactions.

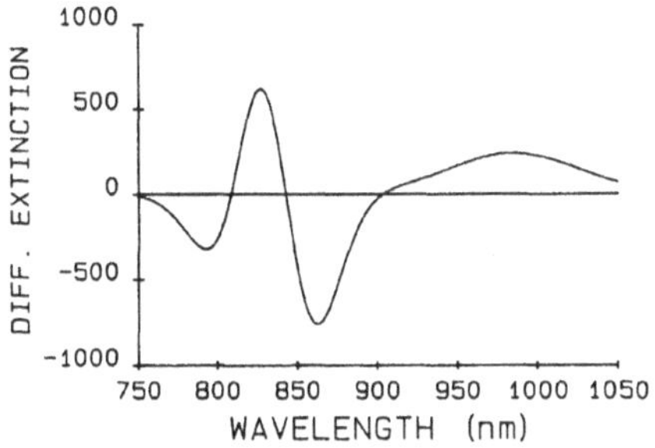

Fig. 3. Theoretical room-temperature CD spectrum. Ordinate is difference molar extinction per RC.

The 870-cm^{-1} value of the pair matrix element calculated here also places the longest-wavelength absorption ("P") band at its fully red-shifted (low temperature) position, ~ 990 nm.[29] In contrast, the parameterized value chosen in EHS-A, 650 cm^{-1}, places the P-band at its room temperature position, 960 nm. In KSF, the value 1000 cm^{-1} is chosen, which places the P-band at 990 nm. The low temperature band position is more nearly indicative of the zero-phonon electronic transition, because the increased band broadening at higher temperature is skewed to shorter wavelength.

It is more difficult to compare the modified-CI model with the model of Parson and Warshel (PW).[7-9] PW find the exciton matrix element of the special pair to be only ~ 350 cm^{-1}, but attribute most of the red shift of the P-band to the effects of charge transfer, for which they calculate some matrix elements $\geq$ 1500 cm^{-1} (W.W. Parson, Personal Communication). The PW model, which relies on extensive molecular orbital calculations, nonetheless places the P-band at ~ 960 nm. While none of the other models mentioned above, including the modified-CI model, takes charge transfer into account at all, neither does any of these models exclude it as a possibility. In PW charge transfer is a large effect, in the other models it is at most a small one. There is as yet no experimental basis for deciding the magnitude of the charge transfer contribution to the P-band.

It should be noted that only the PW and the modified-CI models so far have presented their calculated spectra with absolute extinctions plotted. This facilitates comparison with experiment because extinction and energy splittings simultaneously scale with the magnitude of exciton matrix elements. In the case of the modified-CI model, this makes it apparent that *too much* intensity is transferred to the P-band, i.e. the calculated ε_{990} is too large by ~ 30%. This can presumably be corrected by an improved calculation (see below).

No attempt was made here to include dielectric effects, as was done in EHS-A. However, the neglect of such effects for the non-special-pair matrix elements (the only ones for which they were included in EHS-A), is approximately offset by the use of BChl *a* monopoles -- either procedure leads to smaller matrix elements.

All of the theoretical models discussed here, including the modified-CI, share a peculiar inconsistency. The experimental low-temperature LD spectrum clearly shows that in the Q_Y band, as one moves through wavelengths shorter than the positive peak at ~ 830 nm, one encounters first a strong positive shoulder, and then two negative peaks.[29] Theory finds only three negative features in this region. There is at present no explanation for this anomaly.

All of the models so far mentioned also share the drawback of not calculating the widths and shapes of the various bands on any dynamical basis, but rather just impose Gaussian skirts. Friesner and co-workers have attempted to correct this situation, but in their model the positions of the various spectral features are simply imposed as parameters.[17] Clearly, a fully detailed theory should not impose either spectral line shapes or peak positions.

The current modified-CI model needs to be corrected and extended in several respects. As already noted, the correct distribution of transition monopoles is required, as is some allowance for dielectric effects. In addition, the model itself needs more detail, including the asymmetry of aromatic-amino-acid interactions with L- versus M-chain pigments, and probably more realistic values of the underlying parameters (e.g., starting values of the 4-orbital parameters). Finally, the model should be extended to include other optical spectra, notably those of P^+, borohydride-modified RCs[30,31], and various triplet-minus-singlet spectra[12].

ACKNOWLEDGMENTS

The author thanks J. Deisenhofer and H. Michel for providing the 1985 set of atomic coordinates of the RC bacteriochlorins. He also wishes to thank J. Deisenhofer, B. Honig, and A. Scherz for useful discussions.

REFERENCES

1. K. Sauer, E. A. Dratz, and L. Coyne, *Proc. Natl. Acad. Sci. USA* 61:17 (1968).
2. R. E. Fenna and B. W. Matthews, *Nature (London)* 258:573 (1975).
3. J. Deisenhofer, O. Epp, K. Miki, R. Huber, and H. Michel, *J. Mol. Biol.* 180:385 (1984).
4. T. Schirmer, W. Bode, R. Huber, W. Sidler, and H. Zuber, *J. Mol. Biol.* 184:257 (1985).
5. J. Deisenhofer, O. Epp, K. Miki, R. Huber, and H. Michel, *Nature (London)* 318:618 (1985).

6. W. W. Parson, A. Scherz, and A. Warshel, *in*: "Antennas and Reaction Centers of Photosynthetic Bacteria," M. E. Michel-Beyerle, ed., Springer-Verlag, Berlin (1985), p. 122.
7. A. Warshel and W. W. Parson, *J. Am. Chem. Soc.* 109:6143 (1987).
8. W. W. Parson and A. Warshel, *J. Am. Chem. Soc.* 109:6152 (1987).
9. W. W. Parson and A. Warshel, *in*: "Structure of the Photosynthetic Bacterial Reaction Center" (this volume), J. Breton and A. Vermeglio, eds., Plenum Press, New York (1988).
10. E. W. Knapp and S. F. Fischer, *in*: "Antennas and Reaction Centers of Photosynthetic Bacteria," M. E. Michel-Beyerle, ed., Springer-Verlag, Berlin (1985), p. 103.
11. E. W. Knapp, S. F. Fischer, W. Zinth, M. Sander, W. Kaiser, J. Deisenhofer, and H. Michel, *Proc. Natl. Acad. Sci. USA* 82:8463 (1985).
12. E. W. Knapp, P. O. J. Scherer, and S. F. Fischer, *Biochim. Biophys. Acta* 852:295 (1986).
13. P. O. J. Scherer and S. F. Fischer, *in*: "Structure of the Photosynthetic Bacterial Reaction Center" (this volume), J. Breton and A. Vermeglio, eds., Plenum Press, New York (1988).
14. J. Eccles, B. Honig, and K. Schulten, *Biophys. J.* (in press).
15. B. Honig, *in*: "Structure of the Photosynthetic Bacterial Reaction Center" (this volume), J. Breton and A. Vermeglio, eds., Plenum Press, New York (1988).
16. Y. Won and R. A. Freisner, *Proc. Natl. Acad. Sci. USA* 84:5511 (1987).
17. R. A. Friesner, *in*: "Structure of the Photosynthetic Bacterial Reaction Center" (this volume), J. Breton and A. Vermeglio, eds., Plenum Press, New York (1988).
18. D. E. Tronrud, M. F. Schmid, and B. W. Matthews, *J. Mol. Biol.* 188:443 (1986).
19. S. T. Daurat-Larroque, K. Brew, and R. E. Fenna, *J. Biol. Chem.* 261:3607 (1986).
20. R. M. Pearlstein and R. P. Hemenger, *Proc. Natl. Acad. Sci. USA* 75:4920 (1978).
21. R. M. Pearlstein, *in*: "Organization and Function of Photosynthetic Antennas," H. Scheer and S. Schneider, eds., Walter de Gruyter Publishers, Berlin (1988), in press.
22. C. Weiss, Jr., *J. Mol. Spectrosc.* 44:37 (1972).
23. M. Gouterman, G. H. Wagniere, and L. C. Snyder, *J. Mol. Spectrosc.* 11:108 (1963).
24. H. Michel, O. Epp, and J. Deisenhofer, *EMBO J.* 5:2445 (1986).
25. J. C. Goedheer, *in*: "The Chlorophylls," L. P. Vernon and G. R. Seely, eds., Academic Press, New York (1966), p. 147.
26. R. M. Pearlstein, *in*: "Photosynthesis," J. Amesz., ed., Elsevier, Amsterdam (1987), p. 299.
27. K. D. Philipson and K. Sauer, *Biochemistry* 12:535 (1973).
28. K. D. Philipson, S. C. Tsai, and K. Sauer, *J. Phys. Chem.* 75:1440 (1971).
29. J. Breton, *Biochim. Biophys. Acta* 810:235 (1985).
30. S. L. Ditson, R. C. Davis, and R. M. Pearlstein, *Biochim. Biophys. Acta* 766:623 (1984).
31. P. Maroti, C. Kirmaier, C. Wraight, D. Holten, and R. M. Pearlstein, *Biochim. Biophys. Acta* 810:132 (1985).

TEMPERATURE DEPENDENCE OF THE LONGWAVELENGTH

ABSORPTION BAND OF THE REACTION CENTER OF *Rhodopseudomonas viridis*

Youngdo Won and Richard A. Friesner

Department of Chemistry
The University of Texas at Austin
Austin, TX. 78712 U. S. A.

I. INTRODUCTION

In a series of previous papers(1,2), we have constructed a vibronic coupling model for the excited states of the chromophores in the reaction center (RC) of photosynthetic bacteria. The model has been used to compute various optical properties (absorption, circular dichroism, polarized absorption, holeburning spectra) of the reaction center pigments; a particular focus has been investigation of the low energy Q_y exciton component of the special pair bacteriochlorophylls. Many of the intramolecular vibrational parameters (frequencies, excited state geometry shifts) are obtained a priori from monomer experiments, reducing the parametric flexibility associated with phenomenological lineshape functions. Good agreement with experiment has been obtained by varying a limited set of electronic parameters and by including an intermolecular mode for the special pair dimer P.

Nevertheless, the theoretical model is rather complicated and contains many uncertainties with regard to the parameter set. One way to address this difficulty is to investigate additional features of the experimental data and attempt to interpret them with the model. Such studies can provide insight into the possible origins of these features, leading to proposals as to how to distinguish alternative explanations experimentally.

In this paper we consider two aspects of the low energy absorption band of P which in our previous work were ignored or treated by phenomenological fitting. These are the long wavelength shoulder which is sometimes observed experimentally and the temperature dependence of the peak position and halfwidth of the band. In the first case, our calculations allow for two distinct interpretations which can in principal be resolved experimentally. In the second, we suggest a physical model to explain the anomalously large temperature effect that is basically in agreement with the ideas of Holton and coworkers(3). An alternate explanation of the tempera-ture dependence based on a very large vibronic coupling constant is shown to lead to an erroneous lineshape for the band at low temperature. Two experiments are proposed which should be able to test the validity of these models.

The paper is organized as follows. Section II briefly reviews the theoretical model and the formalism used to compute the optical lineshapes; the reader is advised to consult Refs. 1, 2, 4, and 5, if further details concerning these topics are desired. Section III introduces a modification of the model used in Ref. 2 which leads to reproduction of the long wavelength shoulder via vibronic effects; spectral simulations for this model are presented and a comparison is made with the previously employed model in which the shoulder was absent. We also analyze the temperature dependence of the P parameters required to reproduce experiment, and it is argued that this dependence can be systematically explained in a simple and plausible manner. In conclusion, the two proposed experiments are presented.

We will restrict ourselves below to an analysis of data from the bacterium *Rps. viridis*. We have in previous papers also considered *Rb. sphaeroides* ; analysis of the two organisms has generally produced analogous results.

II. THEORETICAL MODEL AND COMPUTATIONAL METHODS

Our model for the electronic excited state manifold of the reaction center pigments consists of six coupled Q_y excited states (one for each RC pigment). The effects of the protein environment and of other electronic states of the chromophores are incorporated phenomenologically by adjusting electronic site energies for each state so as to obtain agreement with the experimental spectra. Interpigment exchange interactions are evaluated via a point dipole approximation for all chromophores pairs except for those which comprise P; this interaction, designated J_P , is also adjusted empirically (this will be discussed further below). The electronic site energies of the P molecules are for simplicity taken to be degenerate, in the absence of any experimental effect which requires them to be split; small perturbations of this condition have little influence upon calculated optical spectra. The directions of the transition dipole moments are estimated from the X-ray geometry(6) and assumptions concerning the directions in the monomer(7); the oscillator strengths are scaled to obtain agreement with experimental peak heights.

Each chromophore has associated with it a set of molecular vibrations which change their equilibrium position upon electronic excitation, and hence are coupled to the corresponding optical transition. We estimate the vibrational frequencies of these modes and the magnitude of the geometry displacements (linear vibronic coupling constants) from matrix isolated spectra of Hoff and coworkers(8) of chlorophyll-a and pheophytin-a and from supersonic jet spectra of free base porphin obtained by Even and Jortner(9). The details of the modeling procedure are described in Ref. 2. These parameters are not adjusted during subsequent calculations and hence constitute a set of a priori assumptions.

The lowest energy absorption band of P (principally composed of the lowest exciton component of the dimer) is much broader than the corresponding Q_y bands of the remaining RC chromophores; furthermore, this broadening appears to be mainly homogeneous in hole-burning experiments. For notational simplicity, this band, which will be our primary focus in this paper, will henceforth be referred to as the P band. A simple interpretation of this anomalously large homogeneous linewidth is that there exists an intermolecular mode of the dimer which is strongly coupled to the excited state. Relative motion of the P molecules modulates their exchange interaction and hence is a logical candidate for such a strongly coupled mode; in addition, one would expect a substantial change in the equilibrium intermolecular separation upon excitation. Intermolecular vibrations of this type are therefore included in our simulations; the magnitude of the coupling is obtained by fitting the observed holeburning width at a single burn laser frequency. After this one parameter is fixed, the model quantitatively reproduces the experimental dependence of the peak position and width of the hole on burn laser frequency with no further adjustment (this statement applies to calculations on *sphaeroides*; the signal to noise ratio in the *viridis* data renders such an analysis more difficult). The details of this calculation are presented in Ref. 2. It is important to note that the derived coupling strength also properly reproduces the ordinary absorption spectrum of the P band, including the asymmetry at low temperature.

All remaining vibronic effects (e.g., acoustic medium phonons, quadratic coupling, anharmonic terms) and inhomogeneous broadening effects are incorporated into a phenomenological line broadening factor. At low temperature, a single line broadening parameter quantitatively reproduces the entire experimental lineshape. At higher temperatures, two parameters (one in the P band region, one in the remaining region) were required in Ref. 2. An alternate modeling procedure which is physically much more transparent is described and utilized below.

The final feature of the model is a charge transfer (CT) state which interacts with the P band and destroys the zero-phonon line of that band, thus eliminating the usual sharp zero-phonon hole from the holeburning spectrum. Inclusion of this state is essential in simulating the holeburning results; however, it has little effect on the inhomogeneously broadened peaks

TABLE I. Comparison of Model I and Model II*

Temperature	MODEL I				MODEL II	
	g_1	g_2	γ_1	γ_2	g_p	γ
4.2 K	50.0	100.0	55.0	55.0	112.5	55.0
100 K	50.0	100.0	100.0	200.0	165.0	105.0
297 K	50.0	100.0	110.0	250.0	235.0	115.0

* g_1 and g_2 are vibronic coupling constants for 50 and 100 cm^{-1} intermolecular P-modes of Model I respectively, while g_p is the one for 100 cm^{-1} P-mode of Model II.
* γ is the homogeneous line broadening factor in Model II, and γ_1 for B and H band and γ_2 for P band in Model I.
* Parameters are given in cm^{-1} unit.

obtained with other spectroscopic techniques. As we will not be concerned with the hole-burning spectra in detail below, we ignore the CT state in what follows.

A quantitative formulation of the above model is most easily given in second quantized notation. The excited state Hamiltonian is:

$$H = \sum_{i=1}^{6} E_i |i\rangle\langle i| + \sum_{i \neq j}^{6} V_{ij} |i\rangle\langle j| + \sum_{n=1}^{13} w_n b_n^{\dagger} b_n + \sum_{i,j}^{6} \sum_{n}^{13} g_{ij}^{(n)} |i\rangle\langle j| \, (b_n + b_n^{\dagger}) \qquad (1)$$

Here the E_i are electronic excited state energies, the V_{ij} are electronic coupling parameters, the w_n are vibrational frequencies, and the $g_{ij}^{(n)}$ are vibronic coupling constants. b and $b^{\dagger}$ are the usual boson annihilation and creation operators and $|i\rangle\langle j|$ is a projection operator connecting electronic state i to j. The electronic ground state is assumed to be a single harmonic surface with frequencies w_n.

A complete list of the values of all of the above parameters is given in Ref. 2. The focus of the present paper will be on the site energy and exchange coupling of P, the P intermolecular modes, and the phenomenological damping constants. In Ref. 2, two intermolecular modes were assumed and two damping constants were utilized. Table I presents the values for these parameters at the three temperatures studied (4.2 K, 100 K, and 297 K). These values provide a good reproduction of the experimental spectra with the exception of the long wave-length shoulder (to be discussed below). The model defined by these parameters will henceforth be referred to as Model I; a modified model, investigated below, will be designated Model II. The corresponding parameters for Model II are also displayed in Table I to facilitate comparisons. All parameters not shown explicitly are identical for both models and are those of Ref. 2.

Simulation of optical spectra is carried out via a Green's function formalism which has been described previously(4) and has been extensively tested against converged basis set calculations for smaller model Hamiltonians(5). For inhomogeneously broadened lineshapes, agreement to within a few per cent is uniformly obtained over a wide range of parameter values; we expect the results shown below to have this accuracy. The method is computationally efficient, requiring a few seconds on a Cray X-MP to generate a spectrum for a full simulation of Eq. (1). It is therefore possible to carry out extensive investigations of the model parameter space and to test the effects of altering fundamental assumptions. This type of exploration is initiated below.

An important limitation on our analysis of the temperature dependence of the P band using the experimental results displayed below has been pointed out by Kirmaier and Holton (3). The data used here is obtained from three different *Rps. viridis* RC preparations; most importantly, the medium in which the RCs were immersed in is different in each case. It has been observed that these environmental alterations can produce significant changes in the peak position and width of the P band. For a quantitative analysis, therefore, it is necessary to begin with a series of measurements on RCs prepared in the same medium. On the other hand, the qualitative trends in the temperature behavior are still obtainable with our present data set. Therefore, we present below a discussion along these lines. Because the physical mechanism that is postulated is substantially different from alternate suggestions, and because strong support for this mechanism is available even from approximate results, we believe that this effort is worthwhile. Accurate simulation of a homologous series of experiments (see Kirmaier and Holton, this volume, for a presentation of the relevant experimental data) will appear in subsequent publications.

III. RESULTS

A. Description of Model II

In our alternate model, we make two significant changes in the parameter sets used in our spectral simulations. First, we employ only one intermolecular mode (as opposed to two modes of differing frequencies in model I) , of frequency 100 cm^{-1}. At present, there is no way of determining a priori which of these models is closer to reality, so an investigation of this alternate hypothesis is warranted. The coupling constant of this mode at 4.2 K is obtained as before by fitting the holeburning results. The agreement obtained for the holeburning simulations is comparable to that for model I; they will not be displayed explicitly here.

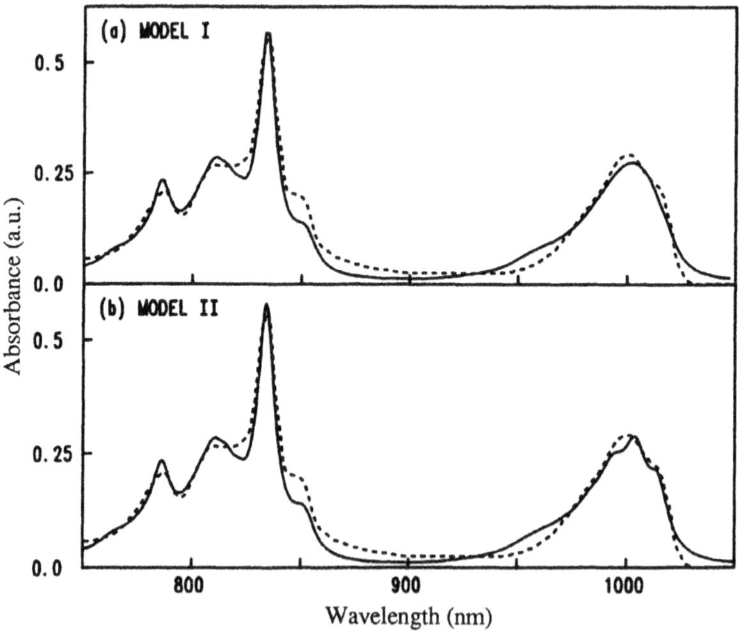

Fig.1 Comparison of the 4.2 K absorption spectra calculated from Model I (a) and Model II (b) (solid line) with the experimental result of Vermeglio et al. (10) (dotted line).

The second modification is to eliminate the use of two phenomenological damping constants at higher temperatures. It seems implausible that the substantial augmentation of the linewidth beyond normal temperature effects for the model are due to acoustic phonon or inhomogeneous interactions. A much more likely hypothesis is that the equilibrium configuration of the P dimer changes as a function of temperature. This idea is supported not only by the analysis which follows but by other, quite different experiments which also suggest substantial configurational changes in the RC as a function of temperature(14). To implement this in our vibronic coupling model, we assume that the vibronic coupling constant g_P of the intermolecular mode changes as a function of temperature. The new value at higher temperatures is obtained by fitting the P band linewidth; only one phenomenological damping constant is now utilized for the entire spectral range. The values of g_P determined by this procedure are listed in Table I.

Fig. 1 compares the experimental and calculated absorption spectrum at 4.2 K for models I and II . The only noticeable difference for Model II is the shoulder on the P band at long wavelength. The significance of this will be discussed below.

Fig. 2 and 3 display the absorption and circular dichroism spectra for Model II (compared to experiments) at 100 K and 297 K respectively. The quality of the fit is roughly equivalent to that for Model I (see Ref. 2 for the corresponding plots for Model I). There are some quantitative discrepancies at short wavelength which need to be explored further. The agreement in the P band region, which will primarily concern us here, is quantitative.

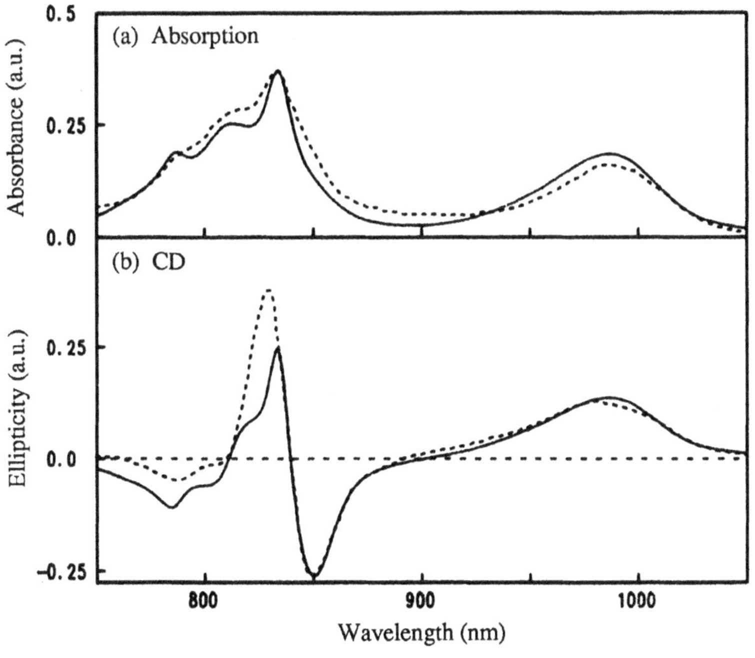

Fig.2 Comparison of 100 K absorption and CD spectra calculated from the model Hamiltonian in Eq. (1) using the parameter values of Model II (solid lines) with the experimental results of Shuvalov and Asadov (11) (dotted lines).

B. Long Wavelength Shoulder of the P Band

A long wavelength shoulder on the P band in *Rps. viridis* has been observed by several experimental groups(10,13). The most detailed and illuminating investigation of this feature to date is reported by Hoff in this volume. Using absorbance detected magnetic resonance (ADMR), he obtains a triplet-minus-singlet (T-S) absorbance profile in the region of the P band. By suitably manipulating the data, a spectrum of the singlet P transition can be extracted.

The most interesting feature of Hoff's results are that the observed ADMR spectrum changes as a function of the applied microwave frequency. The triplet sublevel microwave transitions are inhomogeneously broadened; molecules with differing D and E values are selected by tuning the microwave frequency through the band. The optical results imply that these groups of molecules which differ in their magnetic properties also differ in their optical spectrum. The long wavelength shoulder appears at low microwave frequency and disappears as this frequency is increased. Appearance of the shoulder is also correlated with a shift of the absorption maximum to longer wavelengths.

The most straightforward interpretation of these experiments is that the ADMR experiment is selecting groups of reaction centers with differing dimer configurations. An alteration in the intermolecular distance or orientation would affect the D and E parameters because the dimer wavefunction is strongly delocalized for P. This idea is consistent with the picture that the P potential surface is very "floppy" and anharmonic, as implied by the temperature dependent configurational changes hypothesized above. One can also imagine that different preparation and cooling techniques would "freeze in" different mixtures of dimer configurations, thus leading to discrepancies among experimental groups in measured absorption spectra. Strong variability as function of solvent medium is also consistent with this picture, in that one can imagine forces exerted by different solvents leading to different distributions of dimer geometries.

An analysis of the shoulder suggested by the results of Fig. 1 is as follows. A generalized interpretation of the results of model I is that heterogeneity of the intermolecular

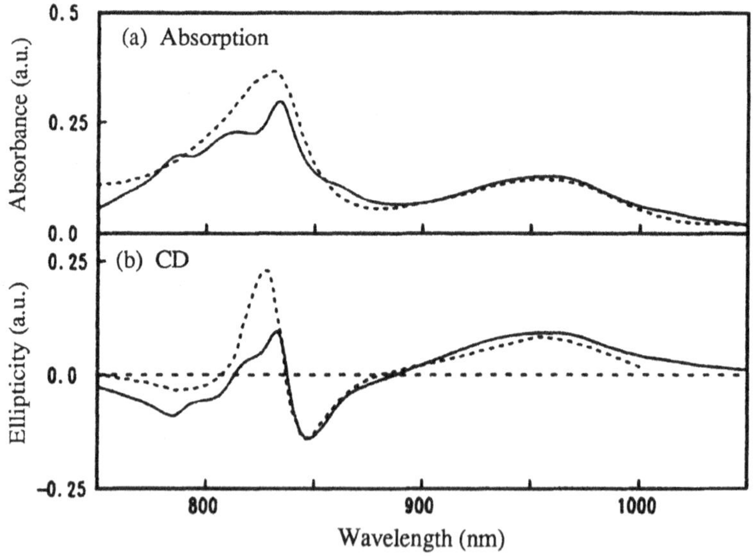

Fig.3 Comparison of 297 K absorption and CD spectra calculated from the model Hamiltonian in Eq. (1) using the parameter values of Model II (solid lines) with the experimental results of Phillipson and Sauer (12) (dotted lines).

vibrational frequency destroys the shoulder by averaging over its position. The coupling strengths used here render the zero phonon line weaker than a one-phonon transition; it is this peak which is underneath the shoulder in Fig. 1(b). In Fig. 1(a), the dispersion induced by having two modes of different frequency is sufficient, in conjunction with the damping constant, to smooth out the sharp feature. Heterogeneity of other parameters in the Hamiltonian could affect the shoulder as well, as has been pointed out by Hoff. The vibrational frequency effect appears to be the most directly connected; one would also expect it to be quite sensitive to the dimer configuration.

The specific behavior of the data of Ref. 15 can be rationalized in the context of this analysis. The key observation is that the shoulder appears when the peak position is shifted to longer wavelengths. A larger shift implies that J_p is larger, and hence that the P molecules are closer together. In other words, the molecules selected by this microwave frequency are the most deeply trapped in the P intermolecular potential well. The well will be most harmonic and uniform at this minimum configuration; consequently, one would expect the vibrational frequency associated with motion in it to be the least dispersed, thus leading to shoulder formation in the optical spectrum (i.e., the description of model II is valid). The frequency would also be expected to be higher, further enhancing its visibility (one needs a frequency larger than the damping width to obtain a visible shoulder).

One point which should be noted is that we have not found support for the idea that the shoulder can be attributed to a charge transfer transition. Simulations in which the CT state is not assigned any oscillator strength (but interacts via vibronic coupling with the P band) fail to produce any feature like that in Fig. 1(b), independent of parameter values. The experimental and theoretical evidence against the CT state possessing oscillator strength of its own appears to be considerable.

C. Temperature Dependence of the P Band

The peak position and width of the P band undergo substantial alteration as a function of temperature. A previous attempt to model these results yielded an extremely large linear vibronic coupling parameter for a low frequency mode of the special pair. Unfortunately, this parameter value leads to an incorrect lineshape for the P band, which is predicted to be completely symmetric at 4.2 K. Fig. 4 displays the 4.2 K absorption spectrum calculated when a Huang-Rhys factor of ~8 and a vibrational frequency of 30 cm^{-1} (per Ref. 16) is assigned to the P intermolecular mode. This result is generic when one assumes linear vibronic coupling and leaves the coupling parameters invariant as a function of temperature; we have been unable to simultaneously reproduce the asymmetric lineshape of the P band and produce a sufficiently large temperature dependence of the linewidth and peak position with any such vibronic coupling model investigated to date without assigning a special and quite large damping constant in the P band region.

If one uses an appropriate set of vibronic coupling constants (i.e., those which give the proper 4.2 K absorption spectrum), neither the experimental linewidth increase nor the peak shift are quantitatively reproduced if one simply raises the temperature without changing any model parameters. This suggests that there are parameters in the Hamiltonian which change as a function of temperature. The equilibrium dimer geometry is the physical parameter which would be likely to be strongly temperature dependent. The excited state equilibrium position of the dimer will affect three energy parameters; the intermolecular vibronic coupling constant g_p, the P site energy E_p, and the P exchange interaction J_p. We already adjust the latter two of these to reproduce the peak position in Model I; Model II completes the picture by also adjusting g_p. Note that E_p and J_p are both rigorously determined from the fitting procedure at all temperatures, at 4.2 K by the (+) exciton component shoulder at 850 nm and at higher temperature by the circular dichroism spectra (these constraints are in addition to the main P band peak position).

As explained above, we will not address the quantitative details of the temperature dependence in this paper. Qualitatively, however, a model attributing the changes in the three parameters to a monotonic increase in the average separation of the P molecules in the excited state explains the data quite well. Both J_p and the red shift of the P band diminish at higher temperatures, consistent with a weaker intermolecular interaction attributable to a larger

TABLE II. Special Pair Parameters at Different Temperatures*

Temperature	E_p	J_p	g_p
4.2 K	11000.0	872.0	112.5
100 K	11075.0	817.0	165.0
297 K	11257.0	700.0	235.0

* Parameters are given in cm^{-1} unit.

separation. An increase in the equilibrium distance in the excited state as compared to the ground state would lead to a corresponding increase in the vibronic coupling constant of the intermolecular mode; it is plausible that the more expanded and diffuse excited state would take advantage of the greater "floppiness" of the protein at higher temperature by expanding the intermolecular separation to a greater degree. Table II lists the three parameters (obtained, as stated above, by fitting to experiment at each temperature) at different temperatures. We again emphasize that the detailed behavior of the three quantities (e.g., linearity or lack thereof) should not be taken seriously. On the other hand, the parallel behavior of E_p and J_p are non-trivial results of the fitting procedure, which provides real support to the model proposed above. The interpretation of the linewidth increase is based on this model; in this case, the fitting procedure merely extracts the consequences of the assumption. A more quantitative study analyzing the detailed temperature dependence may provide confirmation (or denial) of our hypothesis.

This discussion is certainly not definitive; other feasible explanations of the temperature effect can easily be imagined. The model proposed has the advantage of simplicity and consistency with existing experimental data to date.

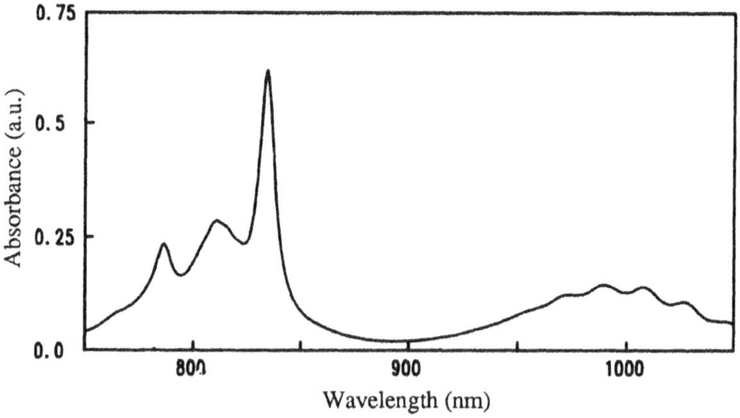

Fig.4 Absorption spectrum calculated from the Hamiltonian in Eq. (1) employing Hayes and Small's model(16) of a 30 cm^{-1} P-mode with Huang-Rhys factor of S=8.

IV. CONCLUSION

The above results both suggest that the P intermolecular potential surface is highly anharmonic, and that motion on this surface plays a key role in generating various aspects of the experimental spectroscopy. The strong coupling of this mode to the excited state would appear to implicate it in controlling the electron transfer dynamics as well. Clearly, further investigation of this subject is warranted.

The most useful experiment which could be carried out in this regard is a resonance Raman study in which the P band is directly excited. The postulated intermolecular mode (as well as all other Franck-Condon active modes) should appear explicitly in the Raman spectrum, and the vibronic coupling constants should be obtainable from the Raman intensities and excitation profiles. The width of the Raman peaks should provide some information about heterogeneity of the vibrational frequencies. Analysis of the temperature dependence of the intensities will allow extraction of a vibronic coupling constant at each temperature; the above model predicts that, for the intermolecular mode, the coupling constant will change substantially. This experiment is difficult because of fluorescence and detector problems in the infrared. Our opinion is that the information it would provide is so important to the issues considered here (and, indeed to assessing the validity of vibronic coupling models in general, and facilitating the construction of better ones) that it ought to become a top priority for experimentalists in this area.

A second experiment of relevance is analysis of X-ray temperature factors of the P dimer. We would predict that relative motion of the P molecules would have a large amplitude at every temperature, and that this amplitude would increase as the temperature increased. Anharmonicity should also manifest itself. The use of molecular dynamics modeling in conjunction with analysis of experimental X-ray data should prove useful here ; indeed, one might be able to design an intermolecular potential surface in part by fitting X-ray data.

ACKNOWLEDGEMENTS

This work was supported by a grant from the National Science Foundation. RAF is an Alfred P. Sloan Foundation Fellow and a Camille and Henry Dreyfus Teacher-Scholar. We thank Dewey Holton, Christine Kirmaier, and Arnold Hoff for useful discussions.

REFERENCES

1. Won, Y. and Friesner, R. A. (1987) *Proc. Natl. Acad. Sci.* USA **84**, 5511-5515.
2. Won, Y. and Friesner, R. A. *J. Phys. Chem.* to be published.
3. Holton, D., private communication.
4. Lagos, R. and Friesner, R. (1984) *J. Chem. Phys.* **81**, 5899-5909.
5. Won, Y., Lagos, R. and Friesner, R. (1986) *ibid.* **84**, 6567-6574.
6. Deisenhofer, J., Epp, O., Miki, K., Huber, R. and Michel, H. (1984) *J. Mol. Biol.* **180**, 385-398.
7. Gouterman, M. (1961) *J. Mol. Spectrosc.* **6**, 138-163.
8. Platenkamp, R. J., Den Blanken, H. J. and Hoff, A. J. (1980) *Chem. Phys. Lett.* **76**, 35-41.
9. Even, U., Magen, J. and Jortner, J. (1982) *ibid.* **88**, 131-134.
10. Vermeglio, A.and Paillotin, G. (1982) *Biochim. Biophys. Acta* **681**, 32-40.
11. Shuvalov, V. A.; Asadov, A. A. (1979) *ibid.* **545**, 296-308.
12. Phillipson, K. D.; Sauer, K. (1973) *Biochemistry* **12**, 535-539.
13. Shuvalov, V. A. and Klevanik, A. V. (1983) *FEBS Lett.* **160**, 51-55.
14. Tiede, D. M., Kellogg, E. and Breton, J. *Biochim. Biophys. Acta*, to be published.
15. Hoff, A. J., this volume.
16. Hayes, J. M. and Small, G. J. (1986) *J. Phys. Chem.* **90**, 4928-4931.

DISCUSSION OF THE LARGE HOMOGENEOUS WIDTH OF THE P-BAND IN BACTERIAL- AND PLANT REACTION CENTERS

Douwe A. Wiersma

Picosecond Laser and Spectroscopy Laboratory
Department of Chemistry
University of Groningen
Nijenborgh 16, 9747 AG Groningen, The Netherlands

Recent hole-burning [1,2,3,4,5,6,7,8,9] and photon echo [2,8,9] experiments on bacterial and plant reaction centers have established that the P-bands in these systems exhibit extremely large 'homogeneous' widths which correspond to ultra-short relaxation times of less than 100 femtoseconds. A second characteristic feature of these transitions in bacterial reaction centers is the large change of dipole moment on optical excitation [10,11]. In photosystem II however, the observed dipole moment change is 'normal' [12], while its P-band still exhibits a large homogeneous width. The question is how these observations should be interpreted and how they relate to the dynamics of the initial photochemical event of charge separation in the reaction center.

Current knowledge of the spatial structure and protein sequence of the bacterial reaction centers [13,14] and protein sequence of the plant photosystem II reaction center strongly suggests [15] that in both type of reaction centers the spatial architecture is similar and that each contains a chlorophyll dimer at the center. The important question is how the reaction proceeds after the energy is trapped at the initial donor. Three different models for charge-separation in bacterial reaction centers have been proposed [16,17,18,19] and are currently under investigation. In each of these models the role of the accessory bacteriochlorophyll on the L-branch ($BChl_L$) is different. As in all models the excited state of the special pair is the initial state of the photochemical reaction, it is important to establish its nature and dynamics.

The present debate on the dynamics of the P-state centers around the question whether vibronic mixing [2,8,20,21] or electron-phonon coupling [5,22] effects are responsible for the observed large homogeneous spectral width of the P-band. The effective hamiltonians that govern these effects are:

$$H_{VM} = \sum_{i,j} \epsilon_{i,j} a_i^\dagger a_j + \sum \hbar\omega_m (b_m^\dagger b_m + 1/2) + \sum_{i,j} \sum_m V_m^{ij} (b_m^\dagger + b_m) a_i^\dagger a_j + \ldots \quad (1)$$

$$H_{LEPC} = \epsilon^p a_p^\dagger a_p + \sum \hbar\omega_\kappa (b_\kappa^\dagger b_\kappa + 1/2) + \sum V_\kappa^p (b_\kappa + b_\kappa^\dagger) a_p^\dagger a_p + \ldots\ldots \quad (2)$$

Here $a_i^\dagger(a_i)$ and $b_m^\dagger(b_m)$ are the creation (destruction) operators for electron and vibrational excitation in (1) and $a_p^\dagger(a_p)$ and $b_\kappa^\dagger(b_\kappa)$ the corresponding electron and phonon operators in (2).

In the vibronic coupling case (1), the *neutral* Frenkel-dimer excitations are vibronically mixed with nearby charge-transfer states (e.g. the intra-dimer $BChl^-$. $BChl^+$ or P^+. $BChl_L^-$ state), leading to a compound state with irregular and dense vibronic structure in its absorption spectrum. Coupling to the phonons is not explicitly considered in this model but their effect can easily be incorporated in terms of a line-width parameter. Won and Friesner [20,21] have recently performed extensive vibronic-coupling calculations, using a Greens function approach, showing that both low-frequency and high-frequency modes are needed to generate a P-transition with a large 'homogeneous' bandwidth. The low-frequency modes in this case are modes that modulate the distance and geometry of the special pair. An important feature of this model-calculation is that a minor displacement along these low-frequency coordinates (a Huang-Rhys factor of about 1) is sufficient to generate a broad homogeneous P-transition. We note that in the case where the charge-transfer state (carrying no oscillator strength) is (in zero order) resonant with or lower in energy than the neutral Frenkel-dimer excitation, the time-evolution of the excited P-state (P*) can be looked at as a (dissipative) decay from a neutral Frenkel excitation into a charge-transfer state on a time-scale concurrent with the large homogeneous width (25 fs for Rb. sphaeroides). Such a description was forwarded in Refs. 2 and 8 for the observed large homogeneous width of the P-band. If the charge-transfer (charge-resonance) state, in zero order, is energetically *higher* than the neutral Frenkel excitation, as seems to be the accepted situation in the reaction center [23], only in a *quantummechanical sense* we can speak of a time-evolution on this time-scale. In both cases however, the underlying physics (breakdown of the Born-Oppenheimer approximation) is identical. At this point it seems important to note that the vibronic mixing model does not necessarily imply the presence of a large change in dipole-moment on optical excitation. Consequently even in the case where a dimer exhibits inversion symmetry, vibronic mixing between two states can lead to a large homogeneous width of a dimer's transition. When this symmetry element is absent vibronic mixing between a neutral and charge-transfer state will generally lead to a polar excited state.

When linear electron-phonon coupling (LEPC, case 2) is assumed to be responsible for the large homogeneous width, the only electronically excited state considered is the P*-state which is assumed to contain charge-transfer character. In fact it is this nature of P* which is held responsible for the strong coupling between the medium phonons and the electronic excitation. In this model the large homogeneous width is due to phonon structure and therefore in a time-domain picture this width corresponds to a "phonon dressing" time of the optical excitation at the special pair. Note that in this model the polarity change on optical excitation is important because it acts as a catalyst for the electron-phonon coupling. This LEPC model has been extremely successfull in explaining spectral features of certain charge-transfer transitions in molecular solids [24]. The active phonon in these systems is thought to be an intra-dimer mode. Another point to be noted is that in the LEPC model the width of the zero-phonon line is directly related to the transfer time of the electron of the special pair to pheophytin. In the vibronic mixing model the width of an observed zero-phonon line may easily be broadened beyond this value [21].

It seems of considerable interest to pose the question whether the different models are related at all and which the criteria are that would enable one to make a choice.

The main difference between the models is the way in which the vibrational-electronic coupling is treated. In the LEPC model the relevant electronically excited state is taken as a *Born-Oppenheimer state* whose charge-transfer nature is determined by electron-nuclear forces of the system. Such an approach seems entirely justified in the case of a *pure* charge-transfer transition as in systems like anthracene-PMDA [25], where the charge-transfer state is far below the neutral Frenkel donor- and acceptor excited states. In the case of the special pair however, we are not dealing with a pure charge-transfer state but with a state that mainly looks like a delocalized neutral Frenkel-dimer excitation with some mixed-in charge-transfer character. When this state is described as a Born-Oppenheimer state, we implicitly assume that the coupling between these different type electronic manifolds is such, that new potentials are formed which are well-behaved in the Born-Oppenheimer sense.

In the VM model we start in the other limit and explicitly consider the possibility of a breakdown of the Born-Oppenheimer approximation. This is expected to occur in case the zero-order neutral Frenkel-dimer and charge-transfer excitation(s) are near-degenerate. In this situation the eigenstates of P* need to be described as linear combinations of *vibronic* functions of the zero-order neutral and charge-transfer states. This vibronic coupling process leads to distortions of the "molecule" along the vibrational coordinates that are most active in the vibronic coupling process. Under certain conditions the energetically lowest potential-well can even attain a double well character (the pseudo Jahn-Teller effect [26,27]) and the nuclear frame work along the coupling coordinate is severely distorted. The coupling of the electronic excitation to the phonons is not included at this level. If needed it can be incorporated by assigning an effective linewidth to the vibronic transitions.

With regards to the question which model is most appropriate for the description of the large homoegeneous width of the P-band, only experiments can tell. So far, in our opinion, the case remains undecided and more detailed information is needed especially on the correlation between the "homogeneous" width and dipole-moment change for different reaction centers and model dimers with and without inversion symmetry.

In view of the above discussion it would be particular important to find a zero-phonon line in these reaction centers and measure its homogeneous linewidth. So far such a transition has not been reported except in the case of photosystem I [6]. Recent experiments on this system [7] and photosystem II [9] strongly suggest however, that the reported zero-phonon line is not due to the reaction center but, most likely, to extraneous chlorphyll.

Returning now to the original question of the relevance of these spectral observations to functioning of the reaction center, it is clear that charge reorganization after optical excitation leading to a polar excited P-state can be considered as a pre-step in the charge-separation process. Calculations by Plato [28] confirm the idea that the excited state charge distribution is asymmetric with an excess electron charge-density on the $BChl_M$-molecule which is closest to the pheophytin electron-acceptor on the L-side.

A final question concerns the relevance of the possible non-Born-Oppenheimer nature of P* to the electron-transfer process. In this context it is important to note that in the field of radiationless processes it has been established [29,30], that a radiationless process becomes more efficient if the nuclear frame-work between the initial and final states is different. The rational for this effect is related to the Franck-Condon principle which only allows vertical optical transitions. Quantum mechanically this can be interpreted as the condition for maximum overlap of vibrational wave functions. When this idea is applied to an *isoenergetic* radiationless process it is clear that the rate is maximized when the zero-point vibrational wave function of the upper potential exhibits maximum overlap with the isoenergetic vibrational wave function of the lower potential. In order to accomplish this the potentials need to be shifted with respect to one another. This suggests that the electron-transfer process in the reaction center, regarded as a super-molecule, depends in the same manner on the difference in nuclear geometry of the initial and final relevant states. In other words a, by vibronic coupling, vibrationally-distorted excited state of the electron donor would increase the rate of electron-transfer in the reaction center. The presence of a dimer in photosynthetic reaction centers would therefore far exceed its function as a light-trap. It would also allow a more efficient non-adiabatic electron-transfer reaction.

It remains a challenge to provide evidence for these speculations.

ACKNOWLEDGEMENT

We are indebted to dr. A.J. Hoff for many discussion pertaining to the optical dynamics of the photosynthetic reaction center.

References

[1] V.G. Maslov, A.S. Chunaev and V.V. Tugarinov, Mol. Biol., **15**, 788 (1981)

[2] S.R. Meech, A.J. Hoff and D.A. Wiersma, Chem. Phys. Lett., **121**, 287 (1985)

[3] S.G. Boxer, D.J. Lockhart and T.R. Middendorf, Chem. Phys. Lett., **123**, 476 (1986)

[4] S.G. Boxer, T.R. Middendorf and D.J. Lockhart, FEBS, **200**, 237 (1986)

[5] J.M. Hayes and G.J. Small, J. Phys. Chem., **90**, 4928 (1986)

[6] J.K. Gillie, B.L. Feary, J.M. Hayes, G.J. Small and J.H. Golbeck, Chem. Phys. Lett., **134**, 316 (1987)

[7] J.M. Hayes, J.K. Gillie, D. Tang and J. Small, submitted for publication in Biochem. Biophys. Acta

[8] S.R. Meech, A.J. Hoff and D.A. Wiersma, Proc. Natl. Acad. Sci., USA, **83**, 9464 (1986)

[9] K.J. Vink, S. de Boer, J.J. Plijter, A.J. Hoff and D.A. Wiersma, Chem. Phys. Lett., to be published (1987)

[10] D. Deleeuv, M. Malley, G. Butterman, M.Y. Okamura and G. Feher, Biophys. Soc. Abstr., **37**, 111a (1982)

[11] D.J. Lockhart and S.G. Boxer, Biochemistry, **26**, 664 (1987)

[12] G. Feher, private communication

[13] J. Deisenhofer, O. Epp, K. Miki, R. Huber and H. Michel, J. Mol. Biol., **180**, 385 (1984)

[14] J. Deisenhofer, O. Epp, K. Miki, R. Huber and H. Michel, Nature (London), **318**, 618 (1985)

[15] A. Trebst, Z. Naturforschung., **C41**, 240 (1986)

[16] S.F. Fischer, I. Nussbaum and P.O.J. Scherer, in: Antennas and reaction centers of photosynthetic bacteria, ed. M.E. Michel-Beyerle (Springer, Berlin, 1985) p. 256

[17] M. Bixon, J. Jortner, M.E. Michel-Beyerle, A. Ogrodnik and W. Lersch, Chem. Phys. Lett., **140**, 626 (1987)

[18] P. Markus, Chem. Phys. Lett., **133**, 471 (1987)

[19] S.F. Fischer and P.O.J. Scherer, Chem. Phys., **115**, 151 (1987), P.O.J. Scherer and S.F. Fischer, Chem. Phys. Lett., **141**, 179 (1987)

[20] Y. Won and R.A. Friesner, Proc. Natl. Acad. Sci. USA, **84**, 5511 (1987)

[21] Y. Won and R.A. Friesner, submitted for publication

[22] P.O.J. Scherer, S.F. Fischer, J.K.H. Horber, M.E. Michel-Beyerle and H. Michel in: Antennas and reaction centers of photosynthetic bacteria, ed. M.E. Michel-Beyerle (Springer, Berlin, 1985) p. 131

[23] W.W. Parson and A. Warshel, J. Am. Chem. Soc., to be published (1987)

[24] D. Haarer and M. Philpott, in: Spectroscopy and Excitation Dynamics of Condensed Molecular Systems, eds. V.M. Agranovich and R.M. Hochstrasser (North Holland Publishing Company, Amsterdam, 1983), p. 27

[25] J. Friedrich, J.D. Swalen and D. Haarer, J. Chem. Phys., **73**, 705 (1980)

[26] R.M. Hochstrasser and C.A. Marzzacco, in: Molecular Luminiscence, ed. E.C. Lim (Benjamin, New York, 1969), p. 631

[27] R.M. Hochstrasser and D.A. Wiersma, Isr. J. Chem., **10**, 517 (1972)

[28] M. Plato, private communication

[29] M. Bixon and J. Jortner, J. Chem. Phys., **48**, 715 (1968)

[30] D.M. Burland and G.W. Robinson, Proc. Natl. Acad. Sci., USA, **60**, 257 (1970)

THEORETICAL MODELS OF ELECTROCHROMIC AND ENVIRONMENTAL EFFECTS ON BACTERIO-CHLOROPHYLLS AND -PHEOPHYTINS IN REACTION CENTERS

Louise Karle Hanson, Mark A. Thompson*, Michael C. Zerner*, and Jack Fajer

Department of Applied Science, Brookhaven National Laboratory, Upton, New York 11973
*Quantum Theory Project, University of Florida
Gainesville, Florida 32611

INTRODUCTION

The primary charge separation in photosynthetic reaction centers (RC) is carried out by an array of (bacterio)chlorophylls arranged in close proximity. The recent x-ray structures of the RC complexes from Rhodopseudomonas viridis[1-3] and Rhodobacter sphaeroides[4,5] reveal two possible electron pathways for purple bacteria, only one of which is active.[6] In order to determine the environmental factors most likely to affect the direction of electron flow in the nascent charge separation, we have calculated the effects of hydrogen bonding and chemical modification (enolization) by nearby residues upon the optical spectra, reduction potentials, unpaired spin density distributions and charge densities of bacterio-chlorophylls (BChl) and -pheophytins (BPh).

In addition, we have calculated the effect of generating the primary cation and anion products on the spectra of the accessory BChl[7] and BPh pigments in the RC. The trends predicted for the electrochromic shifts offer reasonable explanations for some of the optical changes observed during picosecond flash photolysis and in trapping experiments as the photogenerated electron moves to secondary acceptors. At the present levels of refinement, x-ray structures of Rb. sphaeroides[4,5], which contains BChls a, reveal a molecular architecture analogous to that of R. viridis, which is comprised of BChls b. Similar electrochromic shifts are thus predicted for the two species.

Charge iterative extended Hückel (IEH)[8] and spectroscopic INDO[9,10] methods were employed in the calculations. Figure 1 depicts the structure of BChl b, the model used in the INDO calculations, and the numbering scheme.

A. ENVIRONMENTAL EFFECTS

Hydrogen Bonding and Substituent Orientation

The R. viridis RC crystal structure reveals[3] an asymmetric pattern of hydrogen bonding for the $(BChl\ b)_2$ special pair and the two BPhs: the BPh b_L in the active pathway for electron transfer is H-bonded at the

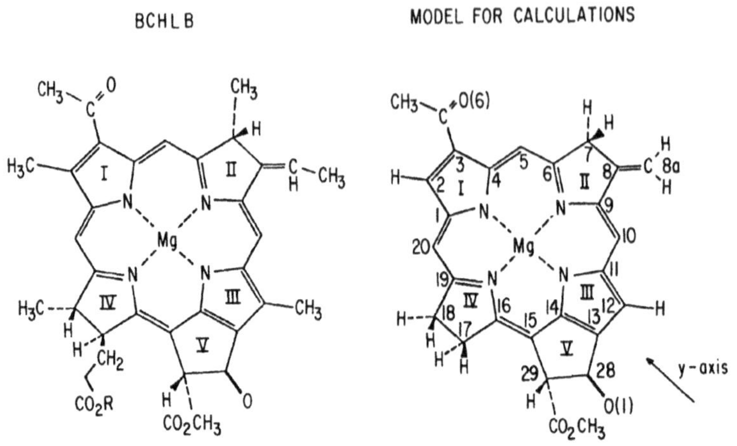

Fig. 1. Structure of BChl b, the chromophore found in R. viridis (left), and of the BChl b model used in the INDO calculations (right). For BPh b, the central Mg was replaced by two protons bound to the nitrogens of rings I and III. The IEH calculations included methyl groups at positions 2, 7, 8a, 12, 17, and 18 (as shown in Figures 3 and 4). Details of the coordinates are given in Ref. 12. H29 was transferred to O1 for the enol calculations, accompanied by appropriate adjustments to bond distances and angles.

TABLE I. Effect of Hydrogen Bonding and Acetyl Orientation on the Q_x and Q_y Bands of BPh b (INDO).

BPh b species	Frequency shifts in cm^{-1} Q_x	Q_y
H_2O H-bonded to O1 (ring V)	55(blue shift)	150(blue shift)*
	relative to non-H-bonded BPh b	
C3 acetyl oriented 90° to macrocycle plane	521(blue shift)	680(blue shift)
	relative to BPh b with C3 acetyl 0°	

*Shifts this small are affected by the precision of the wavefunctions and therefore may not be calculated accurately.

ring V keto oxygen (O1) by a glutamic acid, whereas the BPh b_M in the inactive path is not H-bonded at that position. Furthermore, only one of the BChls comprising the special pair ($BChl_L$) is H-bonded at this keto oxygen. Both IEH and INDO calculations predict that H-bonding at this position will lower the reduction potential of the chromophore by ~40 mV. This calculated effect for the BPh appears to be rather small to be the major factor that determines the path of electron flow. However, the asymmetrical H-bonding of the special pair BChls may result in differences within the donor sufficient to induce vectorial electron transport at the very onset of charge separation.

The optical spectra of BPh_L and BPh_M differ, with the BPh_L Q_x and Q_y bands red-shifted by 10-15nm. Table I presents calculated optical shifts caused by H-bonding and C3 acetyl orientation. These results suggest that the differences in BPh absorption spectra may arise from different C3 acetyl orientations as well as from the presence or absence of H-bonding.

Enolization

Bocian et al.[11] recently suggested on the basis of resonance Raman studies that the glutamic acid H-bonded to the O1 of BPh_L imparts partial enolic character to ring V. We have calculated optical spectra for BPh b, its enol, and their anionic species. The neutral keto and enol forms differ substantially: the enol exhibits a red-shifted Q_y band, loss of oscillator strength in the Soret and Q_y bands, and two strong bands in the 460-590 nm region, whereas the anions are very similar (Figure 2). Unpaired spin density distributions for the keto and enol anions (Figure 3) are again predicted to be very similar. Also shown in Figure 3 is the unpaired spin density distribution for a putative <u>neutral</u> radical species that would result from protonation of the BPh b^- anion at O1. This species does not fit the experimental ESR and ENDOR data as well as the BPh b^- or the BPh b^- enol.[12]

The existence of a BPh_L enol anion cannot be immediately ruled out because of the similarity of the calculated optical spectra and unpaired spin density distributions between the keto and enol anions. However, we do not predict an enol to be preferentially reduced. Both IEH and INDO calculations show the enol to have a slightly higher reduction potential than the keto form. [Wasielewski has found[13] that a model compound, 28-desoxo-28,29-dehydro Pheo a, is ~90 mV easier to reduce than Pheo a. This drop in potential is indeed predicted by IEH calculations for the model but <u>not</u> for the enol.] Note also that the formation of an BPh enol anion from the neutral BPh would require significant and rapid (>200ps) conformational rearrangements of the substituents on ring V or else result in large ring strain.

Charge Densities

Changes in charge densities due to the creation of a BPh b anion are presented in Figure 4. The negative charge is distributed over <u>all</u> the atoms of the macrocycle. The largest changes occur for the atoms that form the conjugated π system; the relative magnitudes roughly parallel the unpaired spin density distribution calculated for BPh b^- (see Figure 3). Also shown in Figure 4 are the effects of hydrogen bonding to the carbonyl oxygens of ring V and of enolization upon the charge densities of BPh b^-. The only significant changes for both species occur in ring V. Enolization affects all the atoms that comprise Ring V and its substituents whereas H-bonding affects only the carbonyl atoms. If these charge density changes translate into bond order shifts, then the effects of enolization and H-bonding on infrared and resonance Raman spectra

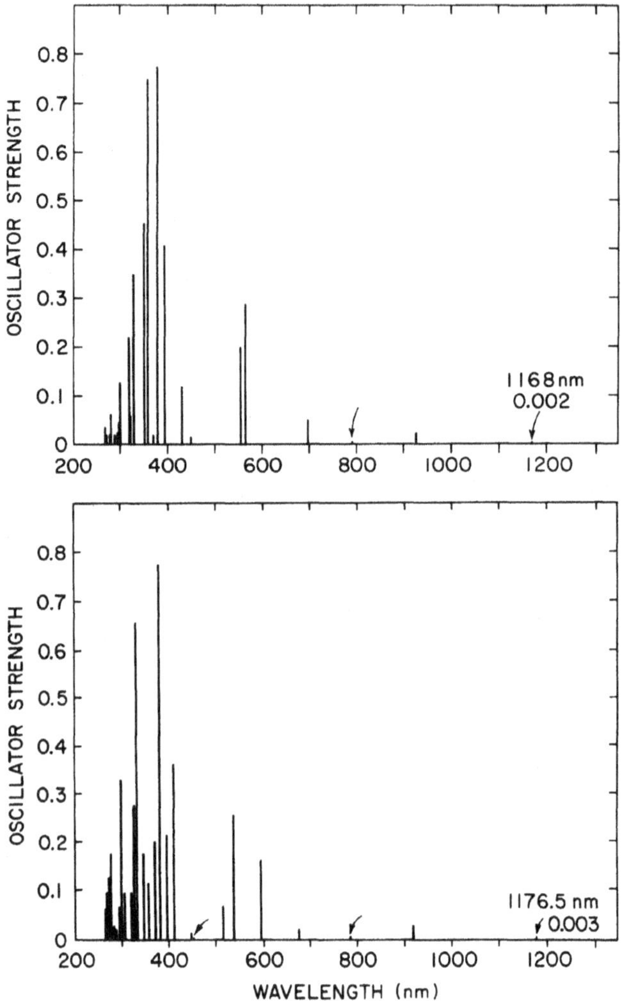

Fig. 2. Calculated absorption spectra (INDO) for anions of BPh b (top) and BPh b enol (bottom).

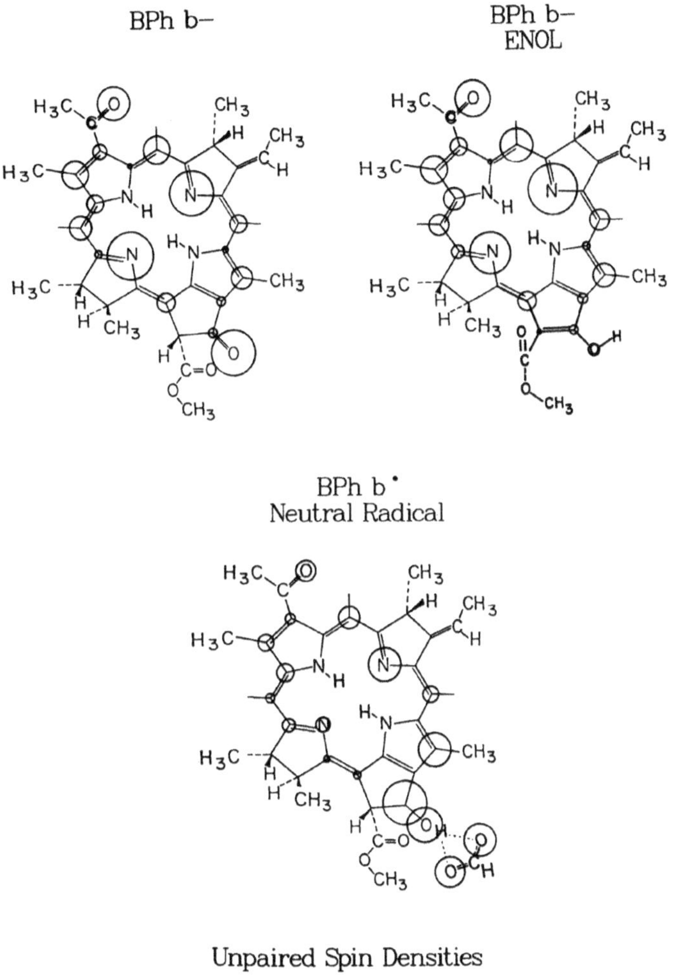

Fig. 3. Unpaired spin density distributions (IEH) for anions of BPh b and BPh b enol, and for a BPh b neutral radical species formed by the transfer of a proton from formic acid to O1 of BPh b^{-}. The C3 acetyl groups are coplanar with the macrocycle for all three species.

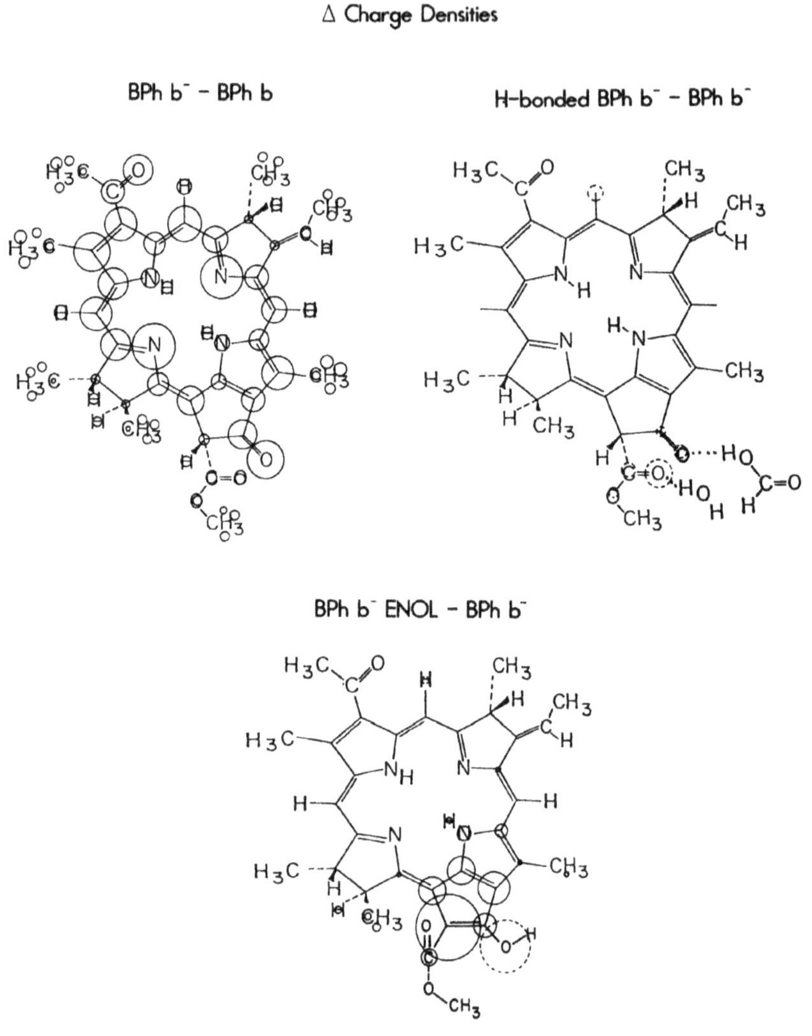

Fig. 4. Differences in calculated charge densities (IEH) for BPh b$^-$ - BPh b (top left), BPh b$^-$ H-bonded at both ring V carbonyl groups - BPh b$^-$ (top right), and BPh b$^-$ enol - BPh b$^-$ (bottom; moving the proton from C29 to O1 increases its charge by +0.163 e$^-$). Solid circles represent an increase in negative charge and broken circles an increase in positive charge at a given position.

should differ. Enolization should alter all those modes that include ring V atoms, whereas H-bonding should shift only the carbonyl modes.

B. ELECTROCHROMIC EFFECTS

The primary charge transfer event in reaction centers of purple bacteria is believed to generate a dimeric BChl cation (P^+) and a BPh anion[14-22] (BPh_L^-). Subsequently, the negative charge migrates to the quinone Q_A. P^+ and BPh_L^- lie within van der Waals contact on opposite sides of a lone accessory BChl "bridging" molecule ($BChl_{LA}$).[1-5] Spectral changes in the red (Q_y) band of the bridge BChls have been observed on picosecond time scales following excitation of the RC[15-24] and have been attributed to the formation of a BChl anion,[23,24] (i.e. P^+ $BChl_{LA}^-$ precedes P^+ BPh_L^-), or to electrochromic effects.[12,14,17,19,22] Other pigments in the RC such as $BChl_{MA}$ and BPh_L, also lie within close proximity of charged species as the charge separation and migration occur. Indeed, shifts in the BPh absorption bands have been observed to accompany quinone reduction.[17-19,25] We present here calculations that examine the electrochromic effects of the charged species P^+, $BChl_{LA}^-$, BPh_L^- and Q_A^- upon the optical spectra of all the accessory BChls and BPhs in the RC.

INDO calculations predict that point charges placed within 3-4Å of BChls can induce significant shifts of the red band.[7] Table II lists values calculated for BChl b. Similar results are obtained for the other BChls found in photosynthetic bacteria, BChls a and g[26]. The magnitude and sign of the optical shifts depend upon the placement and sign of the point charge: positive charges placed near ring I or negative charges situated near ring III are predicted to cause large red shifts whereas reversing the charges should result in large blue shifts.

Point charges representing P^+, $BChl_{LA}^-$, BPh_L^- and Q_A^- in bacterial reaction centers were placed at the centroids of $BChl_{LA}$, BPh_L and the special pair, and between the two oxygens of Q_A using coordinates from the <u>R. viridis</u> structure. The calculated shifts induced by these point charges on the optical spectra of the accessory BChls and the BPhs are presented in Tables III and IV. The signs of the shifts follow the expected pattern and their magnitudes depend upon both the relative orientation of the charge and its distance from the macrocycle (the further away, the smaller the shift). The calculations indicate that the primary charge separation in a RC should induce a significant electrochromic effect on the optical spectrum of $BChl_{LA}$. Shifts were calculated for a positive charge placed at the donor $(BChl\ b)_2$ centroid and a negative charge at the BPh_L center, as well as distributing the point charges over the cores of the donor and primary acceptor (Table III). Individually or together, these charges induce sizeable blue shifts for BChl b. Oxidation of P to P^+ induces a comparable blue shift for the accessory BChl on the inactive M side. However, $BChl_{MA}$ is sufficiently distant from the pigments on the L side that formation of any L anionic species has little additional effect.

In agreement with the trends predicted by the calculations, experimental blue shifts of a few hundred wave numbers in the spectra of the bridging BChl(s) have indeed been observed on picosecond time scales, and when the donor is chemically oxidized or the acceptor is reduced under steady state conditions.[12,15-24] In addition, P^+ BPh_L^- induces a larger blue shift on $BChl_{LA}$ than P^+ Q_A^-.[20]

Small red shifts are calculated for the influence of P^+ upon BPh_M

TABLE II. Effect of Point Charges on the Q_y (red) Band of BChl b (INDO)[a].

Point charge	Position[b] (charge is 3.5A above the atom)	Frequency shift of Q_y relative to no charge[c,d]	
+1	N(I) (y-axis)	-1440 cm^{-1}	red shift
-1	N(I)	2297	blue
+1	N(II) (x-axis)	-239	red
-1	N(II)	109	blue
+1	N(III) (y-axis)	1329	blue
-1	N(III)	-895	red
+1	N(IV) (x-axis)	-564	red
-1	N(IV)	440	blue
+1	C20	-1272	red
-1	C20	1993	blue
+1	C2	-1603	red
-1	C2	2872	blue
+1	O6	-1256	red
-1	O6	1886	blue
+1	C8a	1060	blue
-1	C8a	-1411	red
+1	C12	2579	blue
-1	C12	-1160	red
+1	O1	1796	blue
-1	O1	-1000	red

[a]Data from Ref. 7.
[b]Position numbers are given in Figure 1.
[c]Increasing the dielectric constant from 1 (vacuum) to 2 to simulate a hydrophobic environment roughly halves the values shown.
[d]A point charge affects the optical spectrum by modifying the magnitudes of the state dipole moments of the molecules. The larger the initial difference between the ground and excited state values, the larger the perturbation. The ground state has a considerable dipole moment along the y-axis, which runs through rings I and III, whereas the transition to the Q_y state nearly cancels this moment. Therefore, for a charge at a given distance from the Mg, the calculated optical shifts are largest near the y-axis and fall off towards the x-axis.

Point charges shift the Q_x band in the same direction as the Q_y, but the magnitude of the shift is smaller. For instance, for a positive point charge 3.5Å above N(I), the Q_y red shift of 1440 cm^{-1} is accompanied by a Q_x red shift of 731 cm^{-1}. A negative charge at this position induces blue shifts of 2297 and 483 cm^{-1} for the Q_y and Q_x, respectively. The parallel behavior of the Q_x and Q_y arises because the INDO calculations inpart substantial y character to the Q_x transition.

TABLE III. Effect of Donor and Acceptor Charges on the Q_y (red) Band of the Bridging BChls of R. viridis Reaction Centers (INDO).

Bridging BChl	Point charge	Pt charge Position	Frequency shift of Qy relative to no charge	
L	+1	Center of the (BChl b)$_2$ 3.94, -9.93, 2.54Å*	1063 cm^{-1}	blue
L	-1	Center of the BPh b_L -1.41, 5.54, 8.46Å*	653	blue
L	+1, -1	Cation and anion located as above	1857	blue
L	10 charges of +0.1 a.u. distributed over the 2 Mg and 8N of (BChl b)$_2$		924	blue
L	4 charges of -0.25 a.u. distributed over the 4N of BPh b_L		664	blue
L	+, - Cation and anion distributed as above		1709	blue
L	-1	Midpt. of two O's of Q_A -9.65, 7.47, 19.06Å*	-194	red
L	+1,-1	(BChl b)$_2^+$ and Q_A^-	1063	blue
M	+1	Center of the (BChl b)$^+_2$ 5.20, -9.92, 2.32 Å*	947	blue
M	-1	Center of the BChl b_{LA} 8.01, -17.07, 10.13 Å*	<550*	blue
M	-1	Center of BPh b_L 2.84, -15.27, 18.71Å*	<-150*	red
M	-1	Midpt. of two O's of Q_A 2.03, -7.18, 29.57Å*	-34	red
M	+1,-1	(BChl b)$_2^+$ and Q_A	947	blue

*x, y and z coordinates of the (BChl)$_2$, BChl$_{LA}$, BPh$_L$ or Q_A relative to the center of the bridging BChl. NI and NIII of the bridging BChl define the y-axis.

*The effect of charges at these positions could not be calculated directly. Shifts are estimated from points placed closer to the BChl at the same relative orientations.

and Q_A^- upon both BPhs (Table IV). (Q_A^{2-} should have twice the effect of Q_A^-.) Experimentally, quinone reduction is indeed accompanied by red shifts in the BPh spectra.[17-19,25] The data in Table IV also suggest that if P^+ $BChl_{LA}^-$ preceded P^+ BPh_L^- as a discrete state in the initial charge separation in RCs, then a sizeable blue shift would be expected for the BPh_L prior to its reduction. Subpicosecond measurements[15,16,22] have not revealed such an effect in the BPh_L band.

The calculations correlate well with observed shift patterns for both the accessory BChls and the BPhs in RCs, supporting an electrochromic origin for these shifts. However, the calculations overestimate the magnitudes of the shifts, primarily because they do not take into account shielding by the protein and ring substituents. Extensive delocalization of the charges over $(BChl\ b)^+_2$ and BPh b^-, as is actually observed[12], may also reduce the calculated shifts. The electrochromic effects would be susceptible to additional modulation if changes in distances or orientations between the chromophores follow electron transfer.[27,28]

TABLE IV. Effect of Donor and Acceptor Charges on the Q_y (red) Band of the BPhs of R. viridis Reaction Centers (INDO)

BPh	Point charge	Pt charge Position	Frequency shift of Qy relative to no charge	
L	+1	Center of the $(BChl\ b)_2$ 11.48, 12.92, -2.04Å*	0 cm^{-1}	
L	-1	Center of the BChl b_{LA} 1.42, 10.03, 1.26Å*	1825	blue
L	+1, -1	Cation and anion located as above	1356	blue
L	-1	Midpt. of two O's of Q_A -2.69, -9.63, -9.17Å*	-385	red
M	+1	Center of the $(BChl\ b)_2$ 13.14, 12.44, -0.43Å*	-390	red
M	-1	Center of the BChl b_{LA} 22.02, 8.87, 5.81 Å*	< 60≠	blue
M	-1	Center of the BPh b_L 19.38, 2.00, -12.87 Å*	<-40≠	red
M	-1	Midpt. of the two O's of Q_A 15.79, -10.94, -14.79Å*	<-200≠	red

*x, y and z coordinates of the $(BChl)_2$, $BChl_{LA}$, BPh_L or Q_A relative to the center of the BPh. NI and NIII of the BPh define the y-axis.

≠The effect of charges at these positions could not be calculated directly. Shifts are estimated from points placed closer to the BPh at the same relative orientations.

CONCLUSIONS

The above results lead to the following conclusions:

1. Although it may help stabilize the anion, hydrogen bonding of BPh_L is not likely to be the major driving force that controls vectorial electron transport. Rather, asymmetric hydrogen bonding within the special pair (as well as its environment) may contribute most in determining the direction of electron flow at the very onset of charge spearation.

2. Formation of a putative enol anion would not dramatically affect either the overall unpaired spin density profile nor the optical spectrum of the reduced BPh.

3. Enolization would alter the atomic charge densities of ring V and its substituents. Thus infrared and resonance Raman bands that sample that portion of the molecule should be shifted. In contrast, the perturbations due to hydrogen bonding of a carbonyl oxygen remain localized on the carbonyl group.

4. At least some of the optical shifts observed for the accessory BChls and the BPhs during charge separation and migration in the RCs can be attributed to electrochromic shifts induced by nearby charges. Hence, optical shifts need not necessarily imply actual redox processes by a chromophore such as the bridging $BChl_{LA}$.

ACKNOWLEDGMENT

We thank J. Deisenhofer and H. Michel for the coordinates of the R. viridis chromophores and C. Kirmaier and D. Holten for valuable discussions. This work was supported by the Division of Chemical Sciences, U.S. Department of Energy under contract No. DE-AC02-76CH00016 at BNL and by a University of Florida CRDC Chemical Systems Technology Center award, DAAA15-85-C-0034. M.A.T. is the recipient of an NSF predoctoral fellowship.

REFERENCES

1. J. Deisenhofer, O. Epp, K. Miki, R. Huber, and H. Michel. X-ray structure analysis of a membrane protein complex. Electron density map at 3A resolution and a model of the chromophores of the photosynthetic reaction center from Rhodopseudomonas viridis, J. Mol. Biol., 180:385 (1984).
2. J. Deisenhofer, O. Epp, K. Miki, R. Huber, and H. Michel. Structure of the protein subunits in the reaction centre of Rhodopseudomonas viridis at 3A resolution, Nature, 318:618 (1985).
3. H. Michel, O. Epp, and J. Deisenhofer. Pigment-protein interactions in the photosynthetic reaction centre from Rhodopseudomonas viridis, EMBO J., 5:2445 (1986).
4. C.-H. Chang, D. Tiede, J. Tang, U. Smith, J. Norris, and M. Schiffer. Structure of Rhodopseudomonas sphaeroides R-26 reaction center, FEBS Lett., 205:82 (1986).
5. J. P. Allen, G. Feher, T. O. Yeates, H. Komiya, and D. C. Rees. Structure of the reaction center from Rhodobacter sphaeroides R-26: the cofactors, Proc. Natl. Acad. Sci. USA, 84:5730 (1987).
6. M. E. Michel-Beyerle, ed., "Antennas and Reaction Centers of Photosynthetic Bacteria," Springer-Verlag, Berlin (1985).

7. L. K. Hanson, J. Fajer, M. A. Thompson, and M. C. Zerner. Electrochromic effects of charge separation in bacterial photosynthesis: theoretical models, J. Am. Chem. Soc., 109:4728 (1987).
8. A. M. Schaffer, M. Gouterman, and E. R. Davidson. Porphyrins XXVIII. Extended Hückel calculations on metal phthalocyanines and tetrazaporphins, Theor. Chem. Acta (Berl.), 30:9 (1973).
9. J. Ridley and M. Zerner. An intermediate neglect of differential overlap technique for spectroscopy: pyrrole and the azines, Theor. Chim. Acta (Berl.), 32:111 (1973).
10. A. D. Bacon and M. C. Zerner. An intermediate neglect of differential overlap theory for transition metal complexes: Fe, Co and Cu chlorides, Theor. Chim. Acta (Berl.), 53:21 (1979).
11. D. F. Bocian, N. J. Boldt, B. W. Chadwick, and H. A. Frank. Near-infrared-excitation resonance Raman spectra of bacterial photosynthetic reaction centers. Implications for path-specific electron transfer, FEBS Lett., 214:92 (1987).
12. M. S. Davis, A. Forman, L. K. Hanson, J. P. Thornber, and J. Fajer. Anion and cation radicals of bacteriochlorophyll and bacteriopheophytin b. Their role in the primary charge separation of Rhodopseudomonas viridis, J. Phys. Chem., 83:3325 (1979).
13. M. R. Wasielewski, private communication.
14. J. Fajer, D. C. Brune, M. S. Davis, A. Forman, and L. D. Spaulding. Primary charge separation in bacterial photosynthesis: oxidized chlorophylls and reduced pheophytin, Proc. Natl. Acad. Sci. USA, 72:4956 (1975).
15. J.-L. Martin, J. Breton, A. J. Hoff, A. Migus, and A. Antonetti. Femtosecond spectroscopy of electron transfer in the reaction center of the photosynthetic bacterium Rhodopseudomonas sphaeroides R-26: direct electron transfer from the dimeric bacteriochlorophyll primary donor to the bacteriopheophytin acceptor with a time constant of 2.8±0.2 psec, Proc. Natl. Acad. Sci. USA, 83:957 (1986).
16. J. Breton, J.-L. Martin, A. Migus, A. Antonetti, and A. Orszag. Femtosecond spectroscopy of excitation energy transfer and initial charge separation in the reaction center of the photosynthetic bacterium Rhodopseudomonas viridis, Proc. Natl. Acad. Sci., USA, 83:5121 (1986).
17. C. Kirmaier, D. Holten, and W. W. Parson. Picosecond photodichroism (photoselection) measurements on transient states in reaction centers from Rhodopseudomonas sphaeroides, Rhodospirillum rubrum and Rhodopseudomonas viridis, Biochim. Biophys. Acta, 725:190 (1983).
18. C. Kirmaier, D. Holten, L. J. Mancino, and R. E. Blankenship. Picosecond photodichroism studies on reaction centers from the green photosynthetic bacterium Chloroflexus aurantiacus, Biochim. Biophys. Acta, 765:138 (1984).
19. C. Kirmaier, D. Holten, and W. W. Parson. Picosecond photodichroism studies of the transient states in Rhodopseudomonas sphaeroides reaction centers at 5K. Effects of electron transfer on the six bacteriochlorin pigments, Biochim. Biophys. Acta, 810:49 (1985).
20. P. Maroti, C. Kirmaier, C. Wraight, D. Holten, and R. M. Pearlstein. Photochemistry and electron transfer in borohydride-treated photosynthetic reaction centers, Biochim. Biophys. Acta, 810:132 (1985).
21. C. Kirmaier, D. Holten, and W. W. Parson. The question of the intermediate state P^+ $BChl^-$ in bacterial photosynthesis, FEBS Lett., 185:76 (1985).
22. M. R. Wasielewski and D. M. Tiede. Sub-picosecond measurements of primary electron transfer in Rhodopseudomonas viridis reaction centers using near-infrared radiation, FEBS Lett., 204:368 (1986).

23. V. A. Shuvalov and L. N. M. Duysens. Primary electron transfer reactions in modified reaction centers from Rhodopseudomonas sphaeroides, Proc. Natl. Acad. Sci. USA, 83:1690 (1986).
24. V. A. Shuvalov, J. Amesz, and L. N. M. Duysens. Picosecond charge separation upon selective excitation of the primary electron donor in reaction centers of Rhodopseudomonas viridis, Biochim. Biophys. Acta, 851:327 (1986).
25. A. Vermeglio and R. K. Clayton. Kinetics of electron transfer between the primary and the secondary electron acceptor in reaction centers from Rhodopseudomonas sphaeroides, Biochim. Biophys. Acta, 461:159 (1977).
26. L. K. Hanson, M. A. Thompson, and J. Fajer. Environmental effects on the properties of chlorophylls in vivo. Theoretical models, in: "Progress in Photosynthesis Research," Vol. I, J. Biggins, ed., Martinus Nijhoff, Dordrecht, p. 3.331, (1987).
27. J. Fajer, K. M. Barkigia, K. M. Smith, and D. A. Goff. Consequences of electron transfer in chlorophylls, chlorins, and porphyrins. Structural and theoretical considerations, in: "Porphyrins: Excited States and Dynamics," M. Gouterman, P. M. Rentzepis, and K. D. Straub, eds., ACS Symposium Series No. 321, p. 51 (1986).
28. J. Fajer, K. M. Barkigia, E. Fujita, D. A. Goff, L. K. Hanson, J. D. Head, T. Horning, K. M. Smith, and M. C. Zerner. Experimental, structural and theoretical models of bacteriochlorophylls a, d and g, in: "Antennas and Reaction Centers of Photosynthetic bacteria," M. E. Michel-Beyerle, ed., Springer-Verlag, Berlin, p. 234 (1985).

Electrostatic Control of Electron Transfer in the Photosynthetic Reaction Center of *Rhodopseudomonas viridis*

H. Treutlein *, K. Schulten *, C. Niedermeier *,
J. Deisenhofer +, H. Michel +, and D. DeVault †

* Physik-Department, Technische Universität München, D-8046 Garching, FRG
+ Max-Planck-Institut für Biochemie, D-8033 Martinsried, FRG
† Dept. of Physiology and Biophysics, Univ. of Illinois, Urbana, IL 61801, USA

Abstract

We have investigated electrostatic properties of the photosynthetic reaction center of *Rhodopseudomonas viridis* , the molecular structure of which has been solved recently by X-ray crystallographic analysis[1,2]. Our calculations involved both time-averaged electrostatic properties as can be obtained from the static X-ray structure of the protein as well as fluctuating and time-varying electrostatic properties due to thermal motion of the reaction center and structural rearrangment after electron transfer. The latter properties were derived from a molecular dynamics simulation[3].

1. Introduction

The electrostatic interaction is one of the most important contributions to the total molecular energy and seems to play a major role for structure and function of biological macromolecules[4,5]. Such role of electrostatic interactions can be expected, in particular, for the photosynthetic reaction center which functions as a biological photodiode separating electron and hole charges.

Although Coulomb's law looks rather simple, the calculation of electrostatics inside and outside of macromolecules remains a problem to this day. The difficulty arises because of the the long range of electrostatic interactions and, hence, the large number of atom-pairs carrying fractional charges which contribute. The Coulomb interaction can be evaluated for all these pairs only for the equilibrium structure. Results of such calulation are presented in Sect. 2. Fluctuating and time-varying electrostatic properties originating from thermal motions and structural rearrangements can be described only in an approximate way. In fact, molecular dynamics simulations account only for Coulomb forces between atoms with distances less than a certain cutoff, modifying the force to keep resulting artifical effects small[6].

We have evaluated fluctuating and time-varying electrostatic properties on the basis of a molecular dynamics simulation also presented in these proceedings[3]. The results are presented in Sect. 3. The simulation started from an X-ray structure at 3 Å resolution[1,2]. We investigated the dynamical properties and the response of the protein to the primary electron transfer step simulated by re-charging the apropriate

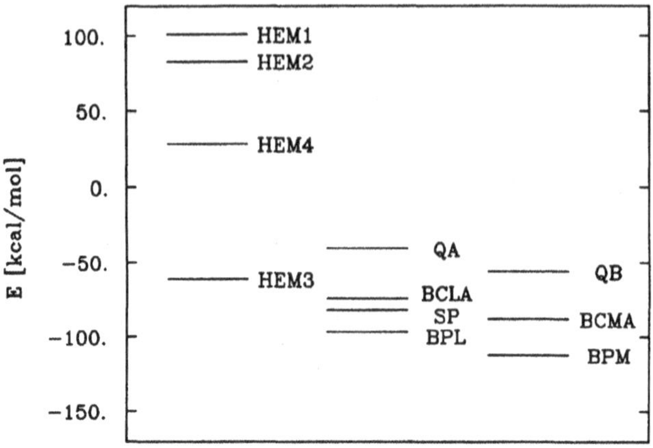

Fig. 1. Mean electrostatic potential levels of reaction center chromophores with neutral special pair and neutral non-heme iron site. The calculations are based on the X-ray structure.

chromophores. The simulation had been carried out in two steps: First, uncharged chromophores were assumed for the dynamics and the reaction center was analyzed in the state before electron transfer. Second, the *special pair* was positively and the bacteriopheophytine of the functional branch (BPL) negatively charged to analyze the state after primary electron transfer.

2. Mean electrostatic potentials of chromophores

As a first step in our investigation we have studied the mean electrostatic potential on the chromophores, i.e. the work done on an electron charge $(-e)$ brought from infinity to the site of the reaction center chromophores. For this purpose we included Coulomb interactions with fractional charges on all reaction center atoms including hydrogens. The fractional charges were those given in the **CHARMM** data base except for the charge distributions of glutamate M232 (see below), the four histidines ligated to the non-heme iron FE1 and the non-heme chromophores. The charge distributions for those residues and chromophores had been obtained by **MNDO** calculations.

The mean electrostatic potentials of the individual chromophores were obtained in two steps:

1. A negative test charge $(-e)$ was brough from infinity (∞) to the central chromophore (tetrapyrol rings and quinone, menaquinone rings) atoms at positions $\vec{r}$ and the energy difference $\Delta V(\vec{r}) = V(\vec{r}) - V(\infty)$ was determined by summing the Coulomb contributions of all charges inside the reaction center, with the exception of the charges on the same tetrapyrol or quinone rings. The influence of the latter should be attributed to the redox energy of a chromophore. In case of the special pair (SP) the Coulomb interactions between the two tetrapyrol moieties were also not counted.
2. The mean potential was then evaluated by averaging over the potential differences $\Delta V(\vec{r})$ of all atoms belonging to the ring systems of the individual chromophores.

We considered four different situations concerning the net charges of the spe-

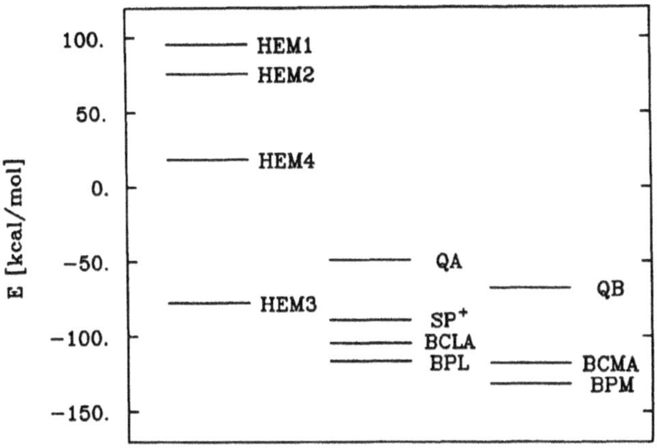

Fig. 2. Mean electrostatic potential levels of reaction center chromophores with positively charged special pair and neutral non-heme iron site.

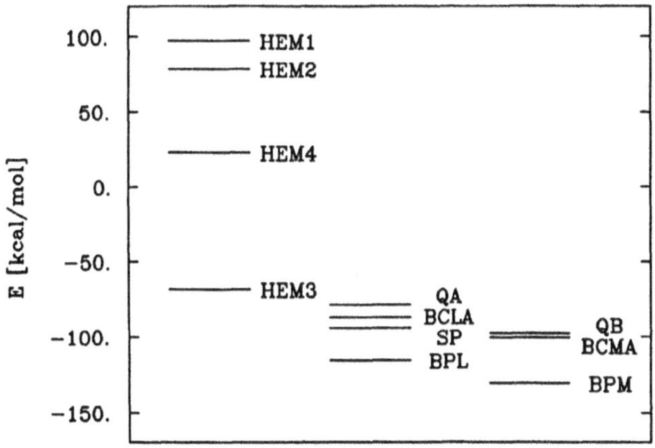

Fig. 3. Mean electrostatic potential levels of reaction center chromophores with neutral special pair and positively charged non-heme iron site.

cial pair and the non-heme iron site FE1: (i) SP-FE1, (ii) SP^+-FE1, (iii) SP-$FE1^+$, (iv) SP^+-$FE1^+$. The different charge states represented in the energy level diagrams Fig. 1–4 were chosen for the following reasons: In case of the special pair the transferred electron actually 'sees' a positively charged special pair (SP^+). Hence, in a description of primary and secondary transfer in terms of electrostatic potentials the *special pair* should be positively charged. Just in order to estimate the influence of localized charges on the chromophore energy levels we have also investigated the case of a neutral special pair. As for the non-heme iron site it is not apparent what the net charge should be. For this reason we investigated both the case of a neutral and of a positively charged ($+e$) state.

Our method of determining energy levels (mean electrostatic potentials) bear

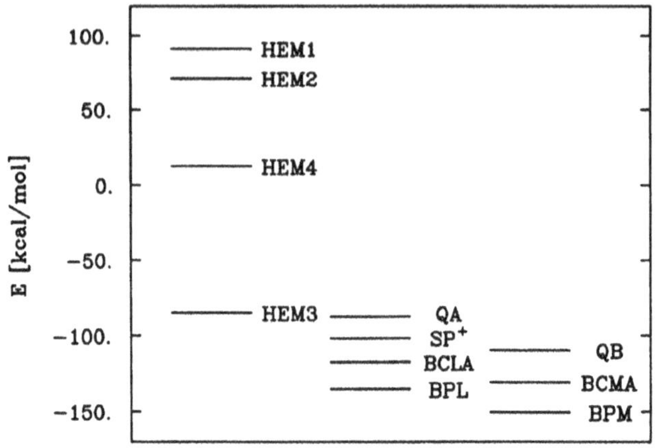

Fig. 4. Mean electrostatic potential levels of reaction center chromophores with positively charged special pair and positively charged non-heme iron site.

some obvious deficiencies. First, we did not include the water and membrane fraction surrounding the reaction center which will exert dielectric screening also on charges inside the reaction center. Also we did not include dielectric screening inside the reaction center due to the infinite frequency dielectric constant (optical density). Second, the H-unit of the reaction center complex carries six net negative charges ($-6e$) which are located at the protein surface and, therefore, can be neutralized partially by H^+ in the water fraction. Third, we cannot judge the accuracy of the (**CHARMM** and **MNDO**) charge distributions assumed in our calculations. Furthermore, there has been some degree of arbitrariness in assigning a protonated state to glutamate L104. Also we assumed that the hemes possess the same charge distributions as in myoglobin, i.e. there were two negatively charged carboxyl groups attached to each heme ring. These charges might explain the high electrostatic potential of the hemes in Fig. 1–4. However, if these carboxyl charges are neutralized, the heme potentials are shifted to values which are unacceptably low. The negative charges of carboxyl groups were also present on the chlorophylls and pheophytins, however, in this case the rings did not carry a net charge except in case (ii) and (iv) defined above.

In our evaluation of energy levels we assumed equal contributions of electrostatic potentials by all ring atoms. This assumption could be replaced by an atom-dependent weighting with weights determined through **MNDO** calculations.

We want to discuss now the energy level diagrams which resulted from our calculations for the charge situations (i) – (iv). The first observation is that the relative ordering of the energy levels is the same in all situations except for the SP–BCLA ordering which is reversed upon charging the special pair (SP $\rightarrow$ SP$^+$). The heme energy levels are found to be more positive than those for the other chromophores in accordance with the function of the hemes as hole carriers. However, the heme energies are spread over a very broad energy range. We attribute this to a neglect of the water fraction surrounding the cytochrome unit and contributing an effective dielectric screening. The energy levels of the M (non-functional) branch chromophores QB, BCMA, BPM are more negative than their L (functional) branch counterparts. This is opposite to the findings of Yeates et al.[7] for the *Rps. sphaeroides* reaction center which also accounted for the effect of a heterogeneous dielectric (water–membrane) surrounding.

The lower position of the energy levels of the non-functional chromophores relative to the levels of the respective functional chromophores does not necessarily imply

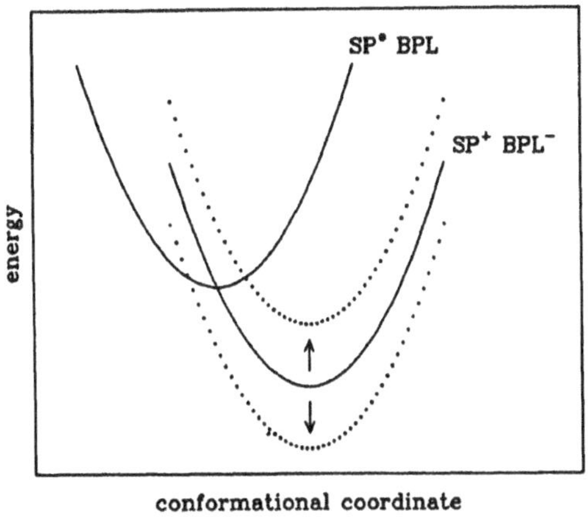

Fig. 5. Nuclear potential energies of neutral (SP-BPL) and charged (SP⁺-BPL⁻) states depending on a generalized conformational coordinate. Solid lines denote an optimum crossing. Dotted lines show hypothetically shifted potential functions, which result in a thermal activation barrier.*

a more favourable, i.e. faster, route for electron transfer along this branch relative to the functional branch. Since observations show that primary electron transfer is temperature independent, in the framework of the Marcus theory[8] depicted in Fig. 5 the potential energy functions of the neutral (SP^*–BPL) and charged (SP^+–BPL^-) states should intersect at the minum of the SP^*–BPL potential. Any shift of the ionic state potential minimum, i.e. to higher as well as to lower values, induces a new crossing of the potential curves at activation barriers $\Delta E > 0$.

The quinone energy levels lie above all the chlorophyll and pheophytine energy levels. We attribute this to Coulomb repulsion with the net six negative charges on the H-unit of the photosynthetic reaction center. This interpretation is corroborated by the fact that a positively charged non-heme iron site shifts the quinone energy levels down. A neutralization of the six negative surface charges of the H-unit through H^+ is likely to lower considerably the quinone energy levels. In this respect it is interesting to note that the H-unit channels protons to the QB binding site, i.e. it is functionally favourable to have a high local concentration of protons near the H-unit.

The remaining ordering of energy levels of the functional branch chromophores BCLA and BPL is in accordance with the known features of primary electron transfer. The accessory chlorophyl (BCLA) energy level lies considerably above the pheophytine energy level. Considering the redox energy difference[9] $E_m(Chl) - E_m(Ph) \approx 5\frac{kcal}{mol}$ in identical solvents (Chlorophyll is more difficult to reduce than Pheophytine) the actual energy level of a reduced accessory Chlorophyll BCLA is not available for thermally assisted light-induced electron transfer.

3. Fluctuating and time-varying electrostatic energy

Because of the weak coupling between covalent and ionic states , i.e. SP^*–BPL and SP^+–BPL^-, compared with vibrational energies in the system, electron transfer is possible only when covalent and ionic states are accidentally degenerate. This corresponds in the theory of Marcus[8] to a transition between covalent and ionic states only at crossings of potential energy surfaces (see Fig. 5). Due to Coulomb interactions with the surrounding protein matrix the energy difference between the covalent and ionic states fluctuates in accordance with thermal motions of the protein. The energy difference can also vary in time when the protein undergoes structural transitions, e.g. those induced by electron transfer between the chromophores. Obviously, the protein can regulate then through its motion the rate of electron transfer by controlling the coincidences of energy values of covalent and ionic states. We have investigated, therefore, how this energy difference for the primary electron transfer $\Delta E = E(\mathrm{SP}^{+}\text{-}\mathrm{BPL}^{-}) - E(\mathrm{SP}^{*}\text{-}\mathrm{BPL})$ varies during the simulated motion of a protein both before and after electron transfer. It should be pointed out, that electron transfer is controlled also by internal degrees of freedom, which can also be described by Marcus theory or by its quantum-mechanical generalizations.

In order to interprete the results of a molecular dynamics simulation regarding the energy $\Delta E(t)$ it is important to realize that **CHARMM** can only account for contributions to $\Delta E(t)$ due to interactions with charges outside the range of atoms which carry the positive and negative charge in the respective chromophores. Included in such calculations are neither the redox energy difference[9] which roughly measures $20\frac{kcal}{mol}$ for the SP^{*}- BPL $\rightarrow$ SP^{+}-BPL^{-}– transfer, nor the Born energy relative to the polar solvents, e.g. DMF, in which redox energies are measured. Assuming a dielectric constant of $\epsilon \approx 2$ inside the protein, the Born energy for moving a single charge from DMF to the protein $\Delta E_B = \frac{e^2}{2r}(\epsilon^{-1} - \epsilon_{DMF}^{-1})$ for $r \approx 5\AA$ measures about $10\frac{kcal}{mol}$, twice that for two charges, and, hence, a total energy of about $40\frac{kcal}{mol}$ needs to be added to the pure external electrostatic contribution. This is, of course, a very rough estimate.

Uncertainties in the redox and Born energies don't allow to determine zero crossings of the energy difference between covalent and ionic states in proteins and, for that reason, we will present below in Fig. 6 only the electrostatic fraction of this energy difference. The electrostatic energy difference presented there has been calculated using **CHARMM**'s electrostatic energy function described above. The cut-off for this energy function is 9 Å. Also Coulomb interactions between nearest neighbours and next nearest neighbours in bonded systems were not included, as these energies were assumed to contribute to the redox energies. It must be emphasized here that the electrostatic potential of the previous section has been calculated differently, using Coulomb's law between all pairs of atoms in the X-ray protein structure, excluding only those atoms belonging to special groups of chromophore ring atoms. This latter type of calculation separates more clearly inter-molecular (electrostatic) from intra-molecular energy contributions in conjugated ring systems. It is therefore not surprising that both calculations yield quantitatively different results.

We calculated (**CHARMM**'s electrostatic) energy-differences both for a simulation before electron transfer and after the transfer. ΔE might also be interpreted as a measure for energy gain or loss due to electron forward or backward transport.

The result for the electrostatic energy difference $\Delta E(t)$ describing a transfer from the *special pair* to the pheophytine BPL is presented in Fig. 6 . The left part of the diagram shows ΔE before the transfer: it fluctuates around an energy value of about $-20\frac{kcal}{mol}$. This value corresponds to the equivalent value of $-27\frac{kcal}{mol}$ for the difference between the energy levels of SP^{+} and BPL in Fig. 2 (case ii). On the basis of a value of $40\frac{kcal}{mol}$ for the further contributions to the SP^{*}–BPL and SP^{+}–BPL^{-} energy difference one estimates an approximate energy of $-20 + 40 = 20\frac{kcal}{mol}$ for the excitation energy of the special pair.

It is suggestive to regard the crossings with the mean value $-20\frac{kcal}{mol}$ as the occurrences of energy degeneracies between the SP^{*}–BPL and SP^{+}–BPL^{-} energy levels. If this analogy holds , primary electron transfer is possible any time a crossing

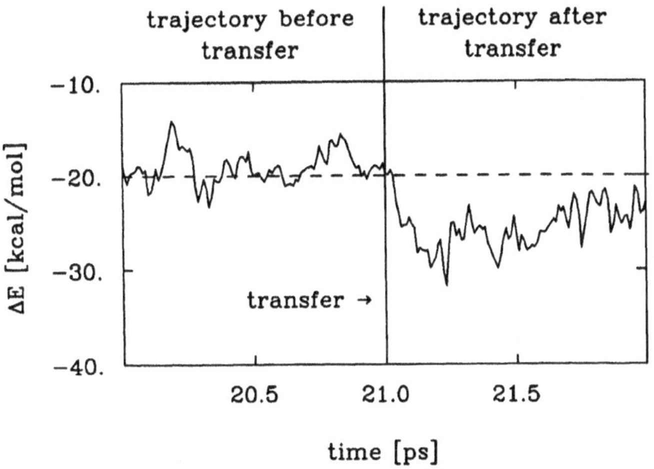

Fig. 6. Energy difference ΔE before and after electron transfer from special pair to BPL. The left part of the diagram displays ΔE obtained from molecular dynamics data (trajectory) before transfer. The instant the electron is transferred by changing the charge distribution on the special pair and BPL is denoted by an arrow. The right part shows ΔE after transfer obtained from molecular dynamics data (trajectory) after transfer.

occurs, i.e. 18 times in the case of Fig. 6. This behaviour demonstrates that protein fluctuations can control primary electron transfer. The situation that crossings with the mean value of $\Delta E(t)$ implies energy degeneracies in the Marcus theory corresponds to the case in this theory that the potential energy functions of the neutral (SP*–BPL) and charged (SP^+–BPL^-) states intersect at the minum of the SP*–BPL potential (see Fig. 5). Obviously, temperature lowering in this situation leads to more frequent crossings with the mean value and larger transfer rates, a behaviour observed for the reaction center.

We want to consider now how electron transfer affects $\Delta E(t)$ and, in particular, if back-transfer can be controlled (hindered) by protein relaxation and fluctuation. We have transferred, therefore, an electron in our simulations (by altering the chromophore charge distributions) at the instant marked by an arrow in Fig. 6 (at the time labeled 21 ps). $\Delta E(t)$ within 0.3 ps after the transfer decreases by about $10\frac{kcal}{mol}$ and then relaxes to a value close to $-25\frac{kcal}{mol}$. This is $5\frac{kcal}{mol}$ below the $-20\frac{kcal}{mol}$ value before transfer which at room temperature is just enough to prevent further crossings with the $-20\frac{kcal}{mol}$ line and, hence, back-transfer. It might be interesting to note that the very rapid initial relaxation appears to be very effective in preventing back-transfer already after about 100 fs. Furthermore, relaxations are sufficient to prevent back-transfer, however, do so with a minimal (exothermic) energy loss.

What processes contribute to the relaxation of $\Delta E(t)$ after electron transfer? Part of the relaxation will certainly be due to the motion of the pheophytine (BPL) ring towards the *special pair* which we discussed extensively in Ref. 3. Because of the large mass of the whole pheophytine ring moiety we expect that this motion will only participate in the slow component of the relaxation. The fast (.3 ps) decrease of $\Delta E(t)$ should be attributed to rearrangements of smaller dipolar side groups of the protein. To test this supposition we have monitored the electrostatic interaction of the tryptophane L100 which lies close to the BPL tetrapyrol ring and can form

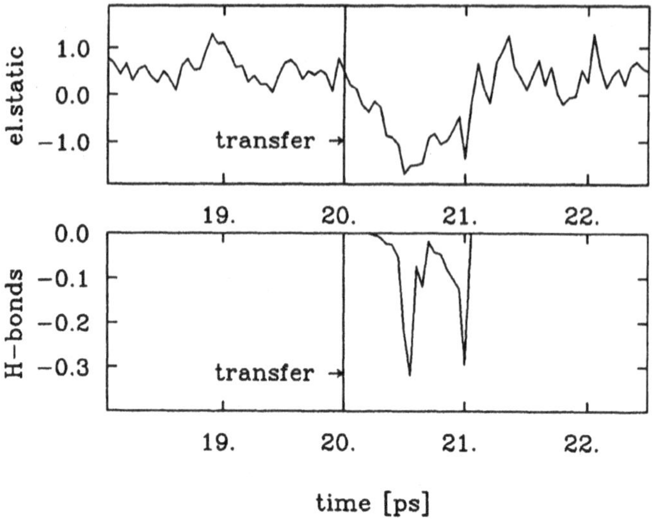

Fig. 7. Electrostatic and hydrogen-bond energy between BPL and a neighboring tryptophane residue (L100) before (time < 20 ps) and after (time > 20 ps) electron transfer to BPL. (Energies are given in $\frac{kcal}{mol}$)

a hydrogen bond with BPL. The behaviour of this group is illustrated in Fig. 7 through its time-dependent electrostatic interaction and through its hydrogen bonding behaviour. One observes that the electrostatic energy of this group decreases very rapidly after electron transfer. Also this group forms a transient hydrogen bond. Of course, this group contributes only a small fraction to $\Delta E(t)$.

Our results suggest that the protein after primary transfer stabilizes the electron to prevent back-transfer.

ACKNOWLEDGEMENTS

The authors like to thank A. Brünger, M. Karplus, Z. Schulten and P. Tavan for advice and help. This work has been supported by the Deutsche Forschungsgemeinschaft (SFB 143-C1) and by the National Center of Supercomputer Applications in Urbana, IL.

REFERENCES

1. J. Deisenhofer, O. Epp, K. Miki, R. Huber, H. Michel, X-ray Structure Analysis of a Membrane Protein Complex J molec Biol 180:385 (1984)
2. J. Deisenhofer, O. Epp, K. Miki, R. Huber, H. Michel, Structure of the protein subunits in the photosynthetic reaction center of Rhodopseudomonas viridis at 3 Å resolution, Nature 318:618 (1985)
3. H. Treutlein et al., Molecular dynamics simulation of the prim ary processes of the photosynthetic reaction center of Rps. viridis, contribution to this workshop

4. A. Warshel, S. T. Russel, Calculations of electrostatic interactions in biological systems and in solution, Q Rev Biophys 17:283 (1984)

5. I. Klapper et al., Focussing of electric fields in the active site of Cu-Zn superoxide dismutase: effects of ionic strength and amino-acid modification, Proteins 1:47 (1986)

6. B. R. Brooks et al. , CHARMM: A Program for Macromolecular Energy Minimization, and Dynamics Calculations, J Comp Chem 4:187 (1983)

7. T. O. Yeates et al., Structure of the reaction center from Rhodobacter sphaeroides R-26: Membrane-protein interactions, PNAS 84:6438 (1987)

8. R. A. Marcus, N. Sutin, Electron transfers in chemistry and biology, Biochim Biophys Acta 811:265 (1985)

9. Fajer et al., JACS 100:1918 (1978)

MOLECULAR ORBITAL STUDIES ON THE PRIMARY DONOR P_{960} IN REACTION CENTERS OF RPS. VIRIDIS

M. Plato*, F. Lendzian*, W. Lubitz+, E. Tränkle# and K. Möbius*

* Institut für Molekülphysik, Freie Universität Berlin Arnimallee 14, 1000 Berlin 33 (FRG)
\+ Institut für Organische Chemie, Freie Universität Berlin Takustr. 3, 1000 Berlin 33 (FRG)
\# Institut für Theorie der Elementarteilchen, Freie Universität Berlin, Arnimallee 14, 1000 Berlin 33 (FRG)

INTRODUCTION

One outstanding question in bacterial photosynthesis research concerns the relevance of the dimeric nature of the primary donor P for the fast electron transfer step from the first excited singlet state $^1P^*H$ to the charge separated state P^+H^-, where H stands for the intermediate acceptor bacteriopheophytin. This problem can only be attacked by performing a detailed study of the electronic structure of the primary donor itself and of the electronic interactions between all involved cofactors in their various active states. This requires the combined effort of high resolution X-ray crystallography, powerful spectroscopic methods and sufficiently advanced quantum-chemical calculations.

X-ray structural data have recently become available for the reaction center (RC) of the bacterium Rhodopseudomonas (Rps.) viridis[1]. These comprise coordinates from all cofactors and amino acid residues gained from a crystallographic refinement at 2.9 Å resolution.

Magnetic resonance experiments (ESR, ENDOR, TRIPLE) have been performed on the radical cation P^+_{960} of the primary donor P_{960} of the same bacterium, yielding ESR linewidths and isotropic hyperfine couplings of the unpaired electron with various protons of the donor[2-5]. Such measurements are particularly well suited for the elucidation of fine details of the electronic structure since they yield local properties of orbital wavefunctions. From these measurements it was concluded that[3-5]:

(i) the density of the unpaired electron (spin density) is distributed over two bacteriochlorophyll b (BChl b) molecules. This is in accordance with the X-ray structural data [1], which clearly show the primary donor to be a BChl b dimer with two overlapping pyrrole rings (distance ca. 3.0 Å);

(ii) the spin density is not equally shared between the two monomeric halves of the dimer but exhibits an asymmetrical distribution of about 2:1 in favor of one (unspecified) half;

(iii) result (ii) differs from the result of similar studies on the primary donor radical cation P^{+}_{865} in RC's of the bacteria Rhodobacter sphaeroides R-26 and Rhodospirillum rubrum G-9[3,6-11]. In this case it has been concluded that the spin density is distributed more evenly over the two BChl a molecules forming this dimer, thus pointing to an approximately symmetrical (C_2) structure of P^{+}_{865}.

This paper will report on molecular orbital calculations on P_{960} in its doublet ground states $^2P^{+}_{960}$ and $^2P^{-}_{960}$ and in its first excited singlet state $^1P^{*}_{960}$ employing the X-ray structural data for the singlet ground state[1]. (The subscript 960 on P will be omitted from here on.) Specifically, we shall

(i) test the consistency of theoretical and experimental hfc's for the cation radical P^+, which probes the electronic charge density distribution in the highest occupied molecular orbital (HOMO) of the ground state $^1P^o$;

(ii) calculate the spin density distribution of the anion radical P^-, which probes the charge density distribution of an electron in the lowest unoccupied molecular orbital (LUMO) of $^1P^o$ - this orbital is occupied by the "excited" electron of the state $^1P^*$;

(iii) examine the influence of polar groups of amino acid residues in the vicinity of P on the spin density distributions of P^+ and P^-.

The aim is to infer from the results of these calculations structure-function relations pertaining to the role of the dimer in the primary charge separation of bacterial photosynthesis.

THEORETICAL METHOD

We have applied the well-known semiempirical self-consistent-field molecular orbital method INDO developed by Pople and Beveridge[12] for closed and open shell ground states and INDO/S developed by Zerner[13] for excited singlet states. These methods use the full valence orbital basis set which is essential for studying molecules of low symmetry. Especially for strong deviations from planarity, as observed for one monomeric half of the dimer (see Fig. 2), separation of π- and σ-electrons is no longer permissible.

We have abandoned the usual Unrestricted Hartree-Fock (UHF) procedure in INDO for open shell doublet ground states for reasons given in ref. 11. Our modified approach, named RHF-INDO/SP, is essentially equivalent to the "half-electron-method" introduced by Dewar et al.[14], which treats open shells similar to the RHF-Roothaan method for closed shells. Since this approach does not take account of spin polarisation effects, these have to be recovered by a subsequent perturbation treatment (SP). The details of this procedure are presented in ref 11. We have recently developed a new SP version which has full rotational invariance for s-spin densities and gives improved results[15]. All calculations were performed on the CRAY X-MP/24 of the Konrad Zuse Rechenzentrum at Berlin.

As a test of the performance of the method RHF-INDO/SP we present a comparison of theoretical and experimental[16] s-spin densities of the BChl b monomer radical cation (see Fig. 1). Except for one position H_{4a}), where the deviation is ca. 50% for reasons not yet understood, it is less than 25% for all large hfc's ($\geq$ 2.5 MHz). The nitrogen hfc's are particularly well reproduced to within 13%.

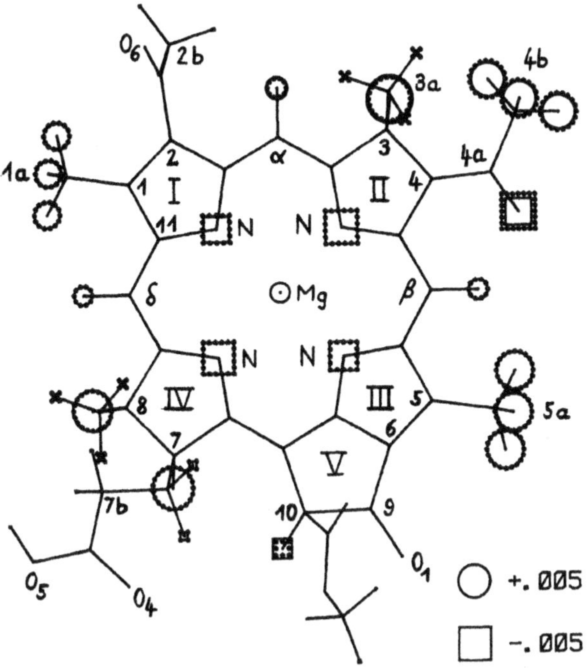

Fig. 1. Comparison of experimental[16] (dotted) and calculated (solid lines) s-spin densities ρ for the BChl b monomer radical cation. Their values are proportional to the area of the squares ($\rho < 0$) and circles ($\rho > 0$). The RHF-INDO/SP method (revised version of SP) was used for the calculations (see text); some improvement over earlier results[16] has been obtained for some positions[4,15]. Theoretical methyl proton hfc's are rotationally averaged. The full molecular structure was taken except for a truncation of the phytyl chain at position O_5. Mostly standard bond lengths and angles were employed[12]; atomic positions for ring V including substituents, for the methyl groups (rings I and III), and for the acetyl group were determined by energy minimization[11]. Experimental values of ρ_i were derived from hfc's a_i by use of $a_i = Q\rho_i$ and Q(H) = 1420 MHz, $Q(^{14}N)$ = 650 MHz [11,15,16].

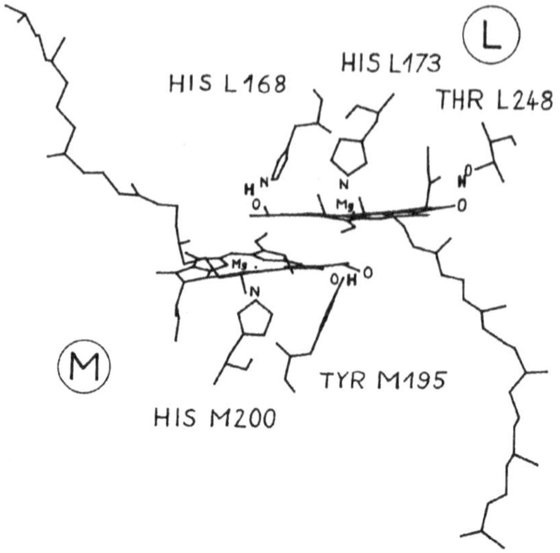

Fig. 2. Polar amino acid residues around the BChl b dimer in Rps. viridis[1]. Hydrogen atoms forming hydrogen bridged bonds between amino acids HIS(L168), THR(L248) and TYR(M195) are also shown. For further details, see references 15 and 19.

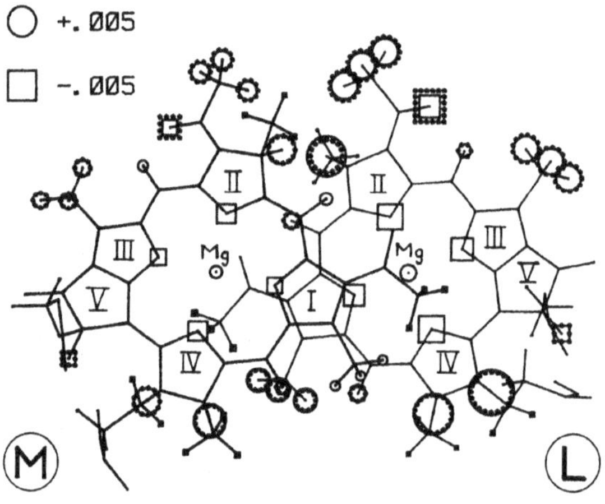

Fig. 3. Theoretical (solid lines) and experimental (dotted) s-spin densities for the BChl b dimer cation radical P^+. Calculations were performed on the structure given by X-ray analysis[1] including surrounding amino acids with polar groups, see Fig. 2 and text. The representation of spin densities is the same as in Fig. 1. Rings are numbered as in Fig. 1. The assignment of experimental s-spin densities to positions corresponds to assignment C of ref. 4.

RESULTS AND DISCUSSION

1) BChl b dimer radical cation P^+

Spin density calculations on the BChl b dimer radical cation in RC's of Rps. viridis were performed on the structure given by the X-ray data[1] with hydrogen atoms attached by standard rules[12]. The final system consisted of 160 atoms, 436 valence orbitals and 463 electrons. The influence of polar groups of neighboring amino acid (AA) residues (see Fig. 2) was taken into account by taking AA-atoms as point charges with net charge values taken from the work of Nemethy et al.[17]. Some minor refinements of $C\text{-}CH_3$ bondlengths were performed by energy minimization resulting in a better agreement between theoretical and experimental methyl hfc's. Detailed numerical results will be published in a forthcoming paper[15].

In Fig. 3 we present a comparison of theoretical and experimental s-spin densities. The most prominent feature of the theoretical spin densities are their asymmetrical distribution over the two monomeric halves, designated "L" and "M" according to their bondage to the respective protein subunits[1]. This asymmetry is clearly seen by comparing, for example, theoretical spin densities on methyl protons at rings I or III on different sides. If the ratio $R = \Sigma\rho_L/\Sigma\rho_M$ of the sums of all s- and p-spin densities is taken as a measure of the asymmetry, one obtains $R = 0.74/0.26$ (0.62/0.38 without AA's). This asymmetry predicted by theory is in good agreement with the results of ESR, ENDOR and TRIPLE measurements[3-5], if the group comprising the larger hfc's is accordingly assigned to the L-half. The corresponding experimental value of R, which is restricted to hyperfine-active positions, is 0.67/0.33 [4]. The chosen assignment appears justified as long as slight changes in the X-ray structure do not reverse the L/M-asymmetry. This has been assured by performing calculations for a (limited) number of geometrical variations (e.g. rotation of acetyl groups, positional changes of amino acids) without rupture of H-bonds between the dimer and the AA's (see Fig. 2).

The agreement between theoretical and experimental s-spin densities, if looked at in detail, is not as good as for the BChl b monomer, which is probably due to inaccuracies of the X-ray structure analysis or small structural changes between the relaxed P^+ and P in its singlet ground state. We do, however, consider the agreement to be sufficiently good to rely on the following predictions concerning the radical anion and the first excited state of P.

2) BChl b dimer radical anion P^- and first excited singlet state P^*

The spin density distribution of P^- is of special interest because it approximately reflects the charge distribution of an electron in the LUMO of the ground state dimer. This orbital is occupied by the "excited" electron of the lowest excited singlet state $^1P^*$. Knowledge of its charge distribution can help to elucidate the route and thus the mechanism by which this electron is transferred to the acceptor. Unfortunately, no magnetic resonance data are available for P^- in RC's of Rps. viridis, so that one has to rely on the present calculations. Fig. 4 shows the π-spin density distribution of P^- in comparison with that of P^+. The presentation of π-spin densities (more accurately p_z-spin densities, where z is perpendicular to a plane passing through N_I^L, N_{II}^L, N_{III}^L)

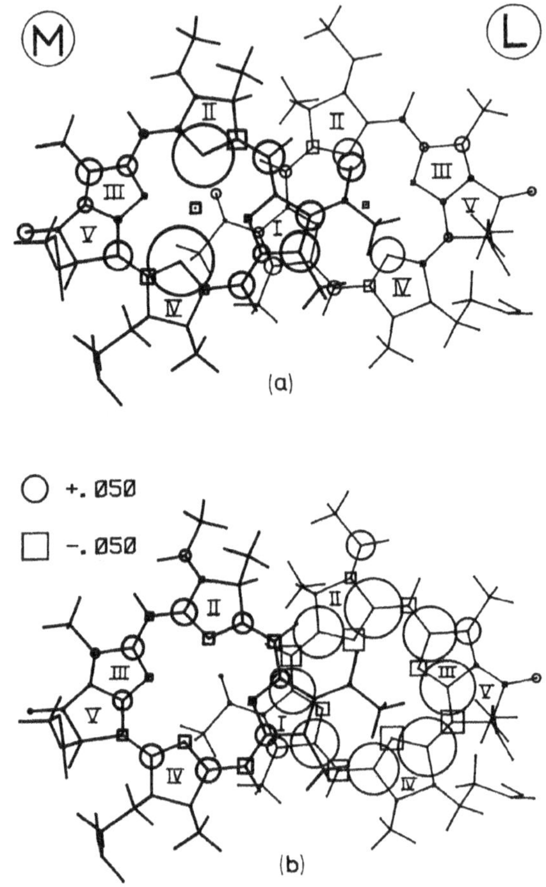

Fig. 4: Theoretical π-spin densities of the BChl b dimer anion radical P^-(a) and dimer cation radical P^+(b). Both calculations were performed on the structure given by X-ray analysis[1] including surrounding amino acids with polar groups. The representation of spin densities is linear in the radius of circles ($\rho > 0$) or side length of squares ($\rho < 0$), in contrast to Fig. 1.

is favored over that of s-spin densities because it shows changes in π-orbital charge distributions more directly. In contrast to P^+ the spin density on P^- is concentrated on the M-half of the dimer, the asymmetry now being R = 0.26/0.74 (0.32/0.68 without AA's). We thus have practically a "mirror-image" reversal of the electronic charge distributions in the HOMO and LUMO of the ground state dimer. This simple picture, which equates the ground state spin density distribution of P^- to the "orbital charge distribution" of the excited electron, can break down on account of configuration mixing of close-lying excited states involving orbitals with strongly differing charge distributions. (This picture also ignores the occurrence of small negative π-spin densities due to spin polarisation effects involving paired π-electrons). We have therefore performed an INDO/S[13] calculation of the first excited singlet state $^1P^*$ including configuration mixing with the 50 nearest singly excited singlet states[15]. These calculations revealed that $^1P^*$ is practically a pure state (ca. 95%), implying that the lowest unoccupied "virtual" molecular orbital of P yields a good representation of the orbital charge distribution of the excited electron. The value of R for this virtual orbital, taking the molecular orbitals of the singlet ground state of P in the INDO-parametrisation of Pople and Beveridge[12], is R = 0.24/0.76 (0.30/ 0.70 without AA's) which is practically identical with R for P^-.

CONCLUSIONS

The main outcome of our calculations - supported by magnetic resonance measurements - is the prediction of a significant asymmetry of the orbital charge density distribution of the "excited" electron in $^1P^*_{960}$ in favor of the M-half P_M by an amount $q_M/q_L \cong 3 : 1$. This asymmetry is caused to the larger part by a structural asymmetry of the dimer itself (2 : 1), whereas the remainder (1.5 : 1) is an enhancement due to the surrounding polar amino acid residues (we are at this point not drawing any conclusions concerning the distribution of the total electronic charge on $^1P^*_{960}$ which would require the summation over all occupied MO's). Our interest is focussed on properties of the excited state $^1P^*$ which determine the vectorial nature and efficiency of the electron transfer (ET) from the dimer P to the intermediate acceptor H. It is generally accepted, that the ET in bacterial RC's is nonadiabatic[20] and therefore governed by an intermolecular electronic interaction term

$$V_{P^*X} = \langle {}^1P^* | \hat{H} | X \rangle \qquad (1)$$

where X is either the acceptor state H^- (in conjunction with P^+) or the state of some intermediate species bridging P and H, and H represents the electronic interaction Hamiltonian.

In a first-order CI-treatment[18] of the electronic interactions between the various cofactors of a bacterial RC, V in eq. 1 is shown to be proportional to the molecular overlap integral

$$V_{P^*X} \propto S_{P^*X} = \Sigma c_i(p^*) c_j(x^*) S_{ij}, \qquad (2)$$

where $c_i(p^*)$ and $c_j(x^*)$ stand for the MO-coefficients of the LUMO's on atomic orbitals i and j of P and X, and S_{ij} is the overlap integral between these atomic orbitals.

Since $|c_i^M(p^*)| > |c_i^L(p^*)|$ from our calculations, we would therefore expect the ET to preferably take its route along that branch (L or M) of the RC which provides a state X localised closely to the M-half P_M of the dimer. Inspection of the X-ray structural data reveals that the closest edge-to-edge distance is between P_M and B_L (see Table 1), where B_L stands for the accessory monomer BChl b on the L-branch[1,19]. This result provides a possible explanation for the observed unidirectionality of the ET along the L-branch[19] and, at the same time, suggests that B_L plays an important mediating role in the primary charge separation process. These ideas can be carried over to the well-known "superexchange" model for the bacterial ET process[20], in which X is assumed to be a virtual ion pair singlet state P^+B^- which effectively mixes into the singlet excited state $^1P^*B$. From recent 50 fs time-resolved studies[21] there is no evidence for a kinetic involvement of B as a distinct electron acceptor. A strong coupling between B and H produces the final state P^+BH^-. In this model

$$k^{ET} \propto (V_{P^*B,P^+B^-} \; V_{B^-H,BH^-})^2 \qquad (3)$$

if equal energy spacings $E_{P^*B} - E_{P^+B^-}$ and equal Franck-Condon factors are assumed in both subunits L and M[19]. Table 1 contains the molecular overlap integrals between all relevant pairs of cofactors in both subunits L and M, calculated from our RHF-INDO wavefunctions with the original SLATER-INDO orbital exponents[18]. Also included are the closest distances r_{IJ} of atoms involved in the overlaps beween cofactors I, J. It should be noted that the largest values of S_{IJ} are obtained for pairwise adjacent methyl groups. This points to an important role of methyl hyperconjugation in the intermolecular electronic couplings[15].

From eqs. 2 and 3 and the values of S_{IJ} from Table 1, we obtain for the relative ET-rates in the L- and M-subunits

$$k_L^{ET}/k_M^{ET} = 34 \qquad (4)$$

with an error of ca. ± 50% on account of uncertainties of ca. ± 0.2 Å in the atomic coordinates[1]. This simple first order estimate of k_L^{ET}/k_M^{ET} on the basis of the superexchange model therefore predicts - at least qualitatively correct - the striking unidirectionality of the ET along the L-branch (experimental value > 5)[19]. The above numerical result can be decomposed into 4 contributions originating from structural asymmetries on the L- and M-branches between P and B (factor 3), between B and H (factor 4) and from the (structure-induced) charge asymmetry on the dimer itself (factor 2) and with (factor 1.4) surrounding polar amino acids. If the structural asymmetry between P and B and between B and H is conserved in the RC's of Rhodobacter sphaeroides, unidirectionality of the ET would still be retained even for a more symmetrical structure of the dimer P_{865} of this species.

ACKNOWLEDGMENT

We thank Dr. C.J. Winscom for many helpful discussions. M. Plato would like to express his gratitude to Prof. J. Jortner for his continuous support in the theoretical aspects of this work. This research project was supported by the Deutsche Forschungsgemeinschaft (Sonderforschungsbereich 337).

Table 1. Molecular overlap integrals and closest distances between pairs of cofactors in Rps. viridis

Pair I, J	S_{IJ} x10^4	r_{IJ}(Å)[a]	
		Heavy Atoms	Protons[b]
P_L, B_L	0.12	3.45 (O_4, C_{4b})	2.08 (H_{7b}, H^m_{4b})
P_L, B_M	0.23	3.75 (O_6, C_{5a})	2.27 (H^m_{2b}, H_{4a})
P_M, B_L	1.00	3.26 (C_{2b}, C_{5a})	1.89 (H^m_{2b}, H^m_{5a})
P_M, B_M	0.17	3.40 (C_9, C_{4b})	2.39 (H_{10}, H^m_{4b})
B_L, H_L	6.82	3.61 (C_{1a}, C_{1a})	1.65 (H^m_{1a}, H^m_{1a})
B_M, H_M	3.28	3.71 (C_{11}, O_6)	2.29 (H^m_{1a}, H^m_{1a})

[a] For numbering of atoms, see Fig. 1.
[b] Protons marked with m belong to methyl groups. Rotational angles of methyl groups were kept at a fixed value[15].

REFERENCES

1. J. Deisenhofer, O. Epp, K, Miki, R. Huber, and H. Michel, J.Mol. Biol., 180:385 (1984); Nature 318:618 (1985); H. Michel, O. Epp, and J. Deisenhofer,EMBO J. 5:2445 (1986).
2. M. S. Davis, A. Forman, L.K. Hansen, J.P. Thornber, and J. Fajer, J.Phys.Chem., 83:3325 (1979).
3. W. Lubitz, F. Lendzian, M. Plato, K. Möbius, and E. Tränkle, in: "Antennas and reaction centers of photosynthetic bacteria - structure, interactions and dynamics," M. E. Michel-Beyerle, ed., Springer, Berlin (1985).
4 F. Lendzian, W. Lubitz, K. Möbius, M. Plato, E. Tränkle, A.J. Hoff, and H. Scheer, Chem.Phys.Lett., to be published.
5 W. Lubitz, F. Lendzian, M. Plato, E. Tränkle, and K. Möbius, Proc.Coll. Ampere XXIII (Rome) 486 (1986).
6. J. R. Norris, R. A. Uphaus, H. L. Crespi and J. J. Katz, Proc.Natl. Acad.Sci. USA, 68:625 (1971).
7. W. Lubitz, F. Lendzian, H. Scheer, J. Gottstein, M. Plato, and K. Möbius, Proc.Natl.Acad.Sci. USA, 81:1401 (1984).
8 W. Lubitz, R. A. Isaacson, E. C. Abresch and G. Feher, Proc.Natl. Acad.Sci. USA, 81:7792 (1984).
9. F. Lendzian, W. Lubitz, H. Scheer, C. Bubenzer, and K. Möbius, J.Am.Chem.Soc. 103:4635 (1981).
10. W. Lubitz, F. Lendzian, H. Scheer, M. Plato, and K. Möbius, in: "Photochemistry and photobiology, Proceedings of the International Conference, University of Alexandria, Egypt, A.H. Zewail, ed., Harwood, New York (1983), p. 1057.
11. M. Plato, E. Tränkle, W. Lubitz, F. Lendzian, and K. Möbius, Chem. Phys. 107:185 (1986).
12. J. A. Pople and D. L. Beveridge, "Approximate molecular orbital theory," McGraw-Hill, New York (1970).

13. J. Ridley and M. Zerner, Theoret.Chim. Acta (Berl.) 32:111 (1973)
14. M. J. S. Dewar, J. A. Hashmall, and C. G. Venier, J.Am.Chem.Soc. 90:1953 (1968).
15. M. Plato, F. Lendzian W. Lubitz, E. Tränkle, and K. Möbius, to be published.
16. F. Lendzian, W. Lubitz, R. Steiner, E. Tränkle, M. Plato, H. Scheer, and K. Möbius, Chem.Phys.Lett., 126:290 (1986).
17. G. Nemethy, M. S. Pottle, and H. A. Scheraga, J.Phys.Chem., 83:1883 (1983).
18. M. Plato and C.J. Winscom, in these Proceedings.
19. M. E. Michel-Beyerle, M. Plato, J. Deisenhofer, H. Michel, M. Bixon, and J. Jortner, Biochim.Biophys. Acta, in press.
20. J. Jortner and M. E. Michel-Beyerle in: "Antennas and reaction centers of photosynthetic bacteria - structure, interactions and dynamics," M. E. Michel-Beyerle, ed., Springer, Berlin (1985).
21. J. Breton, J.-L. Martin, A. Migus, A. Antonetti, and A. Orszag, Proc.Natl.Acad.Sci. USA, 83:5121 (1986).

EARLY STEPS IN BACTERIAL PHOTOSYNTHESIS. COMPARISON OF THREE MECHANISMS

R. A. Marcus

Noyes Laboratory of Chemical Physics
California Institute of Technology
Pasadena, CA 91125

ABSTRACT

Mechanisms for the early electron transfer steps in the bacterial photosynthetic reaction center are discussed. An internal consistency test is described which places a real constraint on the unknown parameters. A resulting possible paradox is described, together with proposed electric field measurements and an alternate mechanism.

INTRODUCTION

The increasingly detailed level of experimental data on the bacterial photosynthetic reaction center is abundantly documented in this conference. In principle there are many unknowns, such as the λ's and ΔG°'s of the individual electron transfer steps (using current notation[1]), and the electron transfer matrix elements of these steps. The apparent presence of both an "unrelaxed" and a "relaxed" state (proposed by Parson and coworkers[2]) of the radical pair that is formed also adds to the complexity. [2]

One puzzle which has arisen, and which we have addressed earlier,[3] concerns the apparent discrepancy between the fast forward rate of electron transfer[4] to form a radical pair $BChl_2{}^{+}BPh^{-}$ and the small singlet-triplet splitting of that pair.[5-7] Here, $BChl_2$ denotes the bacteriochlorophyll dimer and BPh a bacteriopheophytin. A slow recombination of the pair to form a triplet $BChl_2{}^{*}$ and BPh is also pertinent.[5-7] This situation has been discussed by several authors recently.[3,7,8] In the present article I would like to describe an internal consistency test[9] which addresses directly the question of whether or not there is a paradox between the fast forward electron transfer rate and these other observations.

The work presented by Martin *et al.*[10] at this conference provides rather compelling evidence against the chemical intermediate mechanism ($BChl^{-}$ or $BChl^{+}$). This evidence will be taken for granted in the following and it will be used to develop the internal consistency test. A third possible mechanism for the early electron transfer steps in these photosynthetic reaction centers will also be suggested and explored (cf. ref. 9). Experiments which may distinguish between the mechanisms are considered.[9]

INTERNAL CONSISTENCY TEST

We first consider two alternative mechanisms for the role of the accessory bacteriochlorophyll monomer BChl in the initial electron transfer from $BChl_2^*$ to BPh. Three electronic states are particularly relevant to the discussion:

State 1. $BChl_2^*$ BChl BPh
State 2. $BChl_2^+$ $BChl^-$ BPh
State 3. $BChl_2^+$ BChl BPh^- ,

where the asterisk denotes an electronically-excited state.

The free energy curves for these three electronic states are plotted *vs* a reaction coordinate in Fig. 1 for two postulated mechanisms. In the case of a superexchange mechanism the free energy curve for state 2 is relatively high, as in Fig. 1a. In contrast, curve 2 is lower for the case of the $BChl^-$ chemical intermediate mechanism, as in Fig. 1b. We shall frequently refer to the electron

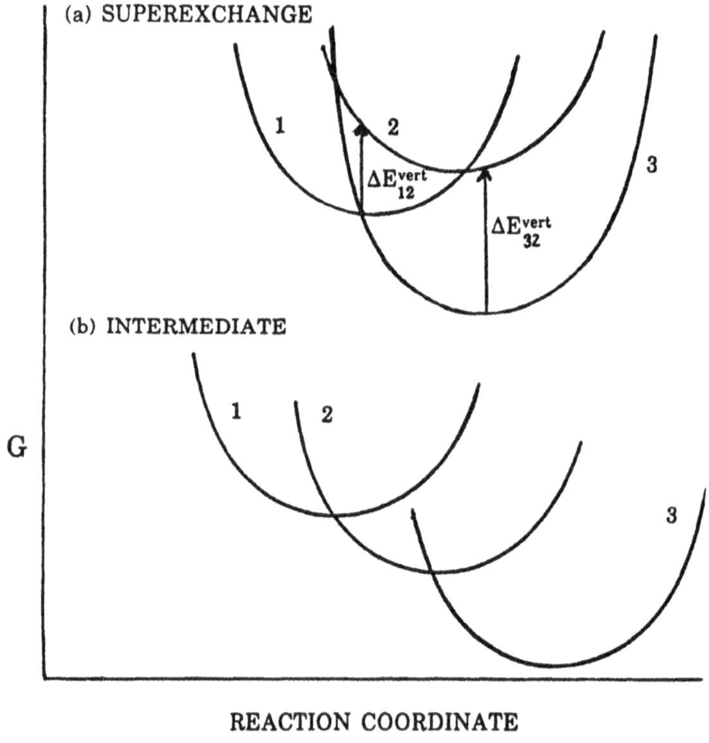

Fig. 1. Plot of free energy *G vs* the reaction coordinate for the various electronic configurations and for two mechanisms, (a) and (b). Curves 1, 2 and 3 refer to the $BChl_2^*$-BChl-BPh, $BChl_2^+$-$BChl^-$-BPh and $BChl_2^+$-BChl-BPh^- electronic states of the system, respectively. Figure (a) is appropriate to a superexchange mechanism and (b) to a chemical intermediate mechanism ($BChl^-$). The reaction coordinate in fig. 1a may differ from that in fig. 1b. Further, in fig. 1a the minimum for curve 2 is not necessarily a global minimum in the space of all reaction coordinates.

transfer steps $1 \rightarrow 2$, $1 \rightarrow 3$, $3 \rightarrow 1$, etc. We first consider the rate expressions for these two mechanisms, before addressing a third possible mechanism described later in Fig. 2.

Using a classical expression for the rate constant of an electron transfer reaction, eq. (1) can be written for the rate constant k_{13}^{super} of formation of $BChl_2^+$ BPh^- (state 3) from $BChl_2^*$ BPh (state 1) at room temperature T for a superexchange mechanism. When, as is the case here, the reaction is activationless we have[1,3]

$$k_{13}^{super} = \frac{2\pi}{\hbar} \frac{|\overline{H}_{13}|^2}{(4\pi\lambda_{13}RT)^{\frac{1}{2}}} \quad . \tag{1}$$

Here, $\overline{H}_{13}$ is the effective (superexchange) electron transfer matrix element and λ_{13} is the reorganizational parameter for this $1 \rightarrow 3$ step.

In a chemical intermediate mechanism the rate constant k_{12} for the formation of $BChl_2^+$ $BChl^-$ from $BChl_2^*$ BChl, where $BChl^-$ is the intermediate, is given by[1]

$$k_{12} = \frac{2\pi}{\hbar} \frac{|H_{12}|^2}{(4\pi\lambda_{12}RT)^{\frac{1}{2}}} e^{-\Delta G_{12}^*/RT} \quad , \tag{2}$$

where H_{12} is the electron transfer matrix element, λ_{12} the reorganization parameter and ΔG_{12}^* the activation free energy, for this $1 \rightarrow 2$ step. The ΔG_{12}^* can be written in terms of λ_{12} and ΔG_{12}°, the standard free energy of reaction for that step[1] :

$$\Delta G_{12}^* = (\lambda_{12} + \Delta G_{12}^\circ)^2 / 4\lambda_{12} \quad , \tag{3}$$

The internal consistency test[9] arises because the $\lambda_{12} + \Delta G_{12}^\circ$ in eq. (3) also occurs as an energy denominator in an expression for $\overline{H}_{13}$. Using the arguments given in Appendix A, $\overline{H}_{13}$ for the superexchange mechanism $1 \rightarrow 3$ can be approximated by

$$\overline{H}_{13} = H_{12} H_{23} / \Delta E_{12}^{vert} \quad , \tag{4}$$

where

$$\Delta E_{12}^{vert} = \lambda_{12} + \Delta G_{12}^o \quad . \tag{5}$$

ΔE_{12}^{vert} is depicted in Fig. 1a. In eq. (4) H_{12} and H_{23} are the matrix elements for the electron transfers $BChl_2^*$ BChl BPh $\rightarrow$ $BChl_2^+$ $BChl^-$ BPh and $BChl_2^+BChl^-$ BPh $\rightarrow$ $BChl_2^+$BChl BPh^-, respectively.

From these equations the ratio of rate constants k_{13}^{super} / k_{12} is given, after some cancellation, by[9]

$$\frac{k_{13}^{super}}{k_{12}} = \frac{|\overline{H}_{13}| \, r e^{\Delta G_{12}^*/RT}}{2(\lambda_{13}\Delta G_{12}^*)^{\frac{1}{2}}} \quad , \tag{6}$$

where

$$r = |H_{23}/H_{12}| \quad . \tag{7}$$

(A main feature of eq.(6) is the manipulation which eliminated the explicit presence in eq. (6) of an additional unknown, the λ_{12} present in eq. (2).)

The $\bar{H}_{13}$ in eq. (6) can be calculated, assuming a superexchange mechanism, by equating the expression in eq. (1) to the observed rate constant for the formation of the radical pair, $BChl_2^+BPh^-$, from $BChl_2^*BPh$. The latter is 3.6 10^{11} s^{-1} at room temperature, taken as 20°C.[4] The value of λ_{13} is about 0.15 eV using the activationless property of the reaction ($\lambda_{13} \simeq -\Delta G_{13}^\circ$)[1] and using a ΔG_{13}° of -0.15 eV for the formation of the (initial) unrelaxed state of the radical pair.[2b] With these values, and assuming a superexchange mechanism, $\bar{H}_{13}$ is found from eq. (1) to be about 25 cm^{-1}.

In order that the contribution of the superexchange mechanism dominate that of the chemical intermediate mechanism, so that the $BChl^-$ concentration at room temperature is small, the ratio k_{13}^{super}/k_{12} should be fairly large. If the ratio is assumed to be larger than some particular value, and if for the r in eq. (7) a value (following Jortner's suggestion)[11] equal to the ratio of the corresponding electronic overlap integrals is used ($\simeq 7$ in the calculation of Plato in this volume), eq. (6) can be solved for ΔG_{12}^*. It is found that ΔG_{12}^* must be equal to or larger than 850 cm^{-1}, if $k_{13}^{super}/k_{12} \geq 5$.

An infinite set of values of the (λ_{12}, ΔG_{12}°) pairs can be found from eq. (3), each pair consistent with this value of ΔG_{12}^*. Examples are (1200 cm^{-1}, 800 cm^{-1}), (850, 850) and (300, 710), the first being in the region where $\lambda_{12} > |\Delta G_{12}^\circ|$ (the "normal" region for the $1 \rightarrow 2$ electron transfer) and the last being in the region where $\lambda_{12} < |\Delta G_{12}^\circ|$ (the "inverted region"). In turn, the corresponding values of H_{23} obtained with these values and eqs. (4), (5) and (7) are 590, 545 and 420 cm^{-1}, respectively. All are rather large and are a consequence of assuming a dominant superexchange mechanism. Before this introduction of an internal consistency test, values of H_{23} and the ΔE_{12}^{vert} in eq. (4) that had been employed were adjustable parameters, without any constraining equation.

SINGLET-TRIPLET SPLITTING

We next consider the implications of the foregoing arguments for the magnitude of the singlet-triplet splitting of the $BChl_2^+$ BPh^- radical pair in the relaxed state of the system. We use for this analysis the three electronic configurations described earlier. The energy of state 3, for the case of a singlet state of the radical pair, is influenced by configuration interaction with states 1 and 2, as in eq. (2) below. The energy of the radical pair triplet state is influenced by a configuration interaction with the corresponding triplet states of 1 and 2. The difference of these interactions gives, in this three-configuration model, a difference in singlet and triplet energies for the radical pair.

The electronic configurations 1,2,3 in a configuration interaction scheme yield a 3×3 Hamiltonian matrix with elements H_{ij}.[3] Using a partitioning argument[3], the corresponding secular determinant for the energy of a state 3 that is modified by the configuration interaction can be easily solved approximately. In this way the energy E of the radical pair is given by[3*]

$$E \simeq \bar{H}_{33} - \frac{|\bar{H}_{13}|^2}{\bar{H}_{11} - \bar{H}_{33}} , \qquad (8)$$

when $|\bar{H}_{13}| << |\bar{H}_{11} - \bar{H}_{33}|$ (as it is in our case) and where values of the barred matrix elements are given in Appendix A. The $\bar{H}_{13}$ in eq. (8) refers to the effective matrix element for the radical pair in its equilibrium nuclear

*In eq. (7) of ref. 3 the sign of the last term, and in eq. (10) that of the rhs, should be changed. The discussion there is unaffected, since only absolute values were considered.

configuration and differs from the $\bar{H}_{13}$ in eq. (4), which is evaluated instead at the equilibrium nuclear configuration of state 1. We denote the $\bar{H}_{13}$ for the radical pair by $\bar{H}_{13}^{rp}$ and the $\bar{H}_{13}$ in eq. (4) by $\bar{H}_{13}^{super}$. Using the argument in Appendix A and assuming that $H_{12}\,H_{23}$ in the radical pair's "relaxed state" is the same as in the initial equilibrium configuration for state 1, we have

$$\frac{\bar{H}_{13}^{rp}}{\bar{H}_{13}^{super}} = \frac{\Delta E_{12}^{vert}}{\Delta E_{32}^{vert}} \tag{9}$$

for the case of the singlet states. In Appendix A we estimate $\Delta E_{32}^{vert} \simeq 4000$ cm^{-1}. From the range of values of the $(\Delta G_{12}^{\circ}, \lambda_{12})$ pairs, the ΔE_{12}^{vert} in eq. (5) would appear to be in the vicinity of 1000 to 2000 cm^{-1}. We shall use a value of 1500 cm^{-1}. $\bar{H}_{13}^{rp}$ can now be calculated from $\bar{H}_{13}^{super}$ and eq. (9) to be about 65 cm^{-1}.

In estimating the last term in eq. (8) for the energy of the radical pair singlet state, a value for $\bar{H}_{11} - \bar{H}_{33}$ is needed at the equilibrium nuclear configuration for state 3. Now, λ_{13} is about 0.16 eV (an average of the 0.15 eV given earlier and the 0.17 eV deduced from the activationless formation of the triplet state of state 1 from the radical pair, state 3[7,12,13]), and ΔG_{13}° is -0.26 eV.[12-14] Since this $\bar{H}_{11} - \bar{H}_{33} = \Delta G_{31}^{\circ} + \lambda_{13} = -\Delta G_{13}^{\circ} + \lambda_{13}$, $\bar{H}_{11} - \bar{H}_{33}$ is approximately 0.4 eV.

The last term in eq. (8) can now be estimated from the above values to be about 0.03 cm^{-1} for the singlet state. Had we used an $r=2$ for the ratio H_{23}/H_{12} instead of $r=7$, the corresponding value would have been 0.05 cm^{-1}.[9]

To calculate the singlet-triplet splitting it is necessary to make a similar calculation for the last term in eq. (8) for the triplet state and to calculate the first term on the right hand side of eq. (8), $\bar{H}_{33}$, for both states. This $\bar{H}_{33}$ is discussed in Appendix B. The singlet-triplet splitting thus consists of several terms, one of which is 0.03 cm^{-1} (or 0.05 cm^{-1} if $r=2$). If we discount the likelihood of cancellation of any large terms (the terms have quite different contributions and there is no reason why there should be such a cancellation) there is seen to be a factor of about 30 discrepancy with the observed (10^{-3} cm^{-1})[5-7] singlet-triplet splitting of the $BChl_2^{+}$ BPh^{-} radical pair. When the ΔE^{vert}'s in eq. (9) become better known an improved estimate of $\bar{H}_{13}^{rp}$ can be made.

It is seen that the above calculated value of $\bar{H}_{13}^{rp}$ rests on the superexchange assumption and on the use of the internal consistency test to narrow the estimates of the ΔE^{vert}'s. The above calculation of a major contribution to the singlet-triplet splitting rests only on this $\bar{H}_{13}^{rp}$ and on the reasonably well known estimate of the denominator $\bar{H}_{11} - \bar{H}_{33}$ in eq. (8).

An important question is whether the value of $H_{12}\,H_{23}$ in eq. (4) for H_{13}^{super} is different from its value in the expression for $\bar{H}_{13}^{rp}$. The effect of energy on H_{12} and H_{23} is very minor.[3] Thus, only if $H_{12}\,H_{23}$ were different for the "unrelaxed" and "relaxed" configurations of the radical pair would there be a difference in $\bar{H}_{13}^{rp}$ and $\bar{H}_{13}^{super}$ due to this source.

TRIPLET RECOMBINATION

The value of $\bar{H}_{13}^{rp}$ estimated from the superexchange mechanism for the $1 \rightarrow 3$ electron transfer and the internal consistency test was seen to yield too high a value of the singlet-triplet radical pair splitting. The value needed for $\bar{H}_{13}^{rp}$ to be consistent with the observed splitting is roughly about $[10^{-3}(\bar{H}_{11} - \bar{H}_{33})]^{\frac{1}{2}}$, where $\bar{H}_{11} - \bar{H}_{33}$ is evaluated at the equilibrium configuration of the radical pair and was determined in the previous section to be about 0.4 eV. Thereby, this $\overline{H}_{13}{}^{rp}$ is about 2 cm^{-1} (1.8 cm^{-1}). It is useful compare this value with a value of $\overline{H}_{13}{}^{rp}$ for the corresponding triplet recombination 3 → 1 step, determined using the rate constant $k_{31}{}^{T}$ for the latter.

The rate constant $k_{31}{}^{T}$ is about 6 10^{8} s^{-1}, and is activationless.[7] The theoretical value of $k_{31}{}^{T}$ is given by eq. (1), but now the $\overline{H}_{13}$ refers to its value $\overline{H}_{13}{}^{rp,T}$ for the triplet system, at the equilibrium nuclear configuration of the radical pair. Calculated from $k_{31}{}^{T}$ using eq. (1) $\overline{H}_{13}{}^{rp,T}$ is found to be about 1.4 cm^{-1}. This value is close to that found above for $\overline{H}_{13}{}^{rp}$. The orbitals of the $BChl_2{}^{*}$ singlet differ somewhat in the asymmetry of the charge distribution from those of the $BChl_2{}^{*}$ triplet, and so some difference in the $\overline{H}_{13}{}^{rp}$ and $\overline{H}_{13}{}^{rp,T}$ is expected.

THE ONE-STEP NONADIABATIC/ADIABATIC MECHANISM

It certainly appears, as already noted, that on the basis of the data of Breton *et al.*[10] the chemical intermediate mechanism can be discarded. We have also seen that using the consistency test the superexchange mechanism encounters problems of a factor of 30 or so. While some other way might possibly be found of avoiding this difficulty it seems appropriate to consider possible alternative mechanisms, such as the one depicted in Fig. 2.

In Fig. 2 the only real barrier to reaction involves the transition 1 → 2 which is a nonadiabatic one. Its matrix element H_{12} is chosen to be 25 cm^{-1}, using the observed rate constant and eq. (2). It is postulated that the system then flows (in Fig. 2) on a new adiabatic curve – the lower curve there, constructed from the original 2 and 3 curves that are split quantum

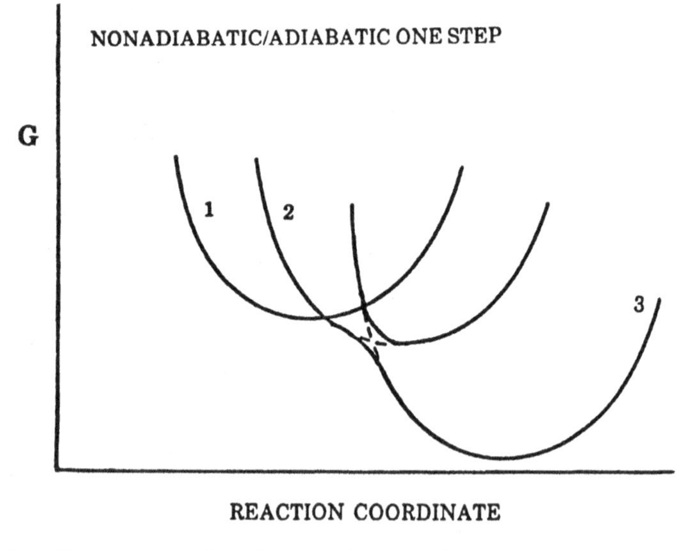

Fig. 2. Plot of free energy G *vs* the reaction coordinate for a mechanism in which there is only one step, namely a curve crossing from curve 1 to the lower solid curve. The latter is an adiabatic curve constructed from the original curves 1 and 2 that are split at their "intersection" by an amount 2 $\overline{H}_{23}$. During the motion of the system on this lower adiabatic curve the $BChl^{-}$ has, in effect, negligible temporal existence.

mechanically at their "intersection" by an amount equal to $2H_{23}$. Using $r = 7$ in eq. (7) we have $H_{23} \approx 175\ \mathrm{cm}^{-1}$. In this mechanism, state 2 in Fig. 2 doesn't even exist for a vibrational period and its concentration (and hence BChl depletion or "bleaching") during the $1 \rightarrow 3$ transition remains small throughout (about 2% or so in a preliminary quantum mechanical wave packet calculation).[15]

A next question is whether such a mechanism would also be consistent with the $10^{-3}\ \mathrm{cm}^{-1}$ singlet-triplet energy difference of the radical pair, since this was the point where the superexchange mechanism had difficulty. ΔE_{32}^{vert} for this mechanism is ~0.4 eV (Appendix B). Using the above values for H_{12} and H_{23}, the value of $\overline{H}_{13}{}^{rp}$ calculated from eq. (4) with ΔE_{12}^{vert} replaced by ΔE_{32}^{vert} is about $1.5\ \mathrm{cm}^{-1}$. Introducing this value into the last term of eq. (8) and using $\overline{H}_{11} - \overline{H}_{33} \simeq 0.4$ eV found earlier for the singlet state at the equilibrium configuration of state 3, the last term in eq. (8) is estimated as $\sim 10^{-3}\ \mathrm{cm}^{-1}$, which is of the same magnitude as the observed singlet-triplet splitting of the radical pair, $10^{-3}\ \mathrm{cm}^{-1}$.

Incidentally, if the situation in Fig. 2 prevails, earlier discussions [8] on the vertical position of curve 2, inferred from an absence of a temperature effect on the singlet-triplet splitting of quinone-free systems, would no longer apply, at least not in their original form.

CHARGE TRANSFER BAND

We have also considered elsewhere the implications of a superexchange mechanism for a $BPh^- \rightarrow BChl$ charge transfer band, if the large values of H_{23} required by the superexchange prevailed.[9] If, indeed, H_{23} were very large, a significant intensity would be found for this hypothetical charge transfer band, which would be located at the ΔE_{32}^{vert} in Fig. 1a, which is very roughly estimated to be around $4000\ \mathrm{cm}^{-1}$ if the superexchange mechanism were valid (Appendix B).

ELECTRIC FIELD EFFECT

Electric field effects on quantum yields have been investigated by Dutton and coworkers[16] using Langmuir-Blodgett monolayer films of close-packed reaction centers. Fields of as much as 150 mV/nm or more were used. The effect of the electric field on the possible bleaching of the BChl monomer and on the rate of the initial electron transfer step would be of particular interest. Using the component of the distances along the C_2-axis and using an assumption regarding the dielectric properties of the system it was deduced that with an adverse field of about 150 mV/nm the energy of state 2 and that of the (unrelaxed) state 3 at their equilibrium configurations would approach each other, if the situation at no field were as depicted in Fig. 2.[9] If there were no "unrelaxed" state or if the situation were, instead, that given by Fig. 1a, the energy of state 2 would still be higher than that of state 3 and there would still be no BChl bleaching.[9]. Thus, with a sufficiently adverse field there would be the possibility of observing $BChl^-$ and a bleaching of BChl, if the mechanism in Fig. 2 prevailed. A direct measurement of the electric field effect on the initial rate, particularly at low temperatures, would also be of interest.

CONCLUSIONS

In summary, we have seen that an internal consistency test can be derived and that it has consequences for the allowed values of the parameters for the superexchange mechanism. The latter mechanism leads, in turn, to a paradox, one possible solution of which is considered in the one-step nonadiabatic/adiabatic mechanism depicted in Fig. 2. Possible experiments, such as the effect of electric fields on the depletion of BChl in the initial reaction steps, are considered and may help distinguish the various mechanisms.

Perhaps it should also be stressed that the mechanisms in Fig. 1a and Fig. 2 have some common features. Both have a substantial H_{23}, though the superexchange mechanism has a larger H_{12} and H_{23}. The main difference is in the vertical position of curve 2. Electric field experiments may shed some light (as might a charge transfer band) on this question, if the dielectric properties are not too complicated.

ACKNOWLEDGMENT

It is a pleasure to acknowledge the support of this research by the Office of Naval Research. This article is Contribution No. 7707 from the Department of Chemistry.

APPENDIX A. H_{ij} TERMS

We have from a partitioning argument[3]

$$\overline{H}_{13} = H_{13} + H_{12} H_{23} / (H_{22} - E) \quad , \tag{A1}$$

where E is replaced by H_{11} when the H_{13} for the activationless $1 \rightarrow 3$ electron transfer is desired. All terms in eq. (A1) are then evaluated at the equilibrium nuclear configuration for state 1. Neglect of the direct interaction term H_{13} then yields eq. (4), upon approximating $H_{22} - H_{11}$ by its thermodynamic average.

When the H_{13} for the radical pair state is desired ($\overline{H}_{13} \equiv \overline{H}_{13}^{rp}$) the E in eq. (A1) is replaced, instead, by H_{33}, and all terms are now evaluated at the equilibrium nuclear configuration for state 3, the radical pair. Once again, the neglect of H_{13} and the replacement of $H_{22} - H_{33}$ by a thermodynamic average leads to eq. (4), with the ΔE_{12}^{vert} replaced by ΔE_{32}^{vert}.

From the same partitioning argument we also have for $\overline{H}_{11}$, the energy of state 1 perturbed by state 2,[3]

$$\overline{H}_{11} \simeq H_{11} - |H_{12}|^2 / (H_{22} - H_{11}) \quad . \tag{A2}$$

Here, H_{11} is the unperturbed energy of state 1, H_{22} that of state 2 and H_{12} the electron transfer matrix element for the $1 \rightarrow 2$, all evaluated at the same nuclear configuration, whichever is the one of interest. Similarly, the energy of a state 3 perturbed by state 2 is[3]

$$\overline{H}_{33} = H_{33} - |H_{32}|^2 / (H_{22} - H_{33}) \quad . \tag{A3}$$

APPENDIX B. SOME ENERGETICS

An estimate of the ΔE_{32}^{vert} in Fig. 1a can be made as follows. It is needed for the "relaxed" configuration of $BChl_2^+$ BPh^- radical pair. ΔE_{32}^{vert} equals $\lambda_{23} + \Delta G_{32}^\circ$, where ΔG_{32}° is the standard free energy of reaction for the $3 \rightarrow 2$ electron transfer, all evaluated at the "relaxed" configuration of state 3. We first consider an estimate made assuming the superexchange mechanism for the $1 \rightarrow 3$ reaction.

We shall approximate λ_{23} crudely by the mean of λ_{12} and λ_{13}. The λ_{13} is estimated in the text to be about 0.16 eV, while a variety of values of λ_{12} were seen to be consistent with the ΔG_{12}^* of 850 cm^{-1} in the text. The ΔG_{32}° equals

$-\Delta G_{23}°$ and the latter, in turn, equals $-\Delta G_{13}° + \Delta G_{12}°$. The value of $\Delta G_{13}°$ for forming the relaxed configuration of state 3 is -0.26 eV,[12-14] while various values of $\Delta G_{12}°$ consistent with $\Delta G_{12}^* = 850\ cm^{-1}$ were given in the text. When various (λ_{12}, $\Delta G_{12}°$) pairs are considered, a value of about 4000 cm^{-1} for $\lambda_{23} + \Delta G_{32}°$ is estimated on the superexchange assumption and used in the text as a consequence of that assumption.

If, instead, the mechanism in Fig. 2 were valid, ΔE_{32}^{vert} would be somewhat smaller, assuming $\Delta G_{12}° \simeq 0$. $\Delta E_{32}^{vert} = \lambda_{23} + \Delta E_{32}° = \lambda_{23} - \Delta E_{23}° \simeq \lambda_{23} - \Delta G_{13}° \simeq 0.4$ eV i.e., 3200 cm^{-1}.

We consider next the first term $\bar{H}_{33}$ in the RHS of eq. (8). It is given, in turn, by eq. (A3). In the latter, H_{33} is essentially the same for the singlet and triplet states of the radical pair, because the radicals are relatively far apart and configuration interaction from electronic configuration 2 doesn't affect $\bar{H}_{33}$. The H_{22} term in (A3) will differ slightly for singlet and triplet states, while H_{23} is expected to be rather similar in the two states. Thus, $\bar{H}_{33}$ will differ somewhat for the singlet and triplet state 3. However, this difference is independent of the differences in the last term in eq. (8), and, as noted earlier, we shall discount the likelihood of a cancellation of the large last term in (8) for the singlet state by other terms, this large term being a consequence of the asumption of a superexchange mechanism for the $1 \rightarrow 3$ reaction.

Configuration interaction may also occur, incidentally, due to electronic states other than 1, 2 and 3, e.g., $BChl_2$ BCl^* BPh.[18]

REFERENCES

1. R. A. Marcus and N. Sutin, *Biochim. Biophys. Acta* 811:265 (1985).
2. (a) N. W. Woodbury and W. W. Parson, *Biochim. Biophys. Acta* 767:345 (1984), *ibid.* 850:197 (1986);
   (b) (quinone-free) N. W. Woodbury, W. W. Parson, M. R. Gunner, R. C. Prince and P. L. Dutton, *Biochim. Biophys. Acta* 851:6 (1986).
3. R. A. Marcus, *Chem. Phys. Letters* 133: 471 (1987).
4. J. Breton, J.-L. Martin, J. Petrich, A. Migus and A. Antonetti, *FEBS Letters* 209:37 (1986);
   J. Breton, J.-L. Martin, A. Migus, A. Antonetti and A. Orzag, *Proc. Natl. Acad. Sci. US* 83:5121 (1986);
   J. -L. Martin, J. Breton, A. J. Hoff, A. Migus and A. Antonetti, *Proc. Natl. Acad. Sci. US* 83:957 (1986).
5. K. W. Moehl, E. J. Lous and A. J. Hoff, *Chem. Phys. Letters* 121:22 (1985); D. A. Hunter, A. J. Hoff and P.J. Hore, *ibid.* 134:6 (1987).
6. J. R. Norris, C.P. Lin and D. E. Budil, *J. Chem. Soc., Faraday Trans.* I. 83:13 (1987);
   J. R. Norris, M. K. Bowman, D. E. Budil, J. Tang, C. A. Wraight and G. L. Closs, *Proc. Natl. Acad. Sci. US* 79:5532 (1982).
7. A. Ogrodnik, N. Remy-Richter, M. E.Michel-Beyerle and R. Feick, *Chem. Phys. Letters* 135:576 (1987).
8. M. Bixon, J. Jortner, M. E. Michel-Beyerle, A. Ogrodnik and W. Lersch, *Chem. Phys. Letters* 140:626 (1987).
9. R. A. Marcus, Chem. Phys. Letters (submitted).
10. J. Breton, G. R. Fleming and J. L.Martin, this volume; G.. R. Fleming, J.-L. Martin and J. Breton, submitted for publication.
11. J. Jortner, private communication;
    J. L. Katz, S. A. Rice, S. -I. Choi and J. Jortner, *J. Chem. Phys.* 39:1683 (1963);
    M. E. Michel-Beyerle, M. Plato, J. Deisenhofer, H. Michel, M. Bixon and J. Jortner, *Biochim. Biophys. Acta*, in press.

12. R. A. Goldstein, L. Takiff and S.G. Boxer, *Biochim. Biophys.Acta*, submitted;
C.E.D. Chidsey, L. Takiff, R. A. Goldstein and S. G. Boxer, *Proc. Natl. Acad. Sci. US* 82:6850 (1985).
13. M.E. Michel-Beyerle and A. Ogrodnik, this volume;
A. Ogrodnik, N. Remy-Richter, and M. E. Michel-Beyerle, *Biochim. Biophys. Act*, submitted.
14. J. K. H. Hörber, W. Göbel, A. Ogrodnik, M. E. Michel-Beyerle and R. J. Cogdell, *FEBS Letters* 198:273 (1986).
15. R. Almeida and R. A. Marcus, unpublished.
16. Z. D. Popovic, G. J. Kovacs, P. S. Vincett, G. Alegria and P. L. Dutton, *Biochim. Biophys. Acta* 851:38 (1986);
Z. D. Popovic, G. J. Kovacs, P. S. Vincett and P. L. Dutton, *Chem. Phys. Letters* 116:405 (1985).
17. J. R. Norris, D. E. Budil, D. M. Tiede, J. Tang, S. V. Kolaczkowski, C. H. Chang and M. Schiffer, *in* "Progress in Photosynthesis Research", Vol 1, ed. J. Biggins (Martinus Nijhoff, 1987), p. 363.

MECHANISM OF THE PRIMARY CHARGE SEPARATION IN BACTERIAL PHOTOSYNTHETIC REACTION CENTERS

M. Bixon and Joshua Jortner, School of Chemistry, Sackler Faculty of Exact Sciences, Tel Aviv University, 69 978 Tel-Aviv (Israel)
M. Plato, Institut für Molekülphysik, Freie Universität Berlin Arnimallee 14, 1000 Berlin 33 (FRG)
M.E. Michel-Beyerle, Institut für Physikalische und Theoretische Chemie Technische Universität München, Lichtenbergstr. 4, Garching (FRG)

ABSTRACT. We present a critical scrutiny of the models for the primary charge separation process in bacterial photosynthesis in the light of recent experimental data and theoretical analysis. The sequential mechanisms are fraught with difficulties due to their incompatibility with magnetic data for the first observable radical pair in conjunction with the activationless nature of the electron transfer process. The implications of the unistep, superexchange mediated nonadiabatic electron transfer mechanism are examined.

I. INTRODUCTION

The primary step in energy conversion in bacterial photosynthesis is the translocation of an electron across a membrane-spanning protein/pigment-complex, the reaction center (RC). The electron is transferred in several discrete steps involving various pigments rigidly arranged in two almost symmetrical protein subunits, L and M. The X-ray structure analysis of the RC of *Rps.viridis* [1,2,3] and *Rb.sphaeroides* [4,5] shows similar spatial arrangement of donors and acceptors together with their protein environment. The primary donor is the singlet excited state of a bacteriochlorophyll dimer (P) [6,7] which transfers an electron with a rate of $3.6 \cdot 10^{11}$ s^{-1} [8-11] at 295K to a bacteriopheophytin molecule (H) on the L-subunit over a center-to-center distance of 17 Å. This rate is sufficiently fast to compete with dissipative radiative and non-radiative pathways and, in particular, it precludes any energy waste due to back transfer to the antenna. The primary charge separation process is too fast to be accounted for simply by a direct non-assisted electron transfer. Several different mechanisms explaining the primary electron transfer step have recently been proposed. It is the goal of this paper to discuss these mechanisms in the light of recent experimental data and theoretical analysis.

II. MECHANISMS

All the mechanisms proposed for the primary charge separation attribute a central role to the accessory monomer bacteriochlorophyll (B), which the X-ray structure revealed to be located between P and H, to act as either a short-lived genuine ionic intermediate, or by the modification of the $^1P^*$-H electronic coupling. Two classes of mechanisms were advanced to account for the role of B in the ultrafast primary charge separation process.

(1) One-step direct electron transfer ${}^1P^*BH \rightarrow P^+BH^-$, which is mediated by superexchange interactions via the virtual states of P^+B^-H (Fig. 1) being located at energy above ${}^1P^*BH$ [12-16].

(2) Two-step sequential electron transfer, which may involve several variants.

2a. Sequential electron transfer via the P^+B^-, which presumably is located at energy between ${}^1P^*$ and P^+H^-, occurring via the following scheme [17,18]

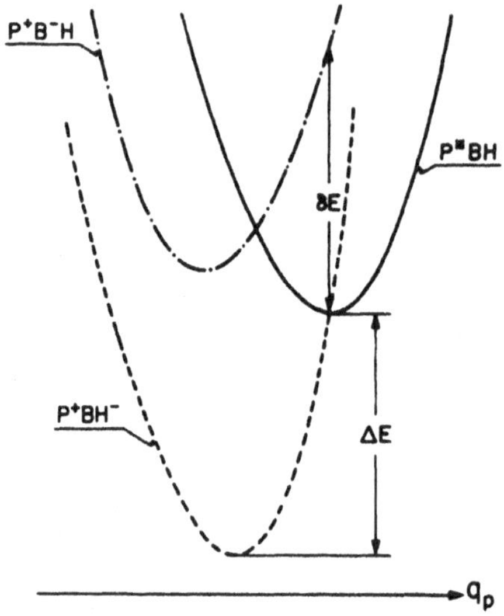

Fig. 1. Nuclear potential energy surface for the superexchange electronic interaction mechanism in the RC.

$$ {}^1P^*BH \xrightarrow{k_1} P^+B^-H \underset{k_{-2}}{\overset{k_2}{\rightleftharpoons}} P^+BH^- \qquad \text{(II.1)}$$

2b. The equilibrium model [19]. This constitutes a special case of the kinetic scheme (2) with the (equilibrated) states P^+B^-H and PB^+H^- being isoenergetic.

2c. The energy transfer model [20], which invokes an intermediate state PB^+H^-, being presumably located at an energy below ${}^1P^*$, and which is effectively populated by the off-resonance interaction with the PB^*H state. The kinetic scheme is

$$^{1}P^{*}\,BH \xrightarrow{k_1} PB^{+}H^{-} \underset{k_{-2}}{\overset{k_2}{\rightleftharpoons}} P^{+}BH^{-} \qquad (II.2)$$

Although no intermediate involving B^+ or B^- was observed in femtosecond time-resolved spectroscopy [8-11], the observation that $^1P^*$ decays and P^+BH^- is built up with the same time constant, and that no bleaching is observed in the absorption region of B, still permits the formation of an intermediate state involving B, according to the sequential kinetic schemes (2a)-(2c). The only requirement is that the formation of the intermediate is the rate determining step, i.e. $k_2 >> k_1$. From recent femtosecond experiments on RCs of *Rb.sphaeroides* and *Rps.viridis* a lower limit $k_2 \gtrsim 50\ k_1$ at 10K was inferred [11].

III. SOME OBSERVABLES

Any acceptable mechanism for the primary charge separation should be compatible with observations on the forward electron transfer $^1P^*BH \rightarrow P^+BH^-$ and the recombination dynamics of P^+BH^-. RCs of *Rb.sphaeroides* and *Rps.viridis* are considered as prototypes for RCs containing bacterio-chlorophylls and -pheophytins a and b in similar spatial distribution within highly homologous protein subunits. For these the relevant kinetic observations on the forward electron transfer are:

(a) The rate $k(P^*BH \rightarrow P^+BH^-)$ exhibits a very weak temperature dependence between 4K and 300K [11,12] with the rate at 4K being enhanced by 2.4 for *Rb.sphaeroides* and by 4 for *Rps.viridis* [11]. The apparent small negative activation energy is compatible with activationless electron transfer processes.

(b) The primary rate is invariant with respect to deuterium isotopic substitution of the protein over the entire temperature domain 4K-300K [11].

(c) The primary electron transfer is strongly unidirectional along the L-protein subunits ($k_L/k_M > 10$ [16,21,22]).

Recombination dynamics can be studied only in RCs in which the electron transfer to the quinone (Q) at the L-branch is blocked. Blocking electron transfer by chemical reduction of Q (ubiquinone or menaquinone in RCs of *Rb.sphaeroides* or *Rps.viridis*, respectively) leads to the formation of a spin-polarized triplet signal of $^3P^*$ observed in ESR-measurements [23,24] and a magnetic field effect on the triplet yield [25-29]. The interpretation of both phenomena within the formalism of Chemically Induced Spin Polarization limits the value of the exchange integral J of the radical pair P^+BH^- to the sum of the hyperfine interactions, $J \leq 10^{-3}\ cm^{-1}$. Since a chemical reduction of Q may complicate the recombination dynamics by the combined effects of electrostatic perturbation and the magnetic coupling of P^+H^- to the Q^--Fe^{2+} complex, the recombination data used in the following refer to RCs of *Rb.sphaeroides*, where both quinones at both protein subunits are absent, yielding:

(d) Small and temperature independent exchange integral $J \simeq 3{\cdot}10^{-4}\ cm^{-1}$ [25, 27-29].

(e) Temperature independent recombination rate k_T for electron transfer from $^3(P^+BH^-) \rightarrow {}^3P^*BH$, k_T being smaller than k ($k/k_T = 600$) [27].

IV. NONADIABATIC OR ADIABATIC PRIMARY ELECTRON TRANSFER?

Electron transfer processes are traditionally seggregated into nonadiabatic and adiabatic processes according to the strength of the electronic coupling. This issue pertains to the distinction between the strong and weak Landau-Zenner (LZ) coupling [30] which is determined by the LZ parameter

$$\gamma_{LZ} = \frac{2\pi V^2}{\hbar v|\Delta F|} \qquad \text{(IV.1)}$$

where V is the electronic coupling between the initial and final states, while v is the velocity in the vicinity of the crossing of the potential surfaces and ΔF is the difference in the slopes of the potential surfaces at the intersection point. The denominator in Eq. IV.1 is given by

$$\hbar v|\Delta F| = \hbar\omega(2\lambda k_B T)^{1/2} \qquad \text{(IV.2)}$$

where ω is the characteristic frequency of the system undergoing ditortion upon electron transfer. For electron transfer in the RC, ω corresponds to the frequency of the medium. λ is the medium reorganization energy. The LZ parameter is then

$$\gamma_{LZ} = \frac{2\pi V^2}{\hbar\omega(2\lambda k_B T)^{1/2}} \qquad \text{(IV.3)}$$

When $\gamma_{LZ} > 1$ the electron transfer process is adiabatic with the rate being given by the Holstein relation

$$k = \frac{\omega}{2\pi} \exp(-E_a/k_B T) \qquad \text{(IV.4)}$$

so that for activationless adiabatic ET

$$k = (\omega/2\pi) \qquad \text{(IV.5)}$$

The maximum value of the electron transfer rate being determined by the characteristic frequency.

When $\gamma_{LZ} < 1$ the electron transfer is nonadiabatic with the rate being given by

$$k = \frac{2\pi}{\hbar} V^2 F \qquad \text{(IV.6)}$$

where F is the thermally averaged Franck-Condon nuclear overlap factor.

Provided that the frequencies of the nuclear modes, $\hbar\omega$, are sufficiently low relative to the thermal energy $k_B T$, the classical high temperature limit of F is applicable, being given by the Marcus relation [31]

$$F = (4\pi\lambda k_B T)^{-1/2} \exp(-E_a/k_B T) \qquad \text{(IV.7)}$$

where the activation energy E_a is given by

$$E_a = (\Delta G - \lambda)^2/4\lambda \qquad \text{(IV.8)}$$

ΔG is the free energy difference between the initial and final states in their equilibrium geometries. For the primary electron transfer process involving porphin prosthetic groups the distortions of nuclear configurations accompanying electron transfer originate mainly from protein modes [32]. For the nuclear protein modes in the RC $\hbar\omega \simeq 100$ cm^{-1} has been estimated [32], so that Eqs. (IV.1) and (IV.8) hold at room temperature.

In order to use this formalism, information on the equilibrium energies, reorganization energies and electronic couplings are needed. The existing data on the energetics [33-37] are summarized in Fig. 2. So far, the energies of the charge transfer states $E(P^+B)$, $E(B^+H^-)$ and $E(P^{\pm})$ are not known. The various models differ in assigning those energies. There is no direct experimental access to the reorganization energy λ and to the electronic couplings. Therefore λ and V are used as adjustable parameters within the frame of the nonadiabatic electron transfer theory. In the case of activationless electron transfer, which constitutes the relevant physical situation for the primary process, $\lambda \simeq \Delta G$ (within a range of 30%) and the following relation emerges for activationless nonadiabatic electron transfer

$$V^2 \simeq (4\pi\Delta G k_B T)^{1/2} (\hbar/2\pi)\, k \tag{IV.9}$$

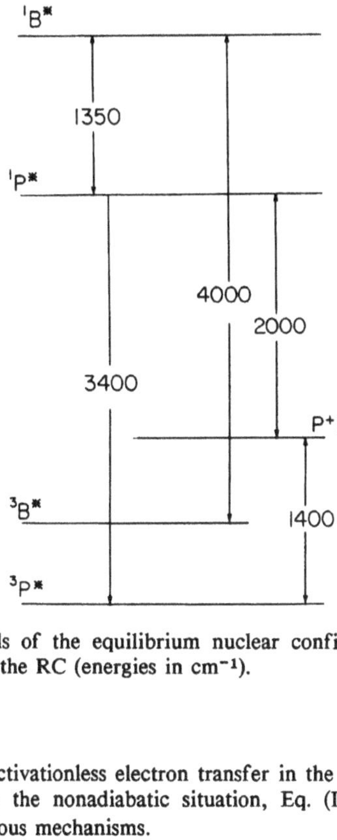

Figure 2. Energy levels of the equilibrium nuclear configurations of singlet and triplet electronic excitations in the RC (energies in cm^{-1}).

Does the primary activationless electron transfer in the RC correspond to the adiabatic limit, Eq. (IV.5), or to the nonadiabatic situation, Eq. (IV.6)? Here we shall have to distinguish between various mechanisms.

(1) One-step direct electron transfer. In this case the experimental value of k and $\lambda \simeq \Delta G = 2000$ cm^{-1} (Fig. 2), together with Eq. (IV.9), result in V = 25 cm^{-1}. Thus, taking $\hbar\omega = 100$ cm^{-1}, we obtain $\gamma_{LZ} = 4\cdot10^{-2}$ and the unistep superexchange mediated process is nonadiabatic.

(2) Two-step sequential electron transfer (mechanisms (2a), (2b) or (2c)). In this case we have to consider the two electron transfer rates k_1 and k_2, Eqs. (I.1) or (I.2), with $k_1 = 3.6\cdot10^{11}$ sec^{-1} and $k_2 > 3.6\cdot10^{12}$ sec^{-1}, both of which being activationless. Consider first the rate k_1. Taking the reasonable value $\Delta G_1 \simeq 800$ cm^{-1} and $\lambda_1 = 800-1200$ cm^{-1} to characterize k_1, we estimate for Eq. (IV.9) $V_1 \simeq 20$ cm^{-1} for the first process which, together with $\hbar\omega = 100$ cm^{-1}, results in the LZ parameter for k_1, $(\gamma_{LZ})_1 \simeq 0.13$, and this process is nonadiabatic. Next, we consider the rate k_2. From recent semi-empirical calculations of the electronic couplings for this process, $V_2/V_1 \simeq 6$ [38] so that the LZ parameter for the second step is $(\gamma_{LZ})_2 \simeq 36\,(\gamma_{LZ})_1 \simeq 4.7$ and the second process is adiabatic. We thus conclude that providing a sequential mechanism prevails, the first step is nonadiabatic, with k_1 being given by Eq. (IV.6), while the second step is adiabatic with $k_2 = \omega/2\pi$. Taking $\omega = 100$ cm^{-1} the fast second step should be characterized by the rate $k_2 \simeq 3.3\cdot10^{12}$ sec^{-1}. The large limit on the ratio $k_2/k_1 \gtrsim 50$ at 10K [11] provides strong evidence against the applicability of the sequential mechanism because of two reasons. First, taking $k_2/k_1 \geq 50$, together with $k_1 = 8.3\cdot10^{11}$ sec^{-1} for *Rb.Sphaeroides* at 10K, results in $k_2 \geq 4\cdot10^{13}$ sec^{-1} at this low temperature. The emerging rate $k_2 \geq 4\cdot10^{13}$ sec^{-1} is considerably larger than the value of $k_2 \simeq 3.3\cdot10^{12}$ sec^{-1} estimated on the basis of Eq. (IV.5) with the appropriate characteristic frequency. Second, the adiabatic rate $k_2 \geq 4\cdot10^{13}$ sec^{-1}, is characteristic of a high frequency of >2000 cm^{-1}, which corresponds to the intramolecular mode of the protein or of the pigments and which is expected to exhibit some deuterium isotope effect, in contrast to experiment (point (b) section III). This argument against the applicability of the sequential mechanism is general, spanning mechanisms (2a), (2b) and (2c).

V. ARE SEQUENTIAL MECHANISMS APPLICABLE?

Evidence against the sequential mechanisms rests also on recent magnetic data for quinone depleted RCs in *Rb.Sphaeroides* [27-29]. The small value of the exchange integral ($J = 3\cdot10^{-4}$ cm^{-1}) of the P^+H^- radical pair and its temperature independence over the temperature domain 300-80K, in quinone depleted RC, together with the activationless nature of the primary charge separation cannot be reconciled with the sequential electron transfer mechanisms [39].

All the sequential models (2a)-(2c) in section III have universal kinetic features although they differ in some of the quantitative aspects of the electronic interactions. It will be appropriate to provide a unified discussion of these sequential models. In the general discussion of sequential mechanisms, both intermediates P^+B^-H and PB^+H^- are denoted by $I^{\pm}$.

The experimental quantity crucial for the discussion of the validity of sequential models is the exchange integral J of the radical pair P^+BH^-. In a sequential mechanism this exchange integral is composed of two contributions:

(i) The exchange integral J_D determining the correlation of the electron spins in the state P^+BH^-, and

(ii) the exchange integral J_C which governs the spin dynamics on the close site, i.e. in the state P^+B^-H ($|J_C| >> |J_D|$).

To correlate J_C and J_D with the experimental exchange interaction J we apply the formalism developed by Kubo [40] to describe the influence of hopping processes on the lineshape of resonance spectra. In the slow hopping limit, i.e. for $k_2+k_{-2} \ll |J_C-J_D|$, two resonances are expected to occur in the MARY (MAgnetic field effects on Reaction Yields)-spectrum corresponding to the two distinct situations of the electron residing on the close site ($J \simeq J_C$) and on the distant site ($J \simeq J_D$), respectively. In the limit of fast hopping ($k_2+k_{-2} \gg |J_C-J_D|$), a single resonance should appear in the MARY-spectrum corresponding to the "effective" exchange interaction

$$J = \frac{k_{-2}}{k_2+k_{-2}} J_C + \frac{k_2}{k_2+k_{-2}} J_D \tag{V.1}$$

which is the average of the two exchange integrals J_C and J_D weighted with the respective site populations in kinetic equilibrium between the two sites. In quinone-depleted reaction centers the lifetime of the radical pair state ($\simeq$10 ns) [26,36,41] is much longer than the time for establishing equilibrium $1/(k_2+k_{-2}) < 600$ fs, and therefore the assumption of an equilibrium situation between the populations on the close and distant site, respectively, is certainly justified.

Under the condition that $k_{-2} < k_2$, Eq. (V.1) reduces to

$$J = (k_{-2}/k_2)\, J_C + J_D = K\, J_C + J_D \tag{V.2}$$

with the equilibrium constant K

$$K = \exp(-\Delta G_2/RT) \tag{V.3}$$

ΔG_2 denoting the free energy difference reaction between the intermediate $I^{\pm}$ state and the radical pair P^+BH^-.

The observations in MARY- [27] and RYDMR- (Reaction Yield Detected Magnetic Resonance) experiments [28-29] indicate the temperature independence of J allowing for a maximal change $\Delta J < 5\ 10^{-4}\ cm^{-1}$ in the temperature range of T_1 = 160K to T_2 = 295K. On the basis of these limitations on the temperature dependence of J we are able to estimate the free energy difference ΔG_2 between the ion pair states P^+B^-H and P^+BH^-.

By inserting Eq. (V.3) into Eq. (V.2) and taking ΔJ between the temperatures T_1 and T_2, the temperature independent contribution from J_D is eliminated according to

$$\Delta J = J(T_2) - J(T_1) = J_C \left(y - y^{T_2/T_1} \right) \tag{V.4}$$

where $y = \exp(-\Delta G_2/RT_2)$.

The exchange integral J_C is estimated from second-order perturbation theory as

$$J_C = \frac{(V_0)^2}{\Delta E_V(^1P^* - I^{\pm})} + \frac{(\eta V_0)^2}{\Delta E_V(I^{\pm} - {}^3P^*)} + (V^*)^2 \left[\frac{1}{\Delta E_V(^1B^* - I^{\pm})} + \frac{1}{\Delta E_V(I^{\pm} - {}^3B^*)} \right] \tag{V.5}$$

where V_0 is the electronic coupling between $^1P^*$ and the intermediate charge-transfer state $I^{\pm}$ and the factor η represents the ratio between the couplings, $^{3,1}I^{\pm}$ and $^3P^*BH$, $^1P^*BH$. V^* is the coupling between B^* and $I^{\pm}$ and assumes the same value for both singlet and triplet states. ΔE_V is the vertical energy difference between the states at the equilibrium medium configuration of the intermediate charge transfer state $I^{\pm}$.

The vertical singlet energy differences can be approximated by the equilibrium energy differences corrected by the reorganization energy λ. The known equilibrium energies are summarized in the level scheme of Fig. 2, and is the equilibrium energy of the intermediate $I^{\pm}$ is a parameter of the model. For the sake of simplicity let us denote the vertical energy difference $\Delta E_V(^1P^*-I^{\pm})$ by $\delta E = \Delta G_1 + \lambda$, where ΔG_1 and λ are the (free) energy gap and the reorganization energy accompanying the formation of $I^{\pm}$. Using the numbers given in the energy scheme (Fig. 2), we rewrite Eq. (V.5) as

$$J_C = \frac{(V_0)^2}{\delta E} + \frac{(\eta V_0)^2}{3400-\delta E} + (V^*)^2 \left[\frac{1}{1350+(\delta E)} + \frac{1}{2650-(\delta E)} \right] \tag{V.6}$$

Now we investigate the two suggested sequential mechanisms (2a) and (2c) separately.

(1) Sequential electron transfer through $I^{\pm} \equiv P^+B^-H$.

The energetic parameters for the formation of $I^{\pm}$ are $\Delta G_1 \simeq \Delta G/2 \simeq 1000$ cm^{-1} and $\lambda_1 \simeq 1000$ cm^{-1}, so that $\delta E = 2000$ cm^{-1}. The matrix element V_0 is taken to account for the magnitude of the first rate determining step. Accordingly, $V_0 \equiv V_1 \simeq 20$ cm^{-1}.

From ESR-experiments on RCs crystals [42] and from the analysis of k_T [43] we can infer that $\eta < 1$. There is no available information on V^*. However, for our analysis it is sufficient to notice that the last term in Eq. (V.6) contributes a positive quantity. The contribution of the last term in Eq. (V.6) is quite large, e.g., for $V^* = V_0$ this term gives 0.69 cm^{-1}. For the sake of an estimate of the lower limit of J_C we shall take the first two terms in Eq. (V.6), setting $\eta = 1$ in the second term, and disregard the substantial positive contribution from the last term . Eq. (V.6) now results in $J_C > 0.48$ cm^{-1}. In order to check the applicability of the fast hopping limit we take $k_2 \gtrsim 4\cdot10^{13}$ sec^{-1} = 200 cm^{-1} (at 10K) as estimated in section IV. Due to the inequality $|J_C| < k_2 \geq 200$ cm^{-1} the fast hopping limit ($k_2+k_{-2} >> |J_C-J_D|$) is indeed applicable. With the value of $J_C > 0.48$ cm^{-1}, using Eq. (V.4) together with $T_2/T_1 = 2$ and $J(T_2) - J(T_1) < 5\cdot10^{-5}$ cm^{-1}, we obtain a lower limit for the (free) energy difference between P^+B^-H and P^+BH^-, which is $\Delta G_2 >$ 1850 cm^{-1}. With the present data for $\Delta G(^1P^*-P^+BH^-) \simeq 2000$ cm^{-1} and the limit on ΔG_2 we have $\Delta G_1 < 150$ cm^{-1} for the (free) energy difference between $^1P^*$ and P^+B^-.

These values of the limits for ΔG_2 and ΔG_1 are expected to result in appreciable temperature dependence of the sequential rates k_1 and/or k_2. In Figure 3 we present a survey of the calculations of the ratios $k_1(300K)/k_1(4K)$ and $k_2(300K)/k_2(4K)$ which were obtained using the full quantum mechanical expression for the electron transfer rates [44] with a characteristic frequency of $\hbar\omega = 100$ cm^{-1}. These calculations were performed for a wide combination of parameters ΔG_j and λ_j (j = 1 or 2). The range of parameters, where both k_1 and k_2 do not vary between 4K and 300K by more than a numerical factor of 3, is marked as an island (to be referred as a "pseudoactivationless island") in Fig.3. Obviously, the limits $\Delta G_2 > 1850$ cm^{-1} and $\Delta G_1 < 150$ cm^{-1} are located outside the "pseudoactivationless island". We thus conclude that the magnetic data imply that the sequential mechanism (2a), Eq. (I.1) is expected to result in a marked temperature dependence of the primary rates which is in contradiction with experiment.

(ii) Sequential electron transfer via $I^{\pm} = PB^+H^-$.

The major contribution to J_C originates now from the interaction between $^1B^*$ and

B^+H^-, i.e., from the third contribution to Eq. (V.6). Using the parameters given by Fischer and Scherer [20] $\delta E = (1133 + \lambda)\ cm^{-1}$ with $\lambda = 800\ cm^{-1}$ together with $V^* = 121\ cm^{-1}$ results in the large value of $J_c \simeq 15\ cm^{-1}$. Now, $|J_c - J_d| \simeq 15\ cm^{-1} < k_2 > 200\ cm^{-1}$, and the system is presumably in the fast hopping limit. Making use of Eq. (V.6) one obtains $\Delta G_2 > 2250\ cm^{-1}$. This analysis implies that the energy of the PB^+H^- intermediate is located by ~250 cm^{-1} above that of $^1P^*$ and that the primary electron transfer process, which presumably proceeds via energy transfer, should be activated. These conclusions regarding mechanism (2c) seem to be incompatible with the experimental facts.

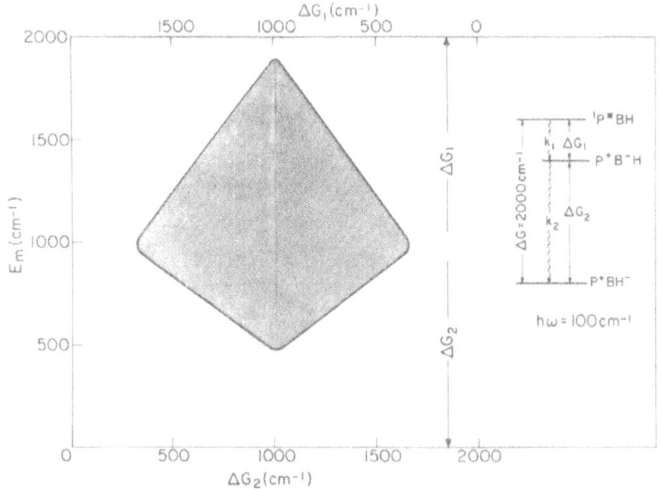

Figure 3. A schematic representation of the range of parameters for the free energies ΔG_2 and ΔG_1 and nuclear reorganization energies λ_1 and λ_2 (denoted by E_m in the figure) for which the electron transfer rates k_1 and k_2 are nearly activationless, i.e. $k_i(300K)/k_i(4K) \leq 3$ for $i = 1,2$. The "pseudoactivationless island" is marked by a shaded region.

VI. SUPEREXCHANGE

On the basis of the foregoing analysis we conclude that the sequential mechanisms are inapplicable and that the unistep, superexchange-mediated electron transfer prevails. This unistep process is nonadiabatic, as discussed in section IV, with the rate being given by Eq. (IV.6). Fig. 1 describes this superexchange interaction between the lowest vibronic level of $^1P^*BH$, with the quasienergetic vibronic manifold of P^+BH^- being mediated by the off-resonance coupling of $^1P^*BH$ with the P^+B^-H state. When the energy differences between $^1P^*BH$ states and the P^+B^-H mediating states effectively coupled to it exceed the average vibrational spacing ($\hbar\omega = 100\ cm^{-1}$) of the protein medium, the electronic coupling, V, assumes the form

$$V = V_{PB}V_{BH}/\delta E \qquad (VI.1)$$

where V_{PB} is the electronic coupling between $^1P^*BH$ and P^+B^-H, while V_{BH} is the electronic coupling between P^+B^-H and P^+BH^-. δE is the vertical energy difference between the potential surfaces for P^*BH and P^+B^-H at the intersection point of the potential surfaces of $^1P^*BH$ and P^+BH^- (Fig. 1).

Information on the electronic coupling terms and the energetic parameter in Eq. (VI.1) are crucial to assess the validity of the superexchange mechanism. We have introduced an intermolecular overlap approximation to calculate the relative magnitudes of the electronic coupling terms V_{PB} and V_{BH} [16,38], demonstrating the dominance of the electronic coupling terms across the L branch, over the corresponding terms in the M subunit, thus establishing that the major ingredient for the unidirectionality of charge separation originates from the electronic coupling. These calculations have also determined the enhancement of the B^-H - BH^- coupling over the $^1P^*B$ - P^+B^- coupling, with

$$\alpha = |V_{BH}/V_{PB}| = 6 \qquad (VI.2)$$

Indirect information on the vertical energy gap δE (Fig. 1) will be inferred from some kinetic constraints on the dynamics of the primary electron transfer.

VII. COMPETITION BETWEEN SUPEREXCHANGE AND THERMALLY ACTIVATED ELECTRON TRANSFER

The prevalence of the superexchange interaction between $^1P^*$ and P^+H^- does not exclude, in principle, the occurrence of activated electron transfer from $^1P^*$ to P^+B^-, which in the classical limit ($\hbar\omega < kT$) proceeds at the crossing of the corresponding potential surfaces (Fig. 1). It was pointed out [38] that a self-consistent framework for the primary activationless superexchange electron transfer process requires that the contribution of the parallel activated channel through the real P^+B^- intermediate is minor.

The rate constant of parallel electron transfer to P^+B^- is

$$k_1 \propto V_{PB}^2 \exp(-E_a/k_BT) \qquad (VII.1)$$

The activation energy can be inferred from the relation $\delta E = \Delta G_1 + \lambda_1$, where ΔG_1 is the free energy gap between the minima of the $^1P^*$ and P^+B^- potential surfaces, δE is the vertical energy difference between the potential surfaces (Fig. 1) and λ_1 is the medium reorganization energy for this process. Accordingly, the activation energy for the parallel process is

$$E_a = \delta E^2/4\lambda_1 \qquad (VII.2)$$

The ratio of the rate constants for the activated channel, Eq. (VII.1) and the superexchange channel, Eqs. (IV.6) and (VI.1), is

$$\frac{k_1}{k} = \left[\frac{\delta E}{V_{BH}}\right]^2 \exp(-E_a/k_BT) \qquad (VII.3)$$

The experimental value of the activationless primary electron transfer $k = 3.6 \cdot 10^{11}$ sec^{-1}, which is attributed to superexchange, implies that the electronic matrix element is according to Eq. (VI.1)

$$V_{PB}\ V_{BH}/\delta E = (25 \pm 5)\ \mathrm{cm}^{-1} \tag{VII.4}$$

At this stage we shall utilize the ratio, Eq. (VI.2), of the electronic coupling terms. Eqs. (VI.2) and (VII.2)-(VII.4) result in

$$\frac{k_1}{k} = \left[\frac{\delta E}{25\alpha}\right] \exp\,[-(\delta E)^2 / 4\lambda_1 k_B T] \tag{VII.5}$$

Eq. (VII.5) has the form $(k_1/k) \propto (\delta E)\exp(-\beta(\delta E)^2)$ where β is temperature dependent, indicating that the increase of δE, although reducing the superexchange rate k, retards the thermally activated channel. In Fig. 4 we present the dependence of k_1/k for several values of λ_1. Obviously, the medium reorganization energy for the direct superexchange rate k ($\lambda \approx 2000\ \mathrm{cm}^{-1}$) and for the thermally activated rate k_1, are not necessarily equal. Eq. (VII.5) implies that the direct process dominates, i.e., $k_1/k \leq 0.2$ for the parameters

$$\lambda_1 \leq (\delta E)^2/[4k_B T\ \ln\,(\delta E/5\alpha)] \tag{VII.6}$$

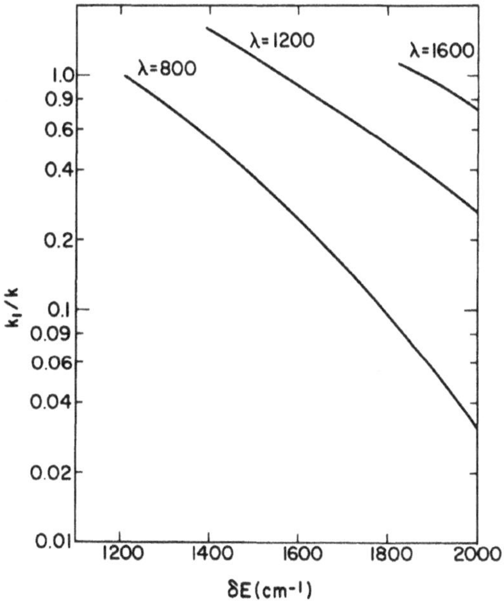

Figure 4. The dependence of k_1/k on λ_1 at 300K.

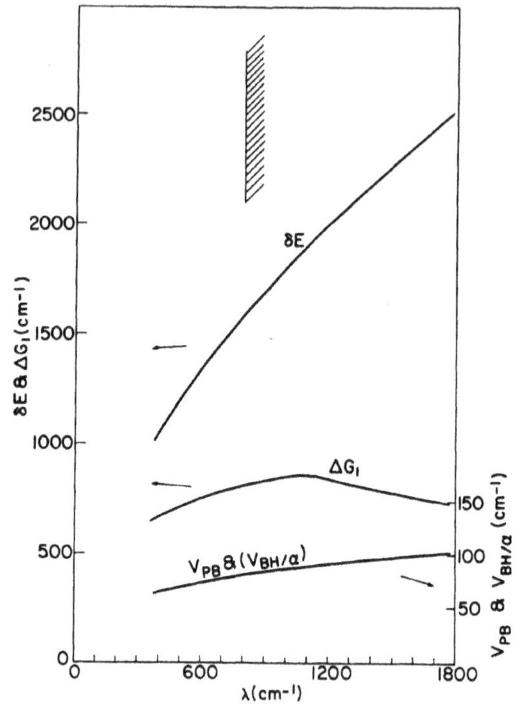

Figure 5. The dependence of the vertical energy δE and the energy gap ΔG_1 between $^1P^*B$ and P^+B^- on the reorganization energy λ_1 at 300K. The dependence of the electronic coupling terms V_{PB} and V_{BH} on λ_1 is also presented. The energetic and coupling parameters are derived from the condition of the dominance of the superexchange mechanism. The lower limit for a "reasonable" value of λ_1 is marked by a dashed line.

Figure 5 displays the interrelationship between the lower limit for the vertical energy δE and the upper limit for λ_1 according to Eq. (VII.6). It is apparent that the medium reorganization energy cannot be too small. For "reasonable" values of λ_1, i.e., $\lambda_1 = 800$ cm^{-1} this analysis implies that $\delta E \geq 1600$ cm^{-1} ($\Delta G_1 \geq 800$ cm^{-1}), while for $\lambda_1 = 1200$ cm^{-1}, i.e. $\delta E \geq 2000$ cm^{-1} ($\Delta G_1 \geq 800$ cm^{-1}). This analysis indicates that the energetics is not very sensitive to the value of λ_1. It should be borne in mind that these estimates of δE rest on the harmonic approximation. Deviations of the potential surfaces at large displacements from harmonicity may result in a considerable decrease of δE. The present analysis within the framework of its inherent limitations sets some rough limits on the electronic coupling matrix elements, which ensure the dominance of the temperature independent superexchange channel (Fig. 5), as required by the experimental activationless nature of the primary process. These matrix elements are given by $V_{PB} = (25\ \delta E/\alpha)^{1/2}$ and $V_{BH} = (25\alpha\ \delta E)^{1/2}$. For example, taking again $\lambda_1 = 800$ cm^{-1} and $\delta E \geq 1600$ cm^{-1} our analysis implies that $V_{PB} = 80$ cm^{-1} and $V_{BH} = 480$ cm^{-1}. We note that in view of the dependence of the electronic matrix elements on $(\delta E)^{1/2}$, the deviations from the harmonic approximation, which lower δE, will result in the decrease of these electronic coupling terms. At present, there is no way to confront these estimates of V_{PB} and V_{BH} with reliable calculations of absolute values of the intermolecular matrix elements in the RC. Calculations of the band structure of excess electron and hole states in organic crystals of aromatic molecules, which include effects of many-electron exchange [45-47], indicate that the largest intermolecular transfer integrals are in the range 100-400 cm^{-1}. Thus the estimates of the matrix elements may be reasonable. The next step in the theoretical analysis should address this problem. The energetic parameters originating from our analysis within the framework of the harmonic approximation imply that the condition for the dominance of the superexchange mediated direct transfer process from $^1P^*$ to P^+H^- dominates over the thermal process from $^1P^*$ to P^+B^- so that δE exceeds 1600 cm^{-1} while $\Delta G_1 \gtrsim 800$ cm^{-1}. Thus the potential surfaces for $^1P^*$ and P^+B^- are well separated in energy, relative to the characteristic (mean) vibrational protein frequency $\hbar\omega = 100$ cm^{-1}, with the energy gap corresponding to $\sim 8\hbar\omega$, while the vertical energy is $\geq 16\hbar\omega$. For such relatively large separation proximity effects between the $^1P^*$ and P^+B^-, potential surfaces will be negligible. Furthermore, small changes between the relative location of these potential surfaces in *Rps.viridis* and *Rb.sphaeroides* are not expected to result in a gross modification of the electron transfer dynamics, providing a rationalization for the similarity of the primary rates in the RCs of these two bacteria.

VIII. ELECTRIC FIELD EFFECTS ON THE PRIMARY CHARGE SEPARATION

The application of strong electric fields across the RC is expected to modify the energetics and the superexchange electronic coupling responsible for the charge separation processes. The kinetics of charge separation in moderately strong (ϵ = 0.1-1 mV/Å) and in strong (ϵ = 1-15 mV/Å) electric fields will be affected by the modification of the energetics of the ion-pair states. From the point of view of general methodology we note that
(1) Electric field effects may be useful in providing a further diagnostic tool for the elucidation of the mechanism for the primary charge separation process.
(2) They provide a test of free-energy relationships for various electron transfer processes in the RC.
(3) They may provide a quantitative estimate of the medium reorganization energy λ and of the relative contributions of medium reorganization and of intramolecular reorganization.

Points (2) and (3) were recently addressed by Dutton et al. [48] in the context of quinone reduction $P^+H^-Q \rightarrow P^+HQ^-$ and by Feher et al. [49] in relation to the $P^+Q^- \rightarrow PQ$ back recombination. The important feature (1) was addressed by Boxer et al. [50] who have observed an increase (of ~20%) of the fluorescence quantum yield at high ($\epsilon \simeq 10$ mV/A) fields in isotropic samples and by Dutton et al. [51] who have observed a substantial drop (to 70%) of the quantum yield of the charge separation process in oriented RCs at high ϵ = -15 mV/Å field along the C2 axis of the RC.

The energetic shifts of ion pair states induced by external electric fields are appreciable in view of the large dipole moments of such states

$$\vec{\mu}(j) = e\vec{R}(j) \qquad \text{(VIII.1)}$$

where j ≡ (P^+B^-,P^+H^- or P^+Q^-) denotes the particular ion-pair state, while $\vec{R}(j)$ is the vector between the components of the jth state. Thus, $|\vec{\mu}(P^+B^-)| = 51D$, $|\vec{\mu}(P^+H^-)| =$

82D and $|\vec{\mu}(P^+Q^-)| = 134D$. The electrostatic energies

$$U(j) = -\vec{\mu}(j) \cdot \vec{\epsilon} \qquad \text{(VIII.2)}$$

can be substantial, modifying the energy gaps. Here and in what follows ϵ is the effective field seen by the dipoles, which will be assumed to be equal to the external field. In Fig. 6 we present the dependence of the energy levels on the electric field, which is directed along the axis of the RC. The superexchange kinetic scheme

$$\xleftarrow{k_d} \ {}^1P^* \underset{k_-}{\overset{k}{\rightleftarrows}} P^+H^- \overset{k_3}{\rightarrow} P^+Q^- \qquad \text{(VIII.3)}$$

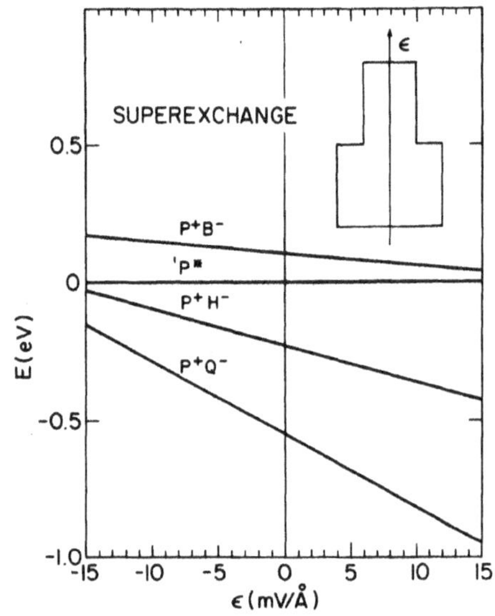

Figure 6. The dependence of the energy levels for the superexchange mechanism on the electric field directed along the C_2 axis of the RC.

has to incorporate the backreaction k_-, due to the energetic proximity of ${}^1P^*$ and P^+H^- (Fig. 6) at large negative fields. k_d represents the combined decay rates of ${}^1P^*$ in all radiative and nonradiative channels except the charge separation channel. The field dependence of the superexchange mediated primary rate is

$$k(\epsilon)/k(0) = |V(\epsilon)/V(0)|^2 \; |F(\epsilon)/F(0)| \qquad \text{(VIII.4)}$$

with the electronic coupling

$$|V(\epsilon)/V(0)|^2 = [\delta E/(\delta E - \mu(P^+B^-)\epsilon \cos\theta_1)]^2 \qquad \text{(VIII.5)}$$

and the Franck Condon factor (FCF)

$$F(\epsilon)/F(0) = \exp[(-(\Delta G(0)-\lambda+\mu(P^+H^-)\,\epsilon\cos\theta_2)^2 + (\Delta G(0)-\lambda)^2) \,/\, 4\lambda k_B T] \qquad \text{(VIII.6)}$$

where θ_1 and θ_2 are the angles between $\vec{\epsilon}$ and $\vec{R}(P^+B^-)$ and between $\vec{\epsilon}$ and $\vec{R}(P^+H^-)$, respectively. Here $\vec{\epsilon}$ is applied along the C_2 axis of the RC. In what follows we shall assert that the process is activationless at zero-field, i.e. $\Delta G(0) = 2000\ cm^{-1} \simeq \lambda$.

The implications of field effects on the superexchange mechanism in oriented RCs are
(1) The electronic coupling is either retarded or enhanced with increasing ϵ depending on the direction of the field (Fig. 7).
(2) The FCF exhibits retardation for all direction of the field (Fig. 7). For strictly activationless processes the decrease of FCF is symmetric for negative and positive values of ϵ.
(3) The rate $k(\epsilon)$ exhibits a combined contribution of $V(\epsilon)$ and of $F(\epsilon)$. For low ϵ the electronic term dominates, while for large ϵ the FCF contribution dominates.
(4) The primary rate $k(\epsilon)$ vs ϵ is asymmetric, decreasing faster at $\epsilon < 0$, (Fig. 7). This rate can be interrogated by time-resolved fsec spectroscopy of $^1P^*$, P^+ or H^-.
(5) The quinone reduction rate, $k_3(\epsilon)$, Eq. (VIII.3), is

$$k_3(\epsilon)/k_3(0) = \exp\,[-(\mu(P^+Q^-)\cos\theta_3 - \mu((P^+H^-)\cos\theta_2)^2\,\epsilon^2/4\lambda_3 k_B T] \qquad \text{(VIII.7)}$$

where θ_3 is the angle between $\vec{\epsilon}$ and $\vec{R}(P^+Q^-)$ and $\lambda_3 \simeq \Delta G_3 = 0.55eV$. $k_3(\epsilon)$ will exhibit a symmetric retardation with increasing $|\epsilon|$ (Fig. 7), due to the contribution of the FCF. Some asymmetry may be exhibited when λ_3 is close to but slightly different from ΔG_3.
(6) A temperature dependence of the rates k (and k_3) is expected to be exhibited at high fields.
(7) The quantum yield for charge separation is $Y(\epsilon) = k_{eff}(\epsilon)/(k_d+k_{eff}(\epsilon))$, where $k_{eff} = k/(1+kK/k_3)$ and $K = k_-/k$. The increase of k_-/k at large negative fields (Fig. 8) due to the energetic proximity effect (Fig. 6) results in appreciable decrease of $k_{eff}(\epsilon)$ at large negative values of ϵ. Thus $Y(\epsilon)$ vs ϵ in oriented samples will decrease faster towards $\epsilon < 0$. Such an asymmetry is exhibited by the experimental quantum yield data for $-15mV/Å < \epsilon < 15mV/Å$ [51]. However, a semiquantitative fit of these experimental data can be accomplished only with an unphysical decay rate or $^1P^*$, $k_d \simeq 9{\cdot}10^9\ sec^{-1}$, which considerably exceeds the radiative ($k_f \simeq 3x10^8\ sec^{-1}$) and intramolecular decay rates of $^1P^*$.

Of considerable interest are experiments in nonoriented RCs, where the field direction is isotropic. Under these circumstances the superexchange model predicts that
(8) The time-resolved dynamics of $^1P^*$ (and of P^+ and H^-) will exhibit a distribution of decay lifetimes, which reflect the effects of dynamics inhomogeneous broadening.
(9) The quantum yield at low fields will be dominated by the field dependence of the superexchange interaction. The fluorescence quantum yield for a certain field direction will be $Y_f(\vec{\epsilon}) = k_d/k_{eff}(\vec{\epsilon})$ where the rates were defined under point (7). The relative fluorescence quantum yield in an isotropic sample at low fields is

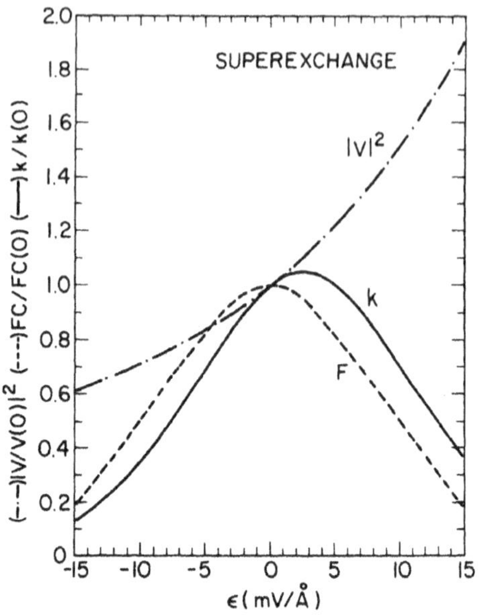

Figure 7. The dependence of the electronic coupling, the FCF and the primary rate k at 300K on the electric field directed across the C_2 axis of the RC for the superexchange mechanism.

$$Y_f(\epsilon) \,/\, Y_f(0) = \left[(1/2)\int_0^{\pi} d\theta \, \sin\theta \, (1-\beta\cos\theta)^{-2}\right]^{-1} \qquad \text{(VIII.8)}$$

where $\beta = \frac{\mu(P^+B^-)\epsilon}{\delta E}$. For low fields $Y_f(\epsilon)/Y_f(0) = (1-\beta^2)$, predicting a small decrease of the fluorescence quantum yield with increasing ϵ.

(10) At moderate and high fields $Y_f(\epsilon)$ for isotropic samples will be determined by the interplay between the field dependence of the superexchange interactions and of the FCF, exhibiting a marked temperature dependence. Numerical calculations using Eqs. (VIII.4)-(VIII.6) at room temperatures and a quantum formalism at low temperatures, with δE = 1600 cm^{-1} and ϵ = 9 mV/Å, result in $Y_f(\epsilon)/Y_f(0)$ = 3.8 at 77 K and $Y_f(\epsilon)/Y_f(0)$ = 1.4 at 298K. The experimental value $Y_f(\epsilon)/Y_f(0)$ = 1.20 for ϵ = 9·mV/Å at 77K, is reported by Boxer at al. [50].

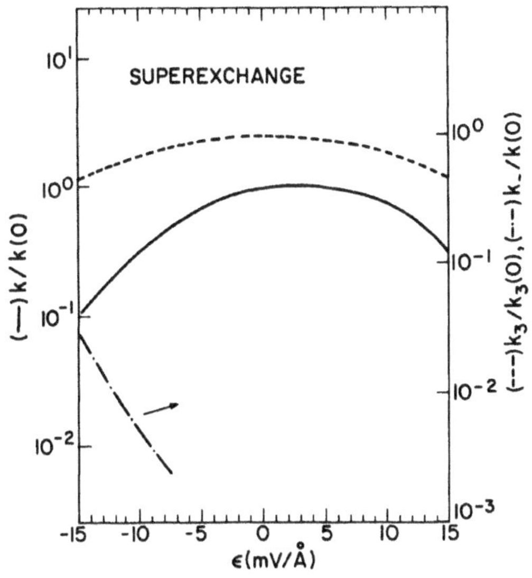

Figure 8. The dependence of the rates k, k_- and k_3 at 300K on the electric field across the C_2 axis of the RC for the superexchange mechanism.

It will be interesting to confront the predictions of the superexchange model, which we endorse, with those of the sequential decay model. A level scheme for the energetics of the sequential model is presented in Fig. 9, where we note the level crossing of $E(P^+H^-)$ and of $E(P^+B^-)$ at large negative fields. The kinetic scheme for the sequential mechanism is

$$\xleftarrow{k_d} {}^1P^* \underset{k_{-1}}{\overset{k_1}{\rightleftarrows}} P^+B^- \underset{k_{-2}}{\overset{k_2}{\rightleftarrows}} P^+H^- \xrightarrow{k_3} P^+Q^- \tag{VIII.9}$$

Assuming that both k_1 and k_2 are activationless, i.e., $\Delta G_i = \lambda_i$ (i = 1.2) the rates are

$$\frac{k_1(\epsilon)}{k_1(0)} = \exp\left[-(\mu(P^+B^-)\,\epsilon\cos\theta_1)^2 / 4\lambda_1 k_B T\right] \tag{VIII.10}$$

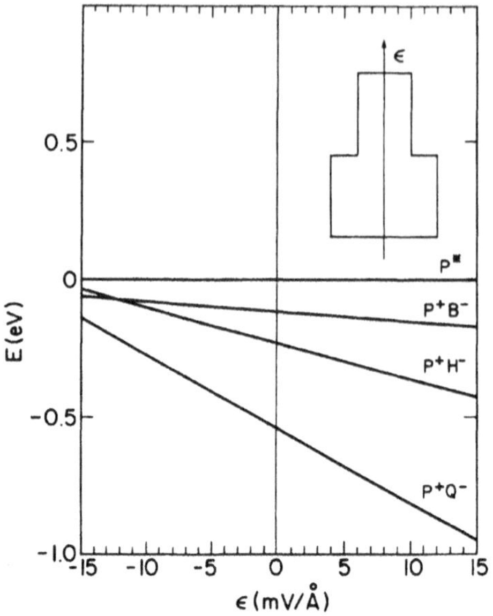

Figure 9. The dependence of the energy levels for the sequantial mechanism on the electric field across the C_2 axis of the RC.

$$\frac{k_2(\epsilon)}{k_2(0)} = \exp\left[-(\mu(P^+H^-)\cos\theta_1 - \mu(P^+B^-)\cos\theta_2)^2\epsilon^2/4\lambda_2 k_B T\right] \quad \text{(VIII.11)}$$

while the equilibrium constants are

$$\frac{k_{-1}}{k_1} = \exp\left[-(\Delta G_1(0) - \mu(P^+B^-)\epsilon\cos\theta_1) / k_B T\right] \quad \text{(VIII.12)}$$

$$\frac{k_{-2}}{k_2} = \exp\left[-(\Delta G_2(0) + (\mu(P^+B^-)\cos\theta_2 - \mu(P^+B^-)\cos\theta_1)\,\epsilon)/k_B T)\right] \quad \text{(VIII.13)}$$

The sequential mechanism will exhibit different qualitative dynamic features. For an oriented sample we have
(1) The rate determining step k_1 (k_1 is affected only through FCF) is expected to exhibit a (25%) symmetric decrease over the entire ϵ domain (Fig. 10). Both the magnitude and the form of the field dependence for the sequential model differ from that predicted for the superexchange model.

(2) k_2 is affected only by the FCF, exhibiting an appreciable symmetric decrease with increasing ϵ (Fig. 10). At the highest field $[k_2(\epsilon)/k_1(\epsilon)]/[k_2(0)/k_1(0)] = 0.25$ at ϵ = 15mV/Å.

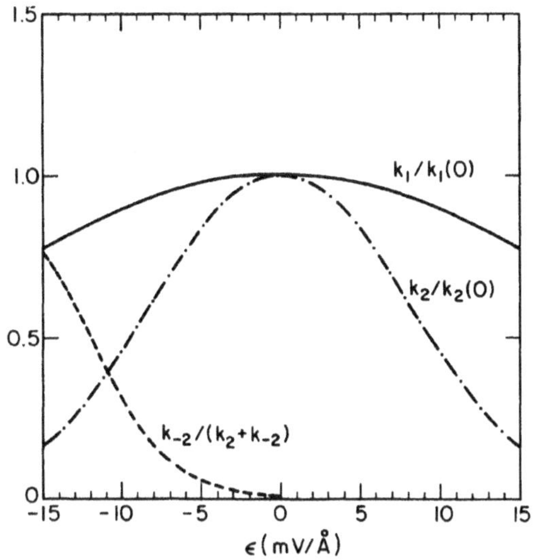

Figure 10. The dependence of k_1, k_2 and $k_2/(k_2+k_{-2})$ at 300K for the sequential mechanism on the electric field across the C_2 axis of the RC.

(3) The level crossing of $E(P^+B^-)$ and $E(P^+H^-)$ at ϵ = 13 mV/A results in an appreciable enhancement of k_{-2}/k_2, with equilibration being achieved at large negative ϵ values. As both k_1 and k_2 are fast while k_3 is slow, creating a kinetic bottleneck effect, equilibration between P^+B^- and P^+H^- is expected to prevail with the relative concentration of P^+B^- which is given by $\frac{[P^+B^-]}{[P^+H^-]} = k_{-2}/(k_2+k_{-2})$. At large fields this ratio approaches unity (Fig. 10). Accordingly, the sequential mechanism predicts the unveiling of the P^+B^- kinetic intermediate at large negative fields in oriented RCs.

(4) The quantum yield for charge separation in oriented RCs is $Y = k_1/(k_d + k_1)$ over the entire ϵ domain, as the equilibrium constant k_{-1}/k_1 is negligible at all fields. In view of the weak ϵ dependence of $k_1(\epsilon)$ (Fig. 10) $Y(\epsilon)$ will exhibit only a small deviation from $Y(0)$, in contrast with the predictions of the superexchange scheme.

Of some interest are also quantitative predictions for the implications of the sequential model for isotropic, randomly oriented RCs.

(5) The low-field fluorescence quantum yield will be

$$Y_f(\epsilon) / Y_f(0) = 1 + \gamma/3$$

where $\gamma = (\mu(P^+B^-)\epsilon)^2/4\lambda_1 k_B T$, predicting the field enhancement of the low-field fluorescence quantum yield in the isotropic sample. This low-field effect for the sequential model is qualitatively different from that predicted (see point (9)) for the superexchange model.

Some of the predictions emerging from the present analysis have to be confronted with experiment in order to provide independent information regarding the nature of the primary charge separation process. Before doing so one has to carefully sort out any alternative field induced processes, e.g., field-induced polarization effects of the pigments and their ions, which may lead to the modification of the charge distribution and of the electron transfer integrals at high electric fields.

REFERENCES

1. J. Deisenhofer, O. Epp, K. Miki, R. Huber, and H. Michel, J. Mol. Biol. **180**, 385 (1984).
2. J. Deisenhofer, O. Epp, K. Miki, R. Huber and H. Michel, Nature **318**, 618 (1985).
3. H. Michel, O. Epp and J. Deisenhofer, EMBO J. **5**, 2445 (1986).
4. J.P. Allen, G. Feher, T.O. Yeates, D.C. Rees, J. Deisenhofer, H. Michel and R. Huber, Proc. Natl. Acad. Sci. USA, **83**, 8589 (1986).
5. C.H. Chang, D. Tiede, J. Tang, U. Smith, J. Norris and M. Schiffer, FEBS Letters **205**, 82 (1986).
6. J.R. Norris, R.A. Uphaus, H.L. Crespi and J.J. Katz, Proc. Natl. Acad. Sci. USA **68**, 625 (1971).
7. G. Feher, A.J. Hoff, R.A. Isaacson and L.C. Ackerson, Ann. N.Y. Acad. Sci. **244**, 239 (1975).
8. J.-L. Martin, J. Breton, A.J. Hoff, A. Migus and A. Antonetti, Proc. Natl. Acad. Sci. USA **83**, 957 (1986).
9. J. Breton, J.-L. Martin, J. Petrich, A. Migus and A. Antonetti, FEBS Lett. **209**, 37 (1986).
10. J. Breton, J.-L. Martin, A. Migus, A. Antonetti and A. Orszag, Proc. Natl. Acad. Sci. USA **83**, 5121 (1986).
11. J.-L. Martin, G. Fleming and J. Breton (this volume).
12. N.W. Woodbury, M. Becker, D. Middendorf and W.W. Parson, Biochem. **24**, 7516 (1985).
13. S.F. Fischer, I. Nussbaum and P.O.J. Scherer, in Antennas and Reaction Centers of Photosynthetic Bacteria, M.E. Michel-Beyerle, ed. p. 256, Springer Berlin (1985).
14. J. Jortner and M.E. Michel-Beyerle, in Antennas and Reaction Centers of Photosynthetic Bacteria, M.E. Michel-Beyerle, ed. p. 345, Springer Berlin (1985).
15. J. Jortner and M. Bixon, in Protein Structure Molecular and Electronic Reactivity, R. Austin, E. Buhks, B. Chance, D. De Vault, P.L. Dutton, H. Frauenfelder and V.I. Gol'danskii, eds. p. 277, Springer-Verlag New York (1987).
16. M.E. Michel-Beyerle, M. Plato, J. Deisenhofer, H. Michel, M. Bixon and J. Jortner, Biochim. Biophys. Acta (in press).
17. R. Haberkorn, M.E. Michel-Beyerle and R.A. Marcus, Proc. Natl. Acad. Sci. USA **70**, 4185 (1979).
18. R. Marcus, Chem. Phys. Lett. **133**, 471 (1987).
19. S.V. Chekalin, Ya.A. Matveetz, A. Ya. Shkuropatov, V.A. Shuvalov and A.P. Yartzev, FEBS Letters **216**, 245 (1987).
20. S.F. Fischer and P.O.J. Scherer, Chem. Phys. **115**, 151 (1987).
21. M.G. Rockley, M.W. Windsor, R.J. Cogdell and W.W. Parson, Proc. Natl. Acad. Sci. USA 72, 2251 (1975).

22. K.J. Kaufman, P.L. Dutton, T.L. Netzel, J.S. Leigh and P.M. Rentzepis, Science **188**, 1301 (1975).
23. A.J. Hoff, Quart. Rev. Biophys. **17**, 153 (1984).
24. P.J. Hore, E.T. Watson, J.Boiden Petersen and A.J. Hoff, Biochim. Biophys. Acta **849**, 70 (1986).
25. A.J. Hoff, Photochem. Photobiol. **43**, 727 (1986).
26. A. Ogrodnik, H.W. Krüger, H Orthuber, R. Haberkorn, M.E. Michel-Beyerle and H. Scheer, Biophys. J. **39**, 91 (1982).
27. A. Ogrodnik, N. Remy-Richter, M.E. Michel-Beyerle and R. Feick, Chem. Phys. Lett. **135**, 576 (1987).
28. K.W. Moehl, E.J. Lous and A.J. Hoff, Chem. Phys. Lett. **121**, 22 (1985).
29. D.A. Hunter, A.J. Hoff and P.J. Hore, Chem. Phys. Lett. **134**, 6 (1987).
30. V.G. Levich, Adv. Electrochem. Eng. **4**, 249 (1965).
31. R.A. Marcus, J. Chem. Phys. **24**, 966,979 (1956).
32. M. Bixon and J. Jortner, J. Phys. Chem. **90**, 3795 (1986).
33. L. Takiff and S.G. Boxer, Photochem. Photobiol. **45**, 61s (1987).
34. J. Breton, Biochim. Biophys. Acta **810**, 235 (1985).
35. C.E.D. Chidsey, L. Takiff, R.A. Goldstein and S.G. Boxer, Proc. Natl. Acad. Sci. USA **82**, 6850 (1985).
36. A. Ogrodnik, N. Remy-Richter and M.E. Michel-Beyerle, (to be published).
37. A. Ogrodnik, M. Volk and M.E. Michel-Beyerle, (this volume).
38. (a) R.A. Marcus, private communication and Chem. Phys. Lett. (in press).
(b) M. Plato, K. Möbius, M.E. Michel-Beyerle, M. Bixon and J. Jortner, (submitted for publication).
39. M. Bixon, J. Jortner, M.E. Michel-Beyerle, A. Ogrodnik and W. Lersch, Chem. Phys. Lett. **140**, 626 (1987).
40. R. Kubo, in Fluctuation, Relaxation and Resonances in Magnetic Systems, ed. D. ter Haar, Oliver and Boyd (1962), p.23.
41. C.E.D Chidsey, C. Kirmaier, D. Holten and S.G. Boxer, Biochim. Biophys. Acta **766**, 424 (1984).
42. J.R. Norris,(private communication).
43. M. Bixon, M.E. Michel-Beyerle, A. Ogrodnik and J. Jortner, (submitted for publication).
44. J. Jortner, J. Chem. Phys. **64**, 4860 (1976).
45. J.L. Katz, S.A.Rice, A.I. Choi and J. Jortner, J. Chem. Phys. **39**, 1683 (1963).
46. R. Silbey, J. Jortner, S.A. Rice and M.T. Vala, J. Chem. Phys. **42**, 733 (1965).
47. R. Silbey, J. Jortner, S.A. Rice and M.T. Vala, J. Chem. Phys. **43**, 2525 (1965).
48. Z.D. Popovic, G.J. Kovacs, P.S. Vincett, G. Algeria and P.L. Dutton, Chem. Phys. **110**, 227 (1986).
49. T. Arno, M.Y. Okamura and G. Feher, (this volume).
50. S.G. Boxer, (this volume).
51. Z.D. Popovic, G.J. Kovacs, P.S. Vincettt, G. Algeria and P.L. Dutton, Biochim. Biophys. Acta **851**, 38 (1986).

A CONFIGURATION INTERACTION (CI) DESCRIPTION OF VECTORIAL ELECTRON TRANSFER IN BACTERIAL REACTION CENTRES

M. Plato and C.J. Winscom

Institut für Molekülphysik, Fachbereich Physik
Freie Universität Berlin
Arnimallee 14, D-1000 Berlin 33, West Germany

INTRODUCTION

The purpose of this communication is to report a CI method for calculating the electronic matrix elements between the various prosthetic groups of a bacterial reaction centre (RC), relevant for estimating the electron transfer (ET) rate in the primary charge-separation step. Although we will focus on just the elements required to estimate the ET rate for the "super-exchange" model[1], the underlying procedure may be used quite generally to determine the energy ordering of the various local and charge-transfer excited states of a molecular aggregate, such as a bacterial RC. It offers a theoretical framework for investigating quite diverse optical properties of the aggregate such as, for example, Stark effects.

The concept of the CI procedure is similar to that adopted by Murrel[2] in 1958 to describe donor-acceptor complexes. It utilises the molecular orbitals (MO's) of semi-empirical SCF calculations performed on the isolated cofactors, and avoids having to consider the overwhelming valence atomic orbital (AO) basis of the entire aggregate in one supermolecule-SCF step. A second strength of this strategy is that medium effects such as those arising from close-lying amino acid residues may be taken into account on the same footing, if required.

METHOD

Only brief details of the full CI procedure[3] will be presented here. The many electron aggregate Hamiltonian under consideration is given by:

$$\hat{H}_{agg} = \sum_i \{T(i)+V_P(i)+V_{B_L}(i)+V_{B_M}(i)+V_{H_L}(i)+V_{H_M}(i)+V_{AA}(i)+\sum_{i<j} r_{ij}^{-1}\} \quad 1)$$

where T(i) represents the kinetic energy of electron i, the V(i) terms represent the potential energy of electron i in the fields of the cofactor-cores P = $(BChl)_2$, B = BChl, H = BPh and the potential of the amino acid residue environment, AA, respectively, and r_{ij}^{-1} accounts for the two-electron repulsion. The additional subscripts, L and M, refer to the corresponding protein sub-units of the RC. Eigenstates are sought as linear combinations of basis configurations, e.g.

$$\Psi_{RC} = c_0\Psi_0(H_M B_M P^{\circ} B_L H_L) + c_1\Psi_1(H_M B_M P^* B_L H_L) + c_2\Psi_2(H_M B_M P^+ B_L^- H_L) + \ldots \qquad 2)$$

Each configuration of the basis is constructed from previously determined SCF MO's of the individual cofactor singlet ground states. For example, the configuration of lowest zero-order energy, Ψ_0, has a single Slater-determinant form, represented by

$$\Psi_0 = |\ \ldots\ h_M\bar{h}_M\ \ldots\ b_M\bar{b}_M\ \ldots\ p\bar{p}\ \ldots\ b_L\bar{b}_L\ \ldots h_L\bar{h}_L|$$

where h_M, b_M, ... represent the highest occupied MO's (HOMO's) in the singlet ground states of the isolated cofactors and α- and β-spin orbitals are distinguished by, e.g. h_M, $\bar{h}_M$, respectively. Excited singlet states may be constructed similarly by taking the symmetric linear combination of α- and β-spin orbital promotions to the relevant virtual orbitals. Specific matrix elements were considered first by application of Slater's rules[4]. Direct use of the Brillouin theorem[5] is excluded since the MO's used are not generally eigenfunctions of the aggregate's SCF Fock operator. Matrix element expressions are rearranged, without neglect, collecting local Fock operators operating on their corresponding eigenfunctions as highest priority. The various components are then collected into intra- and intermolecular contributions and simplified as far as possible. The remaining two-centre exchange integrals are approximated with the help of Mulliken's approximation[6]. Finally, all terms involving intermolecular overlap of S^2 and higher orders are neglected. For the purposes of this communication, the matrix elements of particular interest are:

$$\begin{aligned} V_{PB} &= \langle P^*BH|\hat{H}_{agg}|P^+B^-H\rangle \\ &= \frac{1}{2} S_{p^*b^*}\{\varepsilon_{p^*}+\varepsilon_{b^*}-J_{pp^*}-J_{pb^*} + \sum_{P}^{occ} K_{Pp^*} + \sum_{B}^{occ} K_{Bb^*} \\ &\quad + (U_B+U_H+U_{AA} : p^*p^*) + (U_P+U_H+U_{AA} : b^*b^*)\} \\ V_{BH} &= \langle P^+B^-H|\hat{H}_{agg}|P^+BH^-\rangle \\ &= \frac{1}{2} S_{b^*h^*}\{\varepsilon_{b^*}+\varepsilon_{h^*}-J_{pb^*}-J_{ph^*} + \sum_{B}^{occ} K_{Bb^*} + \sum_{H}^{occ} K_{Hh^*} \\ &\quad + (U_P+U_H+U_{AA} : b^*b^*) + (U_P+U_B+U_{AA} : h^*h^*)\} \end{aligned} \qquad 3)$$

where conventional symbols for the overlap (S), closed shell SCF orbital energies (ε), Coulomb (J) and exchange (K) integrals are used, and, for example, $(U_x : rr) = (V_x : rr) + 2\sum_x J_{xr}$. The subscripts h*, b*, p* refer to the lowest unoccupied MO's (LUMO's) which complement the previously defined HOMO's, and in the ΣK terms P, B, H refer to any doubly occupied orbital in the respective isolated cofactor ground states.

RESULTS AND DISCUSSION

The SCF MO's for each of the cofactors in their ground states were initially obtained using a previously described RHF-INDO procedure[7], and the X-ray structural data[8] of Rps. viridis. The matrix elements V_{PB}, V_{BH} (eq. 3 above) were evaluated; single exponent Slater AO's, whose exponents were identical to those adopted for the INDO procedure, were used in calculating the intermolecular overlap integrals. Both V_{PB} and V_{BH}

are directly proportional to such overlaps, and are thereby critically sensitive to the long-range behaviour of the atomic wavefunctions used. In contrast, the energy factor $1/2\{...\}$ in eq. 3 is less sensitive, varying only slightly over the RC in the range of $10^5 - 2\ 10^5\ cm^{-1}$.

For the L-branch, we obtain:

$$V_{PB} = 5.4\ f_{PB}\ cm^{-1}\ ;\ V_{BH} = 44\ f_{BH}\ cm^{-1},$$

where $f_{PB} = f_{BH} = 1$ for the INDO/SLATER exponent choice. For the M-branch, the crucial overlaps are considerably smaller owing mainly to structural asymmetry[9].

The measured ET rate[10], $k_L^{ET} = 3.6\ 10^{11}\ s^{-1}$, and the theoretically estimated Franck-Condon factor [11], $(2\pi/\hbar)FC = 7.8\ 10^8\ cm^2 s^{-1}$ require an effective

$$V_{PH} = \langle P^*BH|\hat{H}_{agg}|P^+BH^-\rangle \simeq 20\ cm^{-1}$$

In terms of the "super-exchange" model[1], and an estimated energy difference, $\delta E = E_{P^+B^-H} - E_{P^*BH} \simeq 300\ cm^{-1}$ [12], we calculate

$$V_{PH} = V_{PB}\ V_{BH}/\delta E = 0.8\ f_{PB}\ f_{BH}\ cm^{-1}$$

With $f_{PH} = f_{BH} = 1$, the predicted ET rate ($\propto V_{PH}^2$) is too small by a factor of ca. 600. There are two possibilities which can quantitatively account for this discrepancy:

(i) Dielectric polarisation of the protein environment, whereby an electron would move in an effective field of $-e^2/Dr$ at large distances, r, from its counter-charge. Here, D is the dielectric permittivity of the medium. This has the effect of extending the tails of the wavefunctions to larger distances. Other authors[13] have assumed a value of $D \simeq 4$ which would relieve the aforementioned discrepancy. This approach suffers from the rather arbitrary choices of both the magnitude of D and the radius of dielectric onset.

(ii) The Slater/INDO AO exponents are a poor choice for the intermolecular nature of the problem. An alternative set of AO exponents has been proposed by Burns[14], which is generated by optimally fitting Slater-type single exponent orbitals to the Hartree-Fock atomic functions using the method of moments. The outer regions of the Hartree-Fock functions are well-reproduced, and moreover, the INDO-method for tight-binding situations should not be too sensitive to this change. (Indeed, the parameterised one-centre two-electron repulsion integrals of this method are more nearly reproduced theoretically by the Burns set.) Most satisfying is that in the region of maximum overlap between the prosthetic groups, typically 3-4 Å, $S_{2p-2p}(Burns)/S_{2p-2p}(Slater)$ has a range 4-5, yielding a very close agreement for the required $f_{PB}\ f_{BH}$.

In conclusion, the approach adopted in this work demonstrates the feasibility of computing electronic coupling matrix elements in large oligomers with a valence AO basis > 1000. Slater-type single exponent orbitals seem to be an adequate choice provided they reproduce the outer wings of the corresponding multi-exponent Hartree-Fock function. Within the framework of a superexchange model our results reaffirm that unidirectionality of ET in RC's is mainly controlled by structural asymmetry. Finally, we mention that the structure of our CI approach offers a means of quantitatively comparing the two prominently-favoured mechanisms[1,11] of ET on an equal footing. Ongoing work towards this objective is in progress.

Note added in proof. Since presenting this work, a similar CI approach has been reported by Warshel and Parson[15]. Our approach differs in several respects, three of which are relevant here:

(i) P is considered a super-molecule rather than two interacting monomers,
(ii) The SCF MO's of individual cofactors are derived from an all-valence basis,
(iii) the expressions for the interaction matrix elements V_{PB} and V_{BH} (eq. 3) are derived differently. Warshel and Parson equate those typified by V_{PB} to an effective resonance integral, β; those typified by V_{BH} are not explicitly considered and set to zero.

ACKNOWLEDGMENTS

The authors are indebted to Prof. K. Möbius for his stimulation and encouragement of this project, which is supported by the Deutsche Forschungsgemeinschaft (Sonderforschungsbereich 337).

REFERENCES

1. M. Bixon, J. Jortner, M. E. Michel-Beyerle, A. Ogrodnik and W. Lersch, Chem.Phys.Lett. 140:626 (1987).
2. J. N. Murrel, J.Am.Chem.Soc. 81:5037 (1959).
3. C. J. Winscom and M. Plato, to be published.
4. J. C. Slater, Phys.Rev. 34:1293 (1929).
5. L. Brillouin, Actualités sci. et ind. 71 (1933), 159 (1934).
6. R. S. Mulliken, J.Chim.Phys. 46:497 (1949).
7. M. Plato, E. Tränkle, W. Lubitz, F. Lendzian and K. Möbius, Chem. Phys. 107:185 (1986).
8. J. Deisenhofer, O. Epp, K. Miki, R. Huber and H. Michel, J.Mol. Biol. 180:385 (1984); Nature 318:618 (1985).
9. M. Plato, F. Lendzian, W. Lubitz, E. Tränkle, and K. Möbius, in these Proceedings.
10. J. Breton, J.-L. Martin, A. Migus, A. Antonetti and A. Orszag, Proc.Natl.Acad.Sci. USA 83:5121 (1986).
11. S. F. Fischer, P. O. J. Scherer, Chem.Phys. 115:151 (1987).
12. J. Jortner and M. E. Michel-Beyerle, in: "Antennas and reaction centers of photosynthetic bacteria," M. E. Michel-Beyerle, ed., Springer, Berlin (1985) p. 344.
13. H. Kuhn, Phys.Rev. A34:3409 (1986).
14. G. J. Burns, J.Chem.Phys. 41:1521 (1964).
15. A. Warshel and W. W. Parson, J.Am.Chem.Soc. 109:6143 (1987).

CHARGE TRANSFER STATES AND THE MECHANISM OF CHARGE SEPARATION IN BACTERIAL REACTION CENTERS

P. O. J. Scherer and S. F. Fischer

Physikdepartment der Technischen Universität München
D-8046 Garching

ABSTRACT

The six pigments of the reaction center Rps. viridis consisting of the four bacteriochlorophylls BC_{MP}, BC_{LP}, BC_{MA}, BC_{LA} and the two bacteriopheophytins BP_M and BP_L are treated quantum mechanically as a hexamer within a π-electron approximation including extensive configuration interactions. The results indicate that the lowest charge transfer (CT) state consists of the oxydized monomer BC_{MA}^+ and the reduced bacteriopheophytine BP_M^- which are located on the M-branch. We propose that this CT state might induce an efficient quenching mechanism for excess quanta of light absorbed after initiation of the charge separation. Thus a possible function of the so called inactive M-branch may be a protection mechanism against excitation of P^+ which otherwise could damage the special pair.

I. INTRODUCTION

The X-ray structure analysis of the reaction centers Rps. viridis /1, 2/ and also the very recent one of Rb. sphaeroides /3/ have shown, that the arrangements of the prosthetic groups are very similar. The special pair dimer (P) forms a central group from which two branches start: The L-branch, consisting of the accessory monomer BC_{LA} and the bacteriopheophytine BP_L and the M-branch consisting of the accessory monomer BC_{MA} and the pheophytine BP_M which are related to BC_{LA} and BP_L approximately by a C_2 symmetry. Any attempt to interrelate the observed structure and the function of the reaction centers has to deal with the two puzzling questions.

1. In which way is the accessory monomer BC_{LA} involved in the initial charge separation ? and

2. What is the role of the apparently inactive M-branch ?

To shorten the notation we use for states localized on the pigments the symbols P, P^*, P^+, P^-, and B, B^*, B^+, B^- as well as I, I^*, I^+, I^- related to ground, excited, cation or anion states of the dimer, accessory monomer BC_{LA} and the bacteriopheophytine BP_L, respectively. States on the M-branch are denoted as B_M^*, B_M^+, I_M^*, I_M^- etc.

Experimentally it is found, that the initial charge separation starting from P^* to form P^+I^- takes place in 2.8 ps /4,5/ (earlier results /6,7/ gave 4 ps). The formation of P^+ of I^- or shifts of B^* seem to follow the same kinetics of a 2.8 ps decay. The time resolved measurements show also that the excitation transfer from B^*, I^*, B_M^*, and I_M^* to P^*, takes about 50 fs /4,5/. Further more a rapid (400 fs) recovery time for an induced bleaching of the accessory monomer band at 830 nm has been detected /5/.

In this note we suggest that this recovery time of 400 fs may be related to the charge recombination from the $B_M^+I_M^-$ state, which couples efficiently to the monomer band and could act as a trap for excess quanta once P^* or P^+ have been formed. This way the excitation could be deactivated more efficiently than by the cation dimer itself.

II. CHARGE TRANSFER STATES

For a detailed understanding of the initial charge separation it is essential to know the location of low lying CT states. Unfortunately CT states formed between weakly coupled molecules are difficult to detect spectroscopically, since they carry very little oscillator strength /8/ and tend to be very broad. For the six prosthetic groups of the reaction center only the state P^+I^- is clearly identified. It is located about 2000 cm^{-1} below P^* /9, 10/. It is observed as a biradical pair in its singlet or triplet state /11/ but not via direct excitation. Experimental information about other CT states such as P^+B^-, B^+I^-, $P^+I_M^-$ or $B_M^+I_M^-$ as well as internal CT states within the dimer is at best indirect / 8/. Theoretical estimates are not sufficiently accurate to make a priori predictions about their location. One needs to evaluate the ionization potential of the donor, the electron affinity of the acceptor and the polarization of the surrounding protein residues. Any one of these contributions is found with uncertainties of more than 1000 cm^{-1}. This is the reason why different models have been proposed for the electron transfer. These models differ essentially in their assumptions about the location of these CT states relativ to P^*. We hope to show here that one can get interesting information from quantum calculations if not on the absolute values so on the relative ordering of the CT states. Even though our results must be viewed as preliminary we hope that the qualitative picture points towards some essential features of the structure and their importance for the charge separation mechanism.

Since it is at present impossible to perform all valence electron quantum calculations on the hexamer consisting of all six pigments one has to make semiempirical model studies.In Fig. 1 we analyzed the SCF orbitals of an INDO calculation /12/ for BP_L (81 atoms) - most of the phytyl chain is left out- and the BC_{LA}-Histidine complex (99 atoms) with regard to the σ and π-character and compared these with the corresponding π-orbitals of a PPP-model. The highest occupied and lowest unoccupied π-orbitals are similar within the two models. This is the basis for the PPP-model, in which the effect of the σ-system is buried in the semiempirical parameters. In our calculation we used standard π-electron parametrizations /13/ refined by bond distance corrections and by deviations from the planarity /14/. The right side of Fig.1 shows that the lowest excited orbital of BC_{LA} has exclusive σ-character. It is localized on the magnesium atom. Since its location depends sensitively upon the parametrization we will comment on its possible implications for the electron transfer mechanism only briefly at the end. One can further see that the HOMO and LUMO orbitals of BC_{LA} are pushed to higher energies relative to those of BP_L. Without the histidine they were lower than the corresponding orbitals of BP_L. Since the PPP-model cannot account for this effect we will correct our results for the hexamer accordingly.

Next we study the special pair dimer in order to identify the internal CT states. If the dimer would be exactly symmetric the CT states could be characterized by their parity. In a reduced basis set of only one excited state per monomer it becomes clear that the even CT state can couple with the even excitonic state and the odd with the odd. The latter corresponds to P* and carries about 88% of the intensity. However, even a relatively strong coupling between the excitonic and CT states (comparable to the excitonic coupling) does not transfer much oscillator strength into

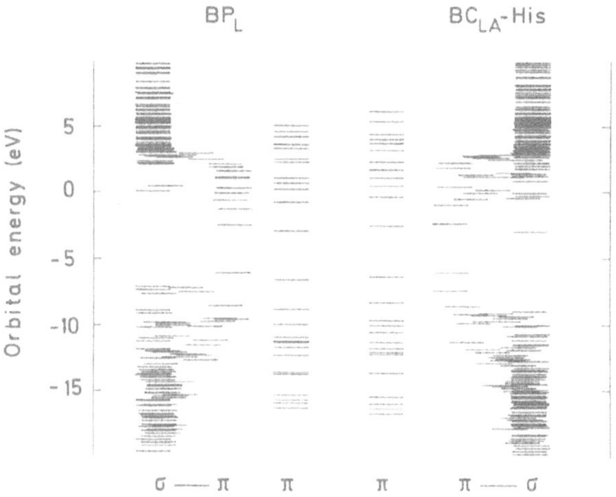

FIGURE 1

Energy locations for SCF- molecular orbitals are shown. On the left side INDO results are analyzed for BP_L in such a way that orbitals with more σ-charatcter are found further to the left. The π-character is defined as a projection on the 26 π-orbitals from a PPP calculation shown as the inner left part. On the right side the orbitals are shown for the BC_{LA}-His complex, with the PPP orbitals at the inner right part and the σ-character of the INDO orbitals increasing towards the right. INDO orbitals above 10eV and below -20eV are not shown. They have exclusive σ- character.

the CT states. It vanishes if the splitting of the two highest occupied states is equal to that of the two lowest unoccupied states . A stronger transfer of oscillator strength can occur if the dimer becomes slightly asymmetric. In Fig. 2 the SCF orbitals based on an PPP-CNDO programm are shown for the isolated pigments BC_{MP} and BC_{LP} and compared with the dimer states. Here the two lowest occupied and the two highest unoccupied states are pretty much delocalized. In the basis of the orbitals of the isolated pigments the CI-interaction is large compared to the asymmetry, defined by the energy difference between corresponding states of P_M and P_L.

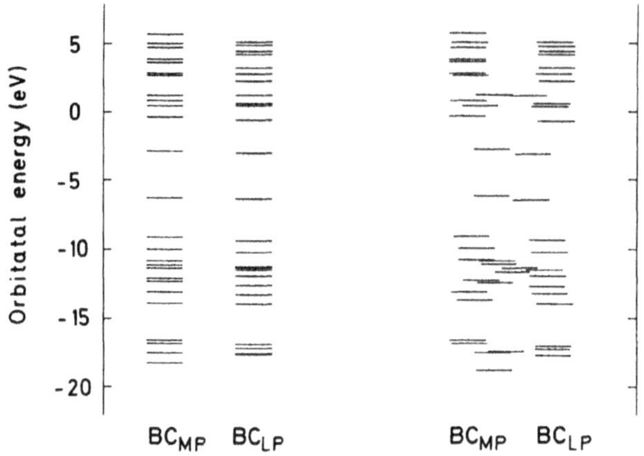

FIGURE 2

PPP orbital energies are shown from left to right for the isolated monomers BC_{LP} and PC_{MP} and for the dimer.

This explains the delocalization. For the CT states the interaction is weaker and consequently they can localize more easily.

The observed strong Stark intensity of the P* state /15/ seems to indicate that P* has admixture of an asymmetric CT state. Such a reasoning might be too simple, since the PPP calculation can not predict permanent dipole changes due to excitation accurately. It does not account properly for the polarization of the σ-system and the location of the CT states. Comparing their energy positions relative to P* with those from INDO calculations /16/ we find that the latter are substantially higher. Again this reflects the fact that the σ system undergoes a complex reorganisation. In the PPP calculations of the hexamer we shifted these internal CT states according to their positions we got from INDO calculations for the dimer. This way the calculated absorption spectrum /17/ agrees also quite well with the experimental.

To learn also about external CT states we studied recently / 8/ the loosely coupled dimers $BC_{LA}BP_L$ and $BC_{MA}BP_M$ and pointed at the extra stabilization of $B_M^+I_M^-$ by the glutamic acid residue. We suggested that $B^+_MI_M^-$ might be higher in energy by about 1000 cm^{-1}, assuming that there are no other differences of importance. In our new calculations, we were able to handle the whole hexamer within the PPP-CNDO model. These studies gave us the following surprising results. Due to certain asymmetries in the dimer and the monomers and due to the Coulombic interactions of all pigments, the HOMO and LUMO orbitals of the monomer B_M are substantially (by 0.3 eV) pushed to higher energies. This change in energy lowers the CT-state $B^+_MI^-_M$ by about the same amount (0.3 eV), relative to B^+I^-. In Fig. 3 the SCF orbitals of the hexamer are shown. The indicated transitions reflect also the relative ordering of the CT states even

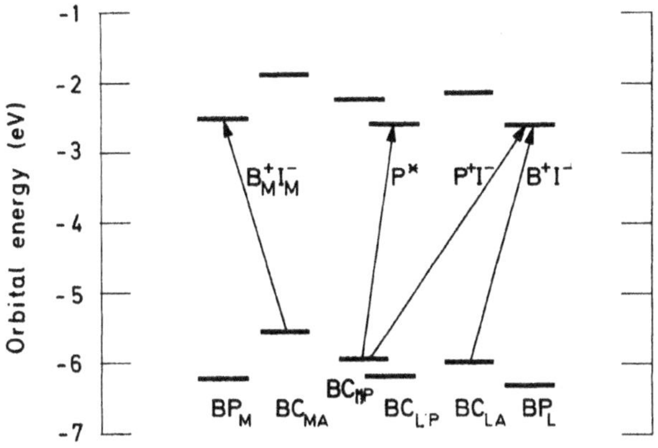

FIGURE 3

The six highest occupied and six lowest unoccupied orbitals are shown for the hexamer. The excited state P^* and a few CT states are represented by the electronic transition between orbitals, which contribute most to the corresponding state.

though CI-interactions and stabilization energies due to medium polarization are not included. A complete analysis of these states including the internal CT state for BC_{MA} and BC_{LA} is in progress.

3. THE ROLE OF THE ACCESSORY MONOMERS BC_{LA} AND BC_{MA} DURING THE INITIAL CHARGE SEPARATION

We mentioned already that the time resolved measurements /4,5/ show that the formation of the state P^+ and I^- follows the same kinetics of 2.8 ps and give no evidence for an intermediate P^+B^- state. Upon excitation into the absorption bands of B^*, B_M^* or I^* and I_M^* a rapid excitation transfer in about 50 fs to P^* has been found. In addition a transient bleaching has been observed after excitation of the B^*, B_M^* band at 830 nm which recovers in 400 fs. There is a stronger intensity dependence of this bleaching than for the shift of the band which follows the 2.8 ps kinetics. In addition it has been seen, that high intensity of absorption into the B^*, B_M^* band lowers the yield of the formation of P^+. Breton et al. / 5/ suggested that this could be due to absorption of reaction centers that have already entered the state P^* during the pulse.

Here we want to offer a microscopic interpretation for these processes. We introduced recently the following model scheme /8,18/. The excited state P^* couples excitonically into the CT state B^+I^- which lies energetically below P^* such that the overlap of the emission from P^* and the absorption of B^+I^- becomes optimal at low temperatures. The state B^+I^- is then converted into the state P^+I^-due to a very rapid (less than 1 ps) electron transfer from P to B^+. The state B^+I^- may thus be looked at as a doorway state for the final state P^+I^- and does not have to lead to resolvable bleaching of B^* prior to the formation of P^+I^-. This scheme gives a coupling of about 5 cm^{-1} for the rate determining step P^* to B^+I^- of the charge separation. This is still too low but better than corresponding results we found for the superexchange model. It is also consistent with the temperature dependence of the rate and is not in conflict with the Stark measurements on P^*.

Our new calculations on the hexamer support this view for the L-branch. However, the study of the full hexamer shows us that there exists a very new effect on the location of the state $B_M^+I_M^-$. It is pushed to lower energies by about 0.3 eV as compared to B^+I^-. To understand the implications of this effect let us consider an excitonic coupling from P^* to $B_M^+I_M^-$. The overlap of its absorption with P^* must be extremely small, since it is far out of resonance. We realize further that it can not act as an intermediate to feed $P^+I_M^-$ since it is lower in energy. This could explain the unidirectional decay along the L-branch. Our calculations show further that the state $B_M^+I_M^-$carries less intensity than B^+I^-since the leading coupling term is smaller. It results from the methyl groups of the pyroll rings I of BC_{MA} and BP_M which are not in so close contact as they are for the corresponding molecules on the L-branch. This shows that not only the energetics but also the electronic coupling support the unidirectionality.

The function of the M-branch is not so easily understood. If $B_M^+I_M^-$ is the lowest CT state it can only decay into the ground state. Such an internal conversion of a CT state can be very efficient. It should be noted that the two charged pigments BC_{LA}^+ and BP_M^- attract each other. The structural analysis shows further that they are less strongly bound to the protein than their mirror images on the L-branch. There the glutamic acid residue forms a hydrogen bond, which stabilizes the CT state B^+I^- but might also hinder the charge recombination by stabilizing the spatial separation. There is a further reason why $B_M^+I_M^-$ is very suited for rapid relaxation. Its energy location should be below P^+I^- at around 7000 cm^{-1}. This energy would match overtones of O-H vibrational bands, which could become good acceptors. With these qualitative arguments in mind we like to propose, that the observed recovery time of 400 fs can be identified with the relaxation time of $B_M^+I_M^-$ to the ground state. Clearly the formation of B_M^+ bleaches partly the monomer band at 830 nm since this is a strong micture of B_M^* and B^*. The bleaching recovers with the relaxation rate of $B_M^+I_M^-$. Also the unusual intensity dependence of the bleaching can be visualized within this model. As P^* is formed the excess excitation of the monomer band can no longer be transferred so efficiently to the special pair and therefore it must decay into the CT-states B^+I^-or $B_M^+I^-$ with a larger yield. The first channel of the two gets blocked as the charge separation proceeds and P^+I^- developes. That means any excitation of P^+ can under these conditions only relax via by $B_M^+I_M^-$. This relaxation should be efficient since the CT state overlaps within our model scheme with the excitation band $(P^+)^*$.

Finally, we like to comment on the possible involvement of an internal CT state B^{+-}(B_M^{+-}) within BC_{LA} (BC_{MA}). If further studies confirm indeed that in the presence of the histidine the lowest excited state becomes the CT state from the HOMO orbital to the magnesium and if this state overlaps with P^* it could act equally well as a doorway state. The charge transfer mechanism could in this case follow the scheme $P^* \rightarrow PB^{+-}I \rightarrow P^+BI^-$. The corresponding couplings are in both cases of the excitonic type. The second involves a cooperative tunneling from P to B^{+-} and from B^{+-} to the acetyl oxygen of BP_L. The state B_M^+ could also be involved in the deactivation to the ground state. It should be located relative to $B_M^+I_M^-$ equally close in energy as B^{+-} to B^+I^-.

4. CONCLUDING REMARKS

In summary we like to point out again that it is not possible to predict the location of the CT states definitely. Even if the calculation for the pigments would be accurate the interaction with the surrounding amino acid residues could have large effects which can not be simply simulated by a homogeneous dielectric medium. By looking at very specific interactions we discovered two effects. The first is due to the histidines, which bind to the Mg atoms and transfer charge of at least 0.13 e. In the INDO calculation the combined system BC_{LA}-His is neutral but it gets a strong dipole with the positive charge on the histidine. For the corresponding molecules of the special pair BC_{LP} and BC_{MP} the charge from the histidines may delocalize further into the bridging water molecules. Such an effect would push the HOMO and LUMO states of the dimer even higher. Also the effect of the hydrogen bonds to the acetyl oxygens must be included. In general we expect that the charge distributions between amino acid groups and the pigments are the most sensitive parameters to control the location of the CT states and thus the charge separation mechanism.

The second effect, which shows up only in the calculation of the full hexamer is even more complex. The BC_{LA}-molecule is somewhat closer to BP_L than BC_{MA} to BP_M. This may be the origin for more relaxed distances within the pyroll ring I of BC_{MA}. That leads to a relatively small shift to higher energies of the HOMO and LUMO orbitals relative to those of the other BC-molecules. This shift becomes strongly enhanced due to the Coulombic interactions with the other pigments. We suggest that this effect could have implications on the function of the M-branch as a protective mechanism. At this point an experimental test seems possible. If the bleaching and 400 fs recovery process is indeed due to population and recombination of the $B_M^+I_M^-$ state one would expect to find the same bleaching and recovery at the band I_M^*. If B_M^{+-} is involved in the charge recombination the corresponding Mg^*-band should show a change.

So far we are aware of two arguments that might speak against our model. First, there is no experimental evidence for a two step mechanism. We already mentioned that the CT-state B^+I^- may be looked at as a doorway state for P^+I^- and is strongly coupled to this state. PPP results show indeed that the positive charge on the special pair distributes partly into the monomers. This underlines the strong coupling between PB^+ and P^+B. Further it should be noted that the states B^* and I^* are effected simultaneously within this model by B^+I^- which exists as a real intermediate only with a very small probability. Under these conditions we can not see a conflict with the experiments so far.

A second argument against the involvement of any intermediate state below P^* is based on the temperature dependence of the exchange interaction J /19/. Theoretically a contribution to J weighted by the factor $\exp(-\Delta G/kT)$ is predicted where ΔG refers to the free enthalpy difference of the states B^+I^- and P^+I^-. Taking the vertical energy difference and neglecting entropy effects this factor would be $\exp(-10)$ at room temperature. ΔG is not substantially different (20 %) for the superexchange model where the state P^+B^- must be placed slightly higher than P^*. The temperature dependence has been looked at experimentally by the MARY-technique in the range between 70 K and room temperature /20/. This technique gives however only an upper limit on J, of the order of $10^{-4} cm^{-1}$ The RYDMR measurements /21/ give values closer to $5x10^{-5} cm^{-1}$ but here the temperature variation spans only the region between 160 K and room temperature. Also here a few percent variations with temperature can not be excluded. So we do not see, that this observation helps to discriminate our model as compared to the superexchange model. On the contrary, the magnitude of J compared to the magnitude of the rate determining coupling forces severe constraints on the models involving P^+B^- as real or virtual intermediate /21,22/. Not so the intermediate B^+I^-, since only P^+B^- acts as intermediate for the superexchange coupling. If this state is indeed much higher than P^* the exchange coupling can become easily as small as $10^{-4}cm^{-1}$ without effecting the coupling between P^* and B^+I^-.

With regard to the temperature dependence of the initial charge separation we have shown that our model gives the right trend /8/. It is related to the decrease of the overlap between the absorption band of B^+I^- and the emission of P^* as P^* moves to higher energies with increasing temperature. Within the superexchange model such a temperature shift must lead to an increase of the rate, due to the decrease of the virtual energy gap between P^* and P^+B^-. In addition an activated contribution coming from P^+B^- is predicted. The activation energy $E_a = E(P^+B^-) - E(P^*)$ can not be much larger than the coupling between P^* and B^+I^- in order to assure a fast forward rate. This coupling is on the other hand limited to about 100 cm^{-1}, since larger values would effect the Stark spectrum and show such a CT state.

ACKNOWLEDGEMENT

We like to thank Dr. J. Deisenhofer, who gave us the coordinates for the atoms of the hexamer. This work has been supported by the Deutsche Forschungsgemeinschaft SFB 143.

REFERENCES

/ 1/ J. Deisenhofer, O. Epp, K. Miki, R. Huber, and H. Michel, J. Mol. Biol. 180 (1984) 385
/ 2/ H. Michel, O. Epp, and J. Deisenhofer, EMBO J. 5 (1986) 2445
/ 3/ J. P. Allen, G. Feher, T. O. Yeates, H. Komiya, and D. C. Rees, Proc. Natl. Acad. Sci. USA 84 (1987) 5730
/ 4/ J. L. Martin, J. Breton, A. J. Hoff, A. Migus, and A. Antonetti, Proc. Natl. Acad. Sci US 83 (1986) 957
/ 5/ J. Breton, J. L. Martin, A. Migus, A. Antonetti, and A. Orsay, Proc. Natl. Acad. Sci USA 83 (1986) 5121
/ 6/ Holten, M. W. Windsor, W. W. Parson, and J. P. Thornber, Biochim. Biophys. Acta 501 (1978) 112
/ 7/ W. Zinth, M. L. Nuss, M. Franz, W. Kaiser, and H. Michel, in: Antennas and reaction centers of photosynthetic bacteria, ed. M. E. Michel-Beyerle (Springer, Berlin, 1985) p. 286
/ 8/ P. O. J. Scherer and S. F. Fischer, Chem. Phys. Lett. 141 (1987) 179
/ 9/ C. E. D. Chidsey, L. Takiff, R. A. Goldstein, and S. G. Boxer, Proc. Natl. Acad. Sci. US 82 (1985) 6850
/10/ N.W.T. Woodbury and W.W.Parson, Biochim. Biophys. Acta 767 (1985) 345
/11/ A. J. Hoff, Quarter, Rev. Biophys. 14 (1981) 599; Photochem. Photobiol. 43 (1986) 424
/12/ QCPE -Program 372
/13/ G. Wagniere, Introduction to elementary molecular orbital theory and to semiempirical methods (Springer, Berlin, 1976)
/14/ I. Fischer-Hjalmas and M. Sundbom, Acta Chem. Scand. 22 (1968) 607
/15/ P. O. J. Scherer and S. F.Fischer, Chem. Phys. Lett. 131 (1986) 153
/16/ unpublished results from M. Plato

/17/ see second contribution by P. O. J. Scherer and S. F. Fischer in this volume
/18/ S. F. Fischer and P. O. J. Scherer, Chem. Phys. 115 (1987) 151

/19/ M. Bixon, J. Jortner, M. E. Michel-Beyerle, A. Ogrodnik, and W. Lersch, Chem. Phys. Lett. 140 (1987) 626
/20/ A. Ogrodnik, N. Remy-Richter, M. E. Michel-Beyerle, and R. Feick, Chem. Phys. Lett. 135 (1987) 576
/21/ D. A. Hunter, A. J. Hoff, and P. J. Hore, Chem. Phys. Lett. 134 (1987) 6
/22/ R. A. Marcus, Chem. Phys. Lett. 133 (1987) 47

LIGHT REFLECTIONS

G. Feher
University of California, San Diego
La Jolla, California 92093

To some Science is a sacred cow;
others make hamburger out of it.

--------------------------------------Ah, the microphone, I forgot; thank you Jacques. It is very important, I wouldn't want to repeat what happened to me a few years ago. In the middle of a lecture, somebody in the last row shouted: ''I didn't hear a single word of what you were saying the last half hour'', upon which a person in the first row shouted back ''I will gladly change seats with you''. Anyway, I like those little microphones that you clip on. In the past you had to hold this large contraption in your hand, which cramped your style. It always reminded me of the sign next to a Tel Aviv bus driver which read: ''don't talk to the driver, he needs his hands for driving''.

Well, by now you must have gotten the spirit of these remarks. This will not be a serious scientific presentation. We had plenty of those during the past week; they were all very interesting and stimulating, albeit occasionally rather demanding and intense. The time has come to depressurize with some laughs and a few frivolous remarks. Arnold (Hoff) had asked me at the banquet yesterday evening to give a light after dinner speech, but it would have been a travesty to spoil such a marvelous banquet, and besides all of us had too much wine to give or receive a coherent rendition of anything. So I said tomorrow. Alas, today is the tomorrow I talked about yesterday.

Where shall I start? At the beginning seems like a good place: Genesis, first page, first paragraph of the Old Testament ''and He commanded, let there be light-and light appeared''. Clearly, the purpose of this act of creation was *photosynthesis*; so ours is the oldest of scientific endeavours - 5748 years, so they say. Well, maybe the estimate of the Creationists is a little off. Let's give the Evolutionists equal time. The Big Bang Boys believe that it all started several billion years ago, i.e. there is a difference of opinion between the two protagonists of ~ 6 orders of magnitude in time. There exists an even more basic disagreement: Evolutionary biologists believe that life began on a very primitive level, probably with photosynthetic bacteria rather than with the Creation of Man. As R. Dickerson put it so well: ''Human beings are the metabolic offspring of defective purple photosynthetic bacteria''. There are, of course, a few species between bacteria and men, plants for instance. Ecologically, the advent of green plant photosynthesis was a disaster. It poisoned the atmosphere with oxygen, which made it possible for parasites like homo sapiens to thrive. But I am digressing; let's get back on track.

At the beginning, progress was very slow; cave men were fully occupied in gathering and eating the fruits of photosynthesis and were not concerned with understanding its mechanism. Besides, they lacked the technology to do serious experiments (Fig. 1). This situation has changed drastically, and the new technology is really mind-boggling. Take for instance femtosecond spectroscopy. Who would ever have dreamed of it just a few years ago? Never mind whether the results are meaningful or even correct, just to work in that time domain!! Besides theorists are a flexible breed and they are always able to explain any results. This reminds me of Joshua's (Jortner) story of the experimentalist who, having just shown that A is larger than B, went to a theorist for an explanation. The theorist, after scratching his head, produced a long and contorted derivation proving the point. The next day the experimentalist went back to the lab and found to his dismay that he had made a mistake, and actually B was larger than A. Somewhat embarrassed he returned to the theorist with his new result. The theorist, unconcerned, replied "Oh, that is even easier to prove". Two more stories related to this subject come to mind: The disciples of two famous rabbis from Minsk and Pinsk (or was it Omsk and Tomsk?) extol the virtues of their respective rabbis. One of them said "My rabbi is for sure the greatest; last week he closed his eyes and with

Fig. 1. *"But we just don't have the technology to carry it out."*
(Modified from a cartoon by S. Harris, American Scientist, May 1971).

utmost concentration proclaimed: I see a great fire in the synagogue of Moscow". Upon which the second asked: "That's strange, I didn't hear any news of this fire; was there really a fire?" To which the first disciple replied: "Actually not, but just to see that far is quite an accomplishment". And then there was the big game hunter who returned from Africa to London where he was dined and wined. At one of the banquets the lady next to him asked him "How many lions did you shoot?". "None" was the answer, "How many leopards?", "None", "How many elephants?", "None", "Did you bag any big game?", "No". "So why are they wining and dining you" asked the lady; "Madam, just to be in this field is an accomplishment", replied the big game hunter.

Coming back to theorists. Their arguments are sometimes bizarre to say the least; sometimes their original assumptions bear no resemblance to reality, although the derivations resulting from them are 100% correct. Take for example the theoretical biophysicist who was asked whether he could explain the low fertility of a certain breed of chickens. His train of reasoning started: "Let's assume for simplicity a spherical chicken....". Then there are those who belabor, in great detail, second order effects with complete abandon of the first order effect. It reminds me of the teacher who explained to his class why two eyes are needed for three dimensional vision. As an example he gave the episode of the cyclops who had only one eye and was therefore such easy prey for Ulysses who slew him. A precocious pupil raised his hand and remarked: "But teacher, somebody had gouged out the cyclop's eye before he was slain by Ulysses". To which the teacher replied "Oh, that's an additional effect".

And then there is a class of theorists who are never in doubt - but often wrong. The louder they proclaim their results the more suspicious one ought to become. How much better science would be if more people would heed the motto, attributed to Will Rogers: "It isn't what we don't know that hurts us, but what we know and ain't so".

But I don't want to knock only theorists. So let me turn now towards my experimental colleagues. Many of them are engaged in performing the proverbial "Lamp Post" experiment. The term has its origin in the story of the guy who was looking for his lost wallet under the lamp post. Not because he lost it there, but because the illumination was better. And then there are experimentalists who get a result that is good to one or two significant places, but when they or their students multiply the result by π, which had been entered in the computer with ten significant digits, the result all of a sudden is quoted to ten places. This phenomenon is perhaps best illustrated by the story of the Scout. An encampment of white settlers was threatened by a group of advancing Indians. The scout was asked to climb up a tree and report on the number of approaching Indians. After a few minutes, he reported that the number of Indians was 1,007. In amazement they asked him how he could get that number so fast and so accurate, to which he replied "I counted 7 Indians on horse back at the head of the group, and from the dust cloud behind, I estimate that there must be about 1,000 more.

There is a rare breed of experimentalists who have the knack of picking the right system or performing the right experiments. When pressed for a rationale, their reasoning is usually pretty cockeyed because by and large, their actions are guided by gut feelings or intuition if you like. Some of these people have made tremendous contributions by this seemingly haphazard procedure. One of them is my late friend and colleague, Bernd Matthias. He discovered more superconductors, ferroelectrics and ferromagnetics than anybody else. Yet, when he tried to explain his reasoning, it made no sense whatsoever. I finally understood him after a joint gambling spree in Las Vegas. He played roulette and bet $1,000 on the number 35. It came up and he left all his winnings again on 35. The number came up again. He repeated the action until he broke the house. As you can imagine, there was a big commotion, and he was surrounded by journalists and TV crews. One of the interviewers asked him: "Professor Matthias, how did you think of putting all your money on the number 35?"; upon which Matthias replied "that's easy gentlemen, I dreamed six times in a row the number six; since six times six is 35, so I bet on 35".

And speaking of numerology, what is the most basic and tentalizing number in science? It clearly is the reciprocal of the fine structure constant, $(\hbar c/e^2)^{-1}$ i.e., the number 137. Generations of scientists have puzzled about the significance of this ubiquitous dimensionless quantity. Did you know that chlorophyll has 137 atoms? It clearly must have a deep significance. And then there is Psalm # 137 that has been haunting me for decades. But that is another story.......

Not everybody is blessed with a gift of intuition like Matthias, and, therefore, most of us have to resort to the "scientific method" and, more often than not, to the trial and error approach. Take for example the case of the isolation of reaction centers. As you know, it all started with the pioneering work of Reed and Clayton who isolated a large complex having a molecular weight of over 1 million that performed the primary photochemistry. The present day reaction center with a molecular weight ~ 10 times smaller than the original complex owes its existence to the choice of the detergent LDAO. We have often been asked in the past what made us pick that particular detergent. Well, it certainly wasn't brains, but a tedious set of experiments based on the trial and error approach illustrated in Fig. 2. A more modern example is H. Michel's breakthrough in crystallizing the RC from *Rp. viridis*. The key to his success was another, smaller detergent, heptane triol and he used the approach just described, namely trial and error (in his case actually error and tri*o*l). Incidentally, we always had great faith that the reaction centers would eventually be crystallized. To keep reminding us of this we had an Escher picture posted in our lab (Fig. 3). After all, if you can produce an ordered array of ducks, you should be able to do the same thing for an integral membrane protein. This optimism was not shared by the establishment, as was brought home to us by the referees of our pre-1980 grant applications. Scientists are like fish, they swim in schools. And what's worse, they are prone to prejudices, which as you know is a disease characterized by the hardening of the categories. Fortunately, it can sometimes be cured by demonstrating that the impossible is possible after all. And what fun doing it!

Are there any take-home lessons from all this? Perhaps that research styles are vastly different and that advances in science can be made by many different approaches. *Vive la difference*! One can, however, divide the approaches into two broad categories: one is the rigorous, analytical, approach which in its extreme form results, unfortunately, in rigor mortis; too much analysis causes paralysis. The other is the simple, intuitive, approach which in its limiting case may result in serious deviation from the truth. But just to make things unrigorous in order to appear simple

Fig. 2. *It worked before - it worked again.*
(Modified from a cartoon by S. Harris, American Scientist, November 1970).

Fig. 3. *If Escher can "crystallize ducks", we should be able to do likewise with integral membrane proteins.*
(Copyright, M. C. Escher, c/o Gordon Art, Baarn - Holland).

won't do. It would be like trying to escape senility by becoming a juvenile delinquent. The best that we can do is to try to find a comfortable compromise between analysis and imagination, reason and fantasy, discipline and fancy. Perhaps, we can summarize the above dilemma by the uncertainty principle

$$(\text{Simplicity}) \times (\text{Truth}) = \text{constant}.$$

I see some uneasiness, and indeed Sighart (Fischer) is violently objecting. No wonder; there is of course an inherent beauty in simplicity for which we don't want to pay with untruth. So take this relation with a grain of salt. It is not as ironclad as the Heisenberg uncertainty principle and there will be many exceptions to it. Incidentally, there is an interesting, irreverent, way of looking at Heisenberg's uncertainty principle: God does not watch over our galaxy all the time. When he looks aside, uncertainties arise. If this is true, it would be very interesting to find out what the value of Planck's constant, $\hbar$ is on other galaxies. If it is smaller than in ours, then He is more interested in them than in us.

So far, we have discussed individual approaches. Time has come to go from the single particle to the many body problem, i.e. to discuss the psychology and sociology of interaction between scientists. The most interesting of these is the reaction of our colleagues when we present them with a new idea. Some of them simply can't take it. Their response is something like: "It is not new, in addition it is wrong, and besides I thought of it before". This is analogous to the response of the guy who, when asked how he liked the restaurant responded: "Terrible, the food is inedible and the portions are too small".

And then there are those who listen to our ideas, ruminate over them and incorporate them into their own thoughts. After a while they may even forget the origin of the idea (not intentionally; it is just a trick of the mind) and explain it to us after a few months with great enthusiasm. I put them into the category of "I thought of it the minute I heard it".

Yet others are trying to find an excuse for not having come up with the right answers, saying: "I thought of it but didn't have the patience to stick to it", or "I didn't get around to doing it" or similar. Didn't we react in this way ourselves when Ted Mar presented his amazing results of getting the pure dimer spectrum after irradiating his sample for many hours? A more historic event of this phenomenon, the discovery of fire, is shown in Fig. 4.

Fig. 4. *"I had the same idea, once, but I just didn't stick to it long enough"*.
(The Saturday Review, Inc. 1972, Reprinted by permission of Ed Fisher).

My time is up, so let me conclude by thanking Jacques and André for having organized this marvelous conference. Jacques agreed at the banquet yesterday to organize another conference in the future if he can get the funding for it. I will end, therefore, with a plea to the granting agencies. Wouldn't you be ashamed to see Jacques hustling at a street corner in Paris?

Fig. 5. *Let that not happen!!*
(Cartoon by Ted Velasquez, UCSD).

P.S. You were such a good audience, laughing at all the appropriate places. As a reward I would like to share with you some good news. In the latest medical breakthrough, it has been shown that lack of humor is carcinogenic.

INDEX

GPSR Compliance

The European Union's (EU) General Product Safety Regulation (GPSR) is a set of rules that requires consumer products to be safe and our obligations to ensure this.

If you have any concerns about our products, you can contact us on ProductSafety@springernature.com

In case Publisher is established outside the EU, the EU authorized representative is:

Springer Nature Customer Service Center GmbH
Europaplatz 3
69115 Heidelberg, Germany

Batch number: 10761599

Printed by Printforce, the Netherlands